Intelligent Solutions for Sustainable Power Grids

L. Ashok Kumar
PSG College of Technology, India

S. Angalaeswari
Vellore Institute of Technology, India

K. Mohana Sundaram
KPR Institute of Engineering and Technology, India

Ramesh C. Bansal
University of Sharjah, UAE & University of Pretoria, South Africa

Arunkumar Patil
Central University of Karnataka, India

A volume in the Advances in Computer and
Electrical Engineering (ACEE) Book Series

Published in the United States of America by
  IGI Global
  Engineering Science Reference (an imprint of IGI Global)
  701 E. Chocolate Avenue
  Hershey PA, USA 17033
  Tel: 717-533-8845
  Fax: 717-533-8661
  E-mail: cust@igi-global.com
  Web site: http://www.igi-global.com

                    Library of Congress Cataloging-in-Publication Data

CIP DATA PROCESSING

Intelligent Solutions for Sustainable Power Grids
  L. Ashok Kumar, S. Angalaeswari, K. Mohana Sundaram, Ramesh C. Bansal, Arunkumar Patil
  2024 Engineering Science Reference

ISBN: 9798369337356 I ISBN: 9798369346952 I ISBN: 9798369337363

This book is published in the IGI Global book series Advances in Computer and Electrical Engineering (ACEE) (ISSN: 2327-039X; eISSN: 2327-0403)

British Cataloguing in Publication Data
A Cataloguing in Publication record for this book is available from the British Library.

For electronic access to this publication, please contact: eresources@igi-global.com.

Advances in Computer and Electrical Engineering (ACEE) Book Series

Srikanta Patnaik
SOA University, India

ISSN:2327-039X
EISSN:2327-0403

MISSION

The fields of computer engineering and electrical engineering encompass a broad range of interdisciplinary topics allowing for expansive research developments across multiple fields. Research in these areas continues to develop and become increasingly important as computer and electrical systems have become an integral part of everyday life.

The **Advances in Computer and Electrical Engineering (ACEE) Book Series** aims to publish research on diverse topics pertaining to computer engineering and electrical engineering. **ACEE** encourages scholarly discourse on the latest applications, tools, and methodologies being implemented in the field for the design and development of computer and electrical systems.

COVERAGE

- Power Electronics
- Computer Science
- Qualitative Methods
- Computer Architecture
- VLSI Fabrication
- Microprocessor Design
- VLSI Design
- Digital Electronics
- Analog Electronics
- Sensor Technologies

IGI Global is currently accepting manuscripts for publication within this series. To submit a proposal for a volume in this series, please contact our Acquisition Editors at Acquisitions@igi-global.com or visit: http://www.igi-global.com/publish/.

Titles in this Series

For a list of additional titles in this series, please visit: www.igi-global.com/book-series/advances-computer-electrical-engineering/73675

Critical Approaches to Data Engineering Systems and Analysis
Abhijit Bora (Assam Don Bosco University, India) Papul Changmai (Assam Don Bosco University, India) and Mrutyunjay Maharana (Xi'an Jiatong University, China)
Engineering Science Reference • copyright 2024 • 326pp • H/C (ISBN: 9798369322604) • US $315.00 (our price)

Applications and Principles of Quantum Computing
Alex Khang (Global Research Institute of Technology and Engineering, USA)
Engineering Science Reference • copyright 2024 • 491pp • H/C (ISBN: 9798369311684) • US $300.00 (our price)

Emerging Trends in Cloud Computing Analytics, Scalability, and Service Models
Dina Darwish (Ahram Canadian University, Egypt)
Engineering Science Reference • copyright 2024 • 468pp • H/C (ISBN: 9798369309001) • US $300.00 (our price)

Principles and Applications of Quantum Computing Using Essential Math
A. Daniel (Amity University, India) M. Arvindhan (Galgotias University, India) Kiranmai Bellam (Prairie View A&M University, USA) and N. Krishnaraj (SRM University, India)
Engineering Science Reference • copyright 2023 • 237pp • H/C (ISBN: 9781668475355) • US $275.00 (our price)

NeutroGeometry, NeutroAlgebra, and SuperHyperAlgebra in Today's World
Florentin Smarandache (University of New Mexico, USA) and Madeline Al Tahan (Abu Dhabi University, UAE)
Engineering Science Reference • copyright 2023 • 264pp • H/C (ISBN: 9781668447406) • US $270.00 (our price)

Principles and Theories of Data Mining With RapidMiner
Sarawut Ramjan (Thammasat University, Thailand) and Jirapon Sunkpho (Thammasat University, Thailand)
Engineering Science Reference • copyright 2023 • 319pp • H/C (ISBN: 9781668447307) • US $255.00 (our price)

Energy Systems Design for Low-Power Computing
Rathishchandra Ramachandra Gatti (Sahyadri College of Engineering and Management, India) Chandra Singh (Sahyadri College of Engineering and Management, India) Srividya P. (RV College of Engineering, India) and Sandeep Bhat (Sahyadri College of Engineering and Management, India)
Engineering Science Reference • copyright 2023 • 389pp • H/C (ISBN: 9781668449745) • US $270.00 (our price)

IGI Global
PUBLISHER of TIMELY KNOWLEDGE

701 East Chocolate Avenue, Hershey, PA 17033, USA
Tel: 717-533-8845 x100 • Fax: 717-533-8661
E-Mail: cust@igi-global.com • www.igi-global.com

Editorial Advisory Board

Durgesh Pratap Singh, *SRM Institute of Science and Technology, Chennai, India*
P. Somasundaram, *Anna University, Chennai, India*
Vishnu Suresh, *Wrocław University of Science and Technology, Poland*
O.V.Gnana Swathika, *Vellore Institute of Technology, Chennai, India*
V. Thiyagarajan, *SSN College of Engineering, India*
Sowmmiya U, *SRM Institute of Science and Technology, Kattankulathur, India*
Gomathi V, *Anna University, Chennai, India*
Mahesh V, *SR University – Telangana, India*
Niveditha VR, *Sathyabama Institute of Science and Technology, Chennai, India*

Table of Contents

Section 1
AI and ML Applications to Power Systems

Section 4
Cyber Physical Systems and Internet of Things

Detailed Table of Contents

Section 1
AI and ML Applications to Power Systems

> *Divya Prabha Varadharajan, S.A. Engineering College, India*
> *Sathiyasekar Kumarasamy, K.S.R. College of Engineering, India*
> *Ramasamy Murugesan, K.S.R. College of Engineering, India*
> *Akila Muthuramalingam, K.S. Rangasamy College of Technology, India*
> *M. Venkatesan, K.S.R. College of Engineering, India*
> *Loganathan Nachimuthu, University of Technology and Applied Sciences, Oman*

Renewable energy sources plays a vital role in power generation. Renewable solar energy resources abound on a global scale. When employing solar energy as a resource, the output of photovoltaic (PV) modules is determined by radiation and the temperature of the sun. The first PV systems were developed in the mid-nineteenth century. The underlying issue in the solar industry is that PV systems capture just a percentage of the sun's beams. The technique works best when the Solar Panel (SP) is perpendicular to the sun's rays. Solar tracking systems (STS) are used to keep a PV panel orthogonal to the radiation at all times, allowing for optimal energy output from the sun. With the help of advanced technology in tracking systems, numerous tracking techniques are available. The goal of this research is to explore the most recent improvements in solar tracking technology to estimate the best orientation for PV panels. The angular displacement of an STS with a single or dual-axis is possible. This chapter starts with mathematical modeling of the SP and ends with evaluation metrics to evaluate the performance of STS. The pipeline between the start and end points covers the identification of sun position, and Artificial Intelligence (AI) based controller. This survey can help the researchers to create an effective STS.

> *Saurabh Laledia, Chandigarh University, India*
> *Harpreet Kaur Channi, Chandigarh University, India*
> *Ramandeep Sidhu, Lovely Professional University, Phagwara, India*

Green buildings using renewable energy are developing as the world pursues sustainability. This abstract covers distributed renewable power planning and construction for energy efficiency, conservation, and built environment sustainability. Distributed solar, wind, biomass, and geothermal for green buildings. Local generation lowers grid dependency in distributed energy systems. Site study, energy demand assessment, and renewable energy source selection for the building's needs and location are design aims. Smart grids and energy storage boost renewables. Energy-efficient passive design, sustainable materials, and creative architecture complement distributed renewable power sources throughout construction. Cost savings, renewable energy ROI, and green building's environmental advantages are examined. Renewable energy may boost green building safety, security, and ecology. Distributed renewable energy green buildings fight climate change and promote sustainability.

Chapter 3

R. Pitchai, Department of Computer Science and Engineering, B.V. Raju Institute of Technology, India

D. Sengeni, Department of Electronics and Communication Engineering, C.K. College of Engineering and Technology, India

Putchakayala Yanna Reddy, Department of Electrical and Electronics Engineering, Bharath Institute of Engineering and Technology, India

M. Mathiyarasi, Department of Aerospace Engineering, Agni College of Technology, India

Devika Sahu, Department of Computer Science and Engineering, Government Engineering College, India

Sampath Boopathi, Mechanical Engineering, Muthayammal Engineering College, Namakkal, India

The growth of digital infrastructure has necessitated the development of data centers, which require efficient cooling systems for reliable operation. This chapter delves into the complex world of cooling technologies, their challenges, and innovative solutions. It emphasizes the importance of efficient cooling in data centers, balancing efficiency and energy consumption, scalability, and environmental considerations. The chapter discusses various cooling systems, including traditional air cooling and advanced liquid cooling and phase-change solutions. It also highlights the use of advanced materials like graphene and carbon nanotubes for improved heat transfer and thermal management. The chapter also discusses the integration of Artificial Intelligence in cooling systems, enabling real-time monitoring and predictive analytics. The future of data center cooling will see continued innovations like modular data centers, rack-level cooling, and advanced free cooling strategies.

Chapter 4

Aamir Bin Rashid, Chandigarh University, India

Harpreet Kaur Channi, Chandigarh University, India

In current power system analysis, integrating dispersed renewable power sources into the infrastructure is a major difficulty. Technical, economic, and environmental factors are considered in this research to optimise distributed renewable power integration. This study presents new power grid performance and reliability methods using advanced power system analytic techniques such load flow analysis, voltage stability evaluation, and transient stability analysis. This research uses sophisticated control algorithms and advanced optimisation to reduce power fluctuations, stabilise the grid, and maximise distributed renewable energy use. The suggested framework balances intermittent renewable sources with demand variations for a sustainable and resilient power system. This research shows that the proposed optimisation strategies are feasible and effective through comprehensive simulations and case studies, providing policymakers, system operators, and stakeholders with valuable insights for sustainable distributed renewable energy integration in the power system.

Chapter 5

G. Jegan, Sathyabama Institute of Science and Technology, India
M. R. Ebenezar Jebarani, Sathyabama Institute of Science and Technology, India
P. Kavipriya, Sathyabama Institute of Science and Technology, India
S. Lakshmi, Sathyabama Institute of Science and Technology, India

The focus of the study is on developing a system for monitoring moving targets. The study's overarching goal is to create a GPS module-based fleet tracking system capable of outputting position updates to Google Maps. By integrating XIAO ESP32 C3 microcontroller with Neo-6M GPS module and A9G GSM/GPRS module into the Maduino Zero 4G platform, military and naval forces can get versatile and convenient GPS tracking and communication solutions with voice and text transmission. In addition, the proposed device has a push button that serves to activate the SOS feature, allowing for discreet dialing and providing the user with GPS data. The proposed system has the capability to provide enhanced levels of accurate GPS data in both indoor and outdoor environments due to its use of a 4G network as an interface. Due to its variable range, the suggested system can save operating costs and improve asset utilization during short and long-range communication. The proposed system is a comprehensive wireless solution that functions only via wireless data power transmission.

Chapter 6

Chalumuru Suresh, Department of Computer Science and Engineering, VNR VJIET, India
V. Nyemeesha, Department of Computer Science and Engineering, VNR VJIET, India
R. Prasath, Department of Computer Science and Engineering, KCG College of Technology, India
K. Lokeshwaran, Department of Computer Science and Engineering (Data Science),Madanapalle Institute of Technology and Science, India
K. Ramachandra Raju, Department of Mechanical Engineering, Bannari Amman Institute of Technology, India
Sampath Boopathi, Mechanical Engineering, Muthayammal Engineering College, Namakkal, India

This chapter explores the role of artificial intelligence (AI) in the energy sector, focusing on energy forecasting, optimization, and demand management. It highlights the importance of AI technologies in utilities, grid operators, and consumers. AI-driven models accurately predict energy consumption patterns and demand, and how machine learning algorithms, data analytics, and IoT devices can improve forecasting precision. AI also optimizes energy production and distribution processes, reducing costs, enhancing reliability, and promoting sustainability. It also emphasizes its role in demand side management, focusing on consumer engagement strategies and AI-driven demand response programs.

S. Saravanan, Department of Electrical and Electronics Engineering, B.V. Raju Institute of
Technology, India
K. S. Pushpalatha, Department of Information Science and Engineering, Acharya Institute of
Technology, Bengaluru, India
Sanjay B. Warkad, Department of Electrical Engineering, P.R. Pote (Patil) College of
Engineering and Management, Amravati, India
A. Prabhu Chakkaravarthy, School of Computing, College of Engineering and Technology,
SRM Institute of Science and Technology, India
Venneti Kiran, Department of Computer Science and Engineering (AIML), Aditya College of
Engineering, India
Sureshkumar Myilsamy, Bannari Amman Institute of Technology, India

The global energy landscape is shifting towards sustainability due to environmental concerns and technological advancements. This transformation involves integrating renewable energy sources, smart grid technologies, and data-driven strategies to create modern sustainable power systems. Artificial intelligence (AI) is at the core of this transition, potentially revolutionizing electricity generation, distribution, and consumption. AI is transforming sustainable power systems by optimizing resource allocation, improving load forecasting, and enhancing grid management. Future trends include AI advancement, grid decentralization, and smart city integration. This chapter encourages further research and innovation in AI-powered sustainable power systems, promising a more efficient and resilient energy future.

S. Saravanan, Department of Electrical and Electronics Engineering, B.V. Raju Institute of
Technology, India
N. M. G. Kumar, Department of Electrical and Electronics Engineering, Sree Vidyanikethan
Engineering College, Mohan Babu University, India
Putchakayala Yanna Reddy, Department of Electrical & Electronics Engineering, Bharath
Institute of Engineering and Technology, India
R. Ramya Sri, Department of English, Kongu Engineering College, India
M. Ramesh, Department of Aerospace Engineering, SNS College of Technology, Coimbatore,
India
B. Sampath, Mechanical Engineering, Mythayammal Engineering College (Autonomous),
India

Microgrid systems, with diverse energy sources and decentralized control, are revolutionizing energy management. However, integrating renewable energy sources and consumer demands poses challenges to grid stability. Machine learning (ML) is a key tool for optimizing energy management and adaptive control in microgrids, enabling accurate load forecasting, renewable energy output prediction, and efficient resource utilization. This abstract discusses the potential of ML in revolutionizing microgrid systems by enabling adaptive control mechanisms, predictive maintenance, early fault detection, and proactive scheduling. ML also addresses cybersecurity concerns, providing sophisticated solutions for intrusion detection and secure data management. This approach optimizes operations, drives innovation, and ensures resilient, cost-effective, and sustainable energy infrastructures.

Artificial intelligence (AI) has played a pivotal role in developing autonomous vehicles and revolutionizing the automotive industry. AI is at the forefront of the autonomous vehicle revolution, with the potential to transform transportation, enhance safety, and reduce environmental impact. While there are challenges and concerns to overcome, ongoing research, technological advancements, and regulatory efforts are helping to pave the way for a future where autonomous vehicles are an integral part of our transportation landscape. The successful integration of AI in autonomous vehicles will depend on a collaborative effort between the automotive industry, regulators, and the public to ensure the safe and responsible deployment of this transformative technology.

The integration of AI and IoT is revolutionizing the energy sector by improving efficiency and minimizing downtime in power plants. IoT sensors and smart grid technologies enhance transmission efficiency by real-time monitoring and power flow optimization. However, challenges like initial investment costs and skill requirements persist. Ethical considerations, data privacy, and equitable access are crucial for fully harnessing the potential of AI and IoT in the energy sector. This chapter synthesizes successful case studies, lessons learned, and future trends, emphasizing the pivotal role of AI and IoT in fostering innovation, optimizing energy systems, and driving the industry towards a cleaner, more sustainable energy landscape.

This study introduces a novel methodology for estimating the parameters of implantable hydroelectric asynchronous generators, combining field simulations with a modified standard measurements approach.

The research focuses on developing a reliable yet straightforward technique for determining the constant parameters essential for the optimal functioning of these generators. The proposed method is tailored specifically to the unique requirements of implantable hydroelectric generators and is primarily based on specialized testing conducted during various stages of the generator's operation cycle. Initially, the approach involves an independent assessment of the stator inductor's field in these generators. Subsequently, all remaining constant parameters are estimated using stationary tests, specifically designed for the unique operational environment of implantable generators. To validate the effectiveness of this methodology, comprehensive field simulations are conducted.

Chapter 20
  Aditi Singh, Vellore Institute of Technology, India
  Akhil Kodalapuram, Vellore Institute of Technology, India
  Prathamesh Prabhu, Vellore Institute of Technology, India
  Arnava Soni, Vellore Institute of Technology, India
  Vijayapriya Ramachandran, Vellore Institute of Technology, India

Initially the project is started by recording power consumption from a laptop at different performance levels using a software known as "HWInfo." Next cleaned this data by filtering the potential variables needed for the analysis and obtained average values for unique utility percentages. The variables to be used for training different models at the time of model choosing were power and temperature for CPU and GPU individually. Upon training different models with the training data, ExtraTreeRegressor was the ML model that gave the least RMSE error, in turn being used for the prediction. Power consumption of VIT university had two variables and one predicting target, them being date, power units, and maximum demand respectively; for which the authors shifted to an ML model called NeuralProphet. This model takes care of any outliers in the data. To deal with the many inconsistencies and unpredicted data, NeuralProphet performs adaptive analysis which is used for training and prediction. After obtaining results from both data, the model is integrated into a web page's front-end.

Section 2
AI and ML Applications to Renewable Energy Applications

Chapter 8
  Dharani Jaganathan, Computer Science and Engineering, KPR Institute of Engineering and
   Technology, Coimbatore, India
  Vishnu Kumar Kaliappan, Computer Science and Engineering, KPR Institute of Engineering
   and Technology, Coimbatore, India

In the rapidly evolving landscape of smart cities, the deployment of ubiquitous robots holds immense potential for enhancing various aspects of urban living. However, the widespread integration of these robots into smart city infrastructures necessitates a careful consideration of energy efficiency to ensure sustainable and long-term operation. By leveraging advanced algorithms, these robots can adapt their behaviors and decision-making processes, leading to reduced energy consumption and increased operational sustainability. This chapter explores the application of reinforcement learning techniques to optimize the energy efficiency of ubiquitous robots operating in smart cities and also investigates various implementation methods of reinforcement learning in the context of smart cities, focusing on enhancing the energy efficiency of ubiquitous robots like search and rescue robots and contributing to the overall development of energy-conscious urban environments.

Chapter 9

    S. Selvakanmani, RMK Engineering College, India
    Seeniappan Kaliappan, KCG College of Technology, India
    M. Muthukannan, KCG College of Technology, India
    Mohammed, SRM Institute of Science and Technology, India

This chapter presents a novel approach for recognizing and securing cyber-physical systems (CPS) through the use of artificial intelligence in wireless sensor networks. The increasing use of CPS in various fields has led to a growing need for effective methods of identifying and securing these systems. The proposed approach utilizes artificial intelligence techniques to analyse network traffic and identify patterns that indicate the presence of a CPS. Additionally, the proposed approach uses this information to secure the CPS by implementing appropriate security measures to protect against cyber-attacks. This study highlights the importance of recognizing and securing CPS in wireless sensor networks, and the potential of artificial intelligence to meet this need. It also emphasizes the importance of developing secure and resilient systems in the face of cyber-threats and the need for a holistic security approach for CPS.

Section 3
Data Analytics for Energy Management

Chapter 13

    S. Fahira Haseen, College of Engineering Guindy, Anna University, India
    P. Lakshmi, College of Engineering Guindy, Anna University, India

Suspension in a vehicle is provided primarily to improve the passenger comfort and vehicle handling for the automobiles moving under any road conditions. Because of the non-linear characteristics of the vehicle, fuzzy logic controller (FLC) fed active suspension system is proposed for a full car with driver vehicle model. This controller dynamics are optimized by meta-heuristic optimization algorithm namely big bang–big crunch (BBBC) optimization and coyote optimization algorithm (COA). The passive system dynamics are compared with controller fed and optimized controller fed system under bump and random road inputs. The passive and active model is simulated in MATLAB/Simulink environment. The results are compared based upon root mean square values of head acceleration, body acceleration, pitch acceleration, roll acceleration, and power spectrum density of head acceleration. The results indicate that implementation of COA optimized FLC is effective in improving ride quality and road handling of the vehicle.

M. Jayalakshmi, Ravindra College of Engineering for Women, India

G. Malarselvi, SRM Institute of Science and Technology, India

Mohammed Ali, SRM Institute of Science and Technology, India

S. Kaliappan, Lovely Professional University, India

This research study focuses on the development of a communication network for industrial sectors using internet of things (IoT) technology in order to enhance privacy and security in the cyber-physical system. The increasing reliance on cyber-physical systems in industrial sectors has highlighted the need for secure and private communication networks. The proposed network utilizes advanced encryption techniques and secure communication protocols to protect sensitive data and critical infrastructure. The network architecture is designed to detect and prevent cyber-attacks in real-time and also implements secure communication protocols to prevent unauthorized access. In case of any failures, the proposed communication network has self-healing mechanisms to automatically restore normal operation. The findings of the research show that the proposed communication network is able to effectively protect against cyber-attacks and unauthorized access while maintaining the availability and integrity of the system.

Chapter 15

Suresh Vendoti, Godavari Institute of Engineering and Technology, India

Dana Victoria, International School for Technology and Science for Women, India

M. Muralidhar, Sri Venkateswara College of Engineering and Technology, India

R. Kiranmayi, JNTUA College of Engineering, India

Kollati Sivaprasad, Godavari Institute of Engineering and Technology, India

Renewable energy systems serve as a sustainable alternative to fossil fuels, deriving from natural ongoing energy flows in our surroundings. These systems encompass the production, storage, transmission, distribution, and consumption of energy. Renewable energy systems offer numerous advantages, such as reliability, environmental friendliness, absence of harmful emissions or pollutants, low or zero carbon and greenhouse gas emissions, reduced maintenance compared to non-renewable sources, cost savings, job creation, and independence from refueling requirements. This chapter provides an overview of various types of renewable energy systems, with a focus on solar/wind/battery or solar/wind/diesel with battery storage integrated energy systems. This chapter also covers the technical and economic aspects of different types of HRES and their comparative results. Based on the findings of this review, the chapter proposes a novel configuration for an off-grid hybrid renewable energy system designed for electrification in rural areas

Section 4
Cyber Physical Systems and Internet of Things

Chapter 16

Suresh Vendoti, Godavari Institute of Engineering and Technology, India
M. Muralidhar, Sri Venkateswara College of Engineering and Technology, India
R. Kiranmayi, JNTUA College of Engineering, India
Dana Victoria, International School of Technology and Science for Women, India
D. Ravi Kishore, Godavari Institute of Engineering and Technology, India

The exploration of renewable energy resources has gained momentum due to the continuous demand for energy consumption and the depletion of fossil fuel reserves. However, these resources possess an intermittent nature and are only viable in certain geographical locations. To address these challenges, this chapter presents a solution in the form of a hybrid energy system (HES). This system operates in an off-grid mode and is specifically designed for high altitude demographic users who face difficulties in accessing the national grid. This chapter utilizes a well-designed hybrid energy system to enhance the reliability and quality of power generation in rural areas. Also, the design of a linear mathematical model is discussed in this chapter, which aims to determine the optimal working and cost optimization of the hybrid energy generating system. The system comprises a wind-biogas-biomass based power generation system, PV array, fuel cells, a battery bank, and a bidirectional converter. To meet economic constraints and load dispatch, an efficient mathematical modeling is employed.

Chapter 18

M. D. Rajkamal, Velammal Institute of Technology, India
T. Mothilal, KCG College of Technology, India
M. Shanmugapriya, KCG College of Technology, India
M. Saravanan, Hindustan Institute of Technology and Science, India

This study presents a comprehensive failure rate examination for implantable antennas, employing a liability tree-based analysis focused on a clamped double subsystem (CDSM) equipped with DC short voltage protective functionality. The enhanced protective feature of the CDSM aims to improve the security and safety of implantable antennas used in critical applications. However, this subsystem's design necessitates the use of additional IGBTs, diodes, and capacitors compared to standard configurations, consequently increasing the complexity and potential failure rate. Given that demanding converter operation in implantable antennas can escalate the failure rate, conducting a precise reliability analysis becomes vital for the deployment of CDSM in these devices. A failure durability analysis is undertaken to address the operational characteristics of CDSM in implantable antennas. Fault Tree Analysis (FTA) is utilized to evaluate the risk with greater precision than previous methods, which mainly considered component types, quantities, and network connectivity states.

Chapter 19
Efficient Design for Implantable Device Constant Current Induction Doubly Fed Generating
S. Socrates, Velammal Institute of Technology, India
M. Shanmugapriya, KCG College of Technology, India
B. Murugeshwari, Velammal Engineering College, India
S. Angalaeswari, Vellore Institute of Technology, Chennai, India

This research presents an innovative approach to the efficient design of implantable devices, focusing on the development and modeling of a constant current induction doubly fed generator (DFIG) system that incorporates grid connectivity under both sub and hyper synchronization conditions. The core of this study is to establish a physical equation for a power station and a DFIG using a combination of power management and voltage estimation techniques in the context of circuit power. The induction generator (IG) blade in the system is designed to rotate synchronously with the photovoltaic (PV) system frequency. The DFIG is connected to a distribution substation, with synchronization between the active power filter and the grid depot achieved through the use of dual converters: a machine side converter (MSC) on the grid side and a grid side converter (GSC) on the power system. Within the circuits, two applications are implemented to recover the parameter spectrum, aiming to maximize the thermodynamic efficiency delivered to the DFIG rotor.

Preface

The globe is changing rapidly in terms of technical advancements, population growth, and energy demands. To address the nation's energy requirements, a variety of energy producing systems have evolved. To reduce carbon emissions, renewable energy sources are increasingly being used to generate electrical power efficiently. With the advancement of information and communication technology (ICT), several electrical engineering applications have been developed to increase efficiency and reliability electrical power systems. Various artificial intelligence (AI) and machine learning (ML)-based intelligent solutions are being developed to improve the secure operation and control of electrical power systems. As data science being widely used across all the sectors, varieties of algorithms are developed to analyze energy flow and consumption patterns, allowing for more efficient energy management.

This book provides insight for scholars working on the development of AI and ML-based intelligent strategies for power system applications. The integration of renewable energy sources into the conventional grid presents several challenges for the grid. The internet of things (IOT) is a cutting-edge technology that addresses many of today's difficulties in energy efficiency and control. All systems use communication technologies to function effectively; numerous algorithms are being developed in cyber physical systems employing network security for wireless sensor networks and industrial sectors to improve privacy and security. Finally, this book helps the engineering community to understand the multiple technological developments taking place in the application of IOT, Data Science, and ICT for efficient power system operation and control.

Chapters 1-5: These chapters provide a comprehensive review of the applications of Artificial Intelligence and Machine Learning to Sustainable Power Systems. This section provides an overview of the applications of Artificial Intelligence and Machine learning to improve the energy efficiency of cooling systems in data centers with multiple energy sources and a Microgrid.

Chapters 6-10: These chapters present an overview of the most recent AI and IOT operation and control solutions for power grids with renewable energy sources. Challenges associated with integration Renewable Energy Sources and parameter estimation for Hydroelectric Asynchronous Generators is explored in these chapters.

Chapters 11-15: These chapters focus on the development of cutting-edge algorithms for AI-Driven Energy Analytics for Demand Side Management. This section focuses mostly on data science applications for power system analysis and predictive analysis.

Chapters 16-20: These chapters explore solutions for real-time GPS coordinates for fleet tracking and improved performance of active suspension systems. Effective recognition and development of cyber physical systems using network security for wireless sensor networks and industrial sectors to improve privacy and security.

L. Ashok Kumar
PSG College of Technology, India

S. Angalaeswari
Vellore Institute of Technology, India

K. Mohana Sundaram
KPR Institute of Engineering and Technology, India

Ramesh C. Bansal
University of Sharjah, UAE & University of Pretoria, South Africa

Arunkumar Patil
Central University of Karnataka, India

Acknowledgement

Dr. L. Ashok Kumar is thankful to his wife, Y. Uma Maheswari, for her constant support during writing. He is also grateful to his daughter, A. K. Sangamithra, for her support; they helped him a lot in completing this work.

Dr. S. Angalaeswari would like to thank her husband Mr. R. Sendraya Perumal for his motivation and support during this book compilation. She is thankful to her son Mr. S. Dhianesh and daughter Ms. S. Rithika Varsha for their constant encouragement and moral support along with patience and understanding. She would like to extend her gratitude to her parents, in-laws for their persistent support throughout the path of success.

Dr. Mohana Sundaram Kuppusamy is thankful to his wife, Nidhya R, for her constant support during the book preparation. He is also grateful to his daughter and son, M.N. Soumitra and M. N. Muhilsai, for their support to complete the work.

Dr. Ramesh C. Bansal would like to thank his family and friends for their moral support in completion of this book. He would like to express his gratitude to the Almighty for guiding him through this creative journey. His deepest gratitude to all who have been a part of this incredible journey and made this book a reality.

Dr. Arunkumar Patil would like to express his special thanks to all the members of this book writing team for their time and efforts provided throughout the process. In this aspect, he wants to express his gratitude to Prof. Ashok Kumar for the useful advice and suggestions given, which were helpful during this process. He is grateful to all his family members for their constant support and guidance.

The editors would like to extend our gratitude to each of the authors who have contributed their chapters in this book, without them, this would not be possible. We acknowledge and appreciate the effort and time dedication by all the authors to complete their book submission on time with quality and efficient work.

The editors would like to thank and dedicate this book to their management and respective Universities. We like to recognize and thank the reviewers of this book who have done wonderful job by reviewing each chapter carefully and gave their comments for the enhancement of the chapters.

We finally thank the editorial board members in bringing the book in most successful manner in this span of time. The editors would like to extend our heartfelt thanks to all the editorial staff of IGI Global team for their immediate response and assistance throughout this journey.

Section 1
AI and ML Applications to Power Systems

Chapter 1
A Comprehensive Review on Single Axis Solar Tracking System Using Artificial Intelligence

Divya Prabha Varadharajan
S.A. Engineering College, India

Akila Muthuramalingam
K.S. Rangasamy College of Technology, India

Sathiyasekar Kumarasamy
K.S.R. College of Engineering, India

M. Venkatesan
K.S.R. College of Engineering, India

Ramasamy Murugesan
iD https://orcid.org/0000-0001-9292-2528
K.S.R. College of Engineering, India

Loganathan Nachimuthu
University of Technology and Applied Sciences, Oman

ABSTRACT

Renewable energy sources plays a vital role in power generation. Renewable solar energy resources abound on a global scale. When employing solar energy as a resource, the output of photovoltaic (PV) modules is determined by radiation and the temperature of the sun. The first PV systems were developed in the mid-nineteenth century. The underlying issue in the solar industry is that PV systems capture just a percentage of the sun's beams. The technique works best when the Solar Panel (SP) is perpendicular to the sun's rays. Solar tracking systems (STS) are used to keep a PV panel orthogonal to the radiation at all times, allowing for optimal energy output from the sun. With the help of advanced technology in tracking systems, numerous tracking techniques are available. The goal of this research is to explore the most recent improvements in solar tracking technology to estimate the best orientation for PV panels. The angular displacement of an STS with a single or dual-axis is possible. This chapter starts with mathematical modeling of the SP and ends with evaluation metrics to evaluate the performance of STS. The pipeline between the start and end points covers the identification of sun position, and Artificial Intelligence (AI) based controller. This survey can help the researchers to create an effective STS.

DOI: 10.4018/979-8-3693-3735-6.ch001

INTRODUCTION

Energy is regarded as an unbeatable resource. The amount of energy consumed is directly related to the development of a country's people power, industry, and economy (Kumar et al., 2014). As the world's energy scarcity worsens, researchers are driven to seek an alternative energy source to replace conventional fossil fuels. As society progresses, so does the energy demand, as does the exploitation of more traditional sources of supply. Furthermore, using nonrenewable energy sources causes the emission of greenhouse gases which produce harmful effects on our ecology. India accounts for 7% of global emissions, whereas the United States accounts for 15% (Lalit, 2019) of total global emissions. Green technology research and application are critical for limiting and mitigating environmental damage. This means that all conventional energy sources must be replaced with renewable energy sources that do not contaminate the environment while in use. Solar energy is defined as the use of the sun's beams to generate electricity. It is the least environmentally damaging energy source. The sun's output of 1.81011 MW, is millions of times greater than the current consumption rate of all commercially available energy sources on the earth.

PV panel-based solar power plants, street lights, transit, Smart Grid systems, and other autonomous systems based on solar energy conversion processes are all examples. According to one recent study, if only 1% of the sun's rays are captured, solar energy can satisfy the whole world's energy requirements. The benefits of solar energy are limitless, and it is also absolutely free. It emits neither pollutants nor greenhouse gases. A solar cell uses PV technology to create energy (Hsu et al., 2014). A PV cell is a device that translates solar energy into electrical energy. A PV is composed of numerous solar cells that are all built of silicon-like semiconductors. Boron or another trivalent impurity (p-type semiconductor) is frequently doped into one layer of the cell to allow for a positive charge, while phosphorus or another pentavalent impurity (n-type semiconductor) is doped into the opposite layer to allow for a negative charge. When two distinct kinds of semiconductors are combined and fused, PN junctions develops. Electron and hole concentration gradients at the semiconductor's surface promote electron and hole diffusion from p to n and n to p-type (Fan et al., 2018). When sunlight strikes the solar cells, photon energy is transferred to silicon electrons via collisions, and this energy is sufficient to free the electrons from their silicon atom parents. The free electrons are collected and moved by n-type silicon. Electrons must flow through the electrodes and into an external circuit, which is coupled to the external circuit, to generate electric current (Stapleton, 2017).

The existing solar energy generation system's efficiency is inadequate due to factors like stationary SP and climate conditions. The great majority of researchers worldwide are working to develop innovative strategies to enhance the effectiveness of PV power generation. The efficiency of a PV panel is affected by elements such as irradiance, temperature, and the angle at which sunlight reaches the panel. These characteristics can be ascribed to the increased output of the PV panels. Aside from that, the most significant disadvantage of solar power is its diluted nature. Solar radiation flow rarely reaches 1 kW/m, which would be inadequate for technological application in the earth's hottest zones. To address this issue, an ST can be used to ensure that the panel receives as much sunshine as possible from sunrise to sunset. While building autonomous systems, it is necessary to balance the reliability, ease of installation, affordability, and efficiency of solar systems (Rodrigues et al., 2016; Tukymbekov et al., 2019). Several tactics and technologies can be used to improve the efficiency of solar systems. One of these choices is an STS. Hence, this chapter discusses about the different kinds of tracking system and techniques adopted to achieve maximum efficiency.

Figure 1. Classification of STS

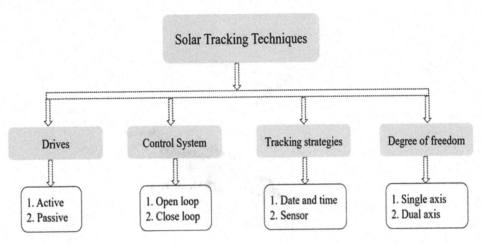

SOLAR TRACKING TECHNIQUES

STS are classified in several ways. It can be classified using control mechanisms, drivers, tracking tactics, and the degree of freedom of travel which is shown in figure 1.

Drives

1. STs (STS) that do not use mechanical motors can orient their sensor units toward the sun's light stream (Ya'u Muhammad et al., 2019). These trackers generally use expandable gas or an alloy with shape memory. The approach employs thermal effects or a pressure differential among two positions at the SP's ends. When the PV panel is orthogonal to the sun, it is in balance. The heated side of the PV panel expands and makes contact with the other as the sun moves across the sky, forcing it to rotate. This method of tracking the sun, however, is very dependent on the weather conditions of the site location, making it less accurate than other approaches. Only if the location where an ST is installed receives enough sunlight, maximum power can be extracted. Because of the emergence of active ST systems, passive ST systems may no longer be required.

2. An active STS (fig. 2) continuously analyses the location of the sun throughout the day using sensors that are already in situ (Razif Hamid et al., 2017). In response to the sun's rays, the sensor will activate the motor attached to the SP. The intensity on one light sensor might change if the sun is not vertical to the tracking system. This variation could be employed to determine where the tracker should be placed vertically to the sun or not.

3. Passive solar tracking systems (Fig. 3) are innovative technologies designed to harness solar energy with minimal or no active mechanical components. Unlike active solar tracking systems that use motors and sensors to follow the sun's movement, passive systems rely on the inherent properties of materials or designs to optimize solar exposure. These systems are more energy-efficient and cost-effective, making them an appealing choice for sustainable and environmentally friendly energy generation. This system comprises two tube tanks attached to the PV panel's side. Passive tracker systems, similar to active trackers, offer various configurations, such as single-axis, dual-axis, and

Figure 2. Active solar system

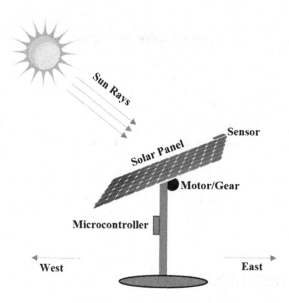

hybrid passive solar tracking systems. Their operational principle is less complex compared to active trackers. Passive trackers are cost-effective, and they exhibit greater resilience to changes in illumination. However, a downside is their slower response, particularly in the morning hours.

Control System

1. Open-loop tracking control schemes determine the sun's location for a specific site rather than actively tracking its position. The tracker estimates the sun's location with help of the current time, date, weekday, month, and year. The tracker uses a stepper motor to track the location of the sun. No sensors are used in this control mechanism. As a result, the open-loop tracker is often used. Time determines the direction of the sun's rays relative to a point on Earth. It is location- or longitudinally-dependent, as opposed to local clock time. The open-loop ST swings the panel from east to west when the sun rises and sets. This activity is only possible if you know the sunrise and sunset times for each day. This is because the open-loop ST must rotate and stop at each dawn and sunset to orient the solar PV panel eastward (Huynh et al., 2013).
2. The closed-loop tracker employs feedback control concepts. Closed-loop tracking control systems use light sensors on the PV panel to trace the location of the sun at any time of day. If the PV panel is not exactly facing the sun, there will be a discrepancy in light intensity between light sensors. We can utilize the difference between the two values to aim the tracker directly toward the sun (Prabha & Mohana, 2018).

Tracking Strategies

1. Microprocessors and other electronic components in date/time tracking systems employ algorithms and basic formulas to determine the location of the sun and then send signals to an electromotor. A

Figure 3. Passive tracking system

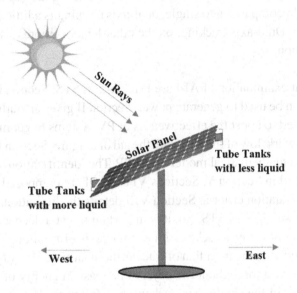

researcher demonstrated a program-based sun-tracking structure for parabolic collectors (Edwards, 1978). A predefined schedule was employed daily to modify the speed of the collection actuators. According to the findings, the system only requires a 500 bit/s data output from the central controller for 10,000 collectors to precisely follow the sun every day.

2. In date/time tracking systems, sensors trace the solar position. The CPU then gets the signal and regulates the motors accordingly. As reported by the author (Kalogirou, 1996), single-axis STS was accomplished with the help of three light sensors, an electrical circuit, and a dc motor. An electronic circuit gets input from sensors to activate the motor. A chronological solar tracking system is designed to adjust the orientation of solar panels or collectors based on the time of day or the position of the sun throughout the year. Unlike traditional solar tracking systems that follow the sun's real-time position in the sky, a chronological tracking system pre-programs and adjusts the angle of the solar panels according to a set schedule or a calendar. This approach takes into account the changing solar angles and positions throughout the year, allowing for optimal energy capture over different seasons and times of the day.

Degree of Freedom

1. The single-axis tracker can follow the sun's horizontal or vertical movements. There appears to be only one axis of rotation on this tracker. Horizontal ST is employed in tropical settings where the sun rises low in the sky and sets swiftly. Vertically oriented STs are employed in areas where the sun does not rise very high yet summer days are long. In concentrated solar power operations, single-axis trackers with flat surface solar modules will be used (Chaysaz et al., 2019).

2. A dual-axis ST appears to move in two directions at the same time. It may be able to track the sun in both horizontal and vertical positions. Dual-axis tracking can be utilized anywhere, allowing solar power to be generated anywhere in the world. It is used in solar power structures and dish systems to concentrate solar power (CSP). This type of tracking is essential in solar power tower

procedures because angle inaccuracy is important for increasing the gap between the reflectors and the tower structure. In comparison to single, dual-axis tracking is a little challenging to design and run (Dhimish, 2019). Dual-axis tracking, on the other hand, is more cost-effective for large-scale solar energy generation.

This paper includes a clear examination of AI-based single-axis STS. Section I provides an overview of the solar cell and how it can be used to generate power. Section II gives an outline of the several types of STS that have been designed to boost the effectiveness of PV systems to get maximum energy from the sun. Section III details the workflow of the STS with the aid of a figure. Section IV describes the available methods for constructing a mathematical model of the SP. The identification technique done to capture the sun position is described in Section V. Sections VI and VII are composed of the advanced controller techniques and their evaluation criteria. Section VIII details the limitation and future scope of STS.

The chapter deals with single-axis STS. So, it is important to get a deeper knowledge of the single-axis system. A single-axis tracker is a tracker with a single axis of rotation. This is because its axis of rotation is typically oriented with the meridian of true northern latitude. They typically cover the motion of the sun between east to west throughout the day. An increase in energy of 25 to 35 percent is to be predicted. The rotating axes of the single-axis tracker are as follows:

- Horizontal Tracker: Smaller latitudes are better suited to the most common single-axis tracker setup. It rotates with an axis of rotation parallel to the ground. They are completely versatile and rotate from east to west direction on a fixed axis horizontal to the earth. Horizontal trackers are often oriented to face the module, which must be parallel to the rotation axis. It brushes a cylinder in the same manner as a module path brushes a symmetrical cylinder around the axis.
- Vertical tracker: These trackers' axes of rotation are aligned vertically concerning the earth's surface. They tend to wander clockwise from east to west on the whole day. These trackers outperform horizontal axis trackers at high latitudes. On vertical single-axis trackers, the module is generally slanted concerning the rotation axis. It, like a module path, brushes a rotationally symmetric cone about the rotation axis.
- Tilt tracker: Trackers featuring both horizontal and vertical axes of rotation are known as "horizontal-to-vertical" trackers. Typically, the tilt angles are restricted to reduce the wind flow and minimize the increased terminal height. They are typically parallel to the rotation axis. For best efficiency and solar tracking capabilities, this sort of ST tracks the sun throughout the day.
- Polar aligned trackers: The employment of these trackers is a well-established way for ascending a telescope's reinforced form. The rotation axis of the polar star is parallel to the tilted single axis. The end product is a polar-aligned single-axis tracker.
3. A hybrid solar tracking system

It combines multiple tracking technologies and approaches to optimize the capture of solar energy. These systems utilize a combination of active and passive tracking mechanisms, as well as various sensors and controls to ensure that solar panels or collectors continuously face the sun for maximum energy output. Hybrid solar tracking systems are designed to improve efficiency and energy generation in a variety of solar applications, making them a promising solution for sustainable power generation.

Figure 4. The workflow of a single-axis SP tracking system

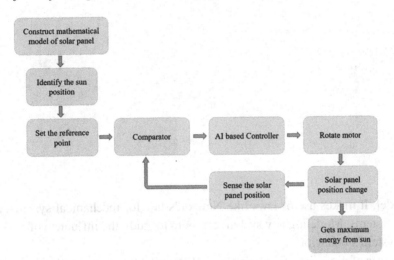

PROCESS FLOW

For precise tracking of the sun by the single-axis SP to get maximum solar energy, the first step is to identify the mathematical model of the SP. Next in this work, the close loop tracking system is taken for research. The set point should be given to the control system, based on the sun's position from the center of the SP. The position of the sun can be calculated by different methods with the help of sensors or cameras. Next to the set point, the comparator is used which helps to identify the error between the given set point and the actual position of the SP. The actual inclined position of the SP is identified using the sensor. The value obtained from the sensor is called the feedback value. The controller generates the control signal based on the error value obtained from the comparator. The control signal is directly given to the motor. The motor is responsible for the inclination of the SP. Hence it is important to design the controller effectively to get the maximum power from the sun. The workflow of the single-axis SP tracking system is detailed with the help of figure 4.

MATHEMATICAL MODEL OF SP

The SP system can be explained using a mathematical model. The process of creating a mathematical model is known as mathematical modeling. Using arithmetic and hypothesized analysis, problems in an application domain are translated into manageable mathematical formulations that provide insight and suggestions for the construction of an application. Mathematical modeling gives reliable techniques for problem-solving and also a comprehensive explanation of the physical process. This technique also facilitates the improvement of the design, management, and utilization of contemporary computational capacity. Mathematical modeling is a crucial transitional step between academic instruction and application-oriented competence; it also assists students in adapting to the difficulties of today's technological society. "Modeling" is the process of creating a differential equation to explain a physical condition. The underlying physical principles that govern the behavior of the system serve as the foundation for a

Figure 5. Variation of sun angle in a day (Aksungur & Koca, 2018)

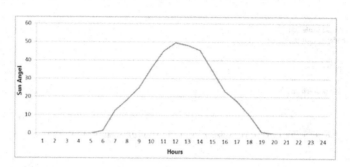

mathematical model. It makes use of laws like Newton's law for mechanical systems and Kirchhoff's law for electrical systems. Modeling any system can help to study the influence of various components and forecast the system's behavior.

There are two sorts of modeling: first-principles (FP) and data-driven (DD) (Czop et al., 2011). The mathematical representation of FP models is developed from a knowledge of the underlying physics of the system. FP models require high-level skill in the field of knowledge to derive mathematical equations from laws, whereas DD models construct their mathematical representation using system test data. The former technique has the advantage of being able to predict system performance, but the latter has the advantage of developing an accurate model quickly and earning trust by using data from the actual system. The latter technique is more difficult to use for determining phenomenological parameters. To achieve the greatest results, FP models are routinely modified via trial-and-error. However, one disadvantage of employing DD models is that many data sets has to be considered for all conceivable operational circumstances.

IDENTIFY SUN POSITION

Due to the permanent and periodic rotation of the Earth, the shifting location of the moving sun varies with the passage of time and the seasons. Consequently, determining the precise location of the sun at a given time has become crucial. The positions are depicted on a Sun Path Diagram. As a function of time, figure 5 displays the daily variation in sun angle. Both light sensors and image processing can be used to determine the location of the sun.

A sensor placed on the sun's position can be employed (Hammoumi, 2018; Kiyak & Gol, 2016; Oladayo & Titus, 2016). Solar tracking devices use two photo resistors that are oriented east-west to monitor the intensity of the sun. A photo resistance sensor, also known as a light-dependent sensor (LDR), changes resistance in response to light. The voltage measurements from the LDR on the east and west sides of the SP are evaluated. The location of the sun is determined based on this.

The image processing technique can be utilised to ascertain the camera's position relative to the sun, and leverage this information to determine its spatial orientation within the image (Syafa'ah et al., 2018). A camera must be used to capture the sun to process the image. Images are handled to determine the location of the sun's centroid in the Cartesian coordinate system of the camera. The original image of the sun will be handled first by the vision assistant. In thresholding, the colours black and red are utilized as a second threshold. In this binary inversion image, the area around the sun appears to be the brightest. The remaining tiny particles can be eliminated using contemporary morphological approaches. The use of a low pass filter causes the image to lose detail that is not immediately visible. The circle detection results show the location of the sun's ring. By applying a coordinate system to the image, the detected circle's coordinate point is recovered from the centre frame.

AI-BASED CONTROLLER

SP tracking systems rely significantly on their controllers. The controller would check to see if the actual position of the SP deviated from the required value and if so, it would make the appropriate adjustments. The controller would detect the fault and conduct the appropriate control function. The controller would be responsible for keeping the process variable in the intended condition. Researchers commonly employ traditional controllers to automate STS. Deep-learning AI controllers, which have recently been presented as a new generation of goal-achieving controllers, have recently been proposed. AI controllers can be a very successful instrument in achieving their specialized operational goals. Through sampling from a target system, AI controllers can transcend the limits of classical controllers such as PID controllers (Lee et al., 2022).

ANN

During the last few decades, AI control has achieved significant advances in practical applications. This strategy can assist both single-input–single-output and single-input–multiple-output systems (Bahman-yar & Karami, 2014). Although linear models have predominated in applied predictive controllers in recent years, the majority of real-world systems are nonlinear. Linear predictive control performs even worse for increasingly advanced nonlinear models. Researchers sometimes utilize predictive control and nonlinear techniques combined to tackle the preceding issue and deal with nonlinear systems regularly. The optimal solution to this problem may be nonconvex, and finding it online will take a long time. A robust mathematical model of the nonlinear system is required for nonlinear predictive controls to work. It became vital during the development of the predictive controller algorithm to discover a simple practical model that accurately explains system behaviour. To address these challenges, Neural Network Predictive Control (NNPC), an advanced control system that can recognize complicated models and overcome parameter uncertainty, was created. In NNPC, NN models the unknown process. Controlling the NNPC is accomplished in two stages:

- NN-based system identification: The offline identification of the system is the first step in predictive control using NN. NN training signals are denoted by NN output and expected error among SP output and required output. The NN model forecasts future output by using the SP's past input and output data. The data from the SP is employed to train the NN offline. The NN can be trained using a variety of ways.
- NN-based predictive control: The control strategy is used in this strategy. A NN can be used to predict the production of SP over a specific period. Numerical optimization is used over the prediction time horizon to regulate the forecasting control signal to reduce the underlying objective function.

Figure 6. ANN controller architecture

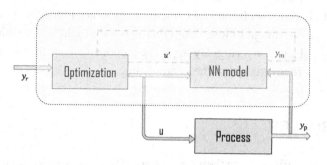

$$J = \sum_{j=N_1}^{N_2} \left(y_r\left(t+j\right) - y_m\left(t+j\right) \right)^2 + \rho \sum_{j=1}^{N_u} \left(u'\left(t+j-1\right) - u'\left(t+j-2\right) \right)^2$$

Here, N_1 represents the minimum and N_2 represents the maximum prediction horizons of the output, N_u represents the control horizon and the control signal u_0 denotes the tentative control signal, y_r denotes the target response, and y_m denotes the network model response. Figure 6 depicts the NNPC architecture. The NN predictive controller and optimization elements in the models are interconnected. When the objective function J is minimized, the u_0 values are modified, and the best control signal u is sent to the SP (Elsisi, 2019).

Fuzzy

Fuzzy logic control (FLC) is a popular AI control technique. It makes use of the functionary's past knowledge of the system. The key functions of the functionary, which establishes decision-based rules for the system, are system behaviour and linguistic input variables. The FLC includes fuzzification interactions, knowledge representation, decision-making algorithms, and defuzzification interactions are shown in figure 7 (Lee, 1990).

1. The fuzzification has the below-mentioned features:
   ○ Input parameters are recorded.
   ○ A scale mapping is used to translate the range of input parameter values to the matching universe of discourse.
   ○ Fuzzification converts input parameters into linguistic values which could assist as labels for fuzzy sets.
2. Understanding the application domain and the accompanying control objectives is an important aspect of developing a robust knowledge base. The system's two components are a "database" and a "Rule basis."
   ○ Using the definitions in the database, an FLC can define language control rules and fuzzy data manipulation.
   ○ To characterize the domain experts' control aims and control policy, the rule base employs a set of language control rules.

Figure 7. FLC architecture

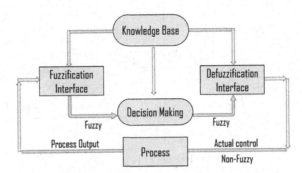

3. FLC employs fuzzy logic to infer control actions and to replicate human decision-making depending on fuzzy ideas and inference procedures.
4. Defuzzification Process includes
    ◦ Scale mapping, which transforms the values of output parameters into appropriate discourse
    ◦ Defuzzification can be used to derive a non fuzzy control action from the fuzzy action.

ANFIS

A network system that receives inputs, processes them and then outputs the desired outcomes. A computer-based NN executes the task at hand by using an algorithm designed for the task at hand. Many inputs are combined to produce one output in a fuzzy logic inference system (FIS). An adaptive neuro-fuzzy inference system (ANFIS) was developed for non-linear applications by combining fuzzy logic and NN. This is because they can substantially alter the membership functions (MF) to achieve the desired results (Mehmet & Erol, 2017). The ANFIS structure employs two basic learning cycles: forward pass and backward pass. As a result, the categorization process can remember earlier observations. ANFIS combines the benefits of NN with fuzzy logic into a single model. The first level of this method is a two-step procedure in which the rules mined from the simulated system's input and output data are employed to create an initial fuzzy model and its input variables. NN is then employed to fine-tune the rules of the early fuzzy model, resulting in the system's ultimate ANFIS model. ANFIS's primary advantage in meeting shifting load needs is its rapid convergence time. Rapid convergence at all levels of the system is required for power supply-demand management. Figure 8 depicts the ANFIS structure, which was created by Jang (Jang, 1993) in 1993. The circle denotes a fixed node and an adaptive node. Tagaki-Sugeno type is constructed, which attempts to simulate a nonlinear model with a range of linearization models. Through training the network with datasets, fuzzy rules and MF are generated. Backpropagation or hybrid methods are employed in the learning stages to find the parameters of the adaptive system.

Reinforcement Learning (RL)

Nonlinear, presumably stochastic, and unknown or severely unpredictable dynamics might benefit substantially from the advanced algorithms of RL (Buşoniu et al., 2018). RL algorithms frequently have two primary objectives. Policy evaluation is one thing, and policy improvement is another. Current policy

Figure 8. ANFIS architecture

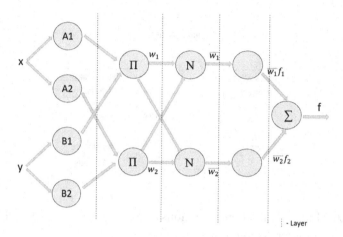

evaluation and improvement are two distinct processes that assess the cost or value function associated with the current policy. The two basic types of RL algorithms utilized to carry out these two processes are policy iteration (PI) and value iteration (VI) (Kiumarsi et al., 2018). PI and VI techniques, which undertake policy assessment and development, can be used iteratively to find a better policy. PI techniques use an arrangement of Bellman equations to estimate the optimal control policy with an accepted control policy. As with PI systems, a stabilizing control policy is not required with VI approaches. Although most RL-based control techniques are PI-based. Concentrate on PI-based RL technique for feedback control design. In the field of control, RL algorithms are commonly employed to resolve the following issues: 1) optimal single-agent system regulation and tracking, and 2) optimal multiagent system coordination. An optimal regulation problem aims to manage system states or outputs to zero or near zero, whereas optimal tracking control seeks to have optimal controllers maintain a suitable reference trajectory between input and output values. The goal of optimal coordination of multiagent is for actors to achieve specified team objectives to create distributed control protocols based purely on local information available to agents. The two types of RL algorithms utilized to address optimal control problems are on-policy and off-policy learning control procedures (Sutton & Barto, 2017). On-policy procedures are used for policy evaluation and improvement. These two obligations are separated through non-policy approaches. When analyzing data, the behavior policy that produced the data may have nothing to do with the estimation or target policies. An offline method can be used to collect information about system dynamics required for online learning of target policy rules. Off-policy approaches leverage the data supplied by a behavior policy efficiently and fast to apprise many value functions for various estimating policies. Offpolicy algorithms also take into account the influence of probing noise required for exploration. Figure 9 depicts a schematic representation of on-policy and off-policy RL.

Optimization

In an automated system, this is the most commonly utilized PID controller. PID parameter setting and optimization are continuously popular research topics in the field of automated control. The PID controller's parameters that influence the overall control impact include proportion, differential, and integral. The controller can employ different weights to calculate the control signal that will drive the controlled

Figure 9. RL architecture a) Off-policy b) On-policy

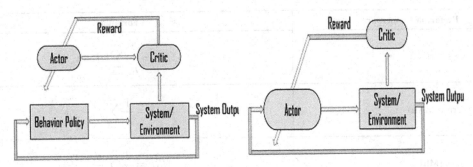

object. If the controller design is appropriate, the inaccuracy in the control signal will be reduced. As a result, optimizing the PID controller's parameters is crucial. Humans typically alter PID controller parameters, which not only takes a long time but also does not provide the best potential results. It is described as an optimization control problem, and different optimization techniques are used to determine the appropriate controller parameters.

The optimization algorithm falls within the category of metaheuristic approaches, particularly belonging to the evolutionary algorithm family. Numerous researchers have drawn inspiration from nature, studying the behavior of various organisms like chromosomes, fish, bees, ants, birds, fireflies, and more. The primary function of an optimization algorithm is to seek the best possible values for unknown parameters in a system, with the objective of either maximizing or minimizing fitness

Evolutionary Programming (EP) (Maddi et al., 2019), Ant Colony Optimization (ACO) (Zhang & Zhang, 2021), Particle swarm optimization (PSO) (Li et al., 2020; Mohanasundaram et al., 2014), cuckoo search optimization (CSO) (Jitwang et al., 2020), moth flame optimization (MFO), Artificial Immune Systems (AIS) (Saleh & Saad, 2016), Genetic Algorithm (GA) (Bhakuni, 2022), Simulated Annealing (SA) (Bhakuni, 2022), water cycle optimization (WCO) (Pachauri, 2020), teaching–learning-based optimization (TLBO) (Shouran & Elgamli, 2020), Bat Algorithm (BA) (Pebrianti et al., 2019), Firefly Algorithm (FA) (Ramyashree et al., 2018), Imperialist Competitive Algorithm (ICA) (Fernandez Cornejo et al., 2020), and hill-climbing (HCO) (Büchi, 2022) have been used. Although optimization algorithms share common characteristics, including three fundamental phases of initialization, main loop, and algorithm outcome, they each exhibit distinctive behaviors. The initialization phase involves generating an initial population with random or uniform values. The main loop is the core of the algorithm, responsible for identifying the optimal value of the unknown parameter through the algorithm's fitness function. Finally, the results obtained from the algorithm reveal the optimized parameter.

EVALUATION METRICS

Evaluation performance criteria like Mean Square Error (MSE), Mean Absolute Error (MAE), Mean Relative Error (MRE), Root Mean Square Error (RMSE), Efficiency, Overall Index (OI), Coefficients of Variance (COV), Coefficients of Determination (COD), Correlation Coefficients (CC) and Coefficient of Residual Mass (CRM) can be used to assess the predictor's quality (Elsheikh et al., 2019; Zangeneh et al., 2012). The variable n_s, d, and y represent the total observation, desired, and expected output val-

Table 1. Performance measure

Performance measure	Formula
Mean square error (MSE)	$$MSE = \frac{1}{n_s}\sum_{i=1}^{n_s}\left(d_i - y_i\right)^2$$
Mean absolute error (MAE)	$$MAE = \frac{1}{n_s}\sum_{i=1}^{n_s}\left|d_i - y_i\right|$$
Mean relative error (MRE)	$$MRE = \frac{1}{n_s}\sum_{i=1}^{n_s}\frac{d_i - y_i}{d_i}$$
Root mean square error (RMSE)	$$RMSE = \sqrt{\frac{1}{n_s}\sum_{i=1}^{n_s}\left(d_i - y_i\right)^2}$$
Efficiency	$$EC = 1 - \frac{\sum_{i=1}^{n_s}\left(d_i - y_i\right)^2}{\sum_{i=1}^{n_s}\left(d_i - d\right)^2}$$
Overall index (OI)	$$OI = \frac{1}{2}\left(1 - \left(\frac{RMSE}{d_{max} - d_{min}}\right) + EC\right)$$
Coefficient of variance (COV)	$$COV = \frac{RMSE}{\frac{1}{n_s}\sum_{i=1}^{n_s}y_i}*100$$
Coefficient of determination (COD)	$$R^2 = \frac{\frac{1}{n_s}\sum_{i=1}^{n_s}\left(\left(d_i - \bar{d}\right)\left(y_i - \bar{y}\right)\right)^2}{\sum_{i=1}^{n_s}(d_i - \bar{d})^2\sum_{i=1}^{n_s}(y_i - \bar{y})^2}$$
Correlation coefficient (CC)	$$CC = \frac{n_s\sum_{i=1}^{n_s}d_i y_i - (\sum_{i=1}^{n_s}d_i)(\sum_{i=1}^{n_s}y_i)}{[n_s\sum_{i=1}^{n_s}d_i^2 - (\sum_{i=1}^{n_s}d_i)^2][n_s\sum_{i=1}^{n_s}y_i^2 - (\sum_{i=1}^{n_s}y_i)^2]}$$
Coefficient of residual mass (CRM)	$$CRM = \frac{\sum_{i=1}^{n_s}y_i - \sum_{i=1}^{n_s}d_i}{\sum_{i=1}^{n_s}d_i}$$

ues for the metrics, as described in Table 1. The d_{max} and d_{min} reflect the maximum and minimum value of desired output, respectively. Next to the $\bar{y}$ and $\bar{d}$ are averaging of expected and observed values. The lower the MSE, MRE, MAE, and RMSE value, the better the algorithm, and zero is the best value for all of these metrics. The Efficiency coefficient is used to determine how well the method fits between the anticipated and observed values, with 1 being the best number. An algorithm with an OI value of 1 is a perfect match between the observed and projected data sets, as a rule of thumb. Meanwhile, a CRM

Table 2. Summary of previous work

Reference	Aim	Results
(Al-Rousan et al., 2020)	ANFIS models containing five MF might improve the efficiency of ST by perfectly anticipating the sun's path across the sky and reducing error, hence enhancing the power generated from the STS.	A solitary MF produced a prediction rate of 37.91% and an MSE of $0.30*10^{-2}$. And 3 MF, a 100% success rate was achieved, with the lowest MSE of $1.60*10^{-6}$
(Zakariah, 2015)	Solar tracking is based on light intensity analysis using an FL estimating technique for two axes of tracking. An FLC employs a power window motor to drive the SP on its axis, which gathers data from LDR sensors and moves accordingly.	In comparison to the stationary panel system, the system achieves a small increase in the output power efficiency of 18.13 percent.
(Larico & Canales, 2022)	The PV effectiveness of a dual-axis ST depending on PV cell incidence angles should be investigated at elevations greater than 3800 meters over mean sea level. Principal component analysis was used to identify factors influencing solar system performance.	The proposed method yields 37.63% efficiency more than the fixed PV system, and the energy generation was 10.66 kWh/ m²/d more than that in the fixed system.
(Hijawi & Arafeh, 2016)	Throughout the day, keep the PV panels pointed directly at the sun. This study presents a fuzzy inference approach-based STS architecture. The system is modeled using the Mamdani fuzzy logic model and different ANFIS modeling combinations.	A suggested dual-axis STS provided 22% more power than a static PV system.
(Hanwate & Hote, 2018)	The construction of a PID controller for an STS using a quadratic regulator technique with compensating poles (QRAWCP). While current iterative soft computing approaches are time-intensive, computationally expensive, and require previous knowledge of the solution space limit, this methodology avoids these limitations.	Performance of proposed model with 50% perturbation with disturbance ISE=0.0315, IAE=0.0536, ITAE=0.0037 Performance of proposed model with 50% perturbation without disturbance ISE=0.0315, IAE=0.0532, ITAE=0.0024
(Rezoug et al., 2021)	Using the Kalman filter to improve Inertial Measurement Unit (IMU) efficiency in an ST system. The orientation/energy yield ratio of autonomous PV panels could be optimized using the astronomical control of an autonomous device.	The tracking method allowed by IMU substantially gains good advantages over a static PV system, and the inclusion of a Kalman filter boosts both the stability and efficiency of the filter-less system (about 0.8 percent).
(González-Acevedo et al., 2021)	An electronic system for monitoring the output power produced by a single-axis ST is also being designed, in addition to the mechanical components. The SP's position is adjusted by the construction of the PD controller.	The average monthly increase in power produced by ST is 19.5 percent, with a maximum rise of 47.84 percent, with a minimum rise of 19 percent.
(Rawat et al., 2020)	An optimization method can be used to improve the position regulation of the STS. A PID controller adjusted using a range of techniques, such as GA, PSO, and TLBO, controls the position of the system.	In comparison, TLBO-tuned PID controllers outperform other approaches. The rise time is 0.0361 seconds, the settling time is 0.0538 seconds, and the overshoot percentage is 0.0110%.
(Roslan, 2020)	The PI controller is being modified to enhance the power quality of the 3φ grid-connected inverter system. The PSO technique was utilized to minimize the voltage and current controller schemes in the PI controller parameters of the inverter system.	As compared with the traditional method, the PI controller using PSO lowers voltage overshoot by 11.1% and duration to stability state by 32.6%.
(Tharakan et al., 2021)	ML-based dual-axis ST to improve energy harvesting efficiency. This method, which employs the Nave Bayes (NB) algorithm, is capable of reliably forecasting the sun's path.	In comparison to a panel held at a flat angle to the ground when this tracker was used, a power gain of around 19.51% is obtained by the suggested technique. The inclusion of a tracker adds an extra layer of shadow protection, significantly lowering the chance of hot-spot damage.
(Datta et al., 2016)	Soft computing is used to optimize the tilt angle of SP. In the first stage, the tilt angle was optimized monthly using a GA. Based on the results of the GA, the dual-axis ST was created in the second stage.	In comparison to a panel held at a flat angle to the ground when this tracker was used, a power gain of around 19.51% is obtained by the suggested technique. The inclusion of a tracker adds an extra layer of shadow protection, significantly lowering the chance of hot-spot damage.
(Ehiagwina, 2021)	A solar tracking device using an artificial NN improves electricity output. An ST ensures that the device's SP is continually towards the sun. Based on gathered solar energy data, a NN was trained to forecast irradiation, energy, and inclination.	The neural system properly monitors the maximum power point with minimal fluctuations and power dissipation. - The obtained value of error is around 0.08072 to 0.06147 and MSE is 0.0025
(Engin & Engin, 2015)	ST can be controlled as a self-contained device by employing the real-time control technique, that combines the best characteristics of both closed-loop and open-loop control. The algorithm determines solar time by employing astronomical calculations and GPS tracking, then translates it to a PWM control signal for a motor.	The total cost of a sun tracker is 21.1% greater than that of a stationary PV system. On a cloudy day, the two-axis STS generated 32.1% more energy than the fixed panel, respectively.
(Ali et al., 2021)	Both a horizontal and a vertical rotating axis of a dual-axis system are utilized by the system. An AI-based PID controller such as PSO, FA, ACO, ICA, and BA is used to control both forms of movement	The sun's location was slightly better tracked by BA than by the other approaches. BA may point at the fastest place of 0.073s for vertical and 0.241s horizontal axis in 0.073seconds.
(Faraji et al., 2020)	Employs a closed-loop combination of MLP NN and P&O (Perturbation & Observation) can rapidly and precisely find the sun via the dual-axis sun tracker. The NN is trained using the back propagation training algorithm and then optimized using the Modified PSO approach.	The proposed monitoring method takes 42 percent to 49 percent less time than the comparison method. As a result, the sun tracker has improved dynamic performance, minimal power consumption, and higher efficiency.

of 0 shows that the model is highly efficient. Furthermore, the CC is utilized to determine the degree to which the predicted and actual values are similar. CC is considered optimum when it is 1, suggesting that the model performs exceptionally well.

Table 2 summarizes the previous research work done on the solar tracking method using AI techniques. The table consists of three columns. The reference paper is cited in the first column. The proposed work of the journal is described in the second column, and the result and performance improved by the proposed method are detailed in the last column.

LIMITATIONS AND FUTURE SCOPE

According to research, ST boosts the overall efficiency of a tracking system. The majority of the challenges that limit the overall effectiveness of static and tracked SP were discovered to be identical in the study (Verma, 2020). Tracking the location of the sun has been shown to significantly increase efficiency, even on cloudy days. However, different types of ST have different benefits and drawbacks in terms of performance and efficiency. Although the number of axes used by ST influences its overall efficiency, it has also been noticed that temperature conditions and the technology used can have a significant impact on this system's efficiency, particularly in hot weather settings. More research is required in this field because several operational efficiencies, minimal cost, and feasibility questions remain unresolved.

- ST, on the other hand, are slightly more expensive than their fixed counterparts since they require a greater level of technology and moving parts to function properly. This frequently corresponds to an increase of $0.08 - $0.10 per watt of electricity generated, depending on the size and location of the SP.
- While reliability improvements mean that ST requires more maintenance than standard fixed racks, the quality of the ST can impact how much and how frequently this maintenance is needed.
- There are more moving parts in a tracker system than in a fixed tracking system. As a result, as part of the total site preparation, additional trenching for wiring and grading is frequently required.
- A single-axis tracker project's stability and financial feasibility must be given special consideration. Funding for these efforts is more difficult to obtain due to the complexity of the systems and, as a result, the higher risk they offer to lenders.
- ST is a more practical option because they are better suited to warm locations with little or no snow. Tracking systems adjust to harsher environmental conditions more slowly than permanent racking.
- In terms of field adjustment, fixed are more flexible than single-axis STS. Fixed systems from the east to west direction can accommodate slopes of up to 20%, whereas tracking systems from the north to south direction can accommodate slopes of about 10%.

Solar power is now the fastest-growing source of electricity. Because of its little environmental impact, it is regarded as a clean energy resource. Many changes are happening because this field of study is evolving so swiftly. The application of improved control and modeling of various solar energy systems, for example, has resulted in greater accuracy, good generation capabilities, and shorter computing time. Deep learning, low-cost, and open hardware for computer vision have been established by scientists and researchers focusing on energy extraction optimization to reduce operational and financial constraints

(Guduru et al., 2023; Karabiber & Güneş, 2023; Mamodiya & Tiwari, 2023; Mohanasundaram et al., 2014; Okoye et al., 2017; Palomino-Resendiz et al., 2023; Patel, 2021; Prabha & Mohana, 2018).

CONCLUSION

Solar energy, being an abundant, inexhaustible, cost-effective, and environmentally friendly resource, has prompted global exploration into the most efficient methods of harnessing it. This chapter offers a concise overview of Solar Panels (SP) and their role in converting solar energy into electricity. Additionally, it outlines the implementation of effective STS through closed-loop control. Recent advancements in STS technologies have led to the creation of diverse SP applications, deviating from the traditional fixed panels. The single-axis STS, in particular, enables SP to adapt to the sun's changing position throughout the day, significantly enhancing solar energy capture and output power. This chapter serves as a valuable resource for aspiring STS researchers, providing insights and guidance for the implementation of their ideas. The contents of this review article hold significance for future researchers seeking to employ artificial intelligence for the development of more precise and efficient STS.

REFERENCES

Aksungur, S., & Koca, T. (2018). *Solar Tracking System with PID control of solar energy panels using servo motor*. International Journal of Energy Applications and Technologies. doi:10.31593/ijeat.450834

Al-Rousan, N., Mat Isa, N. A., & Mat Desa, M. K. (2020). Efficient single and dual axis solar tracking system controllers based on adaptive neural fuzzy inference system. *Journal of King Saud University. Engineering Sciences, 32*(7), 459–469. doi:10.1016/j.jksues.2020.04.004

Ali, M., Firdaus, A. A., Arof, H., Nurohmah, H., Suyono, H., Putra, D. F. U., & Muslim, M. A. (2021). The comparison of dual-axis PV tracking system using artificial intelligence techniques. *IAES International Journal of Artificial Intelligence, 10*(4), 901–909. doi:10.11591/ijai.v10.i4.pp901-909

Bahmanyar, A. R., & Karami, A. (2014). Power system voltage stability monitoring using artificial neural networks with a reduced set of inputs. *International Journal of Electrical Power & Energy Systems, 58*, 246–256. doi:10.1016/j.ijepes.2014.01.019

Bhakuni, A. S. (2022). Design of SA and GA Optimized PID Controllers for Controlling Blend Chest Level of Paper Mill. *2022 4th International Conference on Smart Systems and Inventive Technology (ICSSIT)*, (pp. 545-550). IEEE. 10.1109/ICSSIT53264.2022.9716460

Büchi, R. (2022). PID Controller Parameter Tables for Time-Delayed Systems Optimized Using Hill-Climbing. *Signals, 3*(1), 146–156. doi:10.3390/signals3010010

Buşoniu, L., Bruin, T. D., Tolic, D., Kober, J., & Palunko, I. (2018). Reinforcement learning for control: Performance, stability, and deep approximators. *Annual Reviews in Control, 46*, 8–28. doi:10.1016/j.arcontrol.2018.09.005

Chaysaz, A., Seyedi, S. R. M., & Motevali, A. (2019). Effects of different greenhouse coverings on energy parameters of a PV–thermal solar system. *Solar Energy, 194*(November), 519–529. doi:10.1016/j.solener.2019.11.003

Czop, P., Kost, G., Slawik, D., & Wszolek, G. (2011). Formulation and identification of First- Principle Data-Driven model. Journal of achievements in materials and manufacturing engineering, 44, 179-186.

Datta, S., Bhattacharya, S., & Roy, P. (2016). Artificial Intelligence-based Solar Panel Tilt Angle Optimization and its Hardware Implementation for Efficiency Enhancement. *International Journal of Advanced Research in Electrical, Electronics and Instrumentation Engineering, 5*(10), 7830–7842. doi:10.15662/IJAREEIE.2016.0510006

Dhimish, M. (2019, December). 70% Decrease of Hot-Spotted Photovoltaic Modules Output Power Loss Using Novel MPPT Algorithm. *IEEE Transactions on Circuits and Wystems. II, Express Briefs, 66*(12), 2027–2031. doi:10.1109/TCSII.2019.2893533

Edwards, B. P. (1978). Computer Based Sun following System. *Solar Energy, 21*(1), 491–496. doi:10.1016/0038-092X(78)90073-7

Ehiagwina, F. (2021). Development of a solar energy tracking mechanism with artificial neural network enhancement. *International Research Journal of Modernization in Engineering Technology and Science, 3*(3).

Elsheikh, A., Sharshir, S. W., & Elaziz, E. A. (2019). Modeling of solar energy systems using artificial neural network: A comprehensive review. *Solar Energy, 180*, 622–639. doi:10.1016/j.solener.2019.01.037

Elsisi, M. (2019). Design of neural network predictive controller based on imperialist competitive algorithm for automatic voltage regulator. *Neural Computing & Applications, 31*(9), 5017–5027. doi:10.1007/s00521-018-03995-9

Engin, M., & Engin, D. (2015). Optimization Controller for Mechatronic Sun Tracking System to Improve Performance. *Advances in Mechanical Engineering, 5*, 146352–146352. doi:10.1155/2013/146352

Fan, S., Yang, W., & Hu, Y. (2018). Adjustment and control on the fundamental characteristics of a piezoelectric PN junction by mechanical – loading. Nano Energy, 52, 416–421. .08.017. doi:10.1016/j.nanoen.2018.08.017

Faraji, J., Khanjanianpak, M., Rezaei, M., Kia, M., Aliyan, E., & Dehghanian, P. (2020). Fast-Accurate Dual-Axis ST Controlled by P&O Technique with Neural Network Optimization. *2020 IEEE International Conference on Environment and Electrical Engineering and 2020 IEEE Industrial and Commercial Power Systems Europe (EEEIC / I&CPS Europe)*. IEEE. 10.1109/EEEIC/ICPSEurope49358.2020.9160843

Fernandez Cornejo, E. R., Diaz, R. C., & Alama, W. I. (2020). PID Tuning based on Classical and Metaheuristic Algorithms: A Performance Comparison. *2020 IEEE Engineering International Research Conference (EIRCON)*, (pp. 1-4). IEEE. 10.1109/EIRCON51178.2020.9253750

González-Acevedo, H., Muñoz, Y., Ospino, A., Serrano, J., Atencio, A., & Saavedra, C. (2021). Design and performance evaluation of a solar tracking panel of single axis in Colombia [IJECE]. *Iranian Journal of Electrical and Computer Engineering, 11*(4), 2889. doi:10.11591/ijece.v11i4.pp2889-2898

Guduru, S., Preetham, C. H., & Vijayan, K. (2023). Smart Solar Tracking System for Optimal Power Generation Using Three LDR's. In *2023 International Conference on Recent Advances in Electrical, Electronics, Ubiquitous Communication, and Computational Intelligence (RAEEUCCI)* (pp. 1-5). IEEE 10.1109/RAEEUCCI57140.2023.10134521

Hammoumi, E. (2018). A simple and low-cost active dual-axis ST. *Energy Science & Engineering, 6.* doi:10.1002/ese3.236

Hanwate, S., & Hote, Y. (2018). Design of PID controller for sun tracker system using QRAWCP approach. *International Journal of Computational Intelligence Systems, 11*(1), 133. doi:10.2991/ijcis.11.1.11

Hijawi, H., & Arafeh, L. (2016). Design of Dual Axis ST System Based on Fuzzy Inference Systems. *International Journal on Soft Computing, Artificial Intelligence, and Applications, 5*(2/3), 23–36. Advance online publication. doi:10.5121/ijscai.2016.5302

Hsu, C. F., Li, R.-K., Kang, H.-Y., & Lee, A. H. (2014). A systematic evaluation model for solar cell technologies. *Mathematical Problems in Engineering, 2014,* 1–16. doi:10.1155/2014/542351

Huynh, D. C., Nguyen, T. M., Dunnigan, M. W., & Mueller, M. A. (2013). Comparison between open- and closed-loop trackers of a solar photovoltaic system. *2013 IEEE Conference on Clean Energy and Technology (CEAT),* (pp. 128-133). IEEE. 10.1109/CEAT.2013.6775613

Jang, J. S. R. (1993, May-June). ANFIS: Adaptive-network-based fuzzy inference system. *IEEE Transactions on Systems, Man, and Cybernetics, 23*(3), 665–685. doi:10.1109/21.256541

Jitwang, T., Hlangnamthip, S., & Puangdownreong, D. (2020). Robust PIDA Controller Design by Cuckoo Search for Liquid-Level Control System. *2020 Joint International Conference on Digital Arts, Media, and Technology with ECTI Northern Section Conference on Electrical, Electronics, Computer, and Telecommunications Engineering (ECTI DAMT & NCON),* (pp. 226-229). IEEE. 10.1109/ECTI-DAMTNCON48261.2020.9090746

Kalogirou, S. A. (1996). Design and construction of one-axis Sun-Tracking system. *Sol. Energy, 57*(6). . doi:10.1016/S0038-092X(96)00135-1

Karabiber, A., & Güneş, Y. (2023). Single-motor and dual-axis solar tracking system for micro photovoltaic power plants. *Journal of Solar Energy Engineering, 145*(5), 051004. doi:10.1115/1.4056739

Kiumarsi, B., Vamvoudakis, K. G., Modares, H., & Lewis, F. L. (2018, June). Optimal and Autonomous Control Using Reinforcement Learning: A Survey. *IEEE Transactions on Neural Networks and Learning Systems, 29*(6), 2042–2062. doi:10.1109/TNNLS.2017.2773458 PMID:29771662

Kiyak, E., & Gol, G. (2016). A comparison of fuzzy logic and PID controller for a single-axis Solar Tracking System. *Renewables: Wind, Water, and Solar, 3*(1), 7. doi:10.1186/s40807-016-0023-7

Kumar, S., Bhattacharyya, B., & Gupta, V. K. (2014). Present and Future Energy Scenario in India. *J. Inst. Eng. India Ser. B, 95*(3), 247–254. doi:10.1007/s40031-014-0099-7

Lalit, K. (2019). India projected to be on track to achieve Paris climate agreement target: US expert. *The economic times.* https://energy.economictimes.indiatimes.com/news/renewable/india-projected-to-be-on-track-to-achieve-paris-climate-agreement-target-us-expert/68229428

Larico, E., & Canales, A. (2022). Solar Tracking System with PV Cells: Experimental Analysis at High Altitudes. *International Journal of Renewable Energy Development, 11*(3), 630–639. doi:10.14710/ijred.2022.43572

Lee, C. C. (1990, March-April). Fuzzy logic in control systems: Fuzzy logic controller. I. *IEEE Transactions on Systems, Man, and Cybernetics, 20*(2), 404–418. doi:10.1109/21.52551

Lee, D., Koo, S., Jang, I., & Kim, J. (2022). Comparison of Deep Reinforcement Learning and PID Controllers for Automatic Cold Shutdown Operation. *Energies, 15*(8), 2834. doi:10.3390/en15082834

Li, M., Zhang, Y., & You, D. (2020). Design of fuzzy PID stepping motor controller based on particle swarm optimization. *2020 3rd World Conference on Mechanical Engineering and Intelligent Manufacturing (WCMEIM),* (pp. 449-453). IEEE. 10.1109/WCMEIM52463.2020.00100

Maddi, D., Sheta, A., Davineni, D., & Al-Hiary, H. (2019). Optimization of PID Controller Gain Using Evolutionary Algorithm and Swarm Intelligence. *2019 10th International Conference on Information and Communication Systems (ICICS),* (pp. 199-204). IEEE. 10.1109/IACS.2019.8809144

Mamodiya, U., & Tiwari, N. (2023). Dual-axis solar tracking system with different control strategies for improved energy efficiency. *Computers & Electrical Engineering, 111*, 108920. doi:10.1016/j.compeleceng.2023.108920

Mehmet, S., & Erol, R. (2017). A comparative study of neural networks and ANFIS for forecasting attendance rate of soccer games. *Mathematical & Computational Applications, 43*(4), 22. doi:10.3390/mca22040043

Mohanasundaram, K., Sugavanam, K. R., & Senthilkumar, R. A. (2014). PSO algorithm based pi controller design for soft starting of induction motor. *International Journal of Applied Engineering Research: IJAER, 9*(24), 25535–25542.

Okoye, C., Bahrami, A., & Atikol, U. (2017). Evaluating the solar resource potential on different tracking surfaces in Nigeria. *Renewable & Sustainable Energy Reviews, 81*, 1569–1581. doi:10.1016/j.rser.2017.05.235

Oladayo, B. O., & Titus, A. O. (2016). Development of STS Using IMC-PID Controller [AJER]. *American Journal of Engineering Research, 5*(5).

Pachauri, N. (2020). Water cycle algorithm-based PID controller for AVR. COMPEL. *The international journal for computation and mathematics in electrical and electronic engineering.* . doi:10.1108/COMPEL-01-2020-0057

Palomino-Resendiz, S. I., Ortiz-Martínez, F. A., Paramo-Ortega, I. V., González-Lira, J. M., & Flores-Hernández, D. A. (2023). Optimal Selection of the Control Strategy for Dual-Axis Solar Tracking Systems. *IEEE Access : Practical Innovations, Open Solutions, 11*, 56561–56573. doi:10.1109/ACCESS.2023.3283336

Patel, S. (2021). Review on ST and Comparison on Single Axis ST, Dual Axis ST with Fixed Solar PV System. *International Journal of Innovative Research in Electrical, Electronics, Instrumentation and Control Engineering, 9*(6).

Pebrianti, D., Bayuaji, L., Arumgam, Y., Riyanto, I., Syafrullah, M., & Ann Ayop, N. Q. (2019). PID Controller Design for Mobile Robot Using Bat Algorithm with Mutation (BAM). *2019 6th International Conference on Electrical Engineering, Computer Science and Informatics (EECSI),* (pp. 85-90). IEEE. 10.23919/EECSI48112.2019.8976932

Prabha, G., & Mohana, K. (2018). Design of feasible energy generation using solar panel and control using an IoT. *IACSIT International Journal of Engineering and Technology, 7*(24), 191–196.

Ramyashree, J., Roja, M. R., & Sivagurunathan, G. (2018). *Rangasamy, Kotteeswaran. "Firefly algorithm based multivariable PID controller design for MIMO process"* (Vol. 7). International Journal of Engineering and Technology. doi:10.14419/ijet.v7i2.31.13394

Rawat, A., Jha, S. K., & Kumar, B. (2020). "Position controlling of Sun Tracking System using optimization technique. *Energy Reports, 6,* 304–309. doi:10.1016/j.egyr.2019.11.079

Razif Hamid, A., Khusairy Azim, A., & Hafizuddin Bakar, M. (2017). A review on Solar Tracking System. Proceeding National Innovation and Invention Competition Through Exhibition (iCompEx'17), (pp. 1-9).

Rezoug, M. R., Benaouadj, M., Taibi, D., & Chenni, R. (2021). A New Optimization Approach for a Solar Tracker Based on an Inertial Measurement Unit. Engineering, Technology & *Applied Scientific Research, 11*(5), 7542–7550. doi:10.48084/etasr.4330

Rodrigues, S., Torabikalaki, R., Faria, F., Cafôfo, N., Chen, X., Ivaki, A. R., Mata-Lima, H., & Morgado-Dias, F. (2016). Economic feasibility analysis of small scale PV systems in different countries. *Solar Energy, 131,* 81–95. doi:10.1016/j.solener.2016.02.019

Roslan, M. (2020). Particle swarm optimization algorithm-based PI inverter controller for a grid-connected PV system. *PLoS ONE, 15*(12), e0243581. https://doi.org/. pone.0243581 doi:10.1371/journal

Saleh, M., & Saad, S. (2016). Artificial Immune System based PID Tuning for DC Servo Speed Control. *International Journal of Computer Applications, 155*(2), 23–26. doi:10.5120/ijca2016912265

Shouran, M., & Elgamli, E. (2020). Teaching-Learning based Optimization Algorithm Tuned Fuzzy-PID Controller for Continuous Stirred Tank Reactor (Vol. 7). Research Gate.

Stapleton, A. (2017). How solar cells turn sunlight into electricity. *Cosmos magazine.* https://cosmos-magazine.com/technology/how-solar-cells-turn-sunlightinto-electricity

Sutton, R. S., & Barto, A. G. (2017). *Reinforcement Learning: An Introduction* (2nd ed., Vol. 1). MIT Press.

Syafa'ah, L., Fauziyah, L., & Has, Z. (2018). Robust and Accurate Positioning Control of Solar Panel System Tracking based Sun Position Image. *2018 5th International Conference on Electrical Engineering, Computer Science and Informatics (EECSI),* (pp. 324-329). IEEE. 10.1109/EECSI.2018.8752746

Tharakan, R. A., Joshi, R., Ravindran, G., & Jayapandian, N. (2021). Machine Learning Approach for Automatic Solar Panel Direction by using Naïve Bayes Algorithm. *2021 5th International Conference on Intelligent Computing and Control Systems (ICICCS),* (pp. 1317-1322). IEEE. 10.1109/ICICCS51141.2021.9432114

Tukymbekov, D., Saymbetov, A., Nurgaliyev, M., Kuttybay, N., Nalibayev, Y., & Dosymbetova, G. (2019). Intelligent energy efficient street lighting system with predictive energy consumption, *2019 International Conference on Smart Energy Systems and Technologies (SEST)* (pp. 1-5). IEEE. 10.1109/SEST.2019.8849023

Verma, B. (2020). A Review Paper on STS for PV Power Plant. *International Journal of Engineering Research.* . doi:10.17577/IJERTV9IS020103

Ya'u Muhammad, J., Tajudeen Jimoh, M., Baba Kyari, I., Abdullahi Gele, M., & Musa, I. (2019). A Review on Solar Tracking System: A Technique of Solar Power Output Enhancement. *Engineering and Science, 41*(1), 1–11. doi:10.11648/j.es.20190401.11

Zakariah, A. (2015). Dual-axis Solar Tracking System based on fuzzy logic control and Light Dependent Resistors as feedback path elements. *2015 IEEE Student Conference on Research and Development*, (pp. 139-144). IEEE. 10.1109/SCORED.2015.7449311

Zangeneh, M., Omid, M., & Akram, A. (2012). A Comparative Study Between Parametric and Artificial Neural Networks Approaches for Economical Assessment of Potato Production in Iran. *Spanish Journal of Agricultural Research, 9*(3), 661–671. doi:10.5424/sjar/20110903-371-10

Zhang, X.-L., & Zhang, Q. (2021). Optimization of PID Parameters Based on Ant Colony Algorithm. *2021 International Conference on Intelligent Transportation, Big Data & Smart City (ICITBS)*, (pp. 850-853). IEEE. 10.1109/ICITBS53129.2021.00211

Chapter 2
Distributed Renewable Power in Green Building:
Planning and Construction Strategies

Saurabh Laledia
Chandigarh University, India

Harpreet Kaur Channi
Chandigarh University, India

Ramandeep Sidhu
Lovely Professional University, Phagwara, India

ABSTRACT

Green buildings using renewable energy are developing as the world pursues sustainability. This abstract covers distributed renewable power planning and construction for energy efficiency, conservation, and built environment sustainability. Distributed solar, wind, biomass, and geothermal for green buildings. Local generation lowers grid dependency in distributed energy systems. Site study, energy demand assessment, and renewable energy source selection for the building's needs and location are design aims. Smart grids and energy storage boost renewables. Energy-efficient passive design, sustainable materials, and creative architecture complement distributed renewable power sources throughout construction. Cost savings, renewable energy ROI, and green building's environmental advantages are examined. Renewable energy may boost green building safety, security, and ecology. Distributed renewable energy green buildings fight climate change and promote sustainability.

INTRODUCTION

Climate change and the demand for renewable energy have reshaped the building sector. "Distributed Renewable Power in Green Building Planning and Construction Strategies" is becoming more important as the world works towards a greener future. This introduction discusses the importance and goals

DOI: 10.4018/979-8-3693-3735-6.ch002

of incorporating distributed renewable electricity into green building design and construction. "Green building" refers to planning, building, and operating facilities emphasizing environmental sustainability, energy efficiency, and decreased environmental impact. Green buildings and distributed renewable power technologies might transform energy generation, use, and management. Solar, wind, biomass, and geothermal energy are integrated into green construction projects to minimize fossil fuel usage, greenhouse gas emissions, and energy sustainability. Distributed renewable power systems enable buildings to generate some energy on-site, reducing transmission losses and energy resiliency (Singhal, 2019).

Green building design and construction using distributed renewable electricity aims to produce low-carbon, sustainable structures. These buildings may minimize their carbon footprint, alleviate environmental effects, and help fight climate change using renewable energy. This section discusses green building design and construction solutions that optimize distributed renewable power integration. It involves a site study to find the best renewable energy technologies, energy demand assessment, and technology selection based on the building's needs. Green building design and construction promote energy efficiency and environmental responsibility. Sustainable building materials, passive design, and intelligent building automation technologies boost green building energy efficiency and ensure compatibility with distributed renewable power systems. Green building planning requires the economic feasibility of distributed renewable electricity. This section discusses distributed renewable power's economic advantages and cost-effectiveness. It also considers the long-term effects of mass adoption.In conclusion, distributed renewable electricity in green building design and construction is essential to a more sustainable and resilient built environment. Green buildings may help mitigate climate change by integrating renewable energy sources with energy-efficient design practices (Mohanty, 2010).

In India, the IGBC (Indian Green Building Council) whips hand the construction and blueprint layouts of green buildings (Singhal,2019). The increase in population and the ever-growing demand for power have put a lot of strain on conventional power resources. As a result, electricity bills have experienced a major hike. Therefore, to resolve these problems and to protect the environment, green building construction is an ideal solution. Green building helps reduce energy consumption by providing green energy options and construction instead of standard building techniques. India has over one thousand reported projects with sustainable development at its epicenter. The building sectors consume an approximation of 30% power demand in the country (Streimikiene, 2020) . This percentage can be controlled or reduced by introducing green technology and sustainable building development. Before the construction of a green building, a lot of strategic planning has to be done, like the selection of site, waste management, building design, identifying the type of area, weather, plus smart energy-saving strategies that need to be implemented.

"Green building construction" involves designing, building, and operating a building sustainably and efficiently. It aims to increase residents' health, comfort, and productivity without harming the environment. The present generation's wants and future generations' needs are satisfied by this progressive strategy. Green construction approaches aim to make a building as sustainable as feasible by limiting its environmental effect during its lifespan. Green facilities include location, layout, architecture, energy and water efficiency, material choice, IEQ, waste management, and technology improvements (Çiner, 2019). Site selection and design are important in green development. Brownfield areas are perfect for this since they have little environmental effect and may be oriented to take advantage of sun and breezes. Passive design options reduce a building's energy consumption by eliminating the demand for mechanical heating and cooling systems. Energy efficiency is another sustainable building principle. Examples include high-performance insulation, energy-efficient windows, and on-site renewable energy sources like solar

Figure 1. Features of Green Building
(Chen, 2012) [CC BY 4.0]

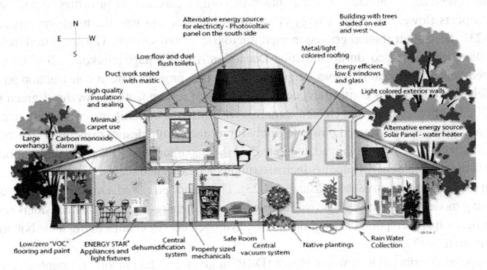

panels or wind turbines. The latest energy management technologies help reduce usage and waste. Water efficiency is important in green building. The building's water-efficient fixtures, rainwater harvesting, and greywater recycling technologies minimize water usage and promote water sustainability, as shown in figure 1 (Chen, 2012). In conclusion, sustainable development reduces global warming, resource depletion, and environmental degradation. We employ sustainable methods and cutting-edge technology to develop buildings that improve residents' health and comfort and demonstrate our commitment to conserving the planet for future generations. It supports responsible development and a greener, brighter future by giving a path to a more sustainable and resilient built environment (Jiang, 2009).

Literature Survey

The design and development of these grids must be integrated to improve efficiency and ensure supply security throughout the transition. Mostafa et al., 2023 addressed these problems and provide unique methods for integrated design of the three grids as one multi-energy grid. For a sustainable future, energy transition with more renewable energy and energy efficiency improvements are essential (Mostafa, 2022).

The researchers used a heuristic approach to assess the feasibility of NZE residential buildings in the KSA using an arid campus case study in 2021–2022. They used a geothermal heat pump (GHP) as a cooling system and photovoltaic thermal collectors (PVT) as a power generation system. The study's findings might be utilized to generate renewable energy system installation recommendations for sustainable and green buildings with comparable features. To coordinate, green, and increase renewable energy use, a new AC/DC power distribution system with low carbonization is needed now (Ismaeil, 2023). Another author combined the connotation and characteristics of renewable energy interconnection and mutual benefit, analyzed the key technology features of AC/DC distribution systems for renewable energy, established a complete overall research framework of key technology for planning and operation from two aspects of system planning and operation, and presented an overview and outlook on each key research direction. This UAE pioneering residential development offers renewable energy storage technologies while meeting the highest green economics and environmental sustainability criteria (Wang, 2023). The researcher examined SSC urban planning components and set priorities. After ranking 15 planning aspects (lower criterion), energy-efficient building design was the most important (0.121) (Jung, 2023). The author studied grid-connected microgrid shared storage. The suggested method was tested in Benguerir, Morocco, during the Solar Decathlon Africa Village (Barkouki, 2023). The author investigated Penang contractors' views on green construction projects. Green construction advantages were widely agreed upon by contractors, demonstrating that respondents are aware of green building (Rahim, 2023). Another researcher created scenarios with different building materials, data center waste heat use, and grid-generated green power to apply three sustainable design principles. The building's construction, with its heavy weight and concrete, had the maximum GWP. The uneven spatiotemporal distribution of renewable energy supplies increases unpredictability and seasonal power imbalance under the global low-carbon objective (Mohtashami, 2023). The author proposed a high-resolution collaborative planning model for electricity-thermal-hydrogen-coupled energy systems that considers renewable energy resource spatiotemporal distribution and the hydrogen energy chain's multi-scale bottom-to-top investment strategy (Yi, 2023).

The scope of Distributed Renewable Power (DRP) in green building involves decentralized, on-site generation from sources like solar, wind, and geothermal. Planning includes site assessments, demand analysis, and hybrid systems. In construction, integrate DRP seamlessly, utilizing Building-Integrated Photovoltaics (BIPV) and smart technologies. Develop microgrids for resilience and employ energy-efficient materials. Operational considerations involve maintenance, real-time monitoring, and compliance with regulations. Leverage government incentives and engage the community through education. Through these strategies, green buildings can enhance sustainability, reduce reliance on centralized power, and contribute to a resilient, energy-efficient built environment.Following are the main objectives:

- Outline a holistic strategy for integrating Distributed Renewable Power (DRP) in green building design.
- Address site assessment, renewable source selection, and energy demand analysis to guide effective planning.
- Provide insights into construction methodologies that seamlessly incorporate DRP technologies.

SITE SELECTION AND DESIGN

Site selection and design are key to green building development. The approach optimizes the building's energy and environmental performance while reducing its ecological effect. Green buildings may effectively utilize resources, minimize energy consumption, and improve tenant well-being by selecting a suitable location and using sustainable design principles (Pol, 2020). Site selection and design for green buildings are described below.

Site Selection

- **Environmental Impact:** The site's closeness to natural habitats, wetlands, and sensitive ecosystems should be considered while assessing its environmental effect. Choose places with little ecological significance or brownfield sites to reduce the influence on natural ecosystems.

- **Access to Amenities:** To increase walkability and minimize single-occupancy transportation, choose locations with convenient access to public transit, attractions, and services.
- **Solar Orientation:** Passive solar design maximizes daylighting and reduces artificial lighting by considering the site's solar orientation.
- **Wind and Ventilation:** Optimize natural ventilation and decrease mechanical cooling using wind patterns.
- **Water Management:** Reduce stormwater runoff and increase water infiltration by considering the site's water flow patterns.
  2.2 Sustainable Site Planning
- **Low-Impact Development:** Reduce soil erosion, protect vegetation, and regulate stormwater flow with low-impact construction.
- **Green Space:** To boost biodiversity, offer recreational activities, and improve the site's microclimate, set aside space for parks, gardens, and green spaces.
- **Permeable Surfaces:** To reduce stormwater runoff, use permeable materials for pathways, driveways, and parking spaces.

Passive Design Strategies

- **Building Orientation:** Optimize energy efficiency by orienting the structure for winter solar gain and summer heat gain.
- **Natural Ventilation:** Allow cross-ventilation using moveable windows and apertures to reduce mechanical cooling.
- **Daylighting:** Maximize natural light by strategically placing windows and skylights (Pan, 2013).

Water Conservation

Reduced water use, preservation of this important natural resource, and lessening the environmental effect of buildings are all achieved via water conservation, an essential component of green building methods. There are long-term advantages to incorporating water conservation measures into green building design and construction. Green buildings may benefit from the following water-saving techniques:

- **Water-Efficient Fixtures and Appliances:** Fixtures like faucets, showerheads, and toilets that consume less water without sacrificing performance may be installed. Adopting high-efficiency equipment, such as washing machines and dishwashers, may also help reduce water use.
- **Greywater and Rainwater Harvesting:** Use greywater (water from sinks, showers, and washing machines) and rainfall for non-drinking functions such as watering plants, flushing toilets, and cooling buildings.
- **Native and Drought-Resistant Landscaping:** Create a landscape that uses less water and is better suited to the local environment by using drought-resistant or native plants. Reduce water waste by using effective irrigation techniques, such as drip irrigation.
- **Smart Irrigation Systems:** Use intelligent irrigation systems that respond to changes in weather, soil moisture, and plant demands by modifying watering times accordingly.
- **Water Leak Detection and Monitoring:** To reduce water waste and possible damage to the building, installing water leak detecting systems and meters is important.

- **Stormwater Management:** To reduce water waste and possible damage to the building, installing water leak-detecting systems and meters is important.
- **Water-Conserving Cooling Towers:** Use water-efficient designs and technology to reduce the water needed for building cooling towers.
- **Water-Efficient Cooling Systems:** Instead of employing water-based cooling systems, you may try air-based ones or a combination (Soni,2013).
- **Educational Initiatives:** Encourage water-saving actions by teaching people in the building and its visitors the value of conserving water and making responsible water use habits.
- **Water Performance Monitoring:** To find places that may be improved and guarantee continued water efficiency, it is important to set up monitoring systems that measure water use in real-time.
- **Greywater Treatment and Recycling:** Advanced greywater treatment systems can recycle and purify greywater for safe usage inside the facility, and they are worth considering for bigger buildings or communities.
- **Water Conservation Policies and Incentives:** Collaborate with regional and national governments to implement water-saving policies, laws, and incentives that boost eco-friendly building water management.

Developers, building owners, and tenants can all do their part to create a more sustainable and ecologically responsible future by using these water-saving methods in green building design and operations (Iravani,2017).

Renewable Energy Integration

One of the most important aspects of sustainable development is the use of renewable energy sources in building design and construction. Solar photovoltaic (PV) systems, wind turbines, and geothermal heat pumps are all examples of renewable energy technology that may be integrated into buildings to provide clean, sustainable energy. Solar photovoltaic (PV) systems, for example, may be set up on the roof or the walls of a building to convert sunlight into electricity and lower the structure's energy needs. Similarly, in regions with stable wind patterns, wind turbines may be incorporated into the construction of buildings to produce power. As an alternative to conventional HVAC systems, geothermal heat pumps make use of the stable temperature of the soil to deliver heating and cooling (Palermo, 2019). Batteries and thermal energy storage systems are two examples of energy storage options that may be used in green buildings to guarantee smooth integration. These methods make it possible to store renewable energy for later use, either to meet peak demand or to make up for times of low output. The total energy efficiency of a building may be improved with the use of modern energy management systems, which allow building managers to keep tabs on and control renewable energy output, consumption, and storage. When renewable energy production exceeds local demand, grid-tied solutions allow green buildings to keep their connections to the regular power grid and send any extra energy back to the utility company. Buildings may now contribute to the larger mission of fostering a sustainable and resilient energy infrastructure thanks to this streamlined approach to energy management. In general, green buildings may help create a more sustainable and environmentally friendly constructed environment by prioritising the use of renewable energy technology as shown in Figure 2 (Huang,2016).

Green buildings create clean, sustainable energy on-site via renewable energy integration. Green buildings employ renewable energy to minimize fossil fuel consumption, greenhouse gas emissions, and climate change (Reztrie,2018) Green buildings may combine many renewable energy sources, including:

- **Solar Photovoltaic (PV) Systems:** Solar panels generate power from sunshine. PV systems may power lights, appliances, and electrical systems when mounted on roofs, facades, or solar carports.
- **Solar Water Heating:** Solar thermal systems heat water for home usage, space heating, and other uses. They decrease the demand for non-renewable energy-based water heating techniques.
- **Wind Turbines:** Small wind turbines may produce power in areas with steady winds. On foggy days, wind energy supplements solar energy.
- **Geothermal Heat Pumps:** Geothermal systems use the earth's steady temperature to heat and cool. HVAC energy consumption may be reduced using these efficient solutions.
- **Biomass Energy:** Biomass systems create heat or power from wood pellets or agricultural waste. Biomass is carbon-neutral when responsibly generated.
- **Micro-Hydro Power:** Micro-hydro systems may produce power from tiny streams or rivers, providing a steady and continuous renewable energy source.
- **Combined Heat and Power (CHP):** CHP (cogeneration) systems generate electricity and utilize waste heat for space heating or hot water, optimizing energy efficiency.

Figure 2. Renewable energy integration

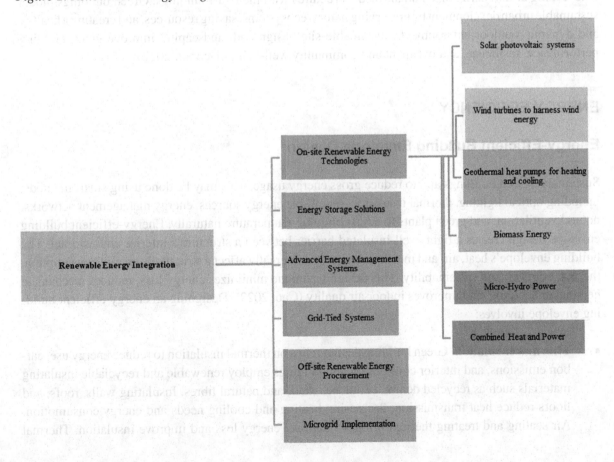

Green buildings must evaluate energy demands and local resources to incorporate renewable energy. Energy audits and feasibility studies determine the best renewable energy options for a place. Smart energy management systems and batteries can optimize energy consumption and offer backup power amid low renewable energy output (Meena, 2022). Green buildings inspire worldwide low-carbon transitions by integrating renewable energy.

Sustainable Site Planning and Landscaping

Green building construction requires sustainable site design and landscaping to create resilient, environmentally friendly buildings. The site's natural features, biodiversity, and climate are considered to reduce the development's ecological impact. Sustainable site design preserves natural features, reduces soil erosion, and manages stormwater runoff through permeable surfaces and rain gardens. Green buildings preserve natural habitats and green spaces, supporting local ecosystems, biodiversity, and community enjoyment. Green building landscaping uses native plants well-adapted to the local environment and needs less water and upkeep. Drought-resistant plants, low-water irrigation systems, green roofs, and vertical gardens increase air quality and thermal performance. Shade, urban heat island reduction, and energy efficiency are achieved by carefully placing trees (Chen, 2022).

Urban agricultural areas encourage local products and community involvement in sustainable site development and landscaping. Green buildings may increase food security and community connection by including edible landscapes and communal gardens. In conclusion, green building site planning and landscaping aim to harmonize human-made structures with their surroundings. Green buildings model sustainable urban development by preserving natural ecosystems, saving resources, and creating a healthy and dynamic outdoor environment. Sustainable site design and landscaping improve green building performance, resilience, and occupant and community well-being (Debrah, 2022).

ENERGY EFFICIENCY

Energy-Efficient Building Envelope Design

Sustainable green buildings aim to reduce gross energy usage. This may be done using smart methods, including optical lighting, thermal insulation, renewable energy sources, energy management networks, natural ventilation, and green plants to adjust building temperature naturally. Energy-efficient building envelope design creates a tight, well-insulated barrier between a structure's interior and external. The building envelope's heat, air, and moisture transmission greatly affect a structure's energy consumption, thermal comfort, and sustainability. This design technique minimizes energy loss, reduces mechanical heating and cooling, and improves indoor air quality (Cao, 2022). Designing an energy-efficient building envelope involves:

- **Thermal Insulation**: Green building design relies on thermal insulation to reduce energy use, carbon emissions, and interior comfort. Green buildings employ renewable and recyclable insulating materials such as recycled denim, cellulose, wool, and natural fibres. Insulating walls, roofs, and floors reduce heat transmission, decreasing heating and cooling needs and energy consumption. Air sealing and treating thermal bridges minimize energy loss and improve insulation. Thermal insulation complements passive design principles like maximizing building orientation and window location for natural heating and cooling to enhance energy efficiency. Green construction projects also include insulating materials' life cycle environmental effects to ensure sustainable, low-impact options. Green buildings favour insulating materials with minimal VOC emissions to improve indoor air quality. Using thermal insulation efficiently, green buildings promote sustainability and occupant comfort (Alshboul, 2022).

- **Air Sealing:** Air sealing minimizes air leakage via building exterior gaps, fractures, and openings. Sealing drafts and minimizing HVAC effort improves energy efficiency.
- **High-Performance Windows:** Installing energy-efficient windows with low-emissivity coatings and many panes reduces heat transmission and maximizes sunshine. It lowers artificial lighting and improves temperature control.
- **Solar Shading:** Strategically positioning overhangs, awnings, or louvres on windows may shade them during the warmest portions of the day, lowering solar heat gain and cooling load.
- **Thermal Mass:** Thermal mass materials like concrete or stone may be included in the building envelope to regulate temperature.
- **Cool Roofs:** Roofing materials that are reflective and light in colour are preferable in hot regions because they absorb less heat and lessen the need for air conditioning.
- **Ventilation Strategies:** Implementing mechanical or natural ventilation systems with heat recovery can regulate indoor air quality while minimizing energy loss.
- **Building Orientation:** The architecture and windows of a structure may be optimized for energy efficiency by using natural light and prevailing breezes.
- **Green Roofing:** The heat island effect is mitigated, extra insulation is provided, and stormwater management is enhanced through green roofs, which are covered with plants.

Energy-efficient building envelope design is essential for high-performance, sustainable buildings that use less energy, decrease greenhouse gas emissions, and provide healthier, more pleasant interior environments. Energy-efficient design and construction may help the building sector fight climate change and create a more sustainable future (Valencia, 2022).

High-Performance Insulation and Windows

High-performance insulation and windows improve energy efficiency, carbon emissions, and occupant comfort in green building design. These components must create a well-insulated building envelope, limit heat transfer, and enhance thermal performance. Insulation reduces heat loss in winter and heat gain in summer (Erten, 2022). Insulation slows heat transfer through walls, roofs, and floors, lowering the need for mechanical heating and cooling.

- **Sustainable Materials:** Cellulose, recycled cotton, and natural fibres are eco-friendly insulating materials used in green buildings.
- **Proper Installation**: Insulation must be installed and sealed to prevent air leaks and maintain the thermal barrier.
- **Continuous Insulation:** Continuous insulation optimizes performance by covering the building envelope without thermal bridging. The heat gained or lost via a building's windows is usually rather high. Modern energy-efficient windows use cutting-edge technology to lessen these impacts

and boost indoor comfort for building occupants (Ur Rehman, 2022). Indicative characteristics include:

- **Low-E Coatings:** Glazing with a low-emissivity (Low-E) coating may reflect infrared light and reduce heat transmission, making the inside of a building more comfortable year-round.
- **Multiple Panes:** Compared to single-pane windows, double- or triple-glazed varieties with air or gas-filled voids between the panes provide superior thermal performance.
- **Framing Materials:** Vinyl, fibreglass, or wood window frames with thermal breaks reduce heat transfer and increase efficiency.
- **Solar Heat Gain Coefficient (SHGC):** Low SHGC windows reduce solar heat, lowering cooling demands.

Energy-efficient windows and high-performance insulation minimize a building's energy consumption, lowering electricity costs and greenhouse gas emissions. Eliminating breezes, temperature variations, and noise transmission improves indoor comfort. These elements make green buildings more resilient, cost-effective, and ecologically beneficial by showing a commitment to sustainability and responsible resource usage (Riadh, 2022).

ENERGY MANAGEMENT SYSTEMS AND AUTOMATION

Implementing effective energy management systems and automation is crucial for integrating distributed renewable power into green buildings. Conduct comprehensive energy audits to understand consumption patterns, and integrate renewable sources such as solar or wind power. Implement smart grid technology for efficient energy distribution and storage, while utilizing advanced energy storage solutions to manage intermittency. Employ building automation systems to regulate energy usage, and prioritize sustainable materials and passive design strategies for improved energy efficiency. Real-time monitoring and analytics can help optimize performance while educating occupants about energy-saving practices fosters a sustainable building community.EMS and automation optimize energy consumption, efficiency, and cost in industrial, commercial, and residential settings. These systems use modern technology to monitor, manage, and optimize energy use, making energy consumption more sustainable and ecologically benign (Kuo, 2016).

Energy Management Systems (EMS)

Green building design and operation rely heavily on energy management technologies. Optimisation of energy consumption, waste minimization, and increased sustainability are the goals of these systems, which integrate technology, data analysis, and control methodologies. Solutions such as smart metres, automatic controls, and renewable energy integration may be part of these systems that work together to save money and help the environment. Implementing thorough monitoring technologies that track energy usage trends in real-time is a crucial part of energy management systems. As a result of the information gleaned from these instruments, building managers may take strategic action to reduce energy use. Building managers may use this information to save costs and improve efficiency in their operations. Energy management systems in sustainable structures also often include automated controls. Occupancy and environmental data are used to automatically adjust the lighting, heating, ventilation, and air conditioning (HVAC) in a building. Reduced operating expenses and a smaller environmental imprint are the results of changing these systems automatically to meet real demands.

By incorporating smart metres into energy management systems, building owners and managers can monitor energy use in real-time and from several locations. Energy use trends may be easily analysed with the help of smart metres, allowing building managers to better allocate resources during peak hours. This not only aids in effective energy management but also makes it possible to use demand response tactics to equalise energy usage at peak hours, easing the load on the power grid. One essential part of green building energy management systems is the use of renewable energy sources. Integrating renewable energy sources like solar panels, wind turbines, and geothermal systems may cut down on the need for external power plants and their associated carbon emissions. By absorbing and storing surplus energy produced during peak production times, energy storage technologies like batteries and thermal energy storage systems supplement the integration of renewable energy sources. A more sus-

Figure 3. Energy management systems

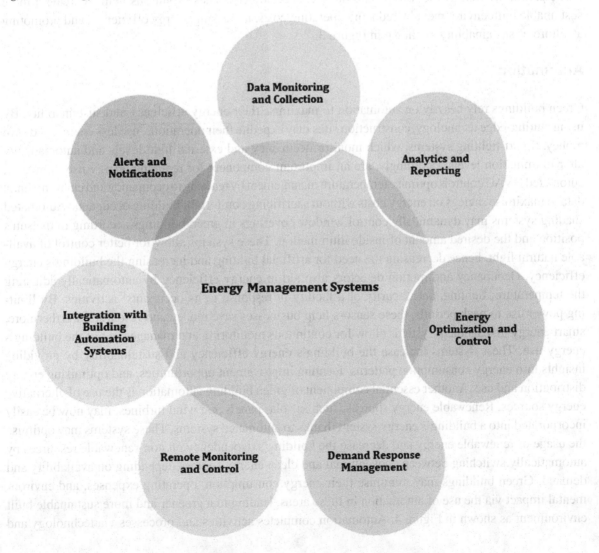

tainable and stable energy supply is achieved when this stored energy is used during peak demand or times of low renewable energy output. Effective energy management is greatly aided by building energy management software. With the help of this programme, building managers may examine energy use, monitor key performance indicators, and identify areas where energy efficiency can be improved. To fine-tune energy use, optimise system performance, and lower total energy costs, building operators may use sophisticated algorithms and data analytics to make data-driven choices (Liu,2021). Effective energy management systems in green buildings often include education and engagement programmes. A culture of sustainability may be developed in a building by teaching its residents the value of energy efficiency and encouraging them to make the switch. Simple things like turning off lights when you're not using them, buying energy-efficient equipment, and being aware of how much energy you use in general may all help. Participation from building occupants in energy conservation measures may help foster a community-wide dedication to sustainability, which in turn can have a multiplied effect on both energy savings and environmental preservation.To sum up, energy management systems in sustainable structures are comprehensive approaches that include not just monitoring and control but also user participation, as well as the use of renewable energy sources. These solutions help to create a more sustainable built environment by reducing operating costs, increasing energy efficiency, and promoting a culture of sustainability as shown in Figure 3.

Automation

Green buildings rely heavily on automation to maximise their energy efficiency and sustainability. By using cutting-edge technology, construction sites may expedite their operations, use less energy, and save money. Smart lighting systems, which monitor occupancy and external light levels and automatically alter illumination levels accordingly, are an important component for reducing energy use. Similarly, automated HVAC controls optimise temperature management by reacting to occupancy and environmental data to maximise savings on energy costs without sacrificing comfort for building occupants. Automated shading systems may dynamically control window coverings in green buildings according to the sun's position and the desired amount of inside illumination. These systems allow for better control of available natural light, hence decreasing the need for artificial lighting and increasing the building's energy efficiency. Occupancy and motion detectors also aid in energy efficiency by automatically adjusting the temperature, lighting, and security of a facility in response to its occupants' activities. By limiting power use to peak periods, these sensors help businesses save money and resources. Furthermore, smart energy management systems allow for continuous monitoring and management of the building's energy use. These systems increase the building's energy efficiency and sustainability by providing insights into energy consumption patterns, locating improvement opportunities, and optimising energy distribution and use. Another essential component of green building automation is the use of alternative energy sources. Renewable energy sources, such as solar panels and wind turbines, may now be easily incorporated into a building's energy system thanks to automated systems. These systems may optimise the usage of renewable energy and decrease the building's dependency on non-renewable resources by automatically switching between conventional and clean energy sources depending on availability and demand. Green buildings may minimise their energy consumption, operating expenses, and environmental impact via the use of automation in these areas, leading to a greener and more sustainable built environment as shown in Figure 4. Automation completes activities and processes via technology and

Figure 4. Automation in green building

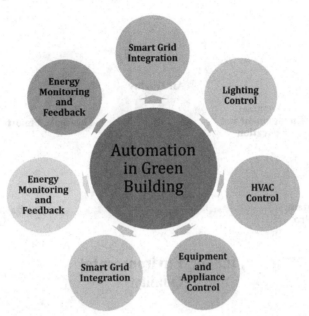

control systems with little human involvement. Automation is used in energy management to facilitate efficiency, effectiveness, and the minimization of human error (Pramanik, 2021).

Indoor Environmental Quality for Green Building

In addition, a conducive interior climate can only be maintained via careful attention to temperature control. Having precise temperature control and thermal zoning made possible by energy-efficient heating, ventilation, and air conditioning (HVAC) systems is a great way to maintain pleasant temperatures inside and make a building habitable year-round. A healthy and pleasant interior atmosphere is largely dependent on the usage of humidifiers and dehumidifiers to maintain the desired relative humidity levels. By lowering chemical emissions and bettering indoor air quality, sustainable and non-toxic construction products including low volatile organic compound (VOC) paints, adhesives, and sealants contribute to IEQ. To maintain a safe and healthy atmosphere for their tenants, building managers need to install sophisticated systems for monitoring and managing indoor air quality (Radziejowska, 2021).In addition, occupant participation and education on healthy lifestyle habits and the correct usage of building systems may greatly contribute to IEQ requirements being maintained at a high level. To preserve the longevity and effectiveness of indoor environmental quality measures in green buildings, regular maintenance and cleaning of ventilation systems, air filters, and building components is also vital. Green buildings emphasize these factors to provide their residents with healthy, pleasant, and productive interior environments that contribute to their happiness and well-being (Jiang,2021).

Creating a healthy, pleasant, and productive interior environment for inhabitants is the primary goal of interior Environmental Quality (IEQ), an essential component of green building design and construction. In addition to improving occupants' happiness and contentment, high IEQ also helps boost productivity

Figure 5. Indoor environmental quality

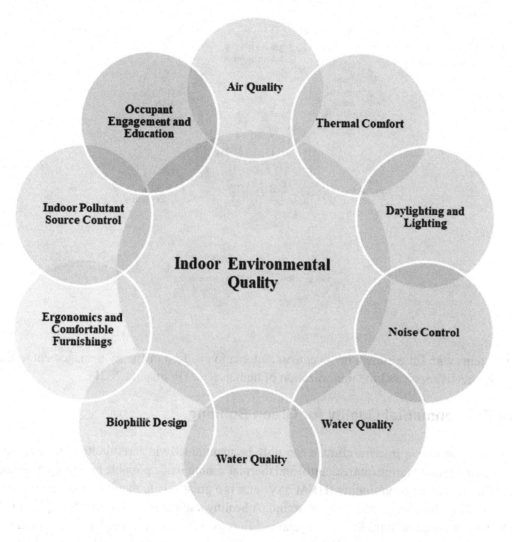

and reduce health problems (Parasher, 2021). High IEQ in environmentally friendly buildings is dependent on many interrelated factors as shown in Figure 5.

GREEN BUILDING CERTIFICATIONS AND STANDARDS

Certifications and standards for green buildings are voluntary grading systems used to assess and recognize buildings for their sustainability and environmental performance. These certifications and standards promote the construction and operation of buildings that are friendly to the environment, efficient in their energy use, and conducive to promoting a healthy living and working environment (Rameshwar, 2020). The following are examples of some of the most well-known green building certifications and standards:

LEED (Leadership in Energy and Environmental Design)

LEED is one of the green building certification methods used most often globally. It was developed by the United States Green Building Council (USGBC). It evaluates the performance of the building in many different areas, including energy efficiency, water consumption, interior environmental quality, the use of sustainable materials, and site selection. Certification levels are available via LEED, including Certified, Silver, Gold, and Platinum.

BREEAM (Building Research Establishment Environmental Assessment Method)

BREEAM is a worldwide sustainability assessment system that analyzes building performance across various environmental and social factors. It was developed in the United Kingdom and has now spread worldwide. It considers energy, water, waste, health and well-being, ecology, pollution, etc. BREEAM offers many grades, ranging from Pass up to Outstanding.

Green Globes

The Green Building Initiative (GBI) is responsible for creating the online evaluation and grading tool Green Globes. It assesses the performance of the building in several different areas, such as energy, water, interior environment, and resource use. Green Globes uses a system ranging from one to four stars with a point value of one thousand. Ratings range from one to four Green Globes.

ENERGY STAR

The Environmental Protection Agency (EPA) in the United States has developed an ENERGY STAR program to encourage energy efficiency. Although it is not a certification for whole buildings, it does give certifications for certain goods (such as appliances, lights, and HVAC systems) and building types (such as schools and hospitals) that exceed stringent requirements for energy efficiency (Casini,2017).

Living Building Challenge (LBC)

One of the most demanding standards for environmentally friendly construction is the Living Building Challenge, designed by the International Living Future Institute (ILFI). It emphasizes regenerative architecture and mandates that buildings satisfy stringent energy, water, material use, equity, and aesthetics requirements. LBC's construction division aims to design and construct environmentally friendly and self-sufficient structures.

WELL Building Standard

The WELL Building Standard, overseen by the International WELL Building Institute (IWBI), emphasizes the health and happiness of the people who live and work in a building. It evaluates aspects of a structure that can benefit human health, such as air, water, food, fitness, and comfort (Chen,2022).

EDGE (Excellence in Design for Greater Efficiencies)

A green building certification system designed specifically for developing nations is called EDGE, established by the International Finance Corporation (IFC). It assesses structures based on resource-efficient design techniques, which may include energy-efficient technology, water-saving measures, and the selection of materials (Circo,2007).

Green Mark (Singapore)

Singapore's Building and Construction Authority (BCA) is responsible for developing and maintaining the Green Mark green building certification scheme. It evaluates buildings based on various environmental factors, such as the amount of energy they save, the amount of water they hold, and the amount of sustainable materials they use. These certifications and standards provide rules and benchmarks for sustainable building practices. They are meant to assist building owners, architects, and developers in creating better buildings for the environment, using less energy, and providing a healthier atmosphere for those living there. Each certification has its own unique set of requirements, as well as a grading system, which is adapted to the various geographical areas and kinds of buildings (Yang, 2022).

CASE STUDIES OF SUCCESSFUL GREEN BUILDING PROJECTS

Case studies of environmentally conscious construction projects that were completed successfully in various regions of the globe are as follows:

One Angel Square, Manchester, United Kingdom

Manchester's Co-operative Group headquarters is One Angel Square. The 2013 BREEAM Outstanding office building was one of the world's most sustainable. The building has an efficient double-skin facade, automated building controls, and on-site renewable energy production. Its unique design and energy-efficient features reduce energy usage by 50% compared to traditional office buildings. The Manchester, UK-based One Angel Square is a model of eco-friendly corporate headquarters that has set the standard for the industry. Built in 2013 as the Co-operative Group's headquarters, this landmark structure exemplifies sustainable design and cutting-edge architecture in perfect harmony. Built by BAM Construct UK and designed by 3DReid, the angular structure of the skyscraper stands out on Manchester's skyline. Nevertheless, its importance goes much beyond mere aesthetics. The most prestigious BREEAM certification, "Outstanding," was bestowed to One Angel Square, revolutionising the way sustainable building techniques are approached. In order to reduce energy consumption and environmental effect, its design contains a number of eco-friendly elements. These include a combined heat and power plant, natural ventilation systems, and rainwater collection. The structure's dedication to sustainability has been recognised with many prizes, one of which being the RIBA North West Award for Architecture. One Angel Square is more than just a beautiful building; it is a symbol of the Co-operative Group's commitment to CSR and environmental protection. Its exemplary sustainable development methods serve as a model for other organisations, encouraging them to have a positive impact on the environment and the future. As a result, One Angel Square is an impressive example of sustainable corporate headquarters architecture in addition to being an outstanding architectural accomplishment (Ismail, 2021).

The Edge, Amsterdam, Netherlands

Deloitte commissioned the design of The Edge, an environmentally friendly office skyscraper in Amsterdam, which was completed in 2015. It was awarded a BREEAM Outstanding certification with a score of 98.36%, making it the office building with the title of being the most environmentally friendly in the whole wide globe. The structure uses cutting-edge technology, such as energy-efficient LED lighting, solar panels, rainwater collecting systems, and intelligent HVAC systems. The outcome is a large reduction in energy consumption due to optimizing energy use depending on occupancy and daylight levels. The Edge, located in the Dutch capital of Amsterdam, is a model of state-of-the-art intelligent design and ecological building. Built with an eye towards efficiency, sustainability, and the comfort of its occupants, this extraordinary building has been home to Deloitte Netherlands since its 2015 completion. A variety of cutting-edge features and technology were integrated into The Edge's design by PLP Architecture in order to lessen its influence on the environment and improve the user experience. Noteworthy features include innovative temperature control systems and LED lighting that optimise energy use, as well as the building's use of solar power via photovoltaic panels covering its roof. Additionally, The Edge is well-known for its use of Internet of Things (IoT) devices and sensors in smart buildings to control and monitor different parts of building operations in real-time. This method not only promotes a more pleasant and productive work atmosphere, but it also guarantees that all environmental factors are at their best. Green walls and interior gardens are only two examples of the biophilic design ideas that the building employs to promote health and a sense of oneness with nature. As a model for environmentally conscious workplace design throughout the globe, The Edge has won a slew of accolades for its cutting-edge sustainability features. Its achievement demonstrates the possibility of constructing high-performance, user-centric buildings that prioritise environmental responsibility and occupant happiness via the integration of cutting-edge technology and sustainable methods (Kuo, 2017).

Bahrain World Trade Center, Manama, Bahrain

The Bahrain World Trade Center (BWTC) is an iconic shopping mall designed with renewable energy sources in mind throughout the construction process. It was finished in 2008 and has three wind turbines incorporated into the design between the two towers. These turbines provide clean energy and supply around 11–15 per cent of the building's overall energy requirements, lowering the structure's carbon footprint. The World Trade Center illustrates how famous designs may benefit from environmentally friendly technologies . An iconic structure in Manama, Bahrain, the Bahrain World Trade Centre showcases cutting-edge sustainable architecture. This famous building, which stood at 240 metres (787 feet) when it was finished in 2008, is known for its twin sail-shaped towers that are linked by three skybridges. One of the most innovative aspects of the Atkins-designed Bahrain World Trade Centre is the use of renewable energy sources. The skyscrapers were among the first in the world to include wind turbines into their construction; each of the two buildings had three enormous turbines, set at a height of 95 metres (312 ft). This building is more sustainable and uses less conventional energy since these turbines take use of the high winds in the area to power a large part of its energy requirements. Use of recycled materials, efficient lighting, and optimised ventilation are just a few of the other sustainable design methods included in the building. The sustainable design and unique architecture of the Bahrain World Trade Centre have earned it several medals and accolades, making it a model for other green buildings in the region and beyond. In addition to lowering the building's carbon footprint, the use of renewable energy sources like wind turbines is a symbolic act of Bahrain's dedication to sustainable development and renewable power. Consequently, the Bahrain World Trade Centre exemplifies how sustainable design may provide aesthetically pleasing and ecologically conscious building solutions (Debrah, 2022).

Pixel Building, Melbourne, Australia

The Pixel Building in Melbourne, Australia, was the first carbon-neutral commercial building in the country when it was finished in 2010. It uses several environmentally friendly components, including a solar photovoltaic system, rainwater collecting, greywater treatment, and natural ventilation, among other things. The building's innovative architecture and energy-saving measures lead to a considerable decrease in the energy used and the amount of carbon emissions produced. The Pixel Building, located in Melbourne, Australia, is a groundbreaking example of sustainable architecture that pushes the boundaries of green design. Completed in 2010, this innovative structure stands as a testament to the city's commitment to environmental sustainability. Designed by renowned architectural firm Studio 505, the Pixel Building incorporates a plethora of eco-friendly features aimed at reducing its carbon footprint and energy consumption. One of its most distinctive features is its striking facade adorned with over 5,000 recycled car windshields, which not only adds a unique aesthetic but also promotes resource reuse and recycling. Additionally, the building utilizes a range of cutting-edge technologies such as solar panels, rainwater harvesting systems, and energy-efficient lighting to minimize its environmental impact. The Pixel Building has achieved the highest Green Star rating, signifying its exceptional sustainability performance. Moreover, its design prioritizes occupant comfort and well-being through features like natural ventilation, ample daylighting, and flexible workspace layouts. As a pioneering green building, the Pixel Building serves as an inspiration for sustainable architecture globally, demonstrating how innovative design strategies and advanced technologies can create buildings that are both environmentally responsible and aesthetically striking (Alagirisamy, 2022).

Bank of America Tower at One Bryant Park, New York City, USA

This commercial skyscraper, finished in 2009 and known as One Bryant Park, was the first commercial high-rise structure to obtain LEED Platinum certification. It is also the tallest building in New York City. It does this by implementing different energy-saving techniques, such as a high-performance glass curtain wall, a sophisticated cogeneration system, and a rainwater harvesting system providing water for cooling and irrigation. The ecological measures implemented in the tower have resulted in an energy efficiency level that is fifty per cent greater than that of conventional office buildings. The Bank of America Tower at One Bryant Park, situated in the heart of New York City, USA, stands as a pinnacle of sustainable skyscraper design and green construction. Completed in 2009, this iconic tower, also known simply as One Bryant Park, is a collaboration between architectural firms Cook+Fox Architects and Adamson Associates Architects. It rises 1,200 feet into the Manhattan skyline and boasts 55 floors of office space. Notably, the tower has achieved LEED Platinum certification, the highest standard for green building, reflecting its commitment to environmental sustainability. One of its most remarkable features is its advanced HVAC system, which utilizes ice storage tanks to cool the building during off-peak hours, significantly reducing energy consumption. Additionally, the tower incorporates a range of eco-friendly technologies, such as rainwater harvesting systems, energy-efficient lighting, and floor-to-ceiling insulated glass windows that maximize natural light while minimizing heat gain. The Bank of America Tower also prioritizes occupant well-being with amenities like indoor air quality monitoring systems and access to outdoor green spaces. Furthermore, the tower's location adjacent to Bryant Park underscores its integration with the surrounding urban environment. As a shining example of sustainable architecture, the Bank of America Tower at One Bryant Park not only redefines the New York City skyline but also sets a new standard for environmentally responsible skyscraper design globally (Ren, 2021).

The Crystal, London, United Kingdom

In 2012, London welcomed the opening of The Crystal, a venue for eco-friendly events and exhibitions. In addition to being one of the most environmentally friendly structures in the world, it has been awarded the BREEAM Outstanding and LEED Platinum certifications. The structure incorporates ground-source heat pump technology, solar panels, and rainwater collection systems. It is built to have a low environmental impact and provides demonstrations of cutting-edge environmentally friendly technology to guests. These eco-friendly construction projects, which have been completed successfully, are examples of how sustainable architecture and building methods may be implemented. They point toward a more sustainable built environment by demonstrating that ecologically responsible structures can simultaneously be visually beautiful and economically feasible. One prime example of environmentally conscious city planning and building design is the Crystal in London, UK. Siemens' sustainable cities programme and an interactive exhibition area focused on urban sustainability and future technology coexist in this landmark edifice, which opened in 2012. A remarkable crystalline shape with a façade composed mostly of glass, The Crystal is a symbol of transparency and sustainability and was designed by the architectural company WilkinsonEyre. Solar panels, rainwater collection systems, energy-efficient lighting, and HVAC systems are just a few of the state-of-the-art features that help reduce the building's negative influence on the environment. The Crystal further improves its ecological credentials by using a Building Management System (BMS) to monitor and optimise energy use in real-time. The Crystal is not just an impressive piece of architecture, but also a centre for sustainability, urban planning, and renewable energy education and innovation, with regular exhibits, seminars, and events. The Crystal, an innovative green building, does more than just show what's possible in green design; it also encourages people to work together and share ideas for future cities that are more sustainable and resilient (Darwish, 2017).

CHALLENGES IN GREEN BUILDING CONSTRUCTION

The green building sector is facing several obstacles despite the increasing interest in eco-friendly building methods. Developers and investors who are only interested in short-term profits may be put off by the higher startup costs of using green building technologies and materials. Additionally, there are obstacles to uniform implementation and compliance due to the absence of standardised green building rules and the variable certification requirements across various areas. The broad adoption of green building practices is hampered by several factors, including the limited availability and high prices of sustainable building materials, as well as a lack of competent labour with knowledge of green construction procedures. There are technical and logistical hurdles to overcome when retrofitting an existing building with renewable energy technology and energy-efficient systems. Policymakers, industry professionals, and financial institutions are just some of the groups that need to work together to find solutions to these problems. They need to develop policies, incentives, and training programmes to boost green building practices and reduce barriers to sustainable building (Amin,2017).

- **High Initial Cost:** Green building construction is known for greater upfront expenditures than traditional structures. Green buildings typically save money in the long run due to decreased energy and water use, operating expenses, and tenant productivity.
- **Limited Awareness and Education:** Developers, contractors, and end-users may not comprehend green construction concepts, preventing implementation. Green building education and training are essential for the sector.
- **Inconsistent Building Codes and Regulations:** Developers may struggle to integrate sustainable elements into their buildings due to uneven or unestablished green building standards and guidelines in various countries (Liu,2021).
- **Fragmented Supply Chain:** Green building encompasses architects, engineers, contractors, and suppliers. Integrating and communicating with these parties might make it difficult to achieve high-performance green buildings.
- **Limited Access to Financing:** Green construction projects may be hard to finance, particularly in areas where lenders don't understand their advantages. This can be addressed with green building certifications and ROI statistics.
- **Performance Verification and Monitoring:** Green buildings must be verified to ensure sustainability. Building owners and operators may struggle with inadequate monitoring systems (Hu,2020).

FUTURE TRENDS IN GREEN BUILDING CONSTRUCTION

Green building practices and renewable energy sources are more likely to merge in the near future. To do this, renewable energy sources including solar panels, wind turbines, and geothermal heat pumps will be included in the design of the buildings. Buildings may store surplus energy for later use via the adoption of energy storage solutions like sophisticated battery technologies and thermal energy storage systems, which further improve the resilience and dependability of renewable energy integration. In addition, it is expected that net-zero energy buildings, which generate as much energy as they use, will become the norm in environmentally friendly construction. For buildings to achieve net-zero energy performance, new energy-efficient systems must be developed. These systems include highly insulated building envelopes, improved glazing technologies, and passive design methods. Incorporating natural materials and green areas into building interiors according to biophilic design principles has been shown to improve the health and happiness of building occupants, leading to a more environmentally friendly and people-focused built environment (Hall,2000).

As climate change and harsh weather become more common, the development of robust and adaptive building designs will become an important goal for sustainable architecture. To ensure the long-term durability and survival of buildings in the face of growing environmental difficulties, disaster-resistant materials, adaptable building structures, and resilient infrastructure systems will be implemented. A more sustainable and environmentally conscious future for the construction industry will be shaped through the collaboration of industry stakeholders, such as architects, engineers, developers, and policymakers, to drive innovation and foster the widespread adoption of sustainable and resilient building practices (Singh, 2021).

CONCLUSION

"Distributed Renewable Power in Green Building Planning and Construction Strategies" combines sustainability with contemporary building strongly and innovatively. Distributed renewable electricity in green construction efforts is essential to addressing climate change and environmental deterioration. Green buildings may minimize greenhouse gas emissions using renewable energy sources, including solar, wind, biomass, and geothermal systems. Distributed renewable power reduces dependency on traditional energy sources and increases energy security due to its decentralization. Green buildings optimize dispersed renewable power resources through careful site study, technology selection, and sustainable design. Innovative architecture and clever automation solutions support energy efficiency, maximizing renewable energy use. Distributed renewable power integration's cost-effectiveness and long-term savings appeal to builders, investors, and occupiers. As technology improves and legislation supports renewable energy usage, economic advantages and general acceptance increase rapidly. Distributed renewable power affects green building design and construction beyond energy efficiency and environmental responsibility. These structures show how smart, networked societies may coexist with nature.

Policymakers, architects, developers, and the community must collaborate to maximize distributed renewable electricity in green construction initiatives. To increase greener buildings, rules, incentives, and education are needed. Distributed renewable electricity in green building design and construction promises a sustainable and resilient future. Green buildings promote good change by using clean, renewable energy sources. Distributed renewable electricity in green buildings brings us closer to a sustainable and prosperous future.

REFERENCES

Alagirisamy, B., & Ramesh, P. (2022). Smart sustainable cities: Principles and future trends. In *Sustainable Cities and Resilience: Select Proceedings of VCDRR 2021* (pp. 301-316). Springer Singapore.

Alshboul, O. (2022). Evaluating the impact of external support on green building construction cost: A hybrid mathematical and machine learning prediction approach. *Buildings, 12*.

Amin, U., Hossain, M. J., Lu, J., & Fernandez, E. (2017). Performance analysis of an experimental smart building: Expectations and outcomes. *Energy, 135*, 740–753. doi:10.1016/j.energy.2017.06.149

Cao, M.-Q., Liu, T.-T., Zhu, Y.-H., Shu, J.-C., & Cao, M.-S. (2022). Developing electromagnetic functional materials for green building. *Journal of Building Engineering, 45*, 103496. doi:10.1016/j.jobe.2021.103496

Casini, M. (2017, August). Green technology for smart cities. []. IOP Publishing.]. *IOP Conference Series. Earth and Environmental Science, 83*(1), 012014. doi:10.1088/1755-1315/83/1/012014

Chen, L., Chan, A. P. C., Owusu, E. K., Darko, A., & Gao, X. (2022). Critical success factors for green building promotion: A systematic review and meta-analysis. *Building and Environment, 207*, 108452. doi:10.1016/j.buildenv.2021.108452

Chen, Y., Huang, D., Liu, Z., Osmani, M., & Demian, P. (2022). Construction 4.0, Industry 4.0, and Building Information Modeling (BIM) for sustainable building development within the smart city. *Sustainability (Basel), 14*(16), 10028. doi:10.3390/su141610028

Chen, Y.-K., Wu, Y.-C., Song, C.-C., & Chen, Y.-S. (2012). Design and implementation of energy management system with fuzzy control for DC microgrid systems. *IEEE Transactions on Power Electronics, 28*(4), 1563–1570. doi:10.1109/TPEL.2012.2210446

Çiner, F., & Doğan-Sağlamtimur, N. (2019, November). Environmental and sustainable aspects of green building: A review. []. IOP Publishing.]. *IOP Conference Series. Materials Science and Engineering, 706*(1), 012001. doi:10.1088/1757-899X/706/1/012001

Circo, C. J. (2007). Using mandates and incentives to promote sustainable construction and green building projects in the private sector: A call for more state land use policy initiatives. *Penn St. L. Rev., 112*, 731.

Darwish, A. S. (2017). Green, smart, sustainable building aspects and innovations. In *Mediterranean Green Buildings & Renewable Energy: Selected Papers from the World Renewable Energy Network's Med Green Forum* (pp. 717-727). Springer International Publishing. 10.1007/978-3-319-30746-6_55

Debrah, C., Albert, P. C. C., & Darko, A. (2022). Green finance gap in green buildings: A scoping review and future research needs. *Building and Environment, 207*, 108443. doi:10.1016/j.buildenv.2021.108443

Debrah, C., Chan, A. P., & Darko, A. (2022). Artificial intelligence in green building. *Automation in Construction, 137*, 104192. doi:10.1016/j.autcon.2022.104192

El Barkouki, B., Laamim, M., Rochd, A., Chang, J. W., Benazzouz, A., Ouassaid, M., Kang, M., & Jeong, H. (2023). An Economic Dispatch for a Shared Energy Storage System Using MILP Optimization: A Case Study of a Moroccan Microgrid. *Energies, 16*(12), 4601. doi:10.3390/en16124601

Erten, D., & Kılkış, B. (2022). How can green building certification systems cope with the era of climate emergency and pandemics? *Energy and Building, 256*, 111750. doi:10.1016/j.enbuild.2021.111750

Hall, R. E., Bowerman, B., Braverman, J., Taylor, J., Todosow, H., & Von Wimmersperg, U. (2000). The vision of a smart city (No. BNL-67902; 04042). Brookhaven National Lab.(BNL), Upton, NY (United States).

Hu, M. (2020). *Smart technologies and design for healthy built environments.* Springer Nature.

Iravani, A. (2017). Advantages and disadvantages of green technology; goals, challenges and strengths. *Int J Sci Eng Appl, 6*(9), 272–284.

Ismaeil, E. M., & Sobaih, A. E. E. (2023). Heuristic Approach for Net-Zero Energy Residential Buildings in Arid Region Using Dual Renewable Energy Sources. *Buildings, 13*(3), 796. doi:10.3390/buildings13030796

Ismail, Z. A. (2021). Maintenance management practices for green building projects: Towards hybrid BIM system. *smart and sustainable. Built Environment, 10*(4), 616–630.

Jiang, Y., & Zheng, W. (2021). Coupling mechanism of green building industry innovation ecosystem based on blockchain smart city. *Journal of Cleaner Production, 307*, 126766. doi:10.1016/j.jclepro.2021.126766

Jung, C., & Awad, J. (2023). Sharjah Sustainable City: An Analytic Hierarchy Process Approach to Urban Planning Priorities. *Sustainability (Basel), 15*(10), 8217. doi:10.3390/su15108217

Kuo, C. F. J., Lin, C. H., & Hsu, M. W. (2016). Analysis of intelligent green building policy and developing status in Taiwan. *Energy Policy, 95*, 291–303. doi:10.1016/j.enpol.2016.04.046

Kuo, C. F. J., Lin, C. H., Hsu, M. W., & Li, M. H. (2017). Evaluation of intelligent green building policies in Taiwan–Using fuzzy analytic hierarchical process and fuzzy transformation matrix. *Energy and Building, 139*, 146–159. doi:10.1016/j.enbuild.2016.12.078

Liu, Z., Chi, Z., Osmani, M., & Demian, P. (2021). Blockchain and building information management (BIM) for sustainable building development within the context of smart cities. *Sustainability (Basel), 13*(4), 2090. doi:10.3390/su13042090

Liu, Z., Chi, Z., Osmani, M., & Demian, P. B. (2021). Building Information Management (BIM) for Sustainable Building Development within the Context of Smart Cities. *Sustainability, 13*.

Meena, C. (2022). Innovation in Green Building Sector for Sustainable Future. *Energies, 15*(18).

Mohanty, S. (2010). *Green technology in construction. Recent Advances in Space Technology Services and Climate Change 2010 (RSTS & CC-2010).* IEEE.

Mohtashami, N., Karuvingal, R., Droste, K., Schreiber, T., Streblow, R., & Müller, D. (2023, November). How to build green substations? An LCA comparison of different sustainable design strategies for substations. []. IOP Publishing.]. *Journal of Physics: Conference Series, 2600*(15), 152022. doi:10.1088/1742-6596/2600/15/152022

Mostafa, M., Vorwerk, D., Heise, J., Povel, A., Sanina, N., Babazadeh, D., & Toebermann, C. (2022, September). Integrated Planning of Multi-energy Grids: Concepts and Challenges. In *NEIS 2022; Conference on Sustainable Energy Supply and Energy Storage Systems* (pp. 1-7). VDE.

Palermo, S. A., Talarico, V. C., & Pirouz, B. (2019). Optimizing rainwater harvesting systems for non-potable water uses and surface runoff mitigation. *International Conference on Numerical Computations: Theory and Algorithms*. Springer, Cham.

Pan, J., Jain, R., & Paul, S. (2013). Nine lessons learned from a green building testbed: A networking and energy efficiency perspective. *2013 World Congress on Sustainable Technologies (WCST)*. IEEE. 10.1109/WCST.2013.6750405

Parasher, Y., Singh, P., & Kaur, G. (2019). Green Smart Town Planning. *Green and Smart Technologies for Smart Cities*, 19-41.

Pol, S., Houchens, B. C., Marian, D., & Westergaard, C. (2020). Performance of AeroMINEs for Distributed Wind Energy. In *AIAA Scitech 2020 Forum* (p. 1241). 10.2514/6.2020-1241

Pramanik, P. K. D., Mukherjee, B., Pal, S., Pal, T., & Singh, S. P. (2021). Green smart building: Requisites, architecture, challenges, and use cases. In *Research anthology on environmental and societal well-being considerations in buildings and architecture* (pp. 25–72). IGI Global. doi:10.4018/978-1-7998-9032-4.ch002

Radziejowska, A., & Sobotka, B. (2021). Analysis of the social aspect of smart cities development for the example of smart sustainable buildings. *Energies*, *14*(14), 4330. doi:10.3390/en14144330

Rahim, N. S. A., Rahman, B. A., Ibrahim, F. A., Ishak, N., & Ayob, A. (2023, September). The Contractors' Perception on the Development of Green Building Projects in Penang. []. IOP Publishing.]. *IOP Conference Series. Earth and Environmental Science*, *1238*(1), 012020. doi:10.1088/1755-1315/1238/1/012020

Rameshwar, R., Solanki, A., Nayyar, A., & Mahapatra, B. (2020). Green and smart buildings: A key to sustainable global solutions. In *Green Building Management and Smart Automation* (pp. 146–163). IGI Global. doi:10.4018/978-1-5225-9754-4.ch007

Ren, X., Li, C., Ma, X., Chen, F., Wang, H., Sharma, A., Gaba, G. S., & Masud, M. (2021). Design of multi-information fusion based intelligent electrical fire detection system for green buildings. *Sustainability (Basel)*, *13*(6), 3405. doi:10.3390/su13063405

Riadh, A. D. (2022). Dubai, the sustainable, smart city. *Renewable Energy and Environmental Sustainability*, *7*, 3. doi:10.1051/rees/2021049

Singh, T., Solanki, A., & Sharma, S. K. (2021). Role of smart buildings in smart city—components, technology, indicators, challenges, future research opportunities. *Digital cities roadmap: IoT-based architecture and sustainable buildings*, 449-476.

Singhal, P. (2019). A Case Study on Energy Efficient Green Building with New Intelligent Techniques Used to Achieve Sustainable Development Goal. *2019 20th International Conference on Intelligent System Application to Power Systems (ISAP)*. IEEE. 10.1109/ISAP48318.2019.9065938

Soni, S. K., & Bartaria, V. N. (2013). An overview of green building control strategies." *2013 International Conference on Renewable Energy Research and Applications (ICRERA)*. IEEE,. 10.1109/ICRERA.2013.6749837

Streimikiene, D., Skulskis, V., Balezentis, T., & Agnusdei, G. P. (2020). Uncertain multi-criteria sustainability assessment of green building insulation materials. *Energy and Building*, *219*, 110021. doi:10.1016/j.enbuild.2020.110021

Valencia, A., Zhang, W., Gu, L., Chang, N.-B., & Wanielista, M. P. (2022). Synergies of green building retrofit strategies for improving sustainability and resilience via a building-scale food-energy-water nexus. *Resources, Conservation and Recycling*, *176*, 105939. doi:10.1016/j.resconrec.2021.105939

Wang, W., Wang, Y., Wang, D., Zou, Y., & Chen, X. (2023, May). A review of key issues in planning AC/DC distribution systems for renewable energy. []. IOP Publishing.]. *Journal of Physics: Conference Series*, *2503*(1), 012055. doi:10.1088/1742-6596/2503/1/012055

Yang, B., Lv, Z., & Wang, F. (2022). Digital Twins for Intelligent Green Buildings. *Buildings*, *12*(6), 856. doi:10.3390/buildings12060856

Yi, X., Lu, T., Li, Y., Ai, Q., & Hao, R. (2023). Collaborative planning and optimization for electric-thermal-hydrogen-coupled energy systems with portfolio selection of the complete hydrogen energy chain. *arXiv preprint arXiv:2311.07891.*

Chapter 3
AI-Integrated Electronic Cooling Systems and Advanced Strategies for Efficient Data Center

R. Pitchai
ⓘ https://orcid.org/0000-0002-3759-6915
Department of Computer Science and Engineering, B.V. Raju Institute of Technology, India

D. Sengeni
Department of Electronics and Communication Engineering, C.K. College of Engineering and Technology, India

Putchakayala Yanna Reddy
Department of Electrical and Electronics Engineering, Bharath Institute of Engineering and Technology, India

M. Mathiyarasi
Department of Aerospace Engineering, Agni College of Technology, India

Devika Sahu
Department of Computer Science and Engineering, Government Engineering College, India

Sampath Boopathi
ⓘ https://orcid.org/0000-0002-2065-6539
Mechanical Engineering, Muthayammal Engineering College, Namakkal, India

ABSTRACT

The growth of digital infrastructure has necessitated the development of data centers, which require efficient cooling systems for reliable operation. This chapter delves into the complex world of cooling technologies, their challenges, and innovative solutions. It emphasizes the importance of efficient cooling in data centers, balancing efficiency and energy consumption, scalability, and environmental considerations. The chapter discusses various cooling systems, including traditional air cooling and advanced liquid cooling and phase-change solutions. It also highlights the use of advanced materials like graphene and carbon nanotubes for improved heat transfer and thermal management. The chapter also discusses the integration of Artificial Intelligence in cooling systems, enabling real-time monitoring and predictive analytics. The future of data center cooling will see continued innovations like modular data centers, rack-level cooling, and advanced free cooling strategies.

DOI: 10.4018/979-8-3693-3735-6.ch003

INTRODUCTION

Electronic devices, which power devices like smartphones and laptops, generate heat as a natural by-product of their operation. Failure to manage this heat can lead to performance degradation and reduced operational lifespan. Electronic cooling systems, such as air, liquid, phase-change, heat sinks, and fans, help mitigate this heat challenge. However, the demand for efficient cooling extends beyond individual devices. In the digital age, data centers house numerous electronic devices, where heat can accumulate quickly, leading to efficiency and reliability concerns. Data center cooling has evolved into a critical science, with challenges and innovative solutions explored. This chapter explores the role of artificial intelligence and advanced materials in shaping the future of electronic cooling systems, as understanding these systems is crucial for the reliability and functionality of our increasingly electronic-dependent world (Capozzoli & Primiceri, 2015).

Cooling systems are crucial in modern technology, ensuring the efficient and reliable operation of various electronic devices. They prevent overheating, which can lead to catastrophic failures and performance degradation. Electronic devices generate heat during operation, which if not managed efficiently can compromise the integrity of sensitive components. Cooling systems, including traditional air cooling and sophisticated liquid cooling systems, help dissipate heat and maintain optimal operating temperature (Hu et al., 2020). These systems, ranging from fans and heat sinks to liquid cooling systems, utilize coolants to transfer heat away from high-performance components. This chapter delves into the complex ecosystem of cooling systems in data centers, which are the backbone of our digital world. These centers house numerous servers, network equipment, and storage systems, generating significant heat. Efficient cooling is crucial for reliability and performance. The chapter explores various cooling methods, unique challenges, and latest innovations like artificial intelligence and advanced materials. Understanding these systems is essential as they ensure the seamless functioning of daily technology (Deng et al., 2018).

Data centers, the digital hubs, house servers, switches, and storage systems that power the internet, cloud services, and online applications. However, they also pose an invisible challenge - heat generation. The operation of data center equipment generates immense heat, which can have significant consequences on performance, reliability, and the bottom line if left unchecked. The vast amount of information processed by data centers is a significant byproduct of the energy required to power and operate the thousands of servers and networking equipment within these facilities. Without efficient cooling solutions, this heat can accumulate, leading to hot spots and elevated temperatures, endangering the equipment they rely on (Xia et al., 2017).

Efficient cooling in data centers is crucial for maintaining the performance and longevity of servers and electronic equipment. It ensures uninterrupted service delivery and reduces hardware failures and downtime. This chapter delves into the complexities of managing heat in data centers, exploring various cooling methods, innovative approaches, and the implications for the digital infrastructure. Understanding this underappreciated aspect is crucial for the resilience and functionality of the digital world we increasingly rely on (Chen et al., 2021). Data centers, the digital backbone of the internet, house vast amounts of critical data and applications. However, they face an invisible battle against heat, as servers, networking equipment, and storage systems generate immense amounts of heat. Effective cooling mechanisms are crucial for maintaining optimal functionality and reliability in these facilities. This chapter delves into the complexities of cooling in data centers (Oró et al., 2015).

Data centers require efficient cooling to prevent overheating, performance issues, and costly downtime. Proper cooling mechanisms are crucial to prevent equipment from overheating, affecting its lifespan and performance. Data centers use various cooling solutions, including precision air conditioning systems, hot/cold aisle containment strategies, and advanced liquid cooling techniques, to maintain a stable temperature, prevent hot spots, and ensure equipment operates within its specified temperature range. This chapter explores the importance of efficient cooling in data centers, focusing on the technologies and strategies that help these facilities run smoothly while minimizing their environmental footprint (Oró et al., 2015). The focus is on energy efficiency, environmental responsibility, and cost-effectiveness, as well as the critical role these facilities play in our digital lives. Understanding the intricacies of cooling in data centers is vital for appreciating their critical role in maintaining a cool, efficient, and reliable environment (Rhee et al., 2017).

The pursuit of efficiency and sustainability has led to significant advancements in electronic cooling systems. Artificial intelligence (AI) has been used for cooling optimization, while advanced materials have been utilized to improve system performance. This chapter explores the intersection of AI and advanced materials in reshaping the electronic cooling system landscape. Artificial intelligence (AI) is revolutionizing the world of electronic cooling systems by analyzing vast datasets and making real-time decisions. AI-driven cooling optimization in data centers can monitor server heat output and direct cooling resources precisely, minimizing energy consumption and reducing hot spots. Predictive analytics powered by AI can forecast temperature trends and suggest proactive adjustments, preventing overheating incidents and equipment failures (Wu et al., 2019). This transforms cooling systems from passive machines into intelligent, adaptive solutions that maintain ideal temperatures, save energy, reduce operational costs, and prolong the lifespan of electronic components. This chapter explores the fascinating world of AI's role in optimizing electronic cooling systems and its potential to revolutionize the field (B et al., 2024; Nishanth et al., 2023; Syamala et al., 2023).

Advanced materials are revolutionizing the cooling systems industry by providing physical means to enhance their cooling capabilities. High-conductivity materials like graphene and phase-change materials are being used to improve heat transfer processes. Miniaturization and microfabrication techniques have enabled the development of intricate structures and surfaces that significantly enhance heat exchange efficiency. These innovations are applicable in consumer electronics, industrial equipment, and data center cooling solutions. The synergy between AI-driven cooling optimization and advanced materials in electronic cooling systems has the potential to unlock new frontiers of sustainability, performance, and reliability in the electronics industry. The convergence of these technologies represents an exciting leap forward in ensuring electronic devices and data centers remain cool, efficient, and resilient in the face of increasing demands.

Objectives

- To elucidate the crucial role of cooling systems in electronic devices and data centers. It provides an overview of why efficient cooling is essential for maintaining performance and reliability.
- To introduce the concept of artificial intelligence (AI) and its applications in optimizing cooling systems. It explores how AI can be harnessed to create smart, adaptive cooling solutions.
- To optimize cooling in electronic devices and data centers. It discusses real-time monitoring, predictive analytics, and intelligent resource allocation for energy-efficient cooling.

- To explore the latest advancements in materials used for heat dissipation in cooling systems. It delves into high-conductivity materials like graphene and novel phase-change materials, showcasing their potential to enhance heat transfer.

Types of Cooling Systems

Cooling systems are crucial for managing heat in various applications, including consumer electronics and industrial settings. They utilize various methods and components to efficiently dissipate heat. Common types include heat pumps, fans, and other essential components (Allouhi et al., 2015; Bahiraei & Heshmatian, 2018; Dincer, 2017).

- Air Cooling System:
  - **Important Components:** Fans, heat sinks, and air ducts.
  - **Functions:** Fans circulate air through the system, while heat sinks absorb and dissipate heat.
  - **Working Principle:** Heat is transferred from electronic components to the heat sink, and the fan blows air over the heat sink to dissipate heat.
  - **Applications:** Consumer electronics like laptops and desktop computers, as well as some industrial equipment.
- Liquid Cooling System:
  - **Important Components:** Coolant (usually water or specialized liquids), pumps, heat exchangers, and a cold plate.
  - **Functions:** The coolant absorbs heat and circulates it to a heat exchanger where it dissipates the heat. A cold plate is in direct contact with the electronic components.
  - **Working Principle:** Liquid cooling systems transfer heat more efficiently than air by using a liquid coolant to absorb and transport heat away from components.
  - **Applications:** High-performance computing, gaming PCs, and overclocked processors.
- Phase-Change Cooling System:
  - **Important Components:** Compressors, evaporators, and condensers.
  - **Functions:** These components work together to change a refrigerant from a liquid to a gas and back, transferring heat efficiently.
  - **Working Principle:** Phase-change systems use the refrigerant's phase transition to absorb and release heat.
  - **Applications:** Medical equipment, semiconductor manufacturing, and specialized cooling applications.
- Heat Sinks:
  - **Important Components:** A heat sink is a passive component made of thermally conductive materials like aluminum or copper.
  - **Functions:** Heat sinks are used to dissipate heat from electronic components through conduction and convection.
  - **Working Principle:** Heat is transferred from the component to the heat sink, which then dissipates the heat to the surrounding environment.
  - **Applications:** CPUs, GPUs, power transistors, and various electronic devices.
- Fans:
  - **Important Components:** Impeller blades, motor, and housing.

Figure 1. Types of cooling systems for electronic devices

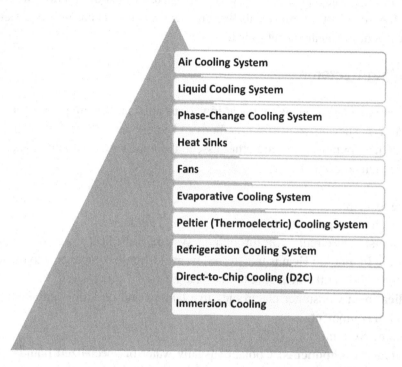

- ○ **Functions:** Fans generate airflow to enhance cooling, either in conjunction with heat sinks or as standalone cooling solutions.
- ○ **Working Principle:** Fans create airflow that helps carry away heat by convection.
- ○ **Applications:** Used in combination with various cooling systems, including air cooling and liquid cooling, in a wide range of electronic and industrial equipment.
- Evaporative Cooling System:
  - ○ **Important Components:** Water reservoir, pump, distribution system, and evaporative media.
  - ○ **Functions:** These systems use the principle of water evaporation to cool the air, which, in turn, cools the equipment.
  - ○ **Working Principle:** Water is pumped over evaporative media, and as it evaporates, it cools the surrounding air, which is then directed over electronic components.
  - ○ **Applications:** Commonly used for cooling data centers, HVAC systems, and in hot, arid climates to improve energy efficiency.
- Peltier (Thermoelectric) Cooling System:
  - ○ **Important Components:** Peltier modules (thermoelectric coolers) and heat sinks.
  - ○ **Functions:** Peltier devices use the Peltier effect to create a temperature differential, transferring heat from one side to the other.
  - ○ **Working Principle:** When a current is applied to the Peltier modules, they transfer heat from one side (the hot side) to the other (the cold side), effectively cooling the cold side.
  - ○ **Applications:** Compact cooling solutions for small electronic devices, such as mini-fridges and portable coolers.

- Refrigeration Cooling System:
  - **Important Components:** Compressor, condenser, evaporator, and refrigerant.
  - **Functions:** These systems use the compression and expansion of a refrigerant to cool the air or liquid in a closed loop.
  - **Working Principle:** A compressor compresses the refrigerant, raising its temperature and pressure, and then the refrigerant is expanded in the evaporator, causing it to cool significantly. This cooled refrigerant is then used to cool the air or liquid in the system.
  - **Applications:** Large industrial cooling systems, commercial refrigeration, and some data center cooling solutions.
- Direct-to-Chip Cooling (D2C):
  - **Important Components:** Microchannel heat sinks and coolant supply.
  - **Functions:** This method provides direct cooling to individual electronic chips or components.
  - **Working Principle:** Coolant is circulated through microchannels in close proximity to electronic chips, effectively cooling them.
  - **Applications:** High-performance computing, supercomputers, and advanced data center cooling, where localized cooling is critical.
- Immersion Cooling:
  - **Important Components:** Liquid coolant and specialized tanks or containers.
  - **Functions:** Entire electronic systems or components are immersed in a dielectric liquid to cool them.
  - **Working Principle:** The liquid coolant absorbs heat, and convection currents circulate it, dissipating heat away from components.
  - **Applications:** High-density data centers, cryptocurrency mining operations, and specialized high-performance computing.

Cooling systems are essential for electronic components to function reliably and optimally, preventing overheating-related issues. The choice of a cooling system depends on factors like heat load, equipment size, energy efficiency requirements, and the environment (Khalaj & Halgamuge, 2017). As devices and data centers become more powerful and heat-intensive, research and innovation continue to focus on developing new cooling technologies and methods to meet the specific needs of the application.

COOLING IN DATA CENTERS

Precision Air Conditioning

Data centers are the lifeblood of our digital world, housing countless servers and networking equipment that power everything from social media to cloud services (Rahamathunnisa, Sudhakar, Murugan, et al., 2023; Reddy et al., 2023; Satav, Hasan, et al., 2024; Vanitha et al., 2023). Efficient cooling in data centers is a mission-critical aspect of ensuring uninterrupted operation and the longevity of electronic components (Wulandari et al., 2020). Precision air conditioning systems are at the forefront of data center cooling, delivering the delicate balance of temperature and humidity control required for optimal performance. In this section, we explore precision air conditioning, its components, working principles, and its pivotal role in data center cooling (Li et al., 2019).

- **Components of Precision Air Conditioning:** Precision air conditioning systems consist of several key components:
  - **Air Handlers:** Air handlers, also known as computer room air conditioners (CRACs), are responsible for cooling the air. They contain components such as filters, cooling coils, and blowers to control the temperature and humidity of the air.
  - **Humidifiers and Dehumidifiers:** Maintaining the right humidity level is crucial in data centers. Humidifiers add moisture, while dehumidifiers remove excess humidity from the air.
  - **Temperature and Humidity Sensors:** Sensors continuously monitor temperature and humidity levels, providing feedback for precise control.
  - **Control Systems:** These systems manage the operation of the air conditioning units, adjusting settings based on real-time data from the sensors.
- **Working Principles of Precision Air Conditioning:** Precision air conditioning systems operate on the principles of temperature and humidity control(Moazamigoodarzi et al., 2019).
  - **Temperature Control:** Air handlers draw warm air from the data center environment. This warm air passes over cooling coils that contain a refrigerant. As the air cools, it's recirculated into the data center to maintain a consistent temperature.
  - **Humidity Control:** To control humidity, precision air conditioning systems use humidifiers to add moisture to the air and dehumidifiers to remove excess moisture. Maintaining the ideal humidity level is crucial for preventing static electricity buildup and protecting sensitive electronic equipment.
  - **Air Filtration:** Air handlers include filters to remove particles and contaminants from the air, ensuring a clean environment for electronic equipment.
- **Applications in Data Centers:** Precision air conditioning systems are specifically designed for data center environments due to the following benefits (Zhang et al., 2021):
  - **Temperature and Humidity Control:** They provide precise control over temperature and humidity levels, creating an optimal environment for electronic components.
  - **Redundancy and Reliability:** Data centers often employ redundant air conditioning units to ensure uninterrupted cooling, even in case of equipment failure.
  - **Scalability:** These systems can be scaled up to accommodate the cooling requirements of larger data centers or additional equipment.
  - **Energy Efficiency:** Advanced models incorporate energy-efficient features, helping data centers reduce their carbon footprint.

In conclusion, precision air conditioning is a cornerstone of efficient data center cooling. Its ability to deliver consistent temperature and humidity control is vital for safeguarding the performance and reliability of the electronic infrastructure that underpins our digital world. As data centers continue to evolve and grow, precision air conditioning systems will play an ever more crucial role in maintaining the optimal operational environment.

Hot/Cold Aisle Containment

In the quest for energy-efficient and effective cooling in data centers, hot/cold aisle containment has emerged as an innovative and practical solution. This approach transforms the traditional open layout of data center racks into a more controlled environment, where hot and cold air are separated to improve cooling efficiency. In this section, we delve into the concept of hot/cold aisle containment, its components, working principles, and the benefits it offers to data center cooling (Zhang et al., 2021).

- **Components of Hot/Cold Aisle Containment:** Hot/cold aisle containment systems consist of various components, including:
  - **Hot Aisle:** This is the aisle where the exhaust air from the servers exits. It is usually enclosed at the end of the racks.
  - **Cold Aisle:** The cold aisle is where the intake air for the servers is supplied. It is typically enclosed, directing cool air to the front of the racks.
  - **Aisle Doors:** Aisle doors enclose the ends of hot and cold aisles to maintain separation and improve airflow control.
  - **Ceiling Panels:** These panels are used to create a physical barrier between the hot and cold aisles, usually extending from the top of the server racks to the ceiling.
  - **Raised Floor:** In some designs, the floor may be raised to accommodate airflow systems or to create an enclosed plenum for directing cold air.
- **Working Principles of Hot/Cold Aisle Containment:** Hot/cold aisle containment is based on the principle of segregating hot and cold air to enhance cooling efficiency:
  - **Cold Aisle Supply:** Cool air is delivered into the cold aisle through perforated tiles in the raised floor or through overhead ducts. This cold air is directed toward the front of the server racks, where the intake of the servers is located.
  - **Server Exhaust:** The servers expel hot air from their rear. This hot air rises naturally and is channeled into the hot aisle.
  - **Aisle Enclosure:** The hot and cold aisles are separated by physical barriers in the form of aisle doors and ceiling panels. This containment minimizes the mixing of hot and cold air.
  - **Return Air:** The hot aisle containment system collects the hot exhaust air and directs it back to the cooling equipment, where it can be cooled and recirculated.
- **Benefits of Hot/Cold Aisle Containment:** Hot/cold aisle containment offers several advantages for data center cooling:
  - **Improved Cooling Efficiency:** By preventing the mixing of hot and cold air, containment ensures that servers receive air at the right temperature, reducing cooling load and energy consumption.
  - **Redundancy and Scalability:** Aisle containment can be designed to accommodate various data center sizes and layouts, making it adaptable for both small and large facilities.
  - **Reduced Operating Costs:** The reduction in energy consumption and more efficient cooling lead to cost savings in the long term.
  - **Enhanced Equipment Lifespan:** Maintaining stable temperatures and reducing thermal stress can extend the lifespan of data center equipment.
  - **Environmental Impact:** Lower energy consumption and increased efficiency contribute to a smaller carbon footprint, aligning with sustainability goals.

In conclusion, hot/cold aisle containment is a proven strategy to enhance cooling efficiency and reduce energy consumption in data centers. It exemplifies how innovative design and engineering can improve the performance and sustainability of the digital infrastructure that underpins our connected world.

Liquid Cooling Solutions

Liquid cooling solutions represent a cutting-edge approach to managing heat in data centers and high-performance computing environments (Vennila et al., 2023). Unlike traditional air cooling, which relies on air circulation, liquid cooling employs specialized coolants to efficiently dissipate heat from electronic components. In this section, we'll explore the concept of liquid cooling, its components, working principles, and applications in data centers (Wulandari et al., 2020).

- **Components of Liquid Cooling Solutions:** Liquid cooling systems for data centers typically comprise the following key components (Boopathi et al., 2023; Boopathi, Sureshkumar, et al., 2023a):
    - **Coolant:** A specialized liquid or fluid designed for its heat-absorbing properties. Common coolants include water, dielectric fluids, and proprietary liquids.
    - **Cold Plates:** These are components in direct contact with electronic devices, such as processors and GPUs. Cold plates are designed to efficiently transfer heat from the components to the coolant.
    - **Pumps:** Pumps are used to circulate the coolant through the system, ensuring a steady flow to carry away heat.
    - **Heat Exchangers:** Heat exchangers transfer heat from the coolant to an external medium (e.g., air or water), dissipating it effectively.
    - **Tubing and Fittings:** These components are used to create a closed-loop system, guiding the flow of coolant from the cold plates to the heat exchangers and back.
    - **Reservoirs and Expansion Tanks:** These containers store and manage the coolant, allowing for system priming and accommodating changes in coolant volume due to temperature variations.
- **Working Principles of Liquid Cooling Solutions:** Liquid cooling systems operate on the principle of heat transfer. Here's how they work:
    - **Heat Absorption:** Cold plates are in direct contact with the heat-producing components, such as CPUs or GPUs. As these components operate, they generate heat that is transferred to the cold plates.
    - **Coolant Circulation:** Pumps circulate the coolant through the system. The warm coolant absorbs heat from the cold plates, which is then transported away.
    - **Heat Dissipation:** Heat exchangers, located in a different environment, transfer heat from the coolant to the external medium, such as air or water.
    - **Coolant Return:** The cooled coolant is returned to the cold plates, where the cycle continues.
- **Applications of Liquid Cooling in Data Centers:** Liquid cooling solutions are well-suited for various applications within data centers, including:
    - *High-Performance Computing (HPC): Liquid cooling is commonly used in HPC clusters where powerful processors and GPUs generate substantial heat. These systems require efficient cooling to maintain peak performance.*
    - *Server Racks: Liquid cooling can be integrated into server racks or applied to specific servers that demand high-performance cooling.*
    - *AI and Deep Learning: Data centers supporting AI and deep learning workloads often rely on liquid cooling to manage the heat generated by dedicated hardware accelerators(Hussain*

et al., 2023; Maguluri, Arularasan, et al., 2023; Pachiappan et al., 2023; Venkateswaran, Kumar, et al., 2023a).

- ○ *Green Data Centers: Liquid cooling is used in energy-efficient and eco-friendly data centers, as it enables more precise temperature control and reduces overall energy consumption.*
- ○ *Overclocked Systems: Enthusiasts and overclocking enthusiasts use liquid cooling to keep their systems at low temperatures during extreme performance demands.*

Liquid cooling offers a highly efficient means of heat management, especially in situations where traditional air-cooling methods may be inadequate. Its ability to maintain lower operating temperatures, reduce energy consumption, and prolong the lifespan of components positions it as a valuable solution for data centers and high-performance computing environments.

Free Cooling Strategies

Data centers are known for their substantial energy consumption, and a significant portion of that energy is dedicated to cooling systems. In pursuit of greater energy efficiency and reduced operational costs, data center operators have turned to free cooling strategies. These techniques leverage external environmental conditions to cool data center facilities without the extensive use of mechanical cooling systems. In this section, we'll explore free cooling strategies, their components, working principles, and the benefits they bring to data center operations (Chu & Huang, 2021; Zhang et al., 2021).

- **Components of Free Cooling Strategies:** Free cooling strategies in data centers don't require many specialized components, but they may include:
  - ○ **Air Dampers:** Dampers control the flow of outside air into the data center. They are often located in air intake systems or on the building's exterior.
  - ○ **Monitoring Systems:** Environmental sensors and control systems are used to assess external weather conditions and adjust the free cooling strategy accordingly.
  - ○ **Filters:** Filters help remove impurities and particulates from incoming air to maintain indoor air quality.
  - ○ **Heat Exchangers:** In some designs, heat exchangers are used to transfer heat from data center air to the incoming outdoor air before it is circulated inside.
  - ○ **Cooling Towers:** In water-based free cooling strategies, cooling towers may be used to reject heat from the data center.
- **Working Principles of Free Cooling Strategies:** The basic premise of free cooling is to use external air, rather than mechanical refrigeration, to cool the data center environment. The working principles vary depending on the specific strategy employed:
  - ○ **Air-Side Economization:** This strategy involves using outdoor air to cool the data center. When external conditions are favorable, air dampers are opened to allow cool outdoor air to enter the data center. Mechanical cooling is only used when necessary, such as during extremely hot or humid conditions.
  - ○ **Water-Side Economization:** Instead of using outdoor air directly, this approach uses a heat exchanger or cooling tower to transfer heat from the data center's water-based cooling system to the incoming outdoor air.

- ◦ **Evaporative Cooling:** In regions with dry climates, evaporative cooling systems use the principle of water evaporation to cool incoming air before it enters the data center. This method is highly energy-efficient.
  - ◦ **Direct and Indirect Free Cooling:** Direct free cooling involves using unconditioned outdoor air, while indirect free cooling uses a heat exchanger to separate indoor and outdoor air. The heat exchanger cools the indoor air using the outdoor air but prevents the two from mixing.
- • **Benefits of Free Cooling Strategies:** Implementing free cooling strategies offers several advantages in data center operations:
  - ◦ **Reduced Energy Consumption:** Free cooling significantly reduces the need for mechanical cooling systems, leading to substantial energy savings.
  - ◦ **Lower Operational Costs:** The reduced energy usage translates into lower operational costs, making data centers more cost-effective.
  - ◦ **Improved Sustainability:** Free cooling aligns with sustainability goals by minimizing the environmental impact of data center operations.
  - ◦ **Enhanced Reliability:** In many cases, free cooling strategies can enhance data center reliability by reducing the reliance on mechanical cooling, which is prone to failures.
  - ◦ **Flexibility:** Free cooling strategies are adaptable and can be fine-tuned to fit various climate conditions and data center requirements.

Free cooling strategies offer a cost-effective and energy-efficient method for cooling data centers, utilizing external environmental conditions to maintain optimal temperatures and reduce their carbon footprint.

USAGE OF AI TO OPTIMIZE COOLING IN DATA CENTERS

Artificial Intelligence (AI) has become a game-changer in the quest for more efficient and reliable data center cooling (Boopathi, Sureshkumar, et al., 2023b; Dhanya et al., 2023; Sampath et al., 2022; Satav, Lamani, et al., 2024). AI-driven cooling optimization leverages machine learning and real-time data analysis to make data centers smarter, greener, and more cost-effective (Van Le et al., 2019). Here's how AI is used to optimize cooling in data centers:

- • **Real-Time Monitoring:** AI continuously monitors various environmental parameters within the data center, including temperature, humidity, airflow, and power consumption. Sensors collect real-time data from servers and cooling systems (Boopathi, 2023b; Kavitha et al., 2023a; Maguluri, Ananth, et al., 2023).
- • **Predictive Analytics:** Machine learning algorithms analyze the historical data and predict temperature trends. By understanding how different variables affect temperature, AI can anticipate overheating risks and adjust cooling settings accordingly.
- • **Dynamic Control:** AI can dynamically adjust cooling settings based on the workload and server utilization. For instance, during periods of high server activity, AI can increase cooling capacity, and during low activity, it can reduce it to save energy.

Figure 2. Usage of AI to optimize cooling in data centers

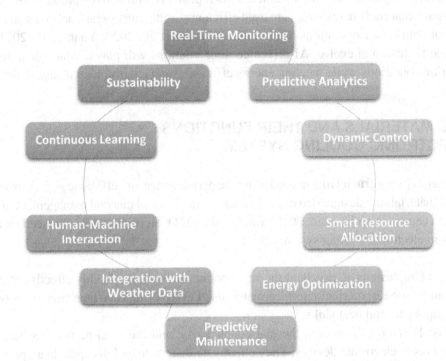

- **Smart Resource Allocation:** AI can intelligently allocate cooling resources to where they are needed the most. For example, if a particular server is generating more heat, AI can direct extra cooling to that server, preventing hot spots.
- **Energy Optimization:** AI-driven systems aim to minimize energy consumption while maintaining optimal operating conditions. They can reduce fan speeds, adjust temperature setpoints, and turn off unnecessary cooling equipment during low-demand periods (Kumar B et al., 2024; Rahamathunnisa, Sudhakar, Padhi, et al., 2023; Satav, Lamani, et al., 2024; Venkateswaran, Vidhya, et al., 2023).
- **Predictive Maintenance:** AI can predict when cooling equipment might fail or require maintenance by analyzing performance data. This proactive approach helps prevent costly downtime and equipment failures.
- **Integration with Weather Data:** AI can integrate external weather data to anticipate temperature fluctuations. For instance, if an AI system predicts a heatwave, it can prepare data center cooling systems to handle increased cooling demands.
- **Human-Machine Interaction:** Some AI systems allow for human interaction through user interfaces. Data center operators can override AI decisions or make manual adjustments when necessary.
- **Continuous Learning:** AI systems can continuously learn and adapt to changing conditions, ensuring that cooling processes become more efficient over time.
- **Sustainability:** AI-optimized cooling often leads to reduced energy consumption, contributing to a data center's sustainability goals and reducing its carbon footprint.

The usage of AI to optimize cooling in data centers represents a proactive approach to cooling management. It ensures that cooling resources are used efficiently, reducing operational costs and improving the reliability of data center operations (Cao et al., 2022; Jin et al., 2020; Yuan et al., 2021). As data centers continue to grow and evolve, AI-driven cooling solutions will play a vital role in maintaining optimal operating conditions while meeting energy efficiency and environmental sustainability targets.

ADVANCED MATERIALS AND THEIR FUNCTIONS IN THE ELECTRONIC COOLING SYSTEM

Advanced materials play a crucial role in enhancing the performance and efficiency of electronic cooling systems. These materials are designed to improve heat dissipation and thermal management in electronic devices and data centers (Van Le et al., 2019; Yang et al., 2021). Here are some advanced materials and their functions in electronic cooling systems:

- **Graphene:** Graphene is an excellent thermal conductor, making it highly effective at dissipating heat. It can be used in thermal interface materials (TIMs) to improve heat transfer between electronic components and heat sinks.
- **Diamonds:** Diamonds have high thermal conductivity, and they can be used as heat spreaders and heat sinks in electronic devices. They effectively distribute and dissipate heat, preventing hot spots.

Figure 3. Advanced materials roles in electronic cooling systems

- Graphene
- Diamonds
- Phase-Change Materials (PCMs)
- Thermally Conductive Polymers
- Carbon Nanotubes (CNTs)
- Gallium Nitride (GaN)
- Boron Nitride (BN)
- Advanced Thermal Interface Materials (TIMs)
- Nanostructured Materials
- Metamaterials
- 3D-Printed Materials

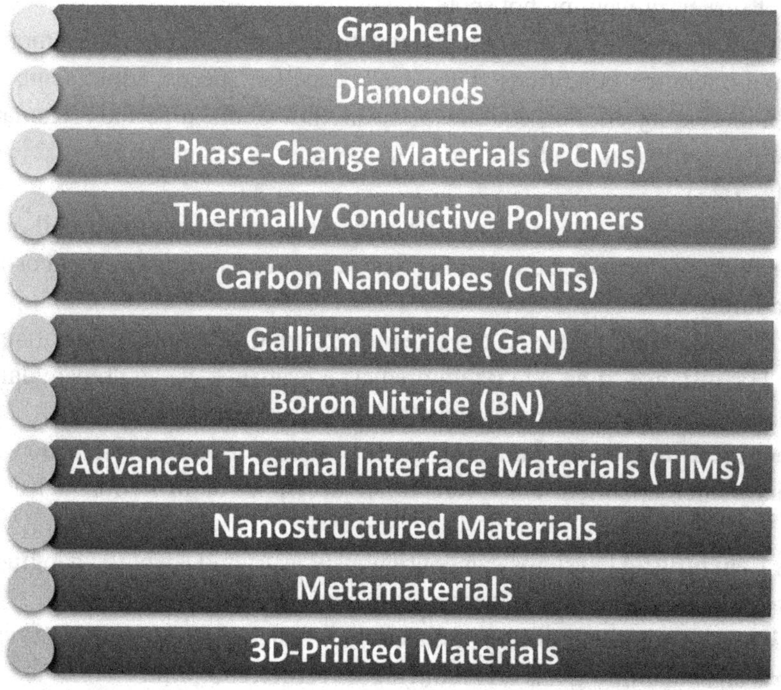

- **Phase-Change Materials (PCMs):** PCMs absorb and release heat during phase transitions. They can be incorporated into electronic cooling systems to store and manage excess heat, providing passive cooling during high-temperature periods.
- **Thermally Conductive Polymers:** These polymers have thermal conductivity properties and can be used as encapsulants or as TIMs in electronic devices to improve heat transfer.
- **Carbon Nanotubes (CNTs):** CNTs are excellent thermal conductors and can be used in composites and coatings to enhance the thermal properties of electronic components. They provide efficient heat dissipation.
- **Gallium Nitride (GaN):** GaN is a semiconductor material with high thermal conductivity. It is used in power electronics and RF devices, where efficient heat dissipation is essential.
- **Boron Nitride (BN):** BN is an excellent thermal conductor and electrical insulator. It can be used as a heat spreader and in thermal interface materials (Boopathi et al., 2022).
- **Advanced Thermal Interface Materials (TIMs):** These materials, often made with a combination of advanced materials, are designed to improve the contact between electronic components and heat sinks. They enhance heat transfer and reduce thermal resistance.
- **Nanostructured Materials:** Nanostructured materials, such as nanostructured metals and ceramics, can be used to create efficient heat sinks and thermal solutions. Their small size and high surface area contribute to improved heat dissipation (Boopathi, 2023a; Boopathi et al., 2023; Boopathi & Davim, 2023; Vijayakumar et al., 2023).
- **Metamaterials:** Metamaterials are engineered materials with unique thermal properties. They can be designed to manipulate heat transfer and control thermal conductivity, making them useful in advanced cooling systems.
- **3D-Printed Materials:** 3D printing allows for the creation of complex geometries and custom designs for heat sinks and cooling components. This customization can improve heat dissipation in specific applications (Mohanty et al., 2023; Senthil et al., 2023).

These advanced materials are essential in addressing the increasing heat challenges in electronic devices and data centers. They help manage thermal issues, reduce hot spots, and enhance the overall performance and reliability of electronic systems by ensuring efficient cooling.

Challenges in Data Center Cooling

Efficient cooling is a critical aspect of data center operations. However, it comes with a set of challenges that data center operators and engineers must overcome to maintain the reliability and performance of electronic equipment. Here are the key challenges in data center cooling (Yang et al., 2021).

- Preventing Hot Spots
  - *Challenge:* Data centers often experience hot spots, localized areas where electronic equipment generates excessive heat. Hot spots can lead to equipment overheating and failures.
  - *Solution:* Cooling systems must be designed to distribute airflow evenly and effectively dissipate heat from all areas, preventing hot spots through measures like proper air circulation and thermal mapping.

Figure 4. Challenges in data center cooling

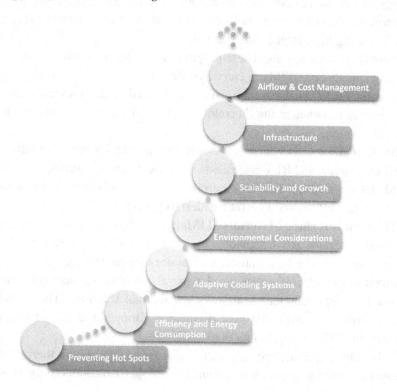

- Balancing Efficiency and Energy Consumption
  - *Challenge:* Striking the right balance between efficient cooling and energy consumption is a constant challenge. Overcooling or undercooling can lead to wasted energy and increased operational costs.
  - *Solution:* Employing adaptive cooling systems and advanced control mechanisms helps data centers dynamically adjust cooling capacity based on real-time conditions to minimize energy usage while maintaining the desired temperature.
- Adaptive Cooling Systems
  - *Challenge:* Traditional cooling systems often operate at a fixed capacity, regardless of server workload. This results in unnecessary energy consumption during periods of lower activity.
  - *Solution:* Adaptive cooling systems use real-time data and AI-driven algorithms to adjust cooling resources in response to server utilization. This reduces energy waste while ensuring consistent cooling.
- Environmental Considerations
  - *Challenge:* Data centers have an environmental impact due to their energy consumption and heat emissions. Balancing operational needs with sustainability goals is a growing concern (Boopathi, 2022b, 2022a; Boopathi et al., 2023; Gowri et al., 2023).
  - *Solution:* Data centers are exploring ways to reduce their carbon footprint by integrating renewable energy sources, implementing free cooling strategies, and enhancing the energy efficiency of cooling systems (Van Le et al., 2019).

- Scalability and Growth
  - *Challenge:* Data centers are constantly evolving and growing, adding more servers and equipment. Maintaining efficient cooling as data centers expand is challenging.
  - *Solution:* Scalable cooling solutions and flexible infrastructure designs are essential. Data centers should plan for future growth and cooling needs to ensure efficient operation as they expand.
- Airflow Management
  - *Challenge:* Inefficiencies in airflow management can lead to hot and cold air mixing, which reduces cooling efficiency.
  - *Solution:* Employing best practices in airflow management, such as hot/cold aisle containment and rack-level cooling, can improve cooling efficiency and prevent recirculation of hot air (Boopathi & Kanike, 2023; Domakonda et al., 2023).
- Cost Management
  - *Challenge:* Investing in advanced cooling technologies and energy-efficient systems can have high upfront costs.
  - *Solution:* Data center operators must consider the long-term benefits of energy savings and reduced downtime when making cooling infrastructure investments.
- Legacy Infrastructure
  - *Challenge:* Many data centers have legacy cooling systems that are less energy-efficient and adaptive.
  - *Solution:* Retrofitting or upgrading legacy infrastructure with modern, energy-efficient cooling solutions can help bring older data centers in line with current best practices.

To address cooling challenges in data centers, innovative solutions such as adaptive cooling systems, optimizing airflow management, and considering environmental sustainability can be implemented to ensure continued efficiency and reliability of facilities.

Innovations in Data Center Cooling

The data center industry continually evolves to address the challenges of increasing heat loads and the need for energy efficiency. Several innovations have emerged to optimize cooling systems and improve the sustainability of data centers. Here are some key innovations (Kavitha et al., 2023b; Rahamathunnisa, Subhashini, et al., 2023; Ugandar et al., 2023; Venkateswaran, Kumar, et al., 2023b):

- **Leveraging AI for Optimization:** Artificial Intelligence (AI) and machine learning are being used to optimize data center cooling. AI-driven systems continuously monitor environmental conditions, predict temperature trends, and dynamically adjust cooling settings to minimize energy consumption while ensuring reliable cooling (Boopathi, 2013, 2022c; Boopathi et al., 2021; Saravanan et al., 2022).
- **Advanced Materials for Enhanced Heat Transfer:** The use of advanced materials, such as graphene, diamond, and carbon nanotubes, has enhanced heat transfer and thermal management in electronic devices and data centers. These materials are integrated into heat sinks, thermal interface materials (TIMs), and other components to improve cooling efficiency.

- **Integrating Renewable Energy Sources:** Data centers are increasingly integrating renewable energy sources, such as solar panels and wind turbines, to power cooling systems and reduce their carbon footprint. This sustainable approach aligns with environmental goals and reduces dependence on fossil fuels.

- **Liquid Cooling Innovations:** Liquid cooling solutions, including direct-to-chip cooling and immersion cooling, have advanced significantly. These innovations offer more efficient heat dissipation and enable data centers to manage higher heat loads generated by powerful servers and GPUs.

- **Rack-Level Cooling:** Cooling at the rack level, as opposed to traditional room-level cooling, is gaining popularity. Rear-door heat exchangers and in-row cooling units allow for more precise cooling at the server level, reducing cooling system energy consumption.

- **Free Cooling Strategies:** Data centers are adopting advanced free cooling strategies that utilize external environmental conditions, such as air-side economization and evaporative cooling, to reduce the reliance on mechanical cooling systems and lower energy consumption.

- **Data Center Infrastructure Management (DCIM):** DCIM software provides real-time monitoring and control of cooling systems, allowing data center operators to make informed decisions, optimize cooling, and identify inefficiencies.

- **Hot/Cold Aisle Containment:** Aisle containment systems have become more sophisticated, providing more effective separation of hot and cold air and improving cooling efficiency. Cold aisle containment solutions, in particular, are widely adopted.

- **Modular and Portable Data Centers:** Modular data centers and containerized solutions come with built-in cooling systems optimized for efficiency and scalability. These solutions offer greater flexibility and can be deployed rapidly.

- **Heat Recycling:** Heat recovery systems capture excess heat generated by data centers and repurpose it for heating nearby buildings or for other industrial processes, enhancing energy efficiency.

The innovations in data center cooling focus on energy efficiency, sustainability, and adaptability to digital infrastructure demands, ensuring efficient and reliable cooling systems as data centers evolve and adapt to evolving needs.

CONCLUSION

The chapter explores the evolution of electronic cooling systems in data centers, highlighting the importance of efficient cooling for maintaining optimal performance and extending the lifespan of critical infrastructure, as data centers house a vast array of electronic equipment. It provides key insights and conclusions.

Data center cooling faces challenges such as preventing hot spots, balancing efficiency and energy consumption, scalability, and environmental considerations. Innovative cooling solutions are crucial. The chapter explores various cooling systems, including traditional air, advanced liquid, and phase-change cooling, each with its advantages. Innovative materials like graphene, carbon nanotubes, and phase-change materials improve heat transfer and thermal management, reducing hot spots and enhancing cooling efficiency.

Artificial Intelligence (AI) is revolutionizing cooling systems by monitoring conditions, predicting trends, and adjusting settings to minimize energy consumption. This aligns with data centers' growing focus on sustainability, incorporating renewable energy sources, free cooling strategies, and heat recycling systems. The future of data center cooling will see continued innovations like modular and portable data centers, rack-level cooling, and advanced free cooling strategies, enhancing efficiency and adaptability.

ABBREVIATIONS

AI: Artificial Intelligence
TIMs: Thermal Interface Materials
PCMs: Phase-Change Materials
DCIM: Data Center Infrastructure Management
GaN: Gallium Nitride
BN: Boron Nitride
CNTs: Carbon Nanotubes

REFERENCES

Allouhi, A., Kousksou, T., Jamil, A., Bruel, P., Mourad, Y., & Zeraouli, Y. (2015). Solar driven cooling systems: An updated review. *Renewable & Sustainable Energy Reviews, 44*, 159–181. doi:10.1016/j.rser.2014.12.014

B, M. K., K, K. K., Sasikala, P., Sampath, B., Gopi, B., & Sundaram, S. (2024). Sustainable Green Energy Generation From Waste Water. In *Practice, Progress, and Proficiency in Sustainability* (pp. 440–463). IGI Global. doi:10.4018/979-8-3693-1186-8.ch024

Bahiraei, M., & Heshmatian, S. (2018). Electronics cooling with nanofluids: A critical review. *Energy Conversion and Management, 172*, 438–456. doi:10.1016/j.enconman.2018.07.047

Boopathi, S. (2013). *Experimental study and multi-objective optimization of near-dry wire-cut electrical discharge machining process* [PhD Thesis, Anna University]. http://hdl.handle.net/10603/16933

Boopathi, S. (2022a). An investigation on gas emission concentration and relative emission rate of the near-dry wire-cut electrical discharge machining process. *Environmental Science and Pollution Research International, 29*(57), 86237–86246. doi:10.1007/s11356-021-17658-1 PMID:34837614

Boopathi, S. (2022b). Cryogenically treated and untreated stainless steel grade 317 in sustainable wire electrical discharge machining process: A comparative study. *Springer :Environmental Science and Pollution Research*, 1–10.

Boopathi, S. (2022c). Experimental investigation and multi-objective optimization of cryogenic Friction-stir-welding of AA2014 and AZ31B alloys using MOORA technique. *Materials Today. Communications, 33*, 104937. doi:10.1016/j.mtcomm.2022.104937

Boopathi, S. (2023a). An Investigation on Friction Stir Processing of Aluminum Alloy-Boron Carbide Surface Composite. In R. V. Vignesh, R. Padmanaban, & M. Govindaraju (Eds.), *Advances in Processing of Lightweight Metal Alloys and Composites* (pp. 249–257). Springer Nature Singapore. doi:10.1007/978-981-19-7146-4_14

Boopathi, S. (2023b). Internet of Things-Integrated Remote Patient Monitoring System: Healthcare Application. In A. Suresh Kumar, U. Kose, S. Sharma, & S. Jerald Nirmal Kumar (Eds.), (pp. 137–161). Advances in Healthcare Information Systems and Administration. IGI Global. doi:10.4018/978-1-6684-6894-4.ch008

Boopathi, S., Alqahtani, A. S., Mubarakali, A., & Panchatcharam, P. (2023). Sustainable developments in near-dry electrical discharge machining process using sunflower oil-mist dielectric fluid. *Environmental Science and Pollution Research International*, 1–20. doi:10.1007/s11356-023-27494-0 PMID:37199846

Boopathi, S., & Davim, J. P. (2023). Applications of Nanoparticles in Various Manufacturing Processes. In S. Boopathi & J. P. Davim (Eds.), (pp. 1–31). Advances in Chemical and Materials Engineering. IGI Global. doi:10.4018/978-1-6684-9135-5.ch001

Boopathi, S., Jeyakumar, M., Singh, G. R., King, F. L., Pandian, M., Subbiah, R., & Haribalaji, V. (2022). An experimental study on friction stir processing of aluminium alloy (AA-2024) and boron nitride (BNp) surface composite. *Materials Today: Proceedings*, 59(1), 1094–1099. doi:10.1016/j.matpr.2022.02.435

Boopathi, S., & Kanike, U. K. (2023). Applications of Artificial Intelligent and Machine Learning Techniques in Image Processing. In B. K. Pandey, D. Pandey, R. Anand, D. S. Mane, & V. K. Nassa (Eds.), (pp. 151–173). Advances in Computational Intelligence and Robotics. IGI Global. doi:10.4018/978-1-6684-8618-4.ch010

Boopathi, S., Myilsamy, S., & Sukkasamy, S. (2021). *Experimental Investigation and Multi-Objective Optimization of Cryogenically Cooled Near-Dry Wire-Cut EDM Using TOPSIS Technique.* IJAMT PREPRINT.

Boopathi, S., Sureshkumar, M., & Sathiskumar, S. (2023a). Parametric Optimization of LPG Refrigeration System Using Artificial Bee Colony Algorithm. In S. Tripathy, S. Samantaray, J. Ramkumar, & S. S. Mahapatra (Eds.), *Recent Advances in Mechanical Engineering* (pp. 97–105). Springer Nature Singapore. doi:10.1007/978-981-19-9493-7_10

Boopathi, S., Sureshkumar, M., & Sathiskumar, S. (2023b). Parametric Optimization of LPG Refrigeration System Using Artificial Bee Colony Algorithm. In S. Tripathy, S. Samantaray, J. Ramkumar, & S. S. Mahapatra (Eds.), *Recent Advances in Mechanical Engineering* (pp. 97–105). Springer Nature Singapore. doi:10.1007/978-981-19-9493-7_10

Boopathi, S., Umareddy, M., & Elangovan, M. (2023). Applications of Nano-Cutting Fluids in Advanced Machining Processes. In S. Boopathi & J. P. Davim (Eds.), (pp. 211–234). Advances in Chemical and Materials Engineering. IGI Global. doi:10.4018/978-1-6684-9135-5.ch009

Cao, Z., Zhou, X., Hu, H., Wang, Z., & Wen, Y. (2022). Toward a systematic survey for carbon neutral data centers. *IEEE Communications Surveys and Tutorials*, 24(2), 895–936. doi:10.1109/COMST.2022.3161275

Capozzoli, A., & Primiceri, G. (2015). Cooling systems in data centers: State of art and emerging technologies. *Energy Procedia*, *83*, 484–493. doi:10.1016/j.egypro.2015.12.168

Chen, Y., Kang, Y., Zhao, Y., Wang, L., Liu, J., Li, Y., Liang, Z., He, X., Li, X., Tavajohi, N., & Li, B. (2021). A review of lithium-ion battery safety concerns: The issues, strategies, and testing standards. *Journal of Energy Chemistry*, *59*, 83–99. doi:10.1016/j.jechem.2020.10.017

Chu, J., & Huang, X. (2021). *Research status and development trends of evaporative cooling air-conditioning technology in data centers*. Energy and Built Environment.

Deng, Y., Feng, C., Jiaqiang, E., Zhu, H., Chen, J., Wen, M., & Yin, H. (2018). Effects of different coolants and cooling strategies on the cooling performance of the power lithium ion battery system: A review. *Applied Thermal Engineering*, *142*, 10–29. doi:10.1016/j.applthermaleng.2018.06.043

Dhanya, D., Kumar, S. S., Thilagavathy, A., Prasad, D. V. S. S. S. V., & Boopathi, S. (2023). Data Analytics and Artificial Intelligence in the Circular Economy: Case Studies. In B. K. Mishra (Ed.), (pp. 40–58). Advances in Civil and Industrial Engineering. IGI Global. doi:10.4018/979-8-3693-0044-2.ch003

Dincer, I. (2017). *Refrigeration systems and applications*. John Wiley & Sons. doi:10.1002/9781119230793

Domakonda, V. K., Farooq, S., Chinthamreddy, S., Puviarasi, R., Sudhakar, M., & Boopathi, S. (2023). Sustainable Developments of Hybrid Floating Solar Power Plants: Photovoltaic System. In P. Vasant, R. Rodríguez-Aguilar, I. Litvinchev, & J. A. Marmolejo-Saucedo (Eds.), (pp. 148–167). Advances in Environmental Engineering and Green Technologies. IGI Global. doi:10.4018/978-1-6684-4118-3.ch008

Gowri, N. V., Dwivedi, J. N., Krishnaveni, K., Boopathi, S., Palaniappan, M., & Medikondu, N. R. (2023). Experimental investigation and multi-objective optimization of eco-friendly near-dry electrical discharge machining of shape memory alloy using Cu/SiC/Gr composite electrode. *Environmental Science and Pollution Research International*, *30*(49), 1–19. doi:10.1007/s11356-023-26983-6 PMID:37126160

Hu, R., Liu, Y., Shin, S., Huang, S., Ren, X., Shu, W., Cheng, J., Tao, G., Xu, W., Chen, R., & Luo, X. (2020). Emerging materials and strategies for personal thermal management. *Advanced Energy Materials*, *10*(17), 1903921. doi:10.1002/aenm.201903921

Hussain, Z., Babe, M., Saravanan, S., Srimathy, G., Roopa, H., & Boopathi, S. (2023). Optimizing Biomass-to-Biofuel Conversion: IoT and AI Integration for Enhanced Efficiency and Sustainability. In N. Cobîrzan, R. Muntean, & R.-A. Felseghi (Eds.), (pp. 191–214). Advances in Finance, Accounting, and Economics. IGI Global. doi:10.4018/978-1-6684-8238-4.ch009

Jin, C., Bai, X., Yang, C., Mao, W., & Xu, X. (2020). A review of power consumption models of servers in data centers. *Applied Energy*, *265*, 114806. doi:10.1016/j.apenergy.2020.114806

Kavitha, C. R., Varalatchoumy, M., Mithuna, H. R., Bharathi, K., Geethalakshmi, N. M., & Boopathi, S. (2023a). Energy Monitoring and Control in the Smart Grid: Integrated Intelligent IoT and ANFIS. In M. Arshad (Ed.), (pp. 290–316). Advances in Bioinformatics and Biomedical Engineering. IGI Global. doi:10.4018/978-1-6684-6577-6.ch014

Kavitha, C. R., Varalatchoumy, M., Mithuna, H. R., Bharathi, K., Geethalakshmi, N. M., & Boopathi, S. (2023b). Energy Monitoring and Control in the Smart Grid: Integrated Intelligent IoT and ANFIS. In M. Arshad (Ed.), (pp. 290–316). Advances in Bioinformatics and Biomedical Engineering. IGI Global. doi:10.4018/978-1-6684-6577-6.ch014

Khalaj, A. H., & Halgamuge, S. K. (2017). A Review on efficient thermal management of air-and liquid-cooled data centers: From chip to the cooling system. *Applied Energy*, *205*, 1165–1188. doi:10.1016/j.apenergy.2017.08.037

Kumar, B. M., Kumar, K. K., Sasikala, P., Sampath, B., Gopi, B., & Sundaram, S. (2024). Sustainable Green Energy Generation From Waste Water: IoT and ML Integration. In B. K. Mishra (Ed.), (pp. 440–463). Practice, Progress, and Proficiency in Sustainability. IGI Global. doi:10.4018/979-8-3693-1186-8.ch024

Li, Y., Wen, Y., Tao, D., & Guan, K. (2019). Transforming cooling optimization for green data center via deep reinforcement learning. *IEEE Transactions on Cybernetics*, *50*(5), 2002–2013. doi:10.1109/TCYB.2019.2927410 PMID:31352360

Maguluri, L. P., Ananth, J., Hariram, S., Geetha, C., Bhaskar, A., & Boopathi, S. (2023). Smart Vehicle-Emissions Monitoring System Using Internet of Things (IoT). In P. Srivastava, D. Ramteke, A. K. Bedyal, M. Gupta, & J. K. Sandhu (Eds.), (pp. 191–211). Practice, Progress, and Proficiency in Sustainability. IGI Global. doi:10.4018/978-1-6684-8117-2.ch014

Maguluri, L. P., Arularasan, A. N., & Boopathi, S. (2023). Assessing Security Concerns for AI-Based Drones in Smart Cities. In R. Kumar, A. B. Abdul Hamid, & N. I. Binti Ya'akub (Eds.), (pp. 27–47). Advances in Computational Intelligence and Robotics. IGI Global. doi:10.4018/978-1-6684-9151-5.ch002

Moazamigoodarzi, H., Tsai, P. J., Pal, S., Ghosh, S., & Puri, I. K. (2019). Influence of cooling architecture on data center power consumption. *Energy*, *183*, 525–535. doi:10.1016/j.energy.2019.06.140

Mohanty, A., Jothi, B., Jeyasudha, J., Ranjit, P. S., Isaac, J. S., & Boopathi, S. (2023). Additive Manufacturing Using Robotic Programming. In S. Kautish, N. K. Chaubey, S. B. Goyal, & P. Whig (Eds.), (pp. 259–282). Advances in Computational Intelligence and Robotics. IGI Global. doi:10.4018/978-1-6684-8171-4.ch010

Nishanth, J., Deshmukh, M. A., Kushwah, R., Kushwaha, K. K., Balaji, S., & Sampath, B. (2023). Particle Swarm Optimization of Hybrid Renewable Energy Systems. In *Intelligent Engineering Applications and Applied Sciences for Sustainability* (pp. 291–308). IGI Global. doi:10.4018/979-8-3693-0044-2.ch016

Oró, E., Depoorter, V., Garcia, A., & Salom, J. (2015). Energy efficiency and renewable energy integration in data centres. Strategies and modelling review. *Renewable & Sustainable Energy Reviews*, *42*, 429–445. doi:10.1016/j.rser.2014.10.035

Pachiappan, K., Anitha, K., Pitchai, R., Sangeetha, S., Satyanarayana, T. V. V., & Boopathi, S. (2023). Intelligent Machines, IoT, and AI in Revolutionizing Agriculture for Water Processing. In B. B. Gupta & F. Colace (Eds.), (pp. 374–399). Advances in Computational Intelligence and Robotics. IGI Global. doi:10.4018/978-1-6684-9999-3.ch015

Rahamathunnisa, U., Subhashini, P., Aancy, H. M., Meenakshi, S., Boopathi, S., & ... (2023). Solutions for Software Requirement Risks Using Artificial Intelligence Techniques. In *Handbook of Research on Data Science and Cybersecurity Innovations in Industry 4.0 Technologies* (pp. 45–64). IGI Global.

Rahamathunnisa, U., Sudhakar, K., Murugan, T. K., Thivaharan, S., Rajkumar, M., & Boopathi, S. (2023). Cloud Computing Principles for Optimizing Robot Task Offloading Processes. In S. Kautish, N. K. Chaubey, S. B. Goyal, & P. Whig (Eds.), (pp. 188–211). Advances in Computational Intelligence and Robotics. IGI Global. doi:10.4018/978-1-6684-8171-4.ch007

Rahamathunnisa, U., Sudhakar, K., Padhi, S. N., Bhattacharya, S., Shashibhushan, G., & Boopathi, S. (2023). Sustainable Energy Generation From Waste Water: IoT Integrated Technologies. In A. S. Etim (Ed.), (pp. 225–256). Advances in Human and Social Aspects of Technology. IGI Global. doi:10.4018/978-1-6684-5347-6.ch010

Reddy, M. A., Reddy, B. M., Mukund, C. S., Venneti, K., Preethi, D. M. D., & Boopathi, S. (2023). Social Health Protection During the COVID-Pandemic Using IoT. In F. P. C. Endong (Ed.), (pp. 204–235). Advances in Electronic Government, Digital Divide, and Regional Development. IGI Global. doi:10.4018/978-1-7998-8394-4.ch009

Rhee, K.-N., Olesen, B. W., & Kim, K. W. (2017). Ten questions about radiant heating and cooling systems. *Building and Environment*, *112*, 367–381. doi:10.1016/j.buildenv.2016.11.030

Sampath, B. C. S., & Myilsamy, S. (2022). Application of TOPSIS Optimization Technique in the Micro-Machining Process. In M. A. Mellal (Ed.), (pp. 162–187). Advances in Mechatronics and Mechanical Engineering. IGI Global. doi:10.4018/978-1-6684-5887-7.ch009

Saravanan, M., Vasanth, M., Boopathi, S., Sureshkumar, M., & Haribalaji, V. (2022). Optimization of Quench Polish Quench (QPQ) Coating Process Using Taguchi Method. *Key Engineering Materials*, *935*, 83–91. doi:10.4028/p-z569vy

Satav, S. D., Hasan, D. S., Pitchai, R., Mohanaprakash, T. A., Sultanuddin, S. J., & Boopathi, S. (2024). Next Generation of Internet of Things (NGIoT) in Healthcare Systems. In B. K. Mishra (Ed.), (pp. 307–330). Practice, Progress, and Proficiency in Sustainability. IGI Global. doi:10.4018/979-8-3693-1186-8.ch017

Satav, S. D., & Lamani, D. K. G., H., Kumar, N. M. G., Manikandan, S., & Sampath, B. (2024). Energy and Battery Management in the Era of Cloud Computing: Sustainable Wireless Systems and Networks. In B. K. Mishra (Ed.), Practice, Progress, and Proficiency in Sustainability (pp. 141–166). IGI Global. doi:10.4018/979-8-3693-1186-8.ch009

Senthil, T. S., Ohmsakthi Vel, R., Puviyarasan, M., Babu, S. R., Surakasi, R., & Sampath, B. (2023). Industrial Robot-Integrated Fused Deposition Modelling for the 3D Printing Process. In R. Keshavamurthy, V. Tambrallimath, & J. P. Davim (Eds.), (pp. 188–210). Advances in Chemical and Materials Engineering. IGI Global. doi:10.4018/978-1-6684-6009-2.ch011

Syamala, M., Komala, C., Pramila, P., Dash, S., Meenakshi, S., & Boopathi, S. (2023). Machine Learning-Integrated IoT-Based Smart Home Energy Management System. In *Handbook of Research on Deep Learning Techniques for Cloud-Based Industrial IoT* (pp. 219–235). IGI Global. doi:10.4018/978-1-6684-8098-4.ch013

Ugandar, R. E., Rahamathunnisa, U., Sajithra, S., Christiana, M. B. V., Palai, B. K., & Boopathi, S. (2023). Hospital Waste Management Using Internet of Things and Deep Learning: Enhanced Efficiency and Sustainability. In M. Arshad (Ed.), (pp. 317–343). Advances in Bioinformatics and Biomedical Engineering. IGI Global. doi:10.4018/978-1-6684-6577-6.ch015

Van Le, D., Liu, Y., Wang, R., Tan, R., Wong, Y.-W., & Wen, Y. (2019). Control of air free-cooled data centers in tropics via deep reinforcement learning. *Proceedings of the 6th ACM International Conference on Systems for Energy-Efficient Buildings, Cities, and Transportation*, (pp. 306–315). ACM. 10.1145/3360322.3360845

Vanitha, S. K. R., & Boopathi, S. (2023). Artificial Intelligence Techniques in Water Purification and Utilization. In P. Vasant, R. Rodríguez-Aguilar, I. Litvinchev, & J. A. Marmolejo-Saucedo (Eds.), (pp. 202–218). Advances in Environmental Engineering and Green Technologies. IGI Global. doi:10.4018/978-1-6684-4118-3.ch010

Venkateswaran, N., Kumar, S. S., Diwakar, G., Gnanasangeetha, D., & Boopathi, S. (2023a). Synthetic Biology for Waste Water to Energy Conversion: IoT and AI Approaches. In M. Arshad (Ed.), (pp. 360–384). Advances in Bioinformatics and Biomedical Engineering. IGI Global. doi:10.4018/978-1-6684-6577-6.ch017

Venkateswaran, N., Kumar, S. S., Diwakar, G., Gnanasangeetha, D., & Boopathi, S. (2023b). Synthetic Biology for Waste Water to Energy Conversion: IoT and AI Approaches. In M. Arshad (Ed.), (pp. 360–384). Advances in Bioinformatics and Biomedical Engineering. IGI Global. doi:10.4018/978-1-6684-6577-6.ch017

Venkateswaran, N., Vidhya, K., Ayyannan, M., Chavan, S. M., Sekar, K., & Boopathi, S. (2023). A Study on Smart Energy Management Framework Using Cloud Computing. In P. Ordóñez De Pablos & X. Zhang (Eds.), (pp. 189–212). Practice, Progress, and Proficiency in Sustainability. IGI Global. doi:10.4018/978-1-6684-8634-4.ch009

Vennila, T., Karuna, M. S., Srivastava, B. K., Venugopal, J., Surakasi, R., & B., S. (2023). New Strategies in Treatment and Enzymatic Processes: Ethanol Production From Sugarcane Bagasse. In P. Vasant, R. Rodríguez-Aguilar, I. Litvinchev, & J. A. Marmolejo-Saucedo (Eds.), *Advances in Environmental Engineering and Green Technologies* (pp. 219–240). IGI Global. doi:10.4018/978-1-6684-4118-3.ch011

Vijayakumar, G. N. S., Domakonda, V. K., Farooq, S., Kumar, B. S., Pradeep, N., & Boopathi, S. (2023). Sustainable Developments in Nano-Fluid Synthesis for Various Industrial Applications. In A. S. Etim (Ed.), (pp. 48–81). Advances in Human and Social Aspects of Technology. IGI Global., doi:10.4018/978-1-6684-5347-6.ch003

Wu, W., Wang, S., Wu, W., Chen, K., Hong, S., & Lai, Y. (2019). A critical review of battery thermal performance and liquid based battery thermal management. *Energy Conversion and Management*, *182*, 262–281. doi:10.1016/j.enconman.2018.12.051

Wulandari, D. A., Akmal, M., Gunawan, Y., & others. (2020). *Cooling improvement of the IT rack by layout rearrangement of the A2 class data center room: A simulation study.*

Xia, G., Cao, L., & Bi, G. (2017). A review on battery thermal management in electric vehicle application. *Journal of Power Sources, 367*, 90–105. doi:10.1016/j.jpowsour.2017.09.046

Yang, J., Zhang, X., Zhang, X., Wang, L., Feng, W., & Li, Q. (2021). Beyond the visible: Bioinspired infrared adaptive materials. *Advanced Materials, 33*(14), 2004754. doi:10.1002/adma.202004754 PMID:33624900

Yuan, X., Zhou, X., Pan, Y., Kosonen, R., Cai, H., Gao, Y., & Wang, Y. (2021). Phase change cooling in data centers: A review. *Energy and Building, 236*, 110764. doi:10.1016/j.enbuild.2021.110764

Zhang, Q., Meng, Z., Hong, X., Zhan, Y., Liu, J., Dong, J., Bai, T., Niu, J., & Deen, M. J. (2021). A survey on data center cooling systems: Technology, power consumption modeling and control strategy optimization. *Journal of Systems Architecture, 119*, 102253. doi:10.1016/j.sysarc.2021.102253

Chapter 4
Power System Analysis:
Optimizing Distributed Renewable Power Integration

Aamir Bin Rashid
Chandigarh University, India

Harpreet Kaur Channi
Chandigarh University, India

ABSTRACT

In current power system analysis, integrating dispersed renewable power sources into the infrastructure is a major difficulty. Technical, economic, and environmental factors are considered in this research to optimise distributed renewable power integration. This study presents new power grid performance and reliability methods using advanced power system analytic techniques such load flow analysis, voltage stability evaluation, and transient stability analysis. This research uses sophisticated control algorithms and advanced optimisation to reduce power fluctuations, stabilise the grid, and maximise distributed renewable energy use. The suggested framework balances intermittent renewable sources with demand variations for a sustainable and resilient power system. This research shows that the proposed optimisation strategies are feasible and effective through comprehensive simulations and case studies, providing policymakers, system operators, and stakeholders with valuable insights for sustainable distributed renewable energy integration in the power system.

INTRODUCTION

Power system analysis has evolved to optimize distributed renewable power source integration in response to climate change and the need for sustainable energy solutions. Electricity system analysis helps integrate dispersed renewable electricity into energy systems effectively. Solar photovoltaics, wind turbines, and small-scale hydropower systems have changed the energy landscape. Power system operators and utilities have distinct problems and possibilities from these decentralized sources. Electricity system analysis is needed to integrate dispersed renewable electricity into the grid. Power system operators must

DOI: 10.4018/979-8-3693-3735-6.ch004

overcome technological and operational difficulties to ensure grid stability, dependability, and resilience. Optimizing distributed renewable power integration requires power system analysis. Load forecasting predicts energy demand and adjusts power production. Grid capacity evaluation guarantees that the infrastructure can accommodate the growing number of distributed energy sources without sacrificing performance. Smart grid technologies improve power system flexibility (D. T. Duong and K. Uhlen, 2018). Distributed renewable power resource monitoring, control, and management allow utilities to adapt to dynamic energy supply and demand. Energy storage devices may mitigate intermittent renewable energy sources. Energy storage reduces dispersed renewable power fluctuation, guaranteeing a stable energy source. Power system analysis considers investment costs, savings, and long-term advantages when assessing distributed renewable power integration. Government subsidies and policies help promote renewable energy technology. Distributed renewable electricity reduces greenhouse gas emissions and fossil fuel consumption. This supports sustainable development and climate objectives. Power system analysis optimizes dispersed renewable power source integration, guiding the energy transition towards sustainability. Smart technology, grid stability issues, and economic possibilities may help utilities integrate distributed renewable power smoothly. This connection improves power grid dependability and worldwide climate change initiatives (S. A. R. Konakalla and R. A. de Callafon, 2017).

Power system analysis is a critical field of study and practice focusing on modelling, optimizing, and operating electrical power systems. It plays a crucial role in ensuring the reliable, efficient, and secure operation of power grids, which are the backbone of modern societies. Power system analysis encompasses various techniques, methodologies, and tools that enable utilities, engineers, and researchers to understand, plan, and manage the complex interactions within a power system. By analyzing various aspects of power generation, transmission, and distribution, power system analysis helps to optimize system performance, mitigate risks, and make informed decisions for future development (S. Ndaba and I. E. Davidson, 2020).

The primary objective of power system analysis is to ensure the balance between electricity supply and demand while maintaining system stability, voltage quality, and frequency regulation. It involves mathematical modelling and simulation of power system components, such as generators, transformers, transmission lines, and loads, to study their behaviour under normal and abnormal operating conditions (C. Baum, 2018). Power system analysis enables assessing power flow, voltage stability, short-circuit faults, and transient stability, among other key parameters, to identify potential issues and optimize system design and operation. Power system analysis also contributes to integrating new technologies and renewable energy sources into the grid. Analyzing their impact on the grid becomes crucial with the increasing deployment of renewable energy, such as solar and wind power. It involves evaluating the variability and intermittency of renewable generation, designing effective control strategies, and assessing the grid's ability to handle these fluctuations (M.Trotignon, C.Counan, *et al*, 1992).

Furthermore, power system analysis is essential for the planning and expansion of power systems. It assists in determining optimal generation capacity, transmission line routes, and substation locations, considering factors such as load growth, environmental considerations, and cost optimization. By conducting comprehensive analyses, utilities can make informed decisions to ensure reliable and cost-effective power supply to consumers. In summary, power system analysis is a multidisciplinary field that combines engineering principles, mathematics, and computer science to analyze, optimize, and operate electrical power systems. Its application ranges from daily system operation and maintenance to long-term planning and development. By employing advanced techniques and tools, power system analysis enables utilities to ensure a resilient, sustainable, and efficient power supply for the ever-evolving needs of society (P. Kundur,1994).

Distributed renewable power is boosting global energy sustainability. Distributed renewable electricity may reduce greenhouse gas emissions and fossil fuel-related environmental challenges, including air pollution and climate change. This decentralized approach empowers communities and increases energy resilience by locally harnessing renewable resources like solar, wind, and hydropower. Distributed renewable power eliminates fossil fuel imports and volatile global energy markets. Energy generating near consumers improves grid stability. This resilience is crucial as increasingly frequent and severe weather events tax centralized electrical infrastructure. Renewable energy and scale have made distributed generation affordable. It can replace conventional energy sources. Policies, financial incentives, and regulatory structures globally encourage its implementation (L. Varga et al., 1999). Distributed renewable power increases local economies and renewable energy jobs. Renewable energy initiatives improve the environment, save power, and generate employment. Distributed renewable power promotes environmental stewardship and CSR. Distributed renewable power helps businesses reduce their carbon impact. Distributed renewable power is essential to a greener energy future. Its energy independence, grid stability, environmental benefits, and economic possibilities help battle climate change. Distributed renewable power will create a cleaner, greener, and more resilient energy landscape, bringing us closer to sustainability (B. Meyer et al.,1992).

Optimizing distributed renewable power integration involves strategic goals. First, it ensures reliable grid integration of intermittent renewable sources to improve grid stability and reduce interruptions. Second, the optimization matches generation and demand to maximize renewable energy use and minimize waste. Efficient grid planning promotes economic viability and affordability by achieving cost-effectiveness. The aim is to speed the transition to a sustainable and resilient energy system, reduce carbon emissions, and meet global climate targets.Sustainable power system analysis guides renewable energy grid integration. It facilitates the easy absorption of clean and renewable energy via rigorous demand forecasting, grid capacity, and transmission planning studies. Power system analysis optimizes grid stability and reduces environmental impact to speed the transition to a sustainable energy landscape, creating a greener and more resilient future (P. Kessel and H. Glavitsch,1986).

Distributed Renewable Power Technologies

Distributed renewable power systems use natural resources to provide clean, sustainable electricity. One of the most popular technologies, solar photovoltaics (PV) converts sunlight into energy using rooftop or ground-mounted PV panels. Onshore or offshore wind turbines convert wind energy into electricity. Biomass energy generates power and heat from agricultural wastes and garbage. Micro-hydro systems produce electricity from tiny streams and rivers. Geothermal systems and small biogas plants that transform organic waste into biogas for energy are alternative distributed renewable technologies. Distributed renewable power solutions offer energy self-sufficiency, greenhouse gas reduction, and climate change mitigation. Their decentralization allows communities, companies, and people to actively produce clean energy, promoting energy resilience and sustainability (B. Gao et al., 1986). Distributed renewable power solutions will play a larger part in greening the global energy landscape as technology improves and prices fall.

Integrating distributed renewable power sources into energy networks ensures a smooth transition. Solar and wind power fluctuate. Weather-dependent renewable sources may not fulfil energy demand.

Grid instability requires grid design and control. Transmission and grid congestion are other issues. Distributed renewable power sources are often in rural areas remote from population centres with heavy electricity demand. Renewable energy transmission infrastructure upgrades are costly and time-consuming. Renewable power's intermittent nature requires energy storage. Batteries and pumped hydro storage can balance supply and demand by storing excess energy during peak generation for low-generation times. Dispersed renewable power integration faces regulatory and legislative obstacles. In certain places, obsolete or confusing regulations may limit renewable energy technology adoption. Streamlining approvals and introducing renewable energy subsidies help overcome these difficulties. Distributed renewable power requires funding (G. Nativel et.al, 1999). Renewable energy systems have substantial setup costs, but the long-term benefits often outweigh them. Low-cost finance and incentives enable distributed renewable power systems. Integrate governments, utilities, investors, and communities. Solving these issues via innovation, policy, and investment may unlock distributed renewable power's full potential and build a sustainable, low-carbon energy future.

LITERATURE REVIEW

(Ahmed et al., 2023) suggested using HOMER Powering Health and the simulation program Hybrid Optimisation Model for Electric Renewables (HOMER) to optimize the healthcare center's renewable energy needs. The load study showed that the hybrid PV system may be better than other power sources for healthcare's daily operations and emergencies. Traditional control system approaches face new problems from renewable distributed producers. (Rodriguez-Martinez et al., 2023) discussed power quality and communication control methods. The gradual integration of renewable energy-based Distribution Generation (DG) using power electronic converters to distribution networks exacerbates this challenge. (Satyanarayana et al., 2023) examined solar dg integrated system power quality enhancement using dc-link fed parallel-vsi-based dstatcom. A typical three-phase distribution system with DG integration and non-linear demand is analyzed. (Alsharif et al., 2023) proposed the Stochastic Monte Carlo Method (SMCM) to assess the system effect of arrival and departure EV uncertainty. The goal is to size system configurations using a metaheuristic algorithm and study how an undetermined number of EVs affect residential power distribution in Tripoli, Libya, to create a cost-effective, dependable, and renewable system. Renewable energy technologies like solar and wind are efficient and eco-friendly. (Shezan et al., 2023) explored the issues of combining a PV plant with a wind power station to provide electricity for the grid or a stand-alone system. The energy interaction between distributed generating units, microgrids, and energy storage systems is growing increasingly complicated as renewable energy becomes more prevalent in the power system. (Shi et al., 2023) proved that the suggested technique can balance virtual power plant component setup and operating costs and construct a reliable, cheap, efficient, and environmentally friendly distribution system. (Lee et al., 2023) examined how multi-terminal direct current (MTDC) works to integrate renewable energy into the Korean electricity grid. Large-scale renewable energy projects planned for the power grid will cause southern line congestion. Therefore, energy transmission and distribution networks must respond quickly. (Kumar et al., 2023) suggested reactive power compensation using STATCOM to reduce voltage sag, swell, fluctuations, and THD. The IEEE 9 bus system's power flow and quality are analyzed to assess the suggested technique. (Dwivedi et al., 2023) introduced resilience frameworks and quantitative power system resilience indicators to measure resilience. Future power system resilience enhancements are considered by analyzing research gaps and obstacles.

The scope of Power System Analysis involves leveraging Artificial Intelligence (AI) to optimize the integration of distributed renewable power sources into existing grids. This encompasses developing advanced algorithms that enhance grid stability, reliability, and efficiency. AI models can predict renewable energy output, optimize energy storage utilization, and dynamically adjust power flow in response to fluctuations. The application of machine learning enables adaptive control strategies, minimizing environmental impact and maximizing the utilization of renewable resources. Through comprehensive analysis and real-time decision-making, this approach aims to address challenges associated with variability, intermittency, and unpredictability in distributed renewable power generation, ultimately contributing to a sustainable and resilient power infrastructure [21]. Following the main objectives of the chapter:

- Provide an overview of distributed renewable energy sources.
- Identify and address stability issues associated with distributed renewables.
- Explore and present AI-driven strategies for optimizing renewable power integration.
- Illustrate successful applications of power system analysis in optimizing distributed renewable integration.
- Discuss emerging trends and advancements in the field for continuous improvement.

Drivers For Change

Customer and Regulatory Drivers

The place of a power system operator as a mediator in market-situated power supply businesses makes it vital that choices can be displayed to showcase members liberated from business predisposition. Since market players are sharp in that reasonable, repeatable, and unquestionable guidelines are kept, there want to share complex, economically delicate choices with strong mathematical calculations which can ensure the outstanding accomplishment of a characterized objective. Similarly, mathematical attributes are progressively requested in framework foundation advancement, particularly when various gatherings should be convinced of the business case for speculation or when a controller intently observes capital consumption (Dwivedi, D., et al.,2023) . In a developing number of spots on the planet, circulation organizations are being boosted to work on the dependability execution of the help they give purchasers. This affects the coming dispersion networks made due and the administration they demand from the transmission. A reaction to this requires types of investigation that utilities are not generally familiar with utilizing. The detachment of various power framework administrations among numerous suppliers has prompted expanding need for information trade. Just as the offices play out the business, this can encourage vulnerability regarding the precision of the information, the administration of which might require new instruments (Y. Xue et al., 1989).

Cost Drivers

Just as bringing vulnerability regarding information, the opening up power supply enterprises to advertise powers has expanded vulnerability in foundation conditions influencing the preparation and activity of transmission frameworks. When a solitary power framework utility has a good vision of things to come, the arrangement of age depends upon a vulnerability regarding the degree of interest; presently age limit is controlled by organizations completely separate from the network proprietor or administrator. Besides, the area of the new limit and the circumstance of retirement of the existing limit are obscure to those answerable for guaranteeing adequate organization ability to empower the certain vehicle of force. Since lead times for transmission support are long and the fortifications themselves are capital serious, there is in this manner a danger both of under-interest in network limit prompting limitations on the exchanging of energy, or then again over-speculation with either greater costs for framework clients or under-recuperation of expenses by the organization financial backer. Such vulnerabilities require new methods to deal with the dangers, thus requiring effective investigation of framework sway and improving healing activities. Comparable market-driven vulnerabilities exist in the exchange of the actual energy. According to the point of view of a framework administrator not involved with reciprocal exchanges yet liable for guaranteeing the safe mass vehicle of force, the dispatch of age is profoundly stochastic, making the upkeep of framework security truly challenging (K.W. Chan et al., 1995). This turns even more so when a few hours' notification of generators' actual positions is accessible. Along these lines, new devices could be significant if greater expenses are not caused. Many power frameworks, particularly in the created world, had their most note-worthy extension time at least 30 years prior. With many plants at first expected to have a valuable existence of 50 years or less, it tends to be seen that a critical time of recharging is going to be entered. Be that as it may, this is incidental with ever more tight controls on expenses and closer working to the furthest reaches of a flawless framework making blackout windows more limited. New devices would also appear to be required here so that resource substitution can be focused on and hazards made due. A last aspect of cost decrease concerns the movement of power framework investigation itself. Investors, clients, furthermore administrative specialists all expect truly lessening costs in the administration of force frameworks, and the expense of data frameworks lingers ever bigger as an extent of total consumption. Pressure in this manner exists to decrease the lifetime expenses of all data frameworks, including power framework investigation instruments. Then, at that point, there is additional pressure to build efficiency. This is frequently deciphered in control framework investigation, which means expanding the proficiency with which a specialist can perform studies (A. Henney, et al., 2001).

Technology Drivers

Developing utilization of strong state power change innovations and computerized gear is generating problems with power quality, especially music and voltage plunges, to be of expanding significance. "Uneven voltages are additionally of concern, and the more noteworthy closeness of low voltage gear to high voltage transmission offices requires more noteworthy regard for potential ascents of earth potential under issue conditions." Similar issues concerning the changing idea of electrical burdens influence the exhibition of short-out examination. That is the thing that is a proper portrayal of an infeed under issue conditions. Distributed generation shows have been followed to limit contrasts of understanding between various network administrators sharing an interface, for example, G74.

Nonetheless, these should be occasionally reexamined and programming corrected in like manner. Few, in any case, exceptionally huge frameworks imploded all over the planet, for example, in the western US in 1996, have roused transmission utilities to explore more modern demonstrating of dynamic peculiarities. Many have looked to broaden the period of such investigations past electro-mechanical homeless people to think about longer-term, slow elements, especially those which impact voltage steadiness, and to create 'safeguard measures' to secure the framework. There has additionally been a proceeding with interest in the viable arrangement of online dynamic security appraisal offices. The developing entrance in power frameworks of HVDC, Realities, and diverse age innovations, for example, various energy sources (like a breeze or new stockpiling innovations) and distinctive control frameworks, has brought concerning the requirement for fusing new models into existing apparatuses. At last, current registering advancements have opened up new freedoms for power framework examination apparatuses, particularly those managed by object direction, versatile improvement instruments for graphical UIs, information bases, and less expensive and all-the-more special equipment (O. Alsac, et al., 1990).

NEW METHODS IN POWER SYSTEM ANALYSIS

Investigation The above drivers propose a solid need to increase the specialist's regular power framework examination toolbox of AC load stream, hamper, and first swing electro-mechanical solidness appraisal. This segment addresses the new instruments by arriving at realization or being embraced for simple applications.

Stability Assessment

Parts of force framework dependability have traditionally been ordered in an accompanying manner. Point security, sub-partitioned int - first-swing (enormous unsettling influence) solidness, little sign security, and voltage security. These perspectives have generally been tended to freely with committed apparatuses, as a rule, 'disconnected,' for example, outside a control room. Nonetheless, generally because of various huge power framework unsettling influences that had not been expected to utilize the techniques accessible at that point, there has been a developing acknowledgement that such differentiations are lacking. Two principal ways to deal with further developing examination offices have been sought after over the most recent ten years. These have been: to present medium and long haul elements into bit-by-bit time space reproduction, to work on the appraisal of an edge of framework static load ability or a sign of the breaking point (H. Wei, et al., 1998).

The amazingly weighty computational burden is the primary inconvenience of a full-time space reenactment. Challenges with voltage dependability markers incorporate the sufficiency of the model's portrayal of how various generators meet expanded interest and the joining of the impacts of discontinuities, for example, generators hitting responsive power cutoff points or transformers hitting tap limits. A compromise approach pertinent to the appraisal of voltage solidness and static security is depicted, where quick elements are considered balanced. Moreover, slow ones are displayed expressly in a semi-consistent state reenactment. The test when furnishing control room staff with online evaluations of framework security has been to do as such rapidly enough. Analysts have proposed any number of alleged 'aberrant' strategies or example acknowledgement in request to 'channel' the possible cases to be concentrated in detail, the framework administrator's prerequisite consistently being that the sifting ought not inaccurately arrange sporadic cases as steady (D. Pudjianto, et al., 2002).

Optimal Power Flow

It has been noticed that a power system proprietor and additionally administrator has both inside and outside needs for finding the ideal answer for some characterized objective. The interior need is for exact expense minimization. The outside, as well, is for cost minimization yet additionally for the exhibition of reasonable treatment of distinctive market players in planning or administration buy choices and, conceivably, in empowering mooted markets in transmission admittance to occur at all. Security-obliged ideal power streams (SC-OPF) guarantee to accomplish such outstanding arrangements while noticing framework security and are, in this way, getting expanding consideration. Be that as it may, the innovation remains complex, and there is a variety of strategies, not many of which show all the ideal presentation highlights. In long-haul framework arranging, especially when considering a venture for framework support, numerous factors exist since there are extensive vulnerabilities; also, innumerable situations should be investigated . Costs would incorporate not just those of age and misfortunes but also of fortifications, such as receptive remuneration, Realities, or higher limit lines. The burden-shedding expense might be generally increased with the extension to add fortifications. Considering the inconceivability of having exact information on conjecture boundaries, it is allowable to determine any mathematical troubles by bothering with specific limitations. In functional preparation, fundamental framework conditions are known with sensible assurance, and applicable principles concerning cutoff points ought to be noticed.

Nonetheless, an arrangement operable on the framework should be found, even at the expense of unwinding, for instance, voltage limits or shedding load. In this manner, punishments might be put on deviations outside working cutoff points and some endeavours made to cost load shedding 'all things considered' to address the social furthermore monetary effect of clients' deficiency of supply; however, to limit the heap shedding to that expected to keep up with the by and large functional trustworthiness of the framework (D. Stroe, et al., 2016). In this last option, to diminish the danger of framework breakdown when some level of vulnerability concerning information is present, some base edge to implode may be seen as a limitation in a perfect world. One could sum these various requirements into 'basic,' 'helpful,' or 'less significant,' as in Table 1.

The fundamental ways to deal with the execution of SC-OPF are direct programming-based and non-straight inside point base. The standing of the previous is for strength as far as assembly, however, with various arrangements in some cases being found from different beginning stages with similar boundaries, while that of the last option is for speed and, assuming it unites, for intermingling to the ideal arrangement paying little mind to begin stage (Y. J. A. Zhang, et al., 2018).

Table 1. Features required from an SC-OPF in different power system planning contexts

Feature	Long-term	Operational
Reliability of convergence to a solution	Less important –can perturb parameters	Critical
Reliability of convergence to the optimal solution	Useful	Critical in a market context; less important in a system emergency
Speed	Useful	Useful
Can handle many variables	Critical	Less important

Statistical and Probabilistic Techniques

Similarly, as a more noteworthy profundity of examination is required, vulnerabilities concerning the conditions which may be found in market-orientated power supply enterprises are likewise prompting significantly more prominent broadness. Many age situations were needed when only a couple of future monetary problems were utilized inside coordinated utilities to design age and bandwidth. Where functional transmission organizers could have a reasonable level of assurance regarding the following day's despatch of dynamic power, presently, there is nothing. At the point when these vulnerabilities are added to the vulnerabilities that power framework engineers are now acquainted with making due, that is possibilities. Vulnerabilities concerning the number of conditions that should be recreated to believe in the operability of the framework turn out to be exceptionally huge. Such vulnerabilities and investigating numerous situations require another way to deal with power framework investigation, for example, in 'Evaluate.' This empowers

- adaptable and efficient examination of stochastic factors
- predictable catch of situations
- programmed reproduction, including dynamic reenactment, of a lot of problems, beginning from sensible starting conditions inferred by an OPF
- data set stockpiling of results
- admittance to strong measurable investigation and information-digging apparatuses for disclosure of basic connections inside the recreation results.

Motivation game plans progressively being set up by administrative experts for circulation organization working organizations are setting a more prominent accentuation on supply unwavering quality than has until now been the situation. This expands existing security rules by considering not simply that supply won't be hindered under certain, characterized conditions subject to certain, characterized occasions but also the likelihood of collection being intruded and for how long, under any doable conditions for any event. Just as being needed by controllers, such data can be of worth to both circulation and transmission proprietors in deciding resource substitution programs. Many such organizations have huge populaces of resources arriving at the finish of their arranged lives. While the difference in disappointment pace of maturing plants is difficult to decide, substitution works should be painstakingly focused on. This can be accomplished with the guide of framework unwavering quality evaluation to figure out what portions of the framework are generally powerless against changes in disappointment and fix rates for various classes of the plant. For a circulation proprietor, weakness would be communicated regarding the number of clients who would have supply hindered and for how long, with what likelihood. For a transmission organization, the attention would be on which matrix supply would experience interference with the most noteworthy recurrence with the probable reclamation time considered straightaway. New programming instruments can make such investigations advantageous by permitting the simple examination of various tomahawks of the issue, displaying the reactions of programmed gear, for example, postponed auto reclose and between outings, and showing the impacts of awful climate when issue occasions are bundled together (Panda S, et al., 2022).

Maintenance of Legacy Software

The interpretation of client necessities into power framework examination programming has generally followed the planned approach of separating every intricate prerequisite into consistent capacities (or systems) that would then be able to be unequivocally coded in a procedural language, for example, FORTRAN or C. Information, the executives' offices, have then, at that point, would, in general, be given later a specific strategy has been created and regularly specially appointed, for example, to provide the information with explicit to one technique. Intricacy or potential duplication is unavoidable in the communication between various methods, and perplexing, firm transformation offices are expected to pass data. The offices so created stay basic to a power framework utility's movement yet are famously troublesome and exorbitant to grow further since with each change, both a technique and a cobweb of information the executives' offices should be altered (J. Cao, et al. 2018).

Distributive Renewable Power Using AI

With the help of AI, distributed renewable power integration is changing the game when it comes to sustainable energy. When it comes to decentralizing renewable energy generation and delivery, AI is crucial. By taking into account ever-changing variables like weather and patterns of renewable resources, AI systems are able to provide precise predictions of future energy generation using predictive analytics. By anticipating future needs, preemptive measures may be taken to balance energy supply and demand. Energy storage is also optimized by AI, which guarantees optimal use of saved power. Figure 1 shows how AI-enabled real-time monitoring and management of distributed energy resources improve grid stability via demand and supply balancing. These resources may react to grid circumstances dynamically with the use of adaptive control algorithms. In addition, AI helps with microgrid optimization, smart grid management, and cybersecurity measures, which all work together to make the distributed renewable power landscape's infrastructure more robust and safe. In addition to resolving issues with fluctuation, this integration lays the groundwork for an intelligent and sustainable energy future (U. Datta, et al. 2020).

Figure 1. Distributive renewable power using AI

THE IMPLEMENTATION OF NEW FACILITIES

In 1965, Dr Gordon Moore, one of the establishing accomplices of Intel, anticipated that the number of semiconductors on a silicon chip will twofold roughly every eighteen months . Today, "Moore's Law" refers to fast advances in figuring power per unit cost, or 'bangs per buck.' This opens up wide, furthermore truly enlarging, new skylines of data framework potential outcomes. The power supply industry is one of the most mind-blowing sets to utilize propels in figuring power. As the US Federal Energy Regulatory Commission highlighted in a report on the formation of a territorial transmission association for the northeastern United States, "the intricacy of an electric framework the executives programming is second just to that of the banking industry". (The creators of the current paper are not satisfied regarding what makes the financial business' programming considerably more perplexing, particularly as it can disregard Ohm's and Kirchoff's laws). A similar report additionally notes that "there are incredibly, scarcely any sellers who plan such programming, and those merchants have a huge monetary motivating force to over-gauge their/their product's capacities to get such worthwhile agreements." The comments in the two going before passages help to delineate that power framework investigation programming is fundamentally significant yet unavoidably both complicated and absurd. With progresses in innovation, a power framework utility looking to acquire the new programming devices expected to meet the challenges illustrated in segment 2 is confronted with attempting to hit a troublesome, costly, and steadily moving objective (M. G. Dozein and P. Mancarella, 2019).

Enterprise Application Reconciliation

Conventional ways to deal with the plan of PC frameworks in light of interaction drove advancements frequently result in a "monolithic" frameworks design by which the client interface, 'calculating' and information/data are all firmly coupled and generally completely held inside a separate single figuring climate for each application. Instances of this in the power business incorporate work-the-board frameworks, energy-the-executives frameworks, disconnected examination frameworks, and resource inventories, all growing altogether freely of one another. Elaborate offices are then regularly evolved to permit frameworks to trade data; however, these are constantly because of the change of information from one appropriate information configuration to another, frequently with strengthening knowledge. This methodology is not generally considered sufficient since an engineer must examine numerous parts of a power framework's presentation, one next to the other, and requirements to ensure the outcomes are reliable and right for the time skyline M. (Zeraati, et al., 2018). All together that a utility's by and large activity can be completely streamlined, data supplementing the electrical qualities, for example, on work furthermore blackout arranging, temporary plant restrictions, upkeep history, and so on, is presently additionally needed in more important detail across an expanding number of timescales. This, too, requires to ever be exact and steady in all the timescales a utility has liability regarding. Such necessities request that these different frameworks be re-engineered (Y. Shi, et al, 2018).

To begin with, they ought to be separated into their central parts, with every aspect executed once with adaptable, characterized interfaces set up. This brings about the idea of programming improvements occurring, not as discrete ventures, but as part of a more extensive undertaking climate where the normal linkage is a steady perspective on the resources being used chasing after business. Generally, everything a power utility jars is related to the resources that make up the power framework. Each interaction that a power utility conveys utilizes a "view" of the resource data that is explicit to the movement referred to,

Figure 2. A power utility software architecture

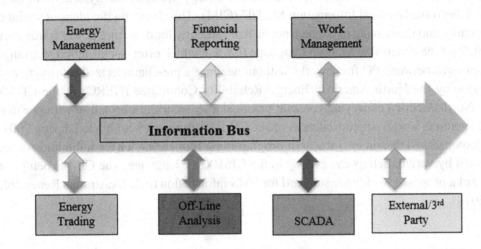

for example, blackout arranging, support action, monetary detailing, security evaluation, and functional survey. It is itself an activity performed on a resource. This normally fits object direction regarding information association and work execution as shown in fig 2. The resource information becomes the normal connection between the various cycles, and a system for correspondence in a standard structure is vital (E. Namor, et al., 2019). That is, a specific language is required that all processes and exercises comprehend and is conveyed using a typical framework known as the 'Data Bus.' This transport ought to be about as free as conceivable from reliance on a singular owner. For example, it ought to be, similarly as possible, 'industry standard.' Transformation to and from the everyday language and the Information Bus is the obligation of each interaction or potential application utilizing it. The colloquial language isn't a data store or information base but an interface. This brings about a product engineering like that outlined in Consider 1, along with which application programming from various merchants can be fitted. In such a model, with a far-reaching and open information worldview, the update of cycles or execution of new mathematical techniques can be accomplished without the need to overhaul information designs (A. Zecchino, et al. 2021).

The Common Language

By and large, data trade inside or between designing substances has occurred through the investigation of explicit configurations. The illustration of a globally concurred fixed arrangement was the IEEE one for the trade of burden stream information . Nonetheless, as examination capacities and the extent of safety appraisal expanded, a further need arose to impart short-out and transient solidness information. In this way, some exclusive information designs have begun to enhance the IEEE arrangement and have acquired overall acknowledgement yet experience the disservice of evolving often. During the 1980s, the IEEE delivered another draft standard that presented the idea of a self-characterizing design – that is, the information sent contained a portrayal of the nature of the report and the basic information.At last, this work never advanced past a draft, yet the idea was taken up by the Electric Power Research Institute (EPRI) in a drive known as the "Control Center Application Program Interface" . This task was initially intended to help integrate multi-seller applications into Energy Management Systems. One of its natural products has been the "Normal Information Model" (CIM). This draws on the ideas of social data sets, object direction, and the craving to create a normalized strategy for depicting power frameworks. It has likewise utilized the eXtensible Markup Language (XML), which offers a cutting-edge arrangement of passing messages between PC frameworks without needing a pre-characterized information structure. The reception by the North American Energy Reliability Committee (NERC) of the CIM/XML for utilities in North America conveying constant practical also security information with one another has shown the "genuine world" appropriateness of the CIM. Preliminaries completed during 2001 have effectively demonstrated that an assortment of programming frameworks on a combination of registering stages created by various sellers can convey in the CIM/XML language. The CIM is being formalized as a feature of a proposed worldwide standard for EMS information trade (Australian Renewable Energy Agency, 2019).

CHALLENGES AND FUTURE PERSPECTIVES

The emerging challenges in power system analysis for utilities include:

- **Renewable Energy Integration:** Integrating renewable energy sources, such as solar and wind power, into the grid presents challenges due to their intermittent nature. Utilities need to ensure the stability and reliability of the grid while managing the variability and uncertainty associated with renewable generation.
- **Grid Resiliency:** With the increasing frequency of extreme weather events and natural disasters, utilities face the challenge of maintaining a resilient power grid. Power system analysis is crucial in assessing vulnerabilities, identifying weak points, and developing strategies to enhance the grid's resilience.
- **Distributed Energy Resources (DERs):** The proliferation of DERs, such as rooftop solar panels and energy storage systems, introduces new complexities for utilities. Power system analysis must address issues related to optimal integration, control, and operation of these decentralized resources to ensure their effective utilization and grid compatibility.
- **Electrification of Transportation:** The electrification of transportation, including electric vehicles (EVs) and charging infrastructure, poses challenges for utilities. Power system analysis is needed to evaluate the impact of increased electricity demand, develop smart charging strategies, and ensure the efficient utilization of the existing grid infrastructure (Channi, H. K. (2023).
- **Cybersecurity:** As power systems become more digital and interconnected, utilities face growing cybersecurity risks. Power system analysis must address vulnerabilities, assess risks, and implement robust security measures to protect critical infrastructure from cyber threats.
- **Ageing Infrastructure:** Many power systems worldwide have ageing infrastructure that requires upgrades and investments. Power system analysis can help utilities identify areas that need modernization, evaluate the impact of infrastructure upgrades, and optimize investment decisions.
- **Grid Modernization:** The transition towards a more modern grid, incorporating advanced technologies like advanced metering infrastructure (AMI), grid automation, and real-time monitoring, presents challenges for utilities. Power system analysis is crucial in designing and optimizing these modernization efforts to improve grid efficiency, reliability, and flexibility.

It's important to note that the power industry is constantly evolving, and new challenges may have emerged since my knowledge cutoff. Utility companies and researchers continue to work on addressing these challenges through advanced power system analysis techniques and innovative solutions. In the future, power system analysis for utilities will face several emerging issues and challenges shaping the industry. One of the primary concerns will be the integration of variable renewable energy sources into the grid. As the share of solar and wind power continues to grow, utilities will need advanced analysis techniques to manage the variability and intermittency of these resources. This will involve developing accurate forecasting models, optimizing resource allocation, and implementing grid operation strategies that ensure a reliable and stable power supply.

Another key issue is enhanced grid flexibility and demand response capabilities. With the rise of distributed energy resources and changing consumer behaviour, utilities must analyze and optimize grid operations in real-time. Power system analysis will play a crucial role in developing advanced optimization algorithms, load management strategies, and demand response programs to ensure the efficient utilization of resources and maintain grid stability. Cybersecurity and grid resilience will be major concerns for utilities as the power system becomes more digitized and interconnected. Power system analysis will focus on identifying vulnerabilities, implementing robust security measures, and developing risk assessment techniques to protect critical infrastructure from cyber threats. Additionally, utilities must analyze and plan for grid resilience in natural disasters and extreme weather events, utilizing power system analysis to assess vulnerabilities and develop contingency plans (kaur Channi, et al., 2020). The increasing electrification of transportation and the rise of decentralized energy systems will also present challenges for utilities. Power system analysis will be essential in optimizing the integration of electric vehicles into the grid, managing bidirectional power flow, and ensuring the reliable operation of charging infrastructure.

Furthermore, the growth of microgrids and local energy communities will require advanced analysis techniques to integrate these decentralized systems into the broader power grid effectively. Lastly, the future of power system analysis will leverage big data analytics and artificial intelligence. The proliferation of smart meters and sensors will generate vast amounts of data that can be harnessed to improve grid operations. Power system analysis will utilize advanced analytics and AI techniques to extract valuable insights, optimize decision-making, and enhance grid planning processes. In conclusion, the future perspective of power system analysis for utilities involves addressing challenges such as renewable energy integration, grid flexibility, cybersecurity, transportation electrification, decentralized energy systems, and leveraging big data analytics and AI. By effectively tackling these emerging issues, utilities can ensure a more sustainable, reliable, and resilient power system for the future (GHD advisory 2020).

AI TECHNIQUES USED IN OPTIMIZING DISTRIBUTED RENEWABLE POWER INTEGRATION

In the pursuit of optimizing the integration of distributed renewable power, artificial intelligence (AI) employs a diverse array of sophisticated techniques. Machine learning algorithms, such as neural networks and decision trees, prove instrumental in predictive modeling for forecasting renewable energy production and demand patterns. Reinforcement learning techniques enable adaptive control strategies, allowing systems to dynamically learn and respond to changing conditions in the energy grid. Optimization algorithms, such as genetic algorithms, contribute to finding the most efficient configuration of distributed renewable resources. Deep learning methodologies enhance the accuracy of forecasting models and aid in intricate pattern recognition. Smart grid technologies leverage AI for real-time monitoring and control, ensuring effective communication and coordination among distributed energy sources. These techniques collectively empower a more intelligent, responsive, and resilient infrastructure for distributed renewable power integration, marking a significant stride towards a sustainable energy future.Several AI techniques are employed in optimizing the integration of distributed renewable power sources into the energy grid. These techniques leverage machine learning, data analytics, and advanced control strategies to enhance efficiency, reliability, and sustainability (North American Electric Reliability Corporation, 2019). Some key AI techniques include:

- **Machine Learning Algorithms:** Various machine learning algorithms, such as neural networks, support vector machines, and decision trees, are used for predictive modeling. These models can forecast renewable energy production, energy demand, and optimize the operation of distributed energy resources.

- **Predictive Analytics:** AI is utilized for predictive analytics to anticipate future energy consumption patterns and renewable energy production. This allows for proactive decision-making and resource allocation to match supply with demand.

- **Reinforcement Learning:** Reinforcement learning techniques enable adaptive control strategies by allowing systems to learn from interactions with the environment. In the context of distributed renewables, this can optimize energy storage, grid management, and demand-side response.

- **Optimization Algorithms:** AI-driven optimization algorithms, including genetic algorithms and particle swarm optimization, are applied to find the optimal configuration of distributed renewable resources, maximizing energy output and grid efficiency.

- **Smart Grid Technologies:** AI is integrated into smart grid systems for real-time monitoring, control, and communication. This facilitates intelligent decision-making to balance and manage distributed energy resources effectively.

- **Deep Learning:** Deep learning techniques, such as deep neural networks, are employed for complex pattern recognition and decision-making tasks. In the context of distributed renewables, this can improve the accuracy of forecasting models and enhance grid management.

- **Edge Computing:** AI at the edge is used for decentralized processing of data from distributed energy sources. This minimizes latency, enhances real-time decision-making, and reduces the need for centralized data processing.

- **Cybersecurity Measures:** AI techniques, including anomaly detection and behavioral analysis, are employed for enhancing the cybersecurity of distributed renewable power systems, protecting against potential cyber threats.

- **Adaptive Control Systems:** AI-based adaptive control systems enable distributed energy resources to dynamically adjust their operation based on real-time conditions, contributing to grid stability and efficiency.

- **Data-driven Decision Support Systems:** AI is utilized to analyze vast amounts of data from distributed energy sources, enabling data-driven decision support systems that guide optimal resource allocation and grid management strategies (L. R. Limongi et al., 2007).

Case Studies

Google's DeepMind and Wind Energy Predictions

DeepMind, a Google subsidiary, used deep learning algorithms to enhance wind farms' ability to forecast future energy production. Improving the dependability and efficiency of wind power production is the central focus of the case study. Historical and real-time data, such as wind speed, direction, and atmospheric conditions, were used to train deep neural networks. When compared to more conventional approaches, the AI models' capacity to predict future wind patterns was striking.In order to effectively prepare for and integrate wind energy into the power system, DeepMind hoped to solve the inherent unpredictability of wind power by using these powerful AI approaches. More accurate scheduling of energy generation and storage was made possible by the enhanced wind energy projections, which maximized the use of wind power resources. In the continuous pursuit of a more sustainable and dependable energy infrastructure, this case study demonstrates how artificial intelligence (AI) may improve renewable energy forecasts (Kumar, R., & Channi, H. K., 2022).

Autonomous Microgrid Control at University of California, Irvine

One example of an AI-powered initiative to improve microgrid management and operation is UCI's Autonomous Microgrid Control Project. A localized energy system that employs renewable energy sources, energy storage, and demand-side resources is the subject of this case study, which aims to maximize its performance via the use of autonomous control mechanisms.The goal of this project was to create a microgrid autonomous control system using optimization and reinforcement learning algorithms developed by academics at UCI (Raut, N., et al.,(2024). The microgrid was able to adapt to its surroundings, including fluctuating energy demand and renewable resource availability, by using reinforcement learning to learn and make choices.Taking into account the intermittent nature of renewable sources, the autonomous control system sought to effectively manage the microgrid's dispersed energy supplies, striking a balance between supply and demand. The microgrid maximized its power generation, storage, and distribution by responding to real-time circumstances by dynamically altering the functioning of its energy assets.This case study shows how AI may be used to build microgrids that are resilient and can provide for themselves. An example of how artificial intelligence is revolutionizing the integration of renewable energy sources is the use of autonomous control techniques, which boost energy efficiency while simultaneously making distributed energy systems more reliable and sustainable (Z. Ali et al., 2018).

Renewable Energy Forecasting with NextEra Energy

In order to improve the accuracy of their renewable energy forecasts, NextEra Energy used machine learning algorithms to analyze past meteorological data and energy production trends. This method, powered by AI, allowed for accurate predictions of future renewable energy production using both current and predicted meteorological conditions. By optimizing resource scheduling, NextEra electricity was able to improve the accuracy of its forecasts, which in turn allowed for smoother integration into the power grid and a more efficient and dependable supply of electricity. This case study demonstrates how artificial intelligence may be used to improve renewable energy forecasts, which in turn helps to build more reliable and sustainable energy infrastructure (Rene, E. A., & Fokui, W. S. T., 2024).

Siemens' Predictive Grid Management

With the use of AI, Siemens introduced Predictive Grid Management, which optimizes the grid in advance. The system uses optimization and predictive analytics to sift through data collected from dispersed energy sources, identify possible disruptions, and redistribute resources on the fly. Improve grid stability with an AI-driven strategy that fixes problems before they happen, makes the system more reliable overall, and reduces the effect of renewable energy swings. To accommodate the growing proportion of distributed renewable energy sources in the dynamic energy environment, forward-thinking strategies like Siemens' Predictive Grid Management are important. By using modern technology, this approach aims to build a power grid that is more robust and responsive (Gundeti, R., et al., 2024).

Pacific Gas and Electric Company's AI-Enhanced Demand Response

The AI-Enhanced Demand Response system was put into place by Pacific Gas and Electric Company. This system makes use of machine learning algorithms to accurately predict future demand. Better coordination of distributed energy resources is made possible by this initiative's enhancement of grid management via the prediction of times of peak energy demand (Hur, J., & Ahn, E.,2024).. In order to reduce stress during peak demand times, the AI-powered system improves demand response programs by proactively modifying energy use. Pacific Gas and Electric showcases the revolutionary power of AI in demand-side management by enhancing grid efficiency, reducing resource requirements, and building a more sustainable and resilient energy infrastructure through the use of machine learning for precise demand predictions (Oyewole, et al.,2024).

These case studies showcase the diverse applications of AI in optimizing distributed renewable power integration, ranging from predictive analytics and demand forecasting to autonomous control systems in microgrids. It's important to note that the field is rapidly evolving, and ongoing research may yield new and innovative applications of AI in renewable energy systems (Mohanty, M., & Sarkar, R., 2024).

CONCLUSION

Finally, Power System Analysis is essential for Optimising Distributed Renewable Power Integration and a sustainable energy future. Distributed renewable power sources seamlessly integrated into energy networks provide a unique chance to cut greenhouse gas emissions, improve energy security, and create a cleaner, more resilient environment. Power system operators and utilities may solve intermittent renewable energy concerns via power system analysis. Accurate load forecasting and grid capacity evaluations allow them to anticipate and handle energy supply and demand variations, guaranteeing grid stability and dependability. Smart grids provide real-time monitoring, control, and management of distributed renewable power supplies, improving system flexibility and adaptability. Energy storage devices allow utilities to store surplus renewable energy for later use and reduce waste. Power system analysis helps determine distributed renewable power integration's economic feasibility. It allows utilities to choose infrastructure improvements, incentives, and cost reductions, enabling more economically efficient energy systems. Optimizing distributed renewable power integration has environmental benefits beyond economic ones. Power system analysis helps meet global climate and environmental objectives by lowering fossil fuel use and increasing sustainable energy.

Integration involves government, policymakers, utilities, and private sector cooperation. Renewable energy adoption and power grid integration need supportive policies, investment incentives, and technological advances. Power system analysis is essential for maximizing distributed renewable power sources in a sustainable energy transition. This strategy will create more robust, adaptable, and efficient energy systems that actively contribute to a greener, more sustainable future for future generations. We harness the revolutionary potential of renewable energy and create a cleaner, more sustainable, and affluent society by continually improving power system analyses. There are huge new drivers for power framework utilities to develop their examination offices further and survey how they are given. The chance exists to use innovation to add esteem and oversee costs. Proposed principles, for example, IEC61970, offer the establishment of means by which the resource-arranged trade of information can be accomplished to such an extent that any investigation capacity should be visible as an activity on a 'genuine world' object. That such powers can be coordinated into a 'data transport utilizing a standard convention in an 'endeavour engineering' offers an opportunity for new scientific techniques to be fused from various merchants, opening up market openings for programming suppliers and adaptability for utilities. Utilities must conduct power system analyses to keep the power grid stable and efficient. Utilities confront additional difficulties as the energy environment changes, such as incorporating renewable energy sources, maintaining grid stability, and making room for charging facilities for electric vehicles. The best ways to improve operations and cater to customers' needs are to use cutting-edge analytics, update the grid, and implement adaptable control systems. Utilities can deal with these new challenges and aid in building a more sustainable and reliable electricity grid by embracing technological advances and planning.

REFERENCES

Ahmed, P., Rahman, M. F., Haque, A. M., Mohammed, M. K., Toki, G. I., Ali, M. H., Kuddus, A., Rubel, M. H. K., & Hossain, M. K. (2023). Feasibility and Techno-Economic Evaluation of Hybrid Photovoltaic System: A Rural Healthcare Center in Bangladesh. *Sustainability (Basel)*, *15*(2), 1362. doi:10.3390/su15021362

Ali, Z., Christofides, N., Hadjidemetriou, L., Kyriakides, E., Yang, Y., & Blaabjerg, F. (2018). Three-phase phase-locked loop synchronization algorithms for grid-connected renewable energy systems: A review. *Renewable & Sustainable Energy Reviews*, *90*, 434–452. doi:10.1016/j.rser.2018.03.086

Alsac, O., Bright, J., Prais, M., & Stott, B. (1990). Further developments in LP-based optimal power flow. *IEEE Transactions on Power Systems*, *5*(3), 697–711. doi:10.1109/59.65896

Alsharif, A., Tan, C. W., Ayop, R., Al Smin, A., Ali Ahmed, A., Kuwil, F. H., & Khaleel, M. M. (2023). Impact of electric Vehicle on residential power distribution considering energy management strategy and stochastic Monte Carlo algorithm. *Energies*, *16*(3), 1358. doi:10.3390/en16031358

Cao, J., Du, W., Wang, H., & McCulloch, M. (2018). Optimal Sizing and Control Strategies for Hybrid Storage System as Limited by Grid Frequency Deviations. *IEEE Transactions on Power Systems*, *33*(5), 5486–5495. doi:10.1109/TPWRS.2018.2805380

Chan, K. W., & Edwards, A. R. (1995). Online dynamic security assessment using a real-time power system simulator with neural network contingency screens. *Proc. IEEE Third International Conference on Advances in Power System Control, Operations and Management (APSCOM '95),* (pp. 461-466). IEEE.

Channi, H. K. (2023). Optimal designing of PV-diesel generator-based system using HOMER software. *Materials Today: Proceedings.*

Datta, U., Kalam, A., & Shi, J. (2020). Battery Energy Storage System Control for Mitigating PV Penetration Impact on Primary Frequency Control and State-ofCharge Recovery. *IEEE Transactions on Sustainable Energy, 11*(2), 746–757. doi:10.1109/TSTE.2019.2904722

Dozein, M. G., & Mancarella, P. (2019). Possible Negative Interactions between Fast Frequency Response from Utility-scale Battery Storage and Interconnector Protection Schemes. *AUPEC, 2019,* 1–6. doi:10.1109/AUPEC48547.2019.211968

Duong, D. T., & Uhlen, K. (2018). *"An empirical method for online detection of power oscillations in power systems,"* 2018 IEEE Innovative Smart Grid Technol. - Asia. ISGT Asia.

Dwivedi, D., Mitikiri, S. B., Babu, K., Yemula, P. K., Srininvas, V. L., Chakraborty, P., & Pal, M. (2023). Advancements in Enhancing Resilience of Electrical Distribution Systems: A Review on Frameworks, Metrics, and Technological Innovations. *arXiv preprint arXiv:2311.07050.*

Gao, B., Morison, G. K., & Kundur, P. (1986). Voltage stability evaluation using modal analysis. *IEEE Transactions on Power Systems, 7*(4), 1529–1542. doi:10.1109/59.207377

Gundeti, R., Vuppala, K., & Kasireddy, V. (2024). The Future of AI and Environmental Sustainability: Challenges and Opportunities. *Exploring Ethical Dimensions of Environmental Sustainability and Use of AI,* 346-371.

Henney, A. (2001, October). Transmission access – a case study in public maladministration? *Power UK,* (92), 18–32.

Hur, J., & Ahn, E. (2024). *An Enhanced Short-term Forecasting of Wind Generating Resources based on Edge Computing in Jeju Carbon-Free Islands.*

Kaur Channi, H., Gupta, S., & Dhingra, A. (2020). Optimization and simulation of a solar–wind hybrid system using HOMER for Rural Electrification. *International Journal of Advanced Science and Technology, 29,* 2108-2116.

Kessel, P., & Glavitsch, H. (1986). Estimating the voltage stability of a power system. *IEEE Transactions on Power Systems, 1*(3).

Konakalla, S. A. R., & de Callafon, R. A. (2017, June). Feature based grid event classification from synchrophasor data. *Procedia Computer Science, 108,* 1582–1591. doi:10.1016/j.procs.2017.05.046

Kumar, A., & Choudhary, J. (2023). Power quality improvement of hybrid renewable energy systems-based microgrid for statcom: Hybrid-deep-learning model and mexican axoltl dingo optimizer (MADO). *Engineering Research Express, 5*(4), 045031. doi:10.1088/2631-8695/ad0287

Kumar, R., & Channi, H. K. (2022). A PV-Biomass off-grid hybrid renewable energy system (HRES) for rural electrification: Design, optimization and techno-economic-environmental analysis. *Journal of Cleaner Production, 349*, 131347. doi:10.1016/j.jclepro.2022.131347

Kundur, P. (1994). *Power System Stability and Control*. McGraw-Hill.

Lee, J., Lee, D., Lee, J., Yoon, M., & Jang, G. (2023). Offshore MTDC Transmission Expansion for Renewable Energy Scale-up in Korean Power System: DC Highway. *Journal of Electrical Engineering & Technology, 18*(4), 2483–2493. doi:10.1007/s42835-023-01513-z PMID:37362030

Mohanty, M., & Sarkar, R. (2024). *The Role of Coal in a Sustainable Energy Mix for India: A Wide-Angle View*.

Namor, E., Sossan, F., Cherkaoui, R., & Paolone, M. (2019). Control of Battery Storage Systems for the Simultaneous Provision of Multiple Services. *IEEE Transactions on Smart Grid, 10*(3), 2799–2808. doi:10.1109/TSG.2018.2810781

Nativel, G., Jacquemart, Y., Sermanson, V., & Gault, J. C. (1999). *Implementation of a voltage stability analysis tool using quasisteady- state time simulation*. Proc 13th PSCC, Trondheim, Norway.

North American Electric Reliability Corporation (NERC). (2019). *Improvements to interconnection requirements for BPS-connected inverter-based resources*. Reliability Guideline.

Oyewole, O. L., Nwulu, N. I., & Okampo, E. J. (2024). Optimal design of hydrogen-based storage with a hybrid renewable energy system considering economic and environmental uncertainties. *Energy Conversion and Management, 300*, 117991. doi:10.1016/j.enconman.2023.117991

Panda, S., Mohanty, S., Rout, P. K., Sahu, B. K., Parida, S. M., Kotb, H., Flah, A., Tostado-Véliz, M., Abdul Samad, B., & Shouran, M. (2022). An Insight into the Integration of Distributed Energy Resources and Energy Storage Systems with Smart Distribution Networks Using Demand-Side Management. *Applied Sciences (Basel, Switzerland), 12*(17), 8914. doi:10.3390/app12178914

Pudjianto, D., Ahmed, S., & Strbac, G. (2002). *"Allocation of VArs Support using LP and NLP based Optimal Power Flows", accepted for publication in IEE Proc*. On Generation, Transmission and Distribution.

Raut, N., Chaudhary, P., Patil, H., & Kiran, P. (2024). 5 Understanding the Contribution of Artificial Intelligence. Handbook of Artificial Intelligence Applications for Industrial Sustainability: Concepts and Practical Examples.

Rene, E. A., & Fokui, W. S. T. (2024). Artificial intelligence-based optimal EVCS integration with stochastically sized and distributed PVs in an RDNS segmented in zones. *Journal of Electrical Systems and Information Technology, 11*(1), 1. doi:10.1186/s43067-023-00126-w

Rodriguez-Martinez, O. F., Andrade, F., Vega-Penagos, C. A., & Luna, A. C. (2023). A Review of Distributed Secondary Control Architectures in Islanded-Inverter-Based Microgrids. *Energies, 16*(2), 878. doi:10.3390/en16020878

Satyanarayana, P. V. V., Radhika, A., Reddy, C. R., Pangedaiah, B., Martirano, L., Massaccesi, A., Flah, A., & Jasiński, M. (2023). Combined DC-Link Fed Parallel-VSI-Based DSTATCOM for Power Quality Improvement of a Solar DG Integrated System. *Electronics (Basel)*, *12*(3), 505. doi:10.3390/electronics12030505

Shezan, S. A., Kamwa, I., Ishraque, M. F., Muyeen, S. M., Hasan, K. N., Saidur, R., Rizvi, S. M., Shafiullah, M., & Al-Sulaiman, F. A. (2023). Evaluation of different optimization techniques and control strategies of hybrid microgrid: A review. *Energies*, *16*(4), 1792. doi:10.3390/en16041792

Shi, Y., Li, W., Tan, B., Yang, F., & Zhang, L. (2023, April). Optimal cost scheduling of virtual power plant for new power system. In *2022 2nd Conference on High Performance Computing and Communication Engineering (HPCCE 2022)* (*Vol. 12605*, pp. 62-70). SPIE. 10.1117/12.2673283

Shi, Y., Xu, B., Wang, D., & Zhang, B. (2018). Using Battery Storage for Peak Shaving and Frequency Regulation: Joint Optimization for Superlinear Gains. *IEEE Transactions on Power Systems*, *33*(3), 2882–2894. doi:10.1109/TPWRS.2017.2749512

Stroe, D., Swierczynski, M., Stroe, A.-I., Laerke, R., Kjaer, P. C., & Teodorescu, R. (2016). Degradation Behavior of Lithium-Ion Batteries Based on Lifetime Models and Field Measured Frequency Regulation Mission Profile. *IEEE Transactions on Industry Applications*, *52*(6), 5009–5018. doi:10.1109/TIA.2016.2597120

Varga, L., Quintana, V. H., & Miranda, R. (1999). Voltage collapse in the Chilean interconnected system. *IEEE Transactions on Power Systems*, *14*(4), 1415–1421. doi:10.1109/59.801905

Wei, H., Sasaki, H., Kubokawa, J., & Yokoyama, R. (1998). An interior point non linear programming for optimal power flow problems with a novel data structure. *IEEE Transactions on Power Systems*, *13*(3), 870–877. doi:10.1109/59.708745

Xue, Y., Van Cutsem, T., & Ribbons-Pavella, M. (1989). Extended equal area criterion justification, generalizations, applications. *IEEE Transactions on Power Systems*, *4*(1), 44–52. doi:10.1109/59.32456

Yuan, W., Yuan, X., Xu, L., Zhang, C., & Ma, X. (2023). Harmonic Loss Analysis of Low-Voltage Distribution Network Integrated with Distributed Photovoltaic. *Sustainability (Basel)*, *15*(5), 4334. doi:10.3390/su15054334

Zecchino, A. (2021). *Optimal provision of concurrent primary frequency and local voltage control from a BESS considering variable capability curves: Modelling and experimental assessment* (Vol. 190). Electric Power Systems Research.

Zeraati, M., Hamedani Golshan, M. E., & Guerrero, J. M. (2018). Distributed Control of Battery Energy Storage Systems for Voltage Regulation in Distribution Networks with High PV Penetration. *IEEE Transactions on Smart Grid*, *9*(4), 3582–3593. doi:10.1109/TSG.2016.2636217

Zhang, Y. J. A., Zhao, C., Tang, W., & Low, S. H. (2018). Profit-Maximizing Planning and Control of Battery Energy Storage Systems for Primary Frequency Control. *IEEE Transactions on Smart Grid*, *9*(2), 712–723. doi:10.1109/TSG.2016.2562672

Chapter 5
Real–Time GPS Coordinates With Voice and Text Transfer Radio Frequency Processing System for Fleet Tracking

G. Jegan
Sathyabama Institute of Science and Technology, India

M. R. Ebenezar Jebarani
Sathyabama Institute of Science and Technology, India

P. Kavipriya
Sathyabama Institute of Science and Technology, India

S. Lakshmi
Sathyabama Institute of Science and Technology, India

ABSTRACT

The focus of the study is on developing a system for monitoring moving targets. The study's overarching goal is to create a GPS module-based fleet tracking system capable of outputting position updates to Google Maps. By integrating XIAO ESP32 C3 microcontroller with Neo-6M GPS module and A9G GSM/GPRS module into the Maduino Zero 4G platform, military and naval forces can get versatile and convenient GPS tracking and communication solutions with voice and text transmission. In addition, the proposed device has a push button that serves to activate the SOS feature, allowing for discreet dialing and providing the user with GPS data. The proposed system has the capability to provide enhanced levels of accurate GPS data in both indoor and outdoor environments due to its use of a 4G network as an interface. Due to its variable range, the suggested system can save operating costs and improve asset utilization during short and long-range communication. The proposed system is a comprehensive wireless solution that functions only via wireless data power transmission.

DOI: 10.4018/979-8-3693-3735-6.ch005

INTRODUCTION

The integration GSM and GPS technologies are driven by the multitude of applications, as well as their widespread adoption by millions of individuals worldwide (Giaglis et al., 2002). GPS/GSM systems are being used by construction and transportation companies in many applications, including road construction and fleet management (Gu & Niemegeers, 2009). The surface information that has been designed is uploaded into the system in a digital format. The computer display and real-time GPS location information enable the operator to ascertain if the right gradient has been achieved. Additionally, it is used for the purpose of monitoring the geographical position and utilization patterns of various devices across many locations. The use of GPS technology allows contractors to optimize the deployment of their equipment in a more effective manner by transmitting relevant information to a centralized location (Koyuncu & Yang, 2010). Furthermore, it is possible to provide effective guidance to vehicle operators in order to facilitate their navigation towards their intended destinations. Currently, individuals use a system for navigation and communication with their control center when using public transportation.

A fleet tracking system involves the installation of an electronic device in one or more vehicles, along with specialized computer software at a central operational base. This system allows the owner or a third party to monitor the location of the vehicle(s), while also collecting relevant data from the field and transmitting it to the operational base (Mautz, 2009). Many businesses and corporations have "field teams" that conduct activities outside from central headquarters. Among them include selling, servicing, maintaining, and delivering products to customers. These human actors may manage a variety of vehicle fleets, including cars, trucks, and others. It may be difficult to keep track of assets and cars efficiently. Fleet and asset tracking software might be useful for businesses. Several companies have made sizable investments in their fleet management solutions and have employed proprietary GPS tracking technology; however, many systems may be overly complex and expensive due to the use of proprietary hardware, software, and customized implementations. To get the most out of a fleet monitoring system, it's important to have a complete picture of the fleet's daily activities and the ability to pinpoint its whereabouts at any moment. Fleet and asset monitoring solutions are becoming more popular in modern businesses due to the many advantages they provide. Companies who use such systems are able to monitor the whereabouts of their fleets in real time and acquire a historical perspective on their vehicles' whereabouts. This has the potential to enhance customer satisfaction, boost earnings, and cut expenses. Satellites are often used inside automotive tracking systems to accurately determine the precise location of the vehicle. The use of internet-accessible electronic maps and specialized software has the potential to facilitate the retrieval of vehicle information. Due to the above-mentioned costs and complexities of existing solutions, in this research, a comprehensive method to investigate and construct a dedicated system to fleet monitoring by combining geo location, satellite, and online technologies.

LITERATURE SURVEY

Akshatha (2017) suggests tracking public transport vehicles using Raspberry Pi and GPS antenna modules. Raspberry Pi processors receive and output values. This method tracks cars from origin to destination. This article tracks the vehicle's location using a GPS receiver module. Car passengers will report their whereabouts between source and destination to the system. If passenger-specified values disagree from current vehicle position information, the Raspberry Pi CPU will alert passengers via display system to prevent drivers from travelling in the incorrect direction. This project created a real-time GPS tracker using Arduino. Monitor salesmen, private drivers, and car safety using this technology. For expensive car owners, the author suggested tracking and monitoring their cars, including their past behavior. Alshamisi and Kepuska (2017) suggested Arduino MEGA-controlled GPS/GSM modules are used. Moving vehicle location will be updated continually. User sends SMS on registered phone for coordinates. Data will remain on the SD card. The website will help folks get there. The method is web-based and works regardless of location during an internet outage.

Jessica Saini et al. (2017) describes an embedded system that uses GSM and GPS technologies to determine vehicle position. A microcontroller and GPS/GSM module must be tightly coupled. Initially, the GPS receiver receives vehicle position data from satellites and stores it in the microcontroller's buffer. To monitor location, the registered mobile phone must make a request. Once authentication is finished, the location will be supplied through SMS. GSM is off and GPS enabled again. SMS contains vehicle latitude and longitude. The Android software can automatically plot the coordinates of the SMS value received. Amol Dhumal et al. (2014) proposes a GPS-based vehicle tracking system to assist organizations in locating car addresses and locations on mobile devices. The author says that the system will provide car position and user-vehicle distance. It will feature one Android phone, GPS, and GSM modems, and a vehicle CPU. When a car is engaged and moves its position is continually updated to a server through GPRS. Monitoring unit checks vehicle position from server database. Google maps will be used to plot location data from the database on the monitoring device. Monitoring units, such as Web or Android apps, provide user with the precise location of the proposed vehicle. Prashant Kokane et al. (2015) discusses a vehicle tracking and accident detection system. Vibration or piezoelectric sensors will detect traffic occurrences. This sensor reports accidents to the microcontroller. When a vehicle crashes, the GPS module determines its latitude and longitudinal position. The ambulance receives the vehicle's latitude and longitude via the GSM module. Automated message transmission alerts the central emergency dispatch server. This traffic accident detection system utilizes Raspberry Pi, GPS, GSM, and vibration sensor.

Tanaka et al. (2017) focuses on the use of LoRa-based bus tracking systems. Results are unrealistic since operational testing was limited to one route and vehicle. LoRa without an appropriate network protocol might cause Gateway message collisions in a multi-vehicle situation. Since LoRa has constraints in this area, the authors concentrated on compressing vehicle-to-Gateway communications. Ganesh et al. (2011) developed the cheapest automobile tracking antitheft system using GSM technology. The car terminal may send the owner's phone the nearest base station's location. This technique is cost-effective, but its location accuracy is poor and the latitude and longitude data is unclear. Lee et al. (2014) developed a smartphone-based automobile tracking system using GPS, GSM, and GPRS. It gives cars a microcontroller to track their whereabouts and speeds. Next, the owner may use Google maps to track their automobile in real time and predict its arrival time. The terminal in this system is costly and has limited functionality.

OVERVIEW OF EXISTING METHODOLOGY

Figure 1 show how an efficient fleet management system helps administrators to monitor and control operational costs, such as vehicle maintenance, and provide appropriate budgets for these charges. Ad-

Figure 1. Essential elements of a fleet management system

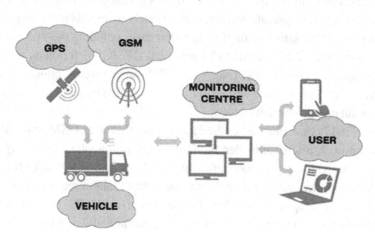

ditionally, it facilitates the ability to monitor and track instances of improper utilization of cars owned by the corporation by drivers.

As it stands, the existing system's Arduino will be used to oversee the whole of the procedure by interfacing with a GPS receiver and GSM module as shown in Figure 2. The vehicle's location can be tracked using a GPS module, and the user's location may be sent through text message with the help of a GSM module. First, it must determine the vehicle's coordinates before continuing to monitor its movement. The GPS module will maintain a constant connection with satellites in order to get location data. After that, Arduino UNO will get the coordinates via GPS. The data received through GPS will be parsed by an Arduino MEGA. When GSM module receives command from user, Arduino MEGA will work with GSM module to respond and transmit vehicle's coordinates to user. GPS will continually connect to satellite to route coordinates. Once the GPS module LED blinks, the position is locked. After establishing connection between mobile networks GSM module's LED will flash. After all LEDs glow, user may text GSM module "START". GSM will respond by Google Maps location and URL. The location will change every minute. Users may stop the system by sending "STOP" to GSM. Then the system stops sending phone messages. Figure 2 depicts overview of fleet management system using GPS and GSM (Jegan et al., 2023).

LIMITATIONS OF EXISTING METHOD

- The system only facilitates text transmission for activation, deactivation, and coordinate acquisition.
- There is a lack of provision for SOS emergency buttons.
- The system does not have voice transfer capabilities.
- Lack of power modes limits low-power mode operation for power saving. If needed, the system may be operated at high power. Two power sources are needed for the gadget. The GSM module needs a 9V DC power source with 2A current from an AC to DC converter. In contrast, the Arduino Mega and GPS module get power from a common USB connection.
- The GSM module operates poorly when powered by a battery, even if the voltage matches the adaptor.

Figure 2. Existing system block diagram

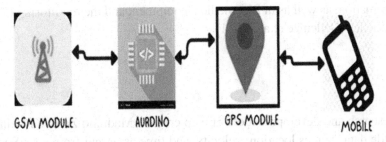

- To track information, separate GPRS or Wi-Fi systems must be activated.
- The maintenance of the Digital Cloud platform is crucial.

OBJECTIVE

This research entails a comprehensive examination of indoor location and navigation methods and technologies, with a particular focus on optimizing positioning accuracy. The successful implementation of this project has the potential to provide fleet owners with a convenient means of tracking their vehicles in the event of theft or loss. The study incorporates many metrics, including A9G, GPS, and other relevant factors. These metrics are derived from measurement attributes such as angle, distance, and signal intensity. The proper implementation of these strategies is crucial inside a system, since they contribute to optimal positioning. The implementation of these strategies in indoor navigation will ascertain the levels of accuracy, scalability, complexity, and performance shown by a wireless system.

INTRODUCTION ABOUT MADUINO ZERO

The Maduino board is a tiny Smart Internet of Things (IoT) device, measuring 55mm x 40mm in size. The Maduino platform combines a microcontroller that is compatible with the Arduino Integrated Development Environment (IDE) and incorporates an Internet of Things (IoT) module, all on a single board. In addition, Maduino offers a comprehensive power control system that includes an integrated battery charging mechanism and a circuit designed to safeguard the battery. The Maduino series incorporates the use of surface-mount device (SMD) microcontroller. An additional noteworthy aspect is the ability to use the Arduino IDE for programming the Maduino Smart IoT modules. The fundamental Arduino Integrated Development Environment (IDE) is enough for initiating the utilization of Maduino boards. The use of Maduino in the Arduino IDE does not need the inclusion of any additional board. The Maduino Series offers a range of GSM and GSM + GPS Integrated modules that may be used according to specific needs and preferences. The GSM and GPS Modules have been used in conjunction with the microcontroller to provide robust and adaptable modules that users can include into their products and facilitate the development of their IoT products. The Maduino with GSM and GPS boards provide the capability to facilitate the development of a personalized, condensed tracking system. The following is a comprehensive enumeration of Maduino boards. Bluetooth and Wi-Fi are often used for the transmission of data over short distances to mobile devices or laptops. These modules are widely used in the field of Home Automation, as well as in Bio-medical equipment and the monitoring of automobiles and Industrial characteristics (Ozdenizc et al., 2011).

SCOPE

The goal of this research and development effort is to create a Maduino Zero 4G Vehicle tracking system that can provide data such as location, velocity, and time as output from a GPS receiver. A viable approach involves the integration of a microcontroller, GPS module, and GSM module into a unified board. The aforementioned board is referred to as the Maduino Zero A9G GPRS/GPS Board. This module is an integration of the low-power A9G GSM+GPS+GPRS Module with the XIAO ESP32 C3 microcontroller. Users will also have the option of utilizing GSM technology to provide commands to the GPS receiver. There are two main sections to this project; the first involves connecting a XIAO ESP32 C3 microcontroller with Neo-6M GPS module. The XIAO ESP32 C3 Microcontroller is a crucial component since it regulates the GPS receiver's behavior. This microcontroller is controlled by C code. Integration of A9G GSM/GPRS module with the microcontroller is second portion. In this stage, we will learn GSM technology. It's necessary for the GSM technology to transport data from the GPS receiver to the computer. To ensure the data is delivered accurately and along the correct route, the proper GSM command set will be required. Once the system design phase is complete, development may begin. This is the stage whereby developers start writing code in the chosen programming language for the whole system. The MADUINO board is programmed using the Arduino IDE.

HARDWARE DESCRIPTIONS

GPS Overview

Early 1970s DoD developed GPS satellite-based navigation system. For the US military, GPS was created. It became a military-civilian dual-use system. The GPS enables global location and timing in any conditions. GPS is a passive, one-way-ranging technology with limitless users and security. User reception is limited to satellite signals (Ruppel & Gschwendtner, 2009). GPS uses satellites, ground-based control and monitoring stations, and receivers. GPS receivers receive satellite signals. GPS receivers then offer Latitude, Longitude, and Altitude time and place. As illustrated in Figure 3, GPS has "space," "control," and "user" segments.

A nominal constellation of 24 operational satellites provides one-way GPS satellite location and time messages. Monitoring and control centers worldwide govern satellite orbits via periodic manoeuvres and clock changes. GPS satellites, navigational data, and constellation health are tracked. User GPS receivers compute three-dimensional location and time using GPS satellite signals. A GPS module shown in Figure 4 receives data from GPS satellites to accurately establish geographical locations. The GPS module provides maps, including street maps, in textual or graphical formats and communicates step-by-step navigation instructions to vehicle operators.

Figure 3. GPS segments

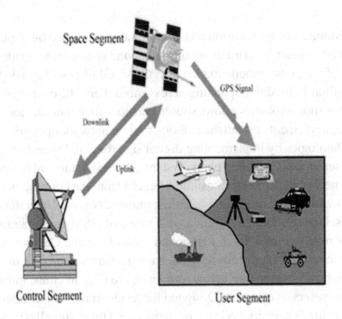

Figure 4. GPS module

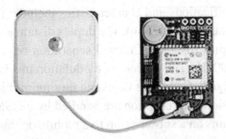

Advantages GPS Tracking Systems

- PS tracking technologies exhibit the potential to monitor numerous items or entities without constraints. Static or flexible may also be beneficial. The right action depends on the purpose. GPS may be used to track the well-being and productivity of employees under supervision.
- GPS tracking equipment and software provide real-time, on-screen information on the exact position of persons, goods, and vehicles globally. This feature is economical. In aviation, shipping, and commercial transportation, GPS tracking systems are widely used and advantageous.
- GPS tracking systems have worldwide applicability regardless of location. GPS technology allows global accessibility due to satellite coverage. Only a trustworthy tracking system and GPS receiver are needed for this capability.

GSM Overview

The Global System for Mobile Communications (GSM), initially developed by Group Special Mobile, is widely regarded as the predominant standard for mobile telephone systems worldwide. The GSM Association estimates that 80% of the global mobile market uses GSM. GSM is widely adopted by a substantial user base of over 1.5 billion individuals spanning across more than 212 countries and territories. The widespread availability of mobile phones allows subscribers to utilize their devices globally, facilitated by international roaming agreements established among mobile network operators. GSM distinguishes itself from previous technologies by implementing digital signaling and speech channels, thereby classifying it as a second generation (2G) mobile phone system. The GSM standard has proven advantageous for both consumers and network operators. Consumers benefit from the convenience of roaming and the ability to switch carriers without the need to replace their phones. Network operators, on the other hand, enjoy the flexibility of selecting equipment from a wide range of GSM equipment vendors.

GSM networks have macro, micro, Pico, femto, and umbrella cells. Cell coverage depends on the implementation environment. 15 Macro cells have the base station antenna on a mast or building above roof level. Micro cells, whose antennas are below roof level, are used in cities. Indoor use of pico cells, which cover a few dozen meters, is common. Designed for residential or small business use, femto cells connect to the service provider's network via broadband internet. Umbrella cells fill gaps between smaller cells and cover shadowed areas (Mehta et al., 2011). GSM uses Gaussian minimum-shift keying (GMSK), a continuous phase frequency shift keying. GMSK effectively reduces interference from neighboring channels by smoothing the signal onto the carrier with a Gaussian low-pass filter before feeding it to a frequency modulator. GSM specifications and characteristics include; the frequency band for GSM is 1,850 to 1,990 MHz (mobile station to base station). The duplex distance is 80 MHz. Uplink and downlink frequencies are separated by duplex distance. Channel separation refers to the separation of adjacent carrier frequencies. GSM uses 200 kHz. The process of modulation involves changing the characteristics of a carrier frequency to send a signal. GSM uses Gaussian minimum shift keying. GSM operates at 270 kbps over-the-air. Data transmission and reception are handled by a GSM module. Similar to a cellular phone, this specialized modem runs on a subscription to a mobile operator and is activated by inserting a SIM card. This is a display less mobile phone. Connecting a GSM modem to a computer enables the computer to access the mobile network through the GSM modem. While the primary function of GSM modems is to enable mobile internet access, many also support SMS and MMS messaging.

GSM/GPRS Module

A GSM/GPRS module refers to a device or integrated circuit that is primarily responsible for facilitating wireless communication between a computer and a GSM-GPRS system as shown in Figure 5. The Global System for Mobile Communication (GSM) is a widely used architectural framework utilized for mobile communication across several nations. Global Packet Radio Service (GPRS) is a technological advancement of the Global System for Mobile Communications (GSM) that facilitates enhanced data transfer speeds. A GSM/GPRS modem is a device that performs the modulation and demodulation of signals sent via a wireless network, hence enabling internet access. A GSM modem typically comprises a GSM module, in addition to supplementary components such as a SIM card, a signal modulation and demodulation device, and a power source. A system, such as a mobile phone, encompasses a comprehensive apparatus consisting of a GSM module, which may be incorporated inside the processor, a

Figure 5. GSM/GPRS module

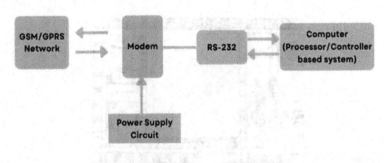

GSM modem, which may also be integrated, and other components such as a processor, screen, keypad, speakers, microphone, and so forth. The GSM/GPRS module is comprised of a GSM/GPRS modem that is integrated with a power supply circuit and several communication interfaces, such as RS-232 and USB, which provide connectivity with computers. The MODEM serves as the central component of these modules.

This GSM/GPRS Module facilitates the execution of the following tasks.

- The user has the ability to initiate, receive, or decline voice calls.
- The SIM card enables the functionality of sending, receiving, or deleting Short Message Service (SMS) messages.
- The user is requesting the ability to add, read, and search contacts stored in the Subscriber Identity Module (SIM) card.
- The transmission and reception of data to and from the GSM/GPRS Network may be facilitated by the use of the GPRS technology.

ATtention Commands, often known as AT Commands, may be used to do any of the aforementioned actions. AT Commands is an integral component of the Hayes Command Set, initially devised for modems. The GSM network also has AT-like instructions for its GSM modules. The responsibility of transmitting AT Commands to the GSM/GPRS Module lies with the processor or controller to which it is linked. The GSM Module carries out task-specific commands such as responding to incoming phone calls and sending SMS messages.

A9G IMPLEMENTATION

The AVP, ESP, and STM32 microcontroller families are often used in IoT devices. Connecting GSM or GPRS requires connecting several sub modules, resulting in a notable increase in the complexity and overall complexity of the device. The device is equipped with a small surface-mount device (SMD) casing that integrates both GSM and GPRS modules is called A9G GPS tracker module. The module has the capability to be manipulated via the use of AT instructions either from an external microcontroller or by means of its own SDK firmware. The A9G Arduino module has been designed by Ai Thinker, a business known for producing ESP microcontrollers. The A9G module is a GSM/GPRS+GPS module that works on the RDA8955 CPU, offering a wide range of functionalities and capabilities. The Low

Figure 6. Development board of A9G module

Bill of Materials (BOM) design of the module offers a cost-effective option for contemporary IoT-based applications.

Similar to other GPRS modules, this module also has the ability to make phone calls and send text messages. In addition to its many features, this module has a low-power mode that is distinguished by a minimum sleep current of 2mA. In addition to the aforementioned functionality, it also facilitates the process of upgrading firmware via the use of the UART connection. The device is equipped with a dual mode positioning feature, which provides compatibility for both the Global Positioning System (GPS) and the BeiDou Navigation Satellite System (BDS). The board is outfitted with a variety of features such as lithium battery charge management, microphone, speaker interface, USB communication interface, multiple user keys/LED, TF card slot, acceleration sensor, SPI interface, I2C2 interface, and ADC interface. These features enable the development of diverse functionalities using this board. The A9G development board has the capacity to be used development and verification of different peripheral prototypes. The prototypes encompass a remote monitoring intercom system that integrates GSM technology, as well as microphone and speaker functionalities. Additionally, the board can be employed for the creation of a remote monitoring camera system that incorporates GPRS, TF, and a camera extension board. The module has the capability to be powered by a 3.7V Lithium Ion Battery, since it necessitates a power supply within the range of 3.5~4.2V, with a usual supply voltage of 4.0V.

This module consumes 1.03mA to 1.14mA, depending on the application. IoT projects including smart-home apps, outside monitoring systems, long-distance monitoring solutions, and GPS trackers benefits from the A9G Module. Using the A9G development board, a slot machine system with We Chat payment capability may be implemented. The A9G module's development board is shown in Figure 6.

The A9G module has GSM functionalities that enable it to initiate and receive phone calls, as well as send SMS messages. When the user dials a phone number associated with the SIM card put into the A9G module, a constant "RING" message will be shown on the monitor. The A9G module supports a set of AT commands that enable users to initiate phone calls and send SMS messages.

The set of instructions for initiating a telephone call:

- ATA: Used to respond to an incoming telephone call. Upon the transmission of the instruction "+CIEV: "CALL", 1 CONNECT", the message is successfully received.
- ATD: s used for the purpose of initiating a telephone call to a certain number. The command is sent in the format "AT+number to be dialled". Upon delivering this command, a message is re-

ceived stating that "ATD+number dialled OK +CIEV: "CALL", 1 +CIEV: "SOUNDER", 1" is received.

- ATH: This command is used for the purpose of terminating a telephone conversation. The command, denoted as "ATH," is sent, and upon transmission, a message is received in the form of "+CIEV: "CALL", 0 OK".The ATH command is used for the purpose of terminating a telephone conversation.

- AT+SNFS=0: This command is used to activate the functionality of any earphones or headphones that are attached to the module. This instruction facilitates their ability to do a certain action.

- AT+SNFS=1: This command is used to activate the Loudspeaker selection feature.

- AT+CHUP: This command is used to terminate the ongoing call on the mobile terminal.

- The set of instructions for sending SMS:

- AT+CMGF=1: This command used for the purpose of selecting the format of the SMS message. Upon issuing his instruction, we obtain a confirmation of "OK." The purpose of this functionality is to enable the interpretation and composition of Short Message Service (SMS) messages in a textual format, rather than relying on hexadecimal representations.

- AT+CMGS: This command is used to transmit Short Message Service (SMS) to a specified mobile phone. The prescribed syntax for transmitting this instruction is as follows: "AT+CMGS=" followed by the cellphone number. Upon executing this command, the monitor will display the prompt "> You may proceed to input the message text and transmit the message by utilizing the designated key combination: -". The user's text is a simple word, "TEST." Upon the lapse of a few seconds, the modem will provide a response in the form of a message identifier (ID), denoted as "+CMGS: 62", therefore confirming the successful transmission of the message. The notification will be received on the mobile device in a brief period of time.

- AT+CMGL: This command is used to retrieve SMS texts from a designated storage location.

XIAO ESP32 C3 MICROCONTROLLER

The Seeded Studio XIAO ESP32C3 is endowed with a highly integrated ESP32-C3 semiconductor, which is based on a 32-bit RISC-V chip processor with a four-stage pipeline that operates at speeds of up to 160 MHz depicts in Figure 5. A highly integrated ESP32-C3 SoC is included on the board. The chip is equipped with a full 2.4 GHz Wi-Fi subsystem, allowing it to function in a variety of Wi-Fi configurations including Station mode, SoftAP mode, SoftAP & Station mode, and promiscuous mode. It operates in a low-power mode and is compatible with Bluetooth 5 and Bluetooth mesh. The chip has 400 KB of static random access memory (SRAM) and 4 MB of flash memory, expanding the possibilities for IoT control applications.

As a member of the Seeded Studio XIAO family, the board maintains the traditional thumb-sized form factor and refined productization of single-sided component mounting. In the meantime, it has been equipped with a battery charge processor and incorporated circuit to increase its carrying capacity. This board incorporates an external antenna to boost the signal strength for wireless applications. Eleven digital I/O can serve as PWM pins, while four analog I/O can serve as ADC pins. It supports the serial communication interfaces UART, IIC, and SPI. Using its compact, exquisite hardware design and potent onboard processor, along with programming by Arduino, it will provide wearable and portable devices and other applications with enhanced functionality. Its elegant form and WiFi+BT connectivity make

Figure 7. Seeded studio XIAO ESP32C3

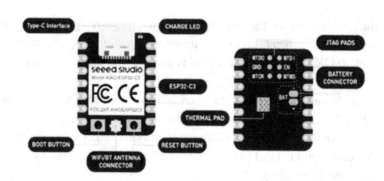

it ideal for a wide range of IoT control use cases and advanced portable software implementations. The board is equipped with an integrated circuit that incorporates a battery charging chip. This particular characteristic significantly improves the product's portability for applications that rely on battery power. The internal construction of the Seeded Studio XIAO ESP32C3 is seen in Figure 7.

The board is furthermore engineered for surface mounting on a printed circuit board (PCB) that has a suitable footprint. With 11 digital IOs, it can be used for PWM, and 4 analogue IOs, it can be used for ADC. Similar to other widely available boards, it supports UART, I2C, and SPI. With existing understanding of Arduino programming, the task of dealing with ESP32C3 becomes relatively easy. The inclusion of an external antenna with this board proves to be very advantageous for wireless applications that need enhanced signal strength.

ARDUINO COMPILER

The Arduino Integrated Development Environment (IDE) is a software application that is platform-independent and implemented in the Java programming language illustrates in Figure 8. It is based on the IDE used for the Processing programming language and the Wiring project. The purpose of this program is to provide an introduction to programming for individuals in artistic fields and other individuals who are not acquainted with software development. The software package encompasses a code editor that incorporates much functionality, including syntax highlighting, brace matching, and automatic indentation. Additionally, it has the capability to compile and upload programs to the board with a single action. In general, it is unnecessary to modify create files or execute programs inside a command-line interface While it is feasible to build on the command-line if necessary using some third-party tools. The Arduino Integrated Development Environment (IDE) is equipped with a C/C++ library known as "Wiring," derived from the project with the same title. This library facilitates certain commonplace input/output tasks, making them more convenient to execute. The programming language used for Arduino programs is C/C++.

Figure 8. Programmed IDE

DESIGN AND CONSTRUCTION PROCEDURE

This chapter focuses on providing a comprehensive and organized overview of the design and construction techniques used in the development of the Maduino Zero 4G Vehicle monitoring system. The A9G board incorporates integrated GSM, GPS, and battery charging circuitry. To control this board, the smallest Seed Studio's Xiao C3 is utilized. The chip also features built-in GPS, allowing for easy location determination, and an on-board microphone for crystal-clear communications. The block diagram of the proposed system is shown in Figure 9.

The NEO-6m GPS module is equipped with a miniature antenna, hence impeding the system's capacity to get precise GPS coordinates. The only information provided to the user consists of geographic coordinates, namely latitude and longitude. A GSM modem is used to transmit the geographical coordinates (Latitude and Longitude) of the vehicle from a distant location to the monitoring center. The GPS modem will consistently provide data, namely the latitude and longitude coordinates, which indicate the precise location of the car. The vehicle tracking system utilizes GPS input and transmits it through the GSM module to a designated mobile device or laptop utilizing mobile communication. The transmitting component of the system is responsible for executing the tracking capability. The fleet is monitored using GPS technology, which allows for the tracking of its current location (Vasireddy et al., 2016). This information is sent to a monitoring center, where different software applications are used to visually represent the vehicle's position on a map. Upon request, the system must provide the precise position of the monitored item at the moment. The monitored object's location should be recorded by the system at regular intervals. Emergency alert features are implemented that activate when a SOS button is pushed. The SOS number will initiate an emergency call and provide a link to a Google Maps page in the text message.

Figure.10 depicts the integration of GPS, GPRS and GSM functionalities with the microcontroller inside a single chip. One side of the chip is visibly connected to a GPS module, while the other side is equipped with a GPRS module.

Figure 9. Block diagram of proposed system

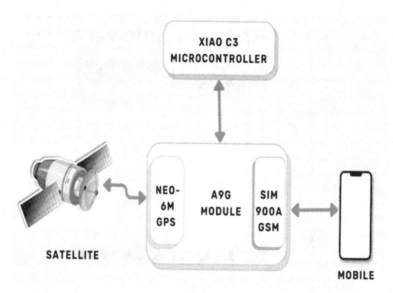

Xiao C3 board is directly mounted on the custom-designed PCB, to bypass the need of eliminating the need for any additional attachments such as voltage regulator entirely. The device has a designated port for connecting external speakers. Additionally, the board is equipped with a dedicated connector for battery attachment and a micro USB port, enabling the board and battery to be powered up. The board may also be equipped with two push buttons, namely a reset button and a power button. The board is equipped with two light-emitting diodes (LEDs), one indicating the presence of GPS and the other indicating the presence of GSM/ GPRS. Additionally, there is a SIM card slot located on the rear of the chip. Both the board and the battery are equipped with rechargeable capabilities, using both a Type-C connector and a USB connection, as seen Figure 11.

Once all the components were successfully mounted, the completed system had a tidy and compact appearance. The wifi.h library is included into the esp32 board. Subsequently, two pins, namely D3 and

Figure 10. Integration of GPS, GPRS, and GSM functionalities with the microcontroller

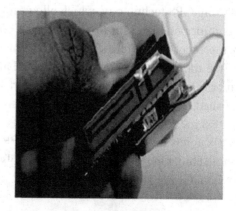

Figure 11. Type-C connector and a USB connection for the proposed system

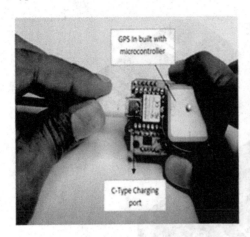
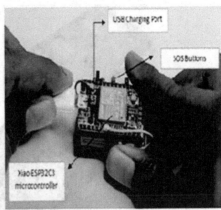

D2, are defined. D3 is assigned as the input pin for a Push Button, serving as an SOS button. On the other hand, D2 is designated as the output pin responsible for regulating the sleep mode of the A9G board. There exists a significant variable referred to be the SOS number, which necessitates the inclusion of the country code. This number is used to transmit latitude and longitude data, as well as to initiate a phone call. A delay of five seconds before sending the coordinates of the device's location in case the user accidentally pressed the SOS button implemented. Only after the user has held the button down for longer than that amount of time will the SOS activity be triggered, at which point the device will send its location data and initiate a phone call. Only when the button is held down for longer than five seconds will the SOS activity be triggered, at which point the coordinates will be sent along with the phone call.

Given the absence of a need for radio modules, Wi-Fi and Bluetooth are disabled to save power. Subsequently, a delay interval of 20 seconds was included, since the A9G device necessitates a brief period to adequately setup itself and establish a connection with the cellular network. The sleep pin may be deactivated by setting it to a low state, and activated by setting it to a high state. It is important to disable the sleep mode before issuing AT instructions. After successfully uploading the code, the next step involves inserting the SIM card into the A9G board which has the flexibility that offers 4G bands or 4G connection. The software is inserted into the SIM card. In order to gather data and ensure prompt retrieval of latitude and longitude coordinates upon pressing the SOS button, the GPS was activated and left enabled. Subsequently, we have activated the low power mode of the GPS, which results in reduced power consumption. This mode is implemented to save power, even while the GPS functionality is active. Efforts will be made to save power in the GPS system, cellular network, and the Xiao C3 board. The loop section is executed when a call is made on the A9G board, resulting in the automated reception of the call.

Tracking position on printed maps may be a time-consuming task. Currently, there is a wide range of websites accessible on the internet that provides online mapping services. Google Maps is a prominent and very advantageous online platform. One may use any of the aforementioned websites to monitor or ascertain the fleet. The location may be determined by using the Longitude and Latitude information obtained from the SMS. Upon receiving GPS data, a Google Maps hyperlink is generated that incorporates latitude and longitude information within the link. This hyperlink enables recipients to conveniently access the location by simply clicking on it, as it automatically directs them to the corresponding position

Figure 12. Getting location via Google Maps

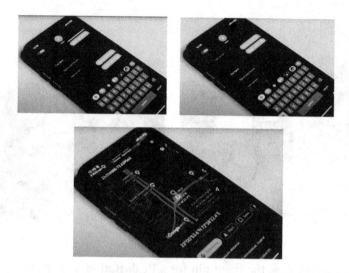

on Google Maps (Karkera et al., 2018). By using these maps, users may get precise geographical coordinates, detailed navigational instructions, and estimated travel durations from their present or preferred location to designated destinations. This approach proves more advantageous than solely transmitting latitude and longitude values, as it also includes a message indicating "I'm here" alongside the Google Maps link as shown in Figure 12.

Subsequently, individual can be contacted instantly, tasked not only with establishing their location, but also with attentively perceiving the ongoing dialogues inside the specific context. This enables us to respond appropriately and effectively address the circumstances at present. After integrating the A9G board onto the PCB, the power button will be pressed. Powering up the board starts by holding down the power button. To send an emergency distress signal, hold the SOS button for more than 5 seconds. After receiving a call on the given number, the user may actively participate in the A9G board debates. The system can transport data utilizing energy-efficient, low-radioactivity technologies in high-moisture settings. The illustration shows how our recommended solution seamlessly integrates an A9G board with a microcontroller for a more compact design.

The board is equipped with GPS and GSM modules, which are visibly shown by a flashing blue light. This light signifies the reception of signals from satellites. The device is equipped with two emergencies SOS buttons that, when pressed for an extended period, enable the user to initiate phone calls and send text a message including GPS coordinates (Hlaing et al., 2019). The system in question is a comprehensive wireless solution that operates using wireless data transmission technology. The device operates based on the fundamental idea of a Tesla wireless charging coil. The integrated circuit has an inherent memory component that may be used for data storage purposes. The update process is limited to a maximum of 32 kilobytes of flash memory. Figure 13 shows overall structure for proposed system.

The tracking system is heavily influenced by the importance of accuracy. Figure 14 depicts the process of measuring the accuracy of the prototype. The blue pin represents the location marker that is placed on a map, indicating a specific longitude and latitude coordinate. This coordinate may be tracked using a GPS module. Table 1 presents the results of the accuracy test for reading.

Figure 13. Proposed structure

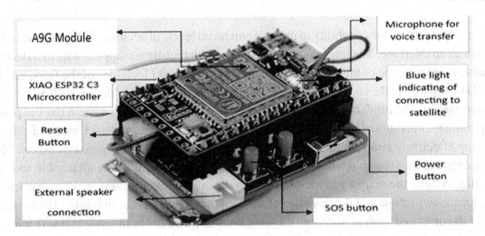

Table 1. Experimental coordinates of GPS coordinates

From	To	Range	Accurate latitude	Accurate longitude
Egmore	Egmore station	100 m	12.013386	80.23254
Sholinganallur	Thorraipakkam	5 km	12.867250	80.221863
Sholinganallur	Koyambedu	10 km	12.941604	80.236214
Sholinganallur	Madhavaram	15 km	12.906361	80.120256

Figure 14. Real-time GPS coordinates

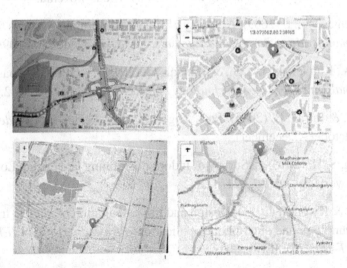

CONCLUSION

The solution presented has the capability to provide enhanced levels of accurate GPS data due to its use of a 4G network as an interface. The integration of GPS data with real-time mapping data in a fleet tracking system enables the provision of up-to-date information on the present locations of the fleet. The Modern Maduino 4G system offers the capability to access and analyze travel history data as well as provide voice calling functionality. The implementation of our suggested fleet monitoring system has the potential to effectively mitigate operational expenses and enhance the utilization of assets. Enhanced driver safety measures may effectively contribute to safeguarding an organization's most precious assets. The GSM module is integrated into the chip, allowing for a compact design that is easily attachable and portable. In the existing system, the primary mode of communication is limited to text transfer, namely via text messages, for the activation and deactivation of the system, as well as for obtaining GPS coordinates. However, in the proposed system, there are additional capabilities for both voice and text transmission, which are integrated inside the device itself and designed with high quality.

REFERENCES

Ganesh, G. S. P., Balaji, B., & Varadhan, T. A. S. (2011). Anti-theft tracking system for automobiles (AutoGSM) *Proceedings of the IEEE International Conference on Anti-Counterfeiting, Security and Identification (ASID '11)*. IEEE.

Giaglis, G. M., Pateli, A., Fouskas, K., Kourouthanassis, P., & Tsamakos, A. (2002). On the Potential Use of Mobile Positioning Technologies in Indoor Environments. In *Proceedings of the 15th Bled Electronic Commerce ConferencevReality: Constructing the Economy*, Bled, Slovenia.

Gu, Y., Lo, A., & Niemegeers, I. (2009). A Survey of Indoor Positioning Systems for Wireless Personal Networks. *IEEE Communications Surveys and Tutorials, 11*(1), 13–32. doi:10.1109/SURV.2009.090103

Karkera, T., Dubey, A., Kamalnakhawa, S., & Mangale, S. (2018). —GPS-GSM based Vehicle Tracking System,‖. *International Journal of New Technology and Research, 4*(3), 140–142.

Kokane, P., Kiran, S., & Imran, B. (2015). Prof. Yogesh Thorat. *Review on Accident Alert and Vehicle Tracking System, 3*(October - December).

Koyuncu, H., & Yang, S. H. (2010). A Survey of Indoor Positioning and Object LocatingvSystems. Int. J. Computer. Sci. Network. *Secur., 10*, 121–128.

Lee, S., Tewolde, G., & Kwon, J. (2014). Design and implementation of vehicle tracking system using GPS/GSM/GPRS technology and smartphone application. *Proceedings of the IEEE World Forum on Internet of Things (WF-IoT '14)*. IEEE.

Mautz, R. (2009). Overview of Current Indoor Positioning Systems. *Geodesy and Cartography (Vilnius), 35*(1), 18–22. doi:10.3846/1392-1541.2009.35.18-22

Mo Khin, J. M., & Nyein Oo, D. N. (2018). —Real-Time Vehicle Tracking System Using Arduino, GPS, GSM and Web-Based Technologies,‖. *International Journal of Science and Engineering Applications, 7*(11), 433–436. doi:10.7753/IJSEA0711.1006

Ozdenizci, B., Ok, K., Coskun, V., & Aydin, M. N. (2011). Development of an IndoorvNavigation System Using A9G Technology. In *Proceedings of the 4th International Conference on Information and Computing*, Phuket Island, Thailand.

People, V. I. (2011). *12th International Conference on Computer Systems and Technologies*, Vienna, Austria.

Ruppel, P., & Gschwendtner, F. (2011). *Spontaneous, and Privacy-Friendly Mobile Indoor Routing and Navigation*. Second Workshop on Services, Platforms, Innovations and Research for New Infrastructures in Telecommunications, Lübeck, Germany.

Tanaka, M. S., Miyanishi, Y., Toyota, M., Murakami, T., Hirazakura, R., & Itou, T. (2017). A study of bus location system using LoRa: Bus location system for community bus Notty. IEEE 6th Global Conference on Consumer Electronics (GCCE), (pp. 1-4). IEEE. 10.1109/GCCE.2017.8229279

Vasireddy, S., Ravipati, V., Ravi, T., & Jegan, G. (2016). Wireless sensor based gps mobile application for blind people navigation. *Journal of Engineering and Applied Sciences (Asian Research Publishing Network)*, *11*(13), 8374–8379.

Chapter 6
AI-Driven Energy Forecasting, Optimization, and Demand Side Management for Consumer Engagement

Chalumuru Suresh

Department of Computer Science and Engineering, VNR VJIET, India

V. Nyemeesha

Department of Computer Science and Engineering, VNR VJIET, India

R. Prasath

Department of Computer Science and Engineering, KCG College of Technology, India

K. Lokeshwaran

Department of Computer Science and Engineering (Data Science),Madanapalle Institute of Technology and Science, India

K. Ramachandra Raju

iD https://orcid.org/0000-0002-1272-7878

Department of Mechanical Engineering, Bannari Amman Institute of Technology, India

Sampath Boopathi

iD https://orcid.org/0000-0002-2065-6539

Mechanical Engineering, Muthayammal Engineering College, Namakkal, India

ABSTRACT

This chapter explores the role of artificial intelligence (AI) in the energy sector, focusing on energy forecasting, optimization, and demand management. It highlights the importance of AI technologies in utilities, grid operators, and consumers. AI-driven models accurately predict energy consumption patterns and demand, and how machine learning algorithms, data analytics, and IoT devices can improve forecasting precision. AI also optimizes energy production and distribution processes, reducing costs, enhancing reliability, and promoting sustainability. It also emphasizes its role in demand side management, focusing on consumer engagement strategies and AI-driven demand response programs.

DOI: 10.4018/979-8-3693-3735-6.ch006

INTRODUCTION

Artificial intelligence (AI) is revolutionizing the energy sector by transforming the generation, distribution, and consumption of energy. AI-driven energy forecasting, optimization, and demand-side management are crucial for creating a sustainable and responsive energy ecosystem. This shift is driven by the need for utilities and energy providers to meet growing demand, integrate renewable sources, and engage consumers in energy efficiency. AI-powered energy forecasting uses sophisticated algorithms and machine learning models to analyze vast datasets, including historical consumption patterns, weather conditions, and demographic trends. This enables accurate predictions of energy demand, enabling grid operators to anticipate fluctuations and optimize resource allocation (Mohammad & Mahjabeen, 2023b).

AI in energy optimization involves dynamic resource management, adjusting energy generation, distribution, and consumption based on real-time data. This ensures optimal resource utilization, reduces waste and environmental impact, and maximizes efficiency. AI also enhances demand side management by providing personalized insights, recommendations, and incentives to consumers. Gamification elements and dynamic pricing models foster a collaborative environment where consumers actively contribute to load balancing and cost savings. This AI-driven energy transformation addresses immediate challenges and lays the foundation for a sustainable, consumer-centric future. The interplay of AI, energy forecasting, optimization, and demand-side management reveals the potential for a greener, more resilient, and interconnected energy ecosystem (Kim et al., 2023).

AI-driven energy systems are transforming the energy industry by improving operational efficiency and adaptability. These systems learn from historical data, improving accuracy and reliability. They can predict energy consumption patterns with precision, enabling utilities to address grid congestion, reduce downtime, and plan for future capacity needs. AI's optimization extends to renewable energy integration, maximizing the potential of clean energy sources like solar and wind. By predicting renewable energy generation patterns, AI ensures seamless grid integration, reducing reliance on non-renewable sources, and promoting a sustainable energy landscape (Mohammad & Mahjabeen, 2023a).

Consumer engagement is crucial in the transformation of energy usage, as AI provides personalized insights and actionable information. Through interactive apps and real-time feedback, consumers gain a deeper understanding of their impact on the grid. This engagement fosters shared responsibility and encourages energy-conscious behaviors. However, challenges arise, such as security and privacy concerns, regulatory compliance, and balancing technological advancement with ethical considerations. Despite these challenges, AI integration remains a significant step forward in promoting sustainable energy use (Yang et al., 2021).

An era of efficiency, sustainability, and consumer involvement in the energy sector is heralded by the use of AI in energy forecast, optimization, and demand-side management. This journey involves technological innovation and collective effort to reshape energy generation, consumption, and thinking, promising a resilient, adaptive, and consumer-centric energy future.

ARTIFICIAL INTELLIGENCE IN ENERGY SECTOR

Artificial Intelligence (AI) is revolutionizing the energy sector by improving efficiency, sustainability, grid management, and energy consumption. Artificial intelligence (AI) developments in machine learning and data analytics facilitate improved decision-making and operational efficiency, driven by global

Figure 1. AI Applied in the various energy sectors

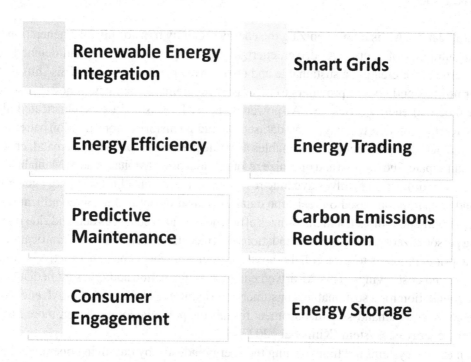

demand and an emphasis on clean and renewable energy sources (Ahmad et al., 2021; Yang et al., 2021). AI is being applied across various energy sectors to generate, distribute, and consume energy (Figure 1).

- **Smart Grids**: With AI, utilities can analyze real-time data from sensors installed throughout the grid to optimize energy distribution. This means adjusting energy flow based on demand fluctuations, weather conditions, and other factors, ensuring that electricity is delivered efficiently and reliably. Predictive analytics help utilities anticipate potential faults or failures in the grid infrastructure, allowing for proactive maintenance to prevent costly outages and improve overall grid reliability. Machine learning algorithms can identify anomalies in energy consumption patterns, potentially indicating issues such as theft or equipment malfunction, enabling utilities to take corrective action promptly.
- **Energy Forecasting**: AI algorithms process large volumes of data, including historical consumption patterns, weather forecasts, economic indicators, and more, to generate accurate predictions of future energy demand and supply. These forecasts enable utilities to optimize their energy generation and distribution strategies, ensuring that sufficient energy is available to meet demand while minimizing excess capacity and associated costs. By better understanding future energy trends, utilities can also make more informed decisions regarding investment in new infrastructure and renewable energy projects.
- **Renewable Energy Optimization**: AI plays a crucial role in maximizing the efficiency and reliability of renewable energy sources like solar and wind power. Predictive models leverage weather data to forecast renewable energy production, allowing utilities to adjust energy generation and distribution accordingly to match demand. AI algorithms can also optimize the placement and

operation of renewable energy assets, such as solar panels and wind turbines, to maximize energy output and minimize intermittency issues.

- **Energy Efficiency**: AI-powered systems monitor energy usage in real-time and identify opportunities for optimization, such as adjusting lighting and HVAC systems based on occupancy patterns and environmental conditions. Advanced analytics provide insights into energy consumption patterns, allowing businesses and consumers to make informed decisions about energy-saving measures and investments in energy-efficient technologies. Smart devices, such as thermostats and appliances, can automatically adjust their settings to minimize energy usage without sacrificing comfort or functionality.

- **Predictive Maintenance**: AI analyzes data from sensors and historical maintenance records to identify patterns indicative of impending equipment failures. By predicting maintenance needs in advance, utilities can schedule repairs during planned downtime, minimizing disruption to operations and reducing the risk of costly unplanned outages. Predictive maintenance also extends the lifespan of critical assets by identifying and addressing issues before they escalate into major problems.

- **Energy Trading and Market Optimization**: Applications driven by AI examine consumer data, including use trends and preferences related to energy in order to offer individualized pricing plans and recommendations for energy-saving strategies. Real-time tracking of energy consumption, goal-setting, and progress monitoring are all made possible for consumers via interactive tools and interfaces. Utilities may promote more collaboration with customers and encourage energy conservation and efficiency by providing them with information and tools to better control their energy use.

- **Grid Security and Resilience**: AI enhances grid security by continuously monitoring network traffic for signs of cyber threats, such as intrusions, malware, and phishing attacks. Machine learning algorithms can detect unusual patterns or anomalies in network behavior, allowing utilities to identify and respond to potential threats in real-time. By proactively addressing cybersecurity risks, utilities can safeguard critical energy infrastructure and ensure the reliability and resilience of the grid.

- **Customer Engagement and Personalization**: Applications driven by artificial intelligence (AI) examine consumer data, including use trends and preferences, to offer individualized pricing plans and advice for energy-saving strategies. Customers may establish energy-saving objectives, monitor their consumption in real time, and get feedback on their success via interactive tools and interfaces. Utilities may encourage a more cooperative relationship with customers and promote greater energy conservation and efficiency by providing customers with the knowledge and resources they need to better control their energy use.

AI is revolutionizing the energy sector by optimizing production, distribution, grid reliability, and enhancing efficiency. As AI technologies evolve, they will play a crucial role in shaping the future of energy. Despite challenges like data security and regulatory compliance, AI integration offers cost savings, environmental impact reduction, and grid resilience, making it a promising field for energy companies, researchers, and policymakers.

Benefits and Challenges

- **Efficiency Improvement:** AI can optimise energy production, distribution, and consumption, resulting in lower energy waste and more operational efficiency. Thus, this decreases operating expenses for both customers and energy businesses (Hussain et al., 2023; Ingle, Swathi, et al., 2023; Mohanty et al., 2023; Ravisankar et al., 2023).
- **Cost Reduction:** Energy infrastructure may have a longer lifespan, less downtime, and lower maintenance costs with predictive maintenance and better asset management made possible by artificial intelligence. Algorithms for energy trading can optimize trading and procurement tactics, which might result in lower energy costs for end users.
- **Sustainability:** AI helps integrate and manage renewable energy sources more effectively, contributing to reduced carbon emissions. It also aids in making energy systems more environmentally friendly by optimizing energy usage and reducing waste.
- **Grid Reliability:** Artificial intelligence (AI)-powered smart grids can identify and address outages and imbalances in real time, improving grid stability and lessening the effects of blackouts.
- **Consumer Empowerment:** Artificial intelligence (AI) apps give customers information on how much energy they use, empowering them to make decisions that will lower their usage, save money, and have a smaller environmental impact.
- **Scalability:** AI is a useful instrument for managing the growing complexity of contemporary energy systems because it can scale and adapt to changing energy demands and sources.
- **Advanced Forecasting:** AI-driven predictive analytics can enhance the accuracy of weather and demand forecasting, enabling better planning and resource allocation.

Challenges

- **Data Quality and Quantity:** AI depends heavily on data, and it can be difficult to get both high-quality and large-volume data. Applications using AI may be hampered by inconsistent, lacking, or out-of-date data.
- **Data Security and Privacy:** The energy sector deals with sensitive data, and maintaining data security and privacy is paramount. Ensuring that AI systems are protected from cyber threats is a critical challenge.
- **Interoperability:** Many energy systems and devices come from different manufacturers and may not easily work together. Integrating AI across these disparate systems can be complex and costly.
- **Regulatory and Policy Issues:** The energy sector is highly regulated, and integrating AI may require navigating complex regulatory environments. Policies and standards must evolve to accommodate AI technologies.
- **Technical Complexity:** Implementing AI systems in the energy sector requires specialized technical expertise, which may be in short supply. Finding and training skilled professionals can be challenging.
- **Initial Investment:** While AI can lead to cost savings over time, there is an upfront cost associated with the implementation of AI technologies, which can be a barrier for some organizations.
- **Ethical and Bias Concerns:** Biases from the data used to train AI may be inherited by them. Particularly in delicate sectors like energy pricing and distribution, it is imperative to ensure fairness and transparency in AI decision-making.

- **Resistance to Change:** The energy industry has traditionally been conservative and slow to adopt new technologies. Overcoming resistance to change and gaining buy-in from stakeholders can be challenging.

In order to meet the growing demand for clean, dependable, and affordable energy solutions, issues like data security, regulation, technical complexity, and societal concerns must be addressed. AI has the potential to improve efficiency, lower costs, promote sustainability, and improve grid reliability (Li & Xu, 2022).

ENERGY FORECASTING WITH AI

Artificial intelligence (AI) is revolutionizing energy forecasting by predicting future energy consumption, generation, and prices. AI models are increasingly used in fields like demand forecasting, supply forecasting, price forecasting, and grid management. They can analyze historical data to identify energy consumption patterns, incorporate weather forecasts for accuracy, and forecast energy production. They can also predict fossil fuel production based on factors like fuel prices, operational constraints, and maintenance schedules (Boopathi, Kumar, et al., 2023; Domakonda et al., 2022; Kumara et al., 2023).

AI models also provide real-time load forecasts to grid operators, enabling them to make immediate adjustments to meet demand efficiently. They can also predict generation from distributed energy resources, such as rooftop solar panels, to better integrate them into the grid. AI models also provide probabilistic forecasts for risk management and decision-making under uncertainty. AI is a powerful tool for energy forecasting, offering improved accuracy due to its ability to handle large datasets and complex relationships. It enables real-time adjustments, optimizing energy companies' operations, reducing costs, enhancing efficiency, and supporting renewable integration by predicting renewable energy generation (Agrawal et al., 2024; Satav, Lamani, G, et al., 2024; Syamala et al., 2023; Vanitha et al., 2023; Vennila et al., 2022).

Energy Consumption Forecasting

Energy consumption forecasting is essential for utilities, grid operators, and energy providers to plan and optimize their resources, anticipate demand fluctuations, and ensure a reliable and efficient energy supply (Kavitha et al., 2023; Satav, Lamani, G, et al., 2024; Venkateswaran, Vidhya, et al., 2023a).

Energy Consumption Forecasting-Methods

Figure 2 depicts the Energy Consumption Forecasting-Methods and explained below.

- **Historical Data Analysis:** Analyzing past energy consumption patterns to identify trends, seasonality, and any long-term changes in demand.
- **Time-Series Analysis:** Employing time-series forecasting models used to predict future energy consumption based on historical data.

Figure 2. Energy consumption forecasting-methods

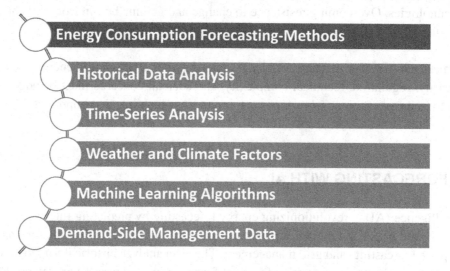

- **Machine Learning Algorithms:** Identifying complex relationships and variables influencing energy usage by using computer learning techniques including decision trees, neural networks, and regression models.
- **Weather and Climate Factors:** Incorporating weather and climate data, as these factors significantly impact energy usage (e.g., heating in cold weather, cooling in hot weather).
- **Demand-Side Management Data:** Integrating data from demand-side management programs, smart meters, and IoT devices to enhance the accuracy of predictions.

Benefits: The project enhanced grid reliability by anticipating peak demand periods, improved energy efficiency through efficient resource allocation, and provided informed decision-making for infrastructure planning and upgrades.

Energy Generation Forecasting

Energy generation forecasting is crucial for optimizing the deployment of energy resources, ensuring a balance between supply and demand, and integrating renewable energy sources effectively (Kumar B et al., 2024; Rahamathunnisa et al., 2023; Satav, Lamani, K. G., et al., 2024; Venkateswaran, Vidhya, et al., 2023b).

- Using specialized models for solar and wind forecasting based on weather conditions, geographic location, and historical renewable energy generation data.
- Employing machine learning algorithms to predict variable energy generation from sources like solar and wind, considering their inherent variability.
- Developing models that consider factors such as reservoir levels, precipitation forecasts, and historical hydroelectric generation patterns.

- Utilizing data on reactor status, fuel availability, and maintenance schedules for nuclear power forecasting. Considering fuel availability, plant efficiency, and maintenance schedules for thermal power forecasting.

Efficient grid management, minimal disruptions in renewable energy integration, and cost savings through optimized resource planning are benefits of renewable energy. Energy consumption and generation forecasting are crucial for creating a balanced, sustainable energy ecosystem.

Grid Demand Prediction

Grid demand prediction involves forecasting the amount of electricity that will be required to meet the needs of consumers at different times. Accurate predictions are crucial for utilities to optimize energy generation, distribution, and demand response programs (Hema et al., 2023; Kavitha et al., 2023; Reddy et al., 2023).

Data Sources and Integration

- Historical consumption data, weather data, calendar and time factors, and economic indicators are essential for predicting energy demand. Historical data helps identify patterns, trends, and seasonality, while weather data enhances predictions in heating and cooling needs. Calendar and time factors consider calendar events, holidays, and time-of-day patterns. Economic indicators, such as GDP and industrial production indices, provide a broader context for demand prediction.
- The integration of demographic data, smart meters, IoT devices, energy efficiency programs, industrial and commercial sector data, and grid operations data can help predict energy consumption patterns. Demographic data provides a sociological dimension to predictions, while real-time data from smart meters and IoT devices offers granular insights into consumer behavior. Energy efficiency programs help adjust predictions by understanding their impact on demand patterns. The integration of these data improves the responsiveness of demand predictions to current grid conditions.

Integration Challenges and Considerations

This emphasizes the importance of maintaining data quality and consistency across sources, implementing efficient real-time data processing mechanisms, ensuring interoperability between data sources and systems, and addressing privacy and security concerns, particularly when handling sensitive consumer data. Effective integration of diverse data sources allows for a holistic approach to grid demand prediction, enabling utilities to make informed decisions and optimize energy resources more effectively.

OPTIMIZING ENERGY PROCESSES

Reducing reliance on conventional fossil fuels and fostering a cleaner, greener energy environment are two major benefits of including renewable energy sources like solar, wind, and hydropower into the energy mix. The efficiency of energy generation is increased by cutting-edge technology such as energy storage systems and smart grids. While energy storage devices like batteries store extra energy during periods of low demand and release it during periods of peak demand, smart grids are able to better regulate the distribution of electricity (Boopathi, Sureshkumar, et al., 2023; Ingle, Swathi, et al., 2023; J. R. Nishanth et al., 2023; Sampath et al., 2022). Predictive analytics and machine learning applications enable accurate forecasting of energy demand patterns, enabling proactive adjustments in production. Predictive maintenance using sensors and data analytics helps identify potential issues in power plants, minimizing downtime and maximizing operational efficiency (Boopathi, Pandey, et al., 2023; Gowri et al., 2023; Kavitha et al., 2023; J. Nishanth et al., 2023; Sampath et al., 2023; Saravanan et al., 2022; Venkateswaran, Kumar, et al., 2023).

Communities can produce their own energy locally using decentralized energy generating models like small-scale solar panels and wind turbines, which lower transmission losses and improve grid resilience. This approach promotes energy independence and sustainability. Optimizing energy generation involves incorporating renewable sources, advanced technologies, predictive analytics, and decentralized models to create a more resilient, sustainable, and efficient energy ecosystem.

Energy Generation Optimization

AI-driven energy generation optimization is revolutionizing power production, distribution, and consumption. This makes it possible for utilities to efficiently satisfy customer requirements by adjusting output levels in real-time. AI-driven algorithms also continuously analyze factors influencing energy generation, optimizing renewable resources like solar and wind by aligning production with peak demand periods, ensuring a sustainable energy mix and maximizing clean, cost-effective sources (B et al., 2024; Boopathi, 2023; Venkateswaran, Vidhya, et al., 2023a).

Demand side management is a crucial aspect of energy generation optimization, using AI algorithms to analyze consumer behavior and usage patterns. This helps balance the grid, reduce system stress, and minimize the need for additional power generation capacity. Because AI technologies offer individualized insights into energy consumption trends and facilitate the making of educated use decisions, consumer interaction plays a pivotal role in this optimization process. This engagement promotes energy efficiency and encourages a more conscious and responsible approach to energy consumption. The integration of AI-driven energy forecasting, optimization, and demand side management improves energy generation efficiency and involves consumers, fostering a sustainable, resilient, and consumer-centric energy ecosystem.

Grid Operations and Management

This document offers a comprehensive procedure for implementing effective grid operations and management, ensuring the reliability, stability, and efficiency of the electrical grid (Agrawal et al., 2024; Ingle, Senthil, et al., 2023; Kavitha et al., 2023; Venkateswaran, Vidhya, et al., 2023a).

a) **Cybersecurity Measures:** Prioritize cybersecurity measures to protect the grid from potential cyber threats. Implement firewalls, encryption, and intrusion detection systems to safeguard critical infrastructure.

b) **Grid Modeling and Simulation:** Develop accurate models of the grid to simulate various operating conditions and scenarios. This helps in optimizing grid performance and identifying potential issues before they occur.

c) **Load Forecasting:** Implement advanced load forecasting techniques using AI and machine learning to predict energy demand patterns accurately. This aids in proactive grid management and resource allocation.

d) **Renewable Integration:** create strategies for efficient use of renewable energy sources, utilizing forecasting technologies to optimize their utilization.

e) **Emergency Response Planning:** Develop comprehensive emergency response plans to address potential disruptions or disasters. This includes contingency plans for equipment failures, natural disasters, and other unforeseen events.

f) **Training and Skill Development:** Train personnel on the use of new technologies and procedures. Continuous skill development is crucial to adapt to evolving grid management practices.

g) **Regulatory Compliance:** Ensure compliance with local and national regulations governing grid operations. Stay informed about regulatory changes and adjust procedures accordingly.

h) **Collaboration and Communication:** Foster collaboration with other utilities, stakeholders, and government agencies. Open communication channels help in coordinating efforts and sharing critical information.

Implement continuous monitoring of grid performance and operations. Regularly review and update procedures to incorporate lessons learned and emerging technologies. The following steps can help utilities establish a robust framework for grid operations and management, thereby improving the overall resilience and efficiency of the electrical grid.

Load Optimization

Load optimization is crucial for achieving sustainability and cost reduction in energy management (Satav, Lamani, K. G., et al., 2024; Venkateswaran, Vidhya, et al., 2023b).

a) To evaluate the effectiveness and utilization of your facility and pinpoint areas with high energy consumption and possible inefficiencies, a thorough energy audit is necessary.

b) Energy management and building automation systems are examples of smart building technologies that optimize energy use by monitoring and managing building systems based on occupancy patterns and real-time data.

c) Use demand response systems, energy storage devices, and scheduling energy-intensive operations for off-peak hours as ways to control and minimize peak energy loads.

d) In order to lessen dependency on non-renewable energy sources and cut long-term energy expenses, renewable energy integration entails integrating renewable sources, such as solar panels or wind turbines.

e) Use load shedding techniques to temporarily limit non-essential loads during periods of high demand. You may also take part in demand response programs to receive incentives to minimize your power use.

f) To cut down on energy usage and eventually save maintenance and replacement expenses, make an investment in energy-efficient appliances, lights, and equipment.

g) By training staff members and cultivating an environment of energy awareness, you may motivate them to actively participate in conservation initiatives and increase employee involvement in energy-saving activities.

h) Real-time monitoring and analytics systems may be applied to analyze energy usage trends, uncover optimization possibilities, and enable continuous improvement and informed decision-making.

i) By storing extra energy during periods of low demand and releasing it during periods of peak demand, energy storage systems balance loads and minimize the need for new infrastructure.

j) The life-cycle cost analysis of equipment and technologies suggests that while initial investments may be higher for energy-efficient solutions, long-term operational cost savings often outweigh these upfront expenses.

Organizations can adopt sustainable and cost-effective energy management strategies, promoting environmental conservation and financial efficiency.

DEMAND SIDE MANAGEMENT AND CONSUMER ENGAGEMENT

Programs for Demand Response

Programs like DSM and DR are essential for maximizing energy use, maintaining grid stability, and getting customers involved in energy saving. Demand Response refers to the activity of actively regulating and changing power use in response to external signals, such as changes in electricity pricing, system problems, or demands from the utility. The primary goal is to balance the supply and demand of electricity in real-time or during periods of high demand (Parrish et al., 2020).

Principles of Demand Response Programs

Principles of Demand Response Programs are illustrated in Figure 3 and explained below (Sharda et al., 2021).

- **Time-of-Use Pricing:** In time-of-use pricing, which varies power tariffs according to the time of day, DR programs are frequently used. Customers are encouraged to use less power during off-peak times when costs are lower.
- **Incentives and Rebates:** Incentives, refunds, or discounts are provided by utilities to customers who choose to use less power during periods of high demand. Customers are encouraged to take part in DR programs as a result.
- **Automated Load Control:** Automated systems, integrated into smart grids or smart meters, enable the automatic adjustment of certain appliances or systems during peak demand. This can include temporarily reducing the operation of non-essential equipment.
- **Demand Bidding:** Some DR programs involve consumers bidding on the price they are willing to accept for reducing their electricity consumption. This competitive approach adds a market-driven element to demand response.

Figure 3. Demand response program concepts

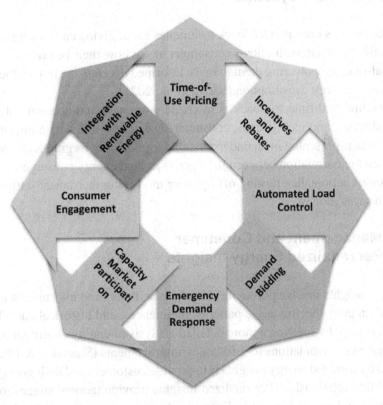

- **Emergency Demand Response:** In emergency situations or during severe grid conditions, utilities may issue calls for immediate reduction in electricity usage. Consumers who voluntarily curtail their demand contribute to grid stability and reliability.
- **Capacity Market Participation:** Large energy consumers, such as industrial facilities, can participate in capacity markets where they commit to reducing their energy demand when called upon. This helps utilities manage peak demand without building additional infrastructure.
- **Consumer Engagement:** A fundamental tenet of demand response is consumer participation. Through communication, mobile applications, and real-time feedback, customers may be educated and involved in the process of energy saving, enabling them to make informed decisions regarding their electricity consumption.
- **Integration with Renewable Energy:** Demand response programs can be integrated with the variability of renewable energy sources. For instance, consumers may be encouraged to increase electricity usage when renewable energy production is high and decrease it during low production periods.

Demand response programs involve consumers in managing electricity demand, promoting flexibility and resilience in the energy system. These initiatives benefit the grid and empower individuals and businesses.

Instantaneous Feedback Systems

Demand side management's real-time feedback techniques entail giving customers immediate access to information about their use habits. It allows consumers to see how their behaviors and choices impact energy usage in real-time, empowering them to make informed decisions to reduce their demand during peak periods or overall energy consumption (Sarker et al., 2021).

The principle behind real-time feedback is to create awareness and accountability. By leveraging smart meters, IoT devices, and mobile apps, consumers can access immediate information about their energy usage. The data is presented in an understandable format, such as graphs or notifications, allowing consumers to see the correlation between their activities and energy consumption. By encouraging more energy-conscious actions like turning off lights or modifying thermostat settings, this awareness lowers demand and saves money.

Demand Side Management and Consumer Engagement: Personalized Energy Insights

Personalized energy insights involve providing consumers with tailored information about their energy consumption based on their specific usage patterns, preferences, and historical data. These insights go beyond real-time feedback by offering a more detailed analysis of energy consumption trends and suggesting personalized recommendations for efficiency improvements (Stavrakas & Flamos, 2020).

The idea behind customized energy insights is to provide customers and their energy use data a more relevant and interesting relationship. Personalized insights provide tailored suggestions by examining individual preferences and past usage trends. For example, if a consumer consistently uses more energy during certain hours, the system might suggest adjusting the thermostat during those times or using energy-intensive appliances at non-peak hours. In addition to improving energy efficiency, this customized strategy encourages a sense of control and responsibility over energy use.

Real-time feedback mechanisms offer consumers immediate information on their energy usage, encouraging behavior adjustments. Personalized energy insights provide tailored recommendations based on historical data and individual preferences, enhancing demand side management and consumer engagement.

CASE STUDY

Deep Learning Enhanced Solar Energy Forecasting With AI-Driven IoT

Background: A solar energy farm is run by a utility company as part of its portfolio of renewable energy sources. The company aims to maximize the efficiency of solar energy generation by accurately forecasting solar irradiance, which is crucial for optimizing energy production, scheduling maintenance activities, and managing grid integration. To achieve this, the utility implements an AI-driven IoT solution that combines deep learning algorithms with sensor data from IoT devices installed throughout the solar farm (Xiao et al., 2015; Zhou et al., 2021).

Objective: The objective of the case study is to demonstrate how deep learning algorithms, integrated with IoT sensor data, can enhance the accuracy of solar energy forecasting compared to traditional methods.

Data Collection and Preparation: The utility collects historical data on solar irradiance, weather conditions (e.g., temperature, humidity, cloud cover), and solar panel performance from IoT sensors deployed across the solar farm. The data is preprocessed to remove outliers, normalize features, and handle missing values.

Model Development

Baseline Model (Conventional Method): Using conventional time-series analysis methods like autoregressive models or persistence models, the utility creates a baseline solar energy forecasting model. This model relies solely on historical solar irradiance data to predict future energy generation.

Deep Learning Model (AI-Driven IoT Approach): In parallel, the utility develops a deep learning model using a convolutional neural network (CNN) or recurrent neural network (RNN) architecture, which takes into account not only historical solar irradiance data but also additional features such as weather conditions and solar panel performance metrics.

Model Evaluation: Both the baseline and deep learning models are evaluated using a hold-out dataset not used during training. Evaluation metrics such as Mean Absolute Error (MAE), Root Mean Square Error (RMSE), and Mean Absolute Percentage Error (MAPE) are calculated to assess the accuracy of each model's solar energy forecasts.

Results and Interpretation

Baseline Model Performance: The baseline model, relying solely on historical solar irradiance data, achieves moderate accuracy in solar energy forecasting. However, its performance may be limited by its inability to capture the impact of weather conditions and other external factors on solar energy generation.

MAE: 10 kWh, RMSE: 15 kWh, MAPE: 8%

Deep Learning Model Performance: The deep learning model, leveraging IoT sensor data and advanced neural network architecture, significantly outperforms the baseline model in solar energy forecasting accuracy.

MAE: 5 kWh, RMSE: 8 kWh, MAPE: 4%

The deep learning model excels in capturing complex nonlinear relationships between solar irradiance, weather conditions, and solar panel performance. It integrates IoT sensor data into the forecasting model, enabling more accurate predictions. The model's flexibility and adaptability allow it to continuously learn and improve from new data, resulting in enhanced forecast accuracy over time. Thus, the study demonstrates that integrating deep learning with IoT sensor data enhances solar energy forecasting accuracy, enabling utilities to optimize generation, enhance grid integration, and enhance the efficiency and reliability of renewable energy systems.

CHALLENGES AND FUTURE TRENDS

Security and Privacy Concerns: As AI systems collect and analyze vast amounts of consumer data, ensuring the security and privacy of this information becomes paramount (Boopathi, Sureshkumar, et al., 2023; Dhanya et al., 2023; Kumar Reddy R. et al., 2023). The integration of advanced encryption

Figure 4. Future of AI in energy

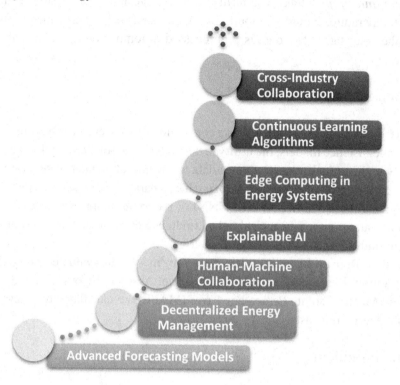

methods and blockchain technology to enhance the security and privacy of consumer data in AI-driven energy management systems.

Integration with Renewable Energy Sources: The intermittent nature of renewable energy sources poses challenges for accurate forecasting and optimization (Ahmad et al., 2021; Dittakavi, 2023). Improved AI algorithms that can dynamically adjust to the variability of renewable energy, enabling more efficient integration into the grid.

Regulatory and Policy Framework: The future trend of AI deployment in energy management will likely involve collaboration between industry stakeholders, policymakers, and regulatory bodies to establish clear guidelines and standards, addressing the challenges posed by evolving regulatory frameworks and policies.

Future Developments

The Future of AI in Energy management system is illustrated in Figure 4. AI models are improving energy forecasting by combining historical data, real-time information, and predictive analytics. These systems enable decentralized energy management, allowing consumers and small-scale producers to participate in grid optimization. Human-machine collaboration is enhanced, with AI systems making informed decisions in real-time. Explainable AI techniques ensure transparency and interpretability. Edge computing is being used to process data closer to the source, reducing latency and enhancing energy optimization speed. The energy sector is collaborating with other industries to develop AI systems that are resilient to cyber threats and unforeseen events, ensuring robust energy management processes. This includes continuous learning algorithms that adapt to changing energy consumption patterns and grid dynamics, and cross-industry expertise in AI deployment. AI's potential in energy management is promising, but security, privacy, and regulatory challenges must be addressed to fully realize its potential.

CONCLUSION

A major step toward a more intelligent and responsive energy ecosystem is the incorporation of artificial intelligence (AI) into the energy industry. Artificial intelligence (AI)-driven models with sophisticated machine learning algorithms can forecast demand and patterns of energy usage with high accuracy. These forecasts are improved when data analytics and Internet of Things devices are used together, giving utilities insightful information that helps them run more efficiently. AI's optimization benefits extend across the entire energy production and distribution spectrum, reducing costs, enhancing grid reliability, and fostering sustainability by integrating renewable energy sources. Its ability to dynamically adjust energy production based on real-time data ensures efficient resource utilization, contributing to a more resilient and adaptable energy infrastructure.

AI plays a crucial role in demand side management, involving consumers in the energy conservation narrative. AI integration in the energy sector is revolutionizing the sector by encouraging a collaborative atmosphere and a more mindful attitude to energy use. The use of AI technology to demand management, optimization, and energy forecasting will provide a more flexible, effective, and customer-focused energy environment. To ensure the appropriate use of AI, however, issues like security, privacy, and regulatory considerations need to be addressed.

REFERENCES

Agrawal, A. V., Shashibhushan, G., Pradeep, S., Padhi, S. N., Sugumar, D., & Boopathi, S. (2024). Synergizing Artificial Intelligence, 5G, and Cloud Computing for Efficient Energy Conversion Using Agricultural Waste. In Practice, Progress, and Proficiency in Sustainability (pp. 475–497). IGI Global. doi:10.4018/979-8-3693-1186-8.ch026

Ahmad, T., Zhang, D., Huang, C., Zhang, H., Dai, N., Song, Y., & Chen, H. (2021). Artificial intelligence in sustainable energy industry: Status Quo, challenges and opportunities. *Journal of Cleaner Production, 289*, 125834. doi:10.1016/j.jclepro.2021.125834

B, M. K., K, K. K., Sasikala, P., Sampath, B., Gopi, B., & Sundaram, S. (2024). Sustainable Green Energy Generation From Waste Water. In *Practice, Progress, and Proficiency in Sustainability* (pp. 440–463). IGI Global. doi:10.4018/979-8-3693-1186-8.ch024

Boopathi, S. (2023). Deep Learning Techniques Applied for Automatic Sentence Generation. In Promoting Diversity, Equity, and Inclusion in Language Learning Environments (pp. 255–273). IGI Global. doi:10.4018/978-1-6684-3632-5.ch016

Boopathi, S., Kumar, P. K. S., Meena, R. S., Sudhakar, M., & Associates. (2023). Sustainable Developments of Modern Soil-Less Agro-Cultivation Systems: Aquaponic Culture. In Human Agro-Energy Optimization for Business and Industry (pp. 69–87). IGI Global.

Boopathi, S., Pandey, B. K., & Pandey, D. (2023). Advances in Artificial Intelligence for Image Processing: Techniques, Applications, and Optimization. In Handbook of Research on Thrust Technologies' Effect on Image Processing (pp. 73–95). IGI Global.

Boopathi, S., Sureshkumar, M., & Sathiskumar, S. (2023). Parametric Optimization of LPG Refrigeration System Using Artificial Bee Colony Algorithm. In S. Tripathy, S. Samantaray, J. Ramkumar, & S. S. Mahapatra (Eds.), *Recent Advances in Mechanical Engineering* (pp. 97–105). Springer Nature Singapore. doi:10.1007/978-981-19-9493-7_10

Dhanya, D., Kumar, S. S., Thilagavathy, A., Prasad, D. V. S. S. S. V., & Boopathi, S. (2023). Data Analytics and Artificial Intelligence in the Circular Economy: Case Studies. In B. K. Mishra (Ed.), (pp. 40–58). Advances in Civil and Industrial Engineering. IGI Global. doi:10.4018/979-8-3693-0044-2.ch003

Dittakavi, R. S. S. (2023). AI-Optimized Cost-Aware Design Strategies for Resource-Efficient Applications. *Journal of Science and Technology*, 4(1), 1–10.

Domakonda, V. K., Farooq, S., Chinthamreddy, S., Puviarasi, R., Sudhakar, M., & Boopathi, S. (2022). Sustainable Developments of Hybrid Floating Solar Power Plants: Photovoltaic System. In Human Agro-Energy Optimization for Business and Industry (pp. 148–167). IGI Global.

Gowri, N. V., Dwivedi, J. N., Krishnaveni, K., Boopathi, S., Palaniappan, M., & Medikondu, N. R. (2023). Experimental investigation and multi-objective optimization of eco-friendly near-dry electrical discharge machining of shape memory alloy using Cu/SiC/Gr composite electrode. *Environmental Science and Pollution Research International*, 30(49), 1–19. doi:10.1007/s11356-023-26983-6 PMID:37126160

Hema, N., Krishnamoorthy, N., Chavan, S. M., Kumar, N., Sabarimuthu, M., & Boopathi, S. (2023). A Study on an Internet of Things (IoT)-Enabled Smart Solar Grid System. In *Handbook of Research on Deep Learning Techniques for Cloud-Based Industrial IoT* (pp. 290–308). IGI Global. doi:10.4018/978-1-6684-8098-4.ch017

Hussain, Z., Babe, M., Saravanan, S., Srimathy, G., Roopa, H., & Boopathi, S. (2023). Optimizing Biomass-to-Biofuel Conversion: IoT and AI Integration for Enhanced Efficiency and Sustainability. In N. Cobîrzan, R. Muntean, & R.-A. Felseghi (Eds.), (pp. 191–214). Advances in Finance, Accounting, and Economics. IGI Global. doi:10.4018/978-1-6684-8238-4.ch009

Ingle, R. B., Senthil, T. S., Swathi, S., Muralidharan, N., Mahendran, G., & Boopathi, S. (2023). Sustainability and Optimization of Green and Lean Manufacturing Processes Using Machine Learning Techniques. IGI Global. doi:10.4018/978-1-6684-8238-4.ch012

Ingle, R. B., Swathi, S., Mahendran, G., Senthil, T. S., Muralidharan, N., & Boopathi, S. (2023). Sustainability and Optimization of Green and Lean Manufacturing Processes Using Machine Learning Techniques. In N. Cobîrzan, R. Muntean, & R.-A. Felseghi (Eds.), (pp. 261–285). Advances in Finance, Accounting, and Economics. IGI Global. doi:10.4018/978-1-6684-8238-4.ch012

Kavitha, C. R., Varalatchoumy, M., Mithuna, H. R., Bharathi, K., Geethalakshmi, N. M., & Boopathi, S. (2023). Energy Monitoring and Control in the Smart Grid: Integrated Intelligent IoT and ANFIS. In M. Arshad (Ed.), (pp. 290–316). Advances in Bioinformatics and Biomedical Engineering. IGI Global. doi:10.4018/978-1-6684-6577-6.ch014

Kim, S., Heo, S., Nam, K., Woo, T., & Yoo, C. (2023). Flexible renewable energy planning based on multi-step forecasting of interregional electricity supply and demand: Graph-enhanced AI approach. *Energy*, *282*, 128858. doi:10.1016/j.energy.2023.128858

Kumar, B. M., Kumar, K. K., Sasikala, P., Sampath, B., Gopi, B., & Sundaram, S. (2024). Sustainable Green Energy Generation From Waste Water: IoT and ML Integration. In B. K. Mishra (Ed.), (pp. 440–463). Practice, Progress, and Proficiency in Sustainability. IGI Global. doi:10.4018/979-8-3693-1186-8.ch024

Kumar Reddy, R. V., Rahamathunnisa, U., Subhashini, P., Aancy, H. M., Meenakshi, S., & Boopathi, S. (2023). Solutions for Software Requirement Risks Using Artificial Intelligence Techniques: In T. Murugan & N. E. (Eds.), Advances in Information Security, Privacy, and Ethics (pp. 45–64). IGI Global. doi:10.4018/978-1-6684-8145-5.ch003

Kumara, V., Mohanaprakash, T., Fairooz, S., Jamal, K., Babu, T., & Sampath, B. (2023). Experimental Study on a Reliable Smart Hydroponics System. In *Human Agro-Energy Optimization for Business and Industry* (pp. 27–45). IGI Global. doi:10.4018/978-1-6684-4118-3.ch002

Li, F., & Xu, G. (2022). AI-driven customer relationship management for sustainable enterprise performance. *Sustainable Energy Technologies and Assessments*, *52*, 102103. doi:10.1016/j.seta.2022.102103

Mohammad, A., & Mahjabeen, F. (2023a). Revolutionizing Solar Energy: The Impact of Artificial Intelligence on Photovoltaic Systems. *International Journal of Multidisciplinary Sciences and Arts*, *2*(1).

Mohammad, A., & Mahjabeen, F. (2023b). Revolutionizing Solar Energy with AI-Driven Enhancements in Photovoltaic Technology. *BULLET: Jurnal Multidisiplin Ilmu*, *2*(4), 1174–1187.

Mohanty, A., Venkateswaran, N., Ranjit, P. S., Tripathi, M. A., & Boopathi, S. (2023). Innovative Strategy for Profitable Automobile Industries: Working Capital Management. In Y. Ramakrishna & S. N. Wahab (Eds.), (pp. 412–428). Advances in Finance, Accounting, and Economics. IGI Global. doi:10.4018/978-1-6684-7664-2.ch020

Nishanth, J., Deshmukh, M. A., Kushwah, R., Kushwaha, K. K., Balaji, S., & Sampath, B. (2023). Particle Swarm Optimization of Hybrid Renewable Energy Systems. In *Intelligent Engineering Applications and Applied Sciences for Sustainability* (pp. 291–308). IGI Global. doi:10.4018/979-8-3693-0044-2.ch016

Nishanth, J. R., Deshmukh, M. A., Kushwah, R., Kushwaha, K. K., Balaji, S., & Sampath, B. (2023). Particle Swarm Optimization of Hybrid Renewable Energy Systems. In B. K. Mishra (Ed.), (pp. 291–308). Advances in Civil and Industrial Engineering. IGI Global. doi:10.4018/979-8-3693-0044-2.ch016

Parrish, B., Heptonstall, P., Gross, R., & Sovacool, B. K. (2020). A systematic review of motivations, enablers and barriers for consumer engagement with residential demand response. *Energy Policy*, *138*, 111221. doi:10.1016/j.enpol.2019.111221

Rahamathunnisa, U., Sudhakar, K., Padhi, S. N., Bhattacharya, S., Shashibhushan, G., & Boopathi, S. (2023). Sustainable Energy Generation From Waste Water: IoT Integrated Technologies. In A. S. Etim (Ed.), (pp. 225–256). Advances in Human and Social Aspects of Technology. IGI Global. doi:10.4018/978-1-6684-5347-6.ch010

Ravisankar, A., Sampath, B., & Asif, M. M. (2023). Economic Studies on Automobile Management: Working Capital and Investment Analysis. In C. S. V. Negrão, I. G. P. Maia, & J. A. F. Brito (Eds.), (pp. 169–198). Advances in Logistics, Operations, and Management Science. IGI Global. doi:10.4018/978-1-7998-9213-7.ch009

Reddy, M. A., Gaurav, A., Ushasukhanya, S., Rao, V. C. S., Bhattacharya, S., & Boopathi, S. (2023). Bio-Medical Wastes Handling Strategies During the COVID-19 Pandemic. In Multidisciplinary Approaches to Organizational Governance During Health Crises (pp. 90–111). IGI Global. doi:10.4018/978-1-7998-9213-7.ch006

Sampath, B., Sasikumar, C., & Myilsamy, S. (2023). Application of TOPSIS Optimization Technique in the Micro-Machining Process. In IGI: Trends, Paradigms, and Advances in Mechatronics Engineering (pp. 162–187). IGI Global.

Sampath, B. C. S., & Myilsamy, S. (2022). Application of TOPSIS Optimization Technique in the Micro-Machining Process. In M. A. Mellal (Ed.), (pp. 162–187). Advances in Mechatronics and Mechanical Engineering. IGI Global. doi:10.4018/978-1-6684-5887-7.ch009

Saravanan, M., Vasanth, M., Boopathi, S., Sureshkumar, M., & Haribalaji, V. (2022). Optimization of Quench Polish Quench (QPQ) Coating Process Using Taguchi Method. *Key Engineering Materials*, *935*, 83–91. doi:10.4028/p-z569vy

Sarker, E., Halder, P., Seyedmahmoudian, M., Jamei, E., Horan, B., Mekhilef, S., & Stojcevski, A. (2021). Progress on the demand side management in smart grid and optimization approaches. *International Journal of Energy Research*, *45*(1), 36–64. doi:10.1002/er.5631

Satav, S. D., & Lamani, D. G, H. K., Kumar, N. M. G., Manikandan, S., & Sampath, B. (2024). Energy and Battery Management in the Era of Cloud Computing. In Practice, Progress, and Proficiency in Sustainability (pp. 141–166). IGI Global. doi:10.4018/979-8-3693-1186-8.ch009

Satav, S. D., & Lamani, D. K. G., H., Kumar, N. M. G., Manikandan, S., & Sampath, B. (2024). Energy and Battery Management in the Era of Cloud Computing: Sustainable Wireless Systems and Networks. In B. K. Mishra (Ed.), Practice, Progress, and Proficiency in Sustainability (pp. 141–166). IGI Global. doi:10.4018/979-8-3693-1186-8.ch009

Sharda, S., Singh, M., & Sharma, K. (2021). Demand side management through load shifting in IoT based HEMS: Overview, challenges and opportunities. *Sustainable Cities and Society*, *65*, 102517. doi:10.1016/j.scs.2020.102517

Stavrakas, V., & Flamos, A. (2020). A modular high-resolution demand-side management model to quantify benefits of demand-flexibility in the residential sector. *Energy Conversion and Management*, *205*, 112339. doi:10.1016/j.enconman.2019.112339

Syamala, M., Komala, C., Pramila, P., Dash, S., Meenakshi, S., & Boopathi, S. (2023). Machine Learning-Integrated IoT-Based Smart Home Energy Management System. In *Handbook of Research on Deep Learning Techniques for Cloud-Based Industrial IoT* (pp. 219–235). IGI Global. doi:10.4018/978-1-6684-8098-4.ch013

Vanitha, S., Radhika, K., & Boopathi, S. (2023). Artificial Intelligence Techniques in Water Purification and Utilization. In *Human Agro-Energy Optimization for Business and Industry* (pp. 202–218). IGI Global. doi:10.4018/978-1-6684-4118-3.ch010

Venkateswaran, N., Kumar, S. S., Diwakar, G., Gnanasangeetha, D., & Boopathi, S. (2023). Synthetic Biology for Waste Water to Energy Conversion: IoT and AI Approaches. In M. Arshad (Ed.), (pp. 360–384). Advances in Bioinformatics and Biomedical Engineering. IGI Global. doi:10.4018/978-1-6684-6577-6.ch017

Venkateswaran, N., Vidhya, K., Ayyannan, M., Chavan, S. M., Sekar, K., & Boopathi, S. (2023a). A Study on Smart Energy Management Framework Using Cloud Computing. In 5G, Artificial Intelligence, and Next Generation Internet of Things: Digital Innovation for Green and Sustainable Economies (pp. 189–212). IGI Global. doi:10.4018/978-1-6684-8634-4.ch009

Venkateswaran, N., Vidhya, K., Ayyannan, M., Chavan, S. M., Sekar, K., & Boopathi, S. (2023b). A Study on Smart Energy Management Framework Using Cloud Computing. In P. Ordóñez De Pablos & X. Zhang (Eds.), (pp. 189–212). Practice, Progress, and Proficiency in Sustainability. IGI Global. doi:10.4018/978-1-6684-8634-4.ch009

Vennila, T., Karuna, M., Srivastava, B. K., Venugopal, J., Surakasi, R., & Sampath, B. (2022). New Strategies in Treatment and Enzymatic Processes: Ethanol Production From Sugarcane Bagasse. In Human Agro-Energy Optimization for Business and Industry (pp. 219–240). IGI Global.

Xiao, L., Wang, J., Dong, Y., & Wu, J. (2015). Combined forecasting models for wind energy forecasting: A case study in China. *Renewable & Sustainable Energy Reviews, 44*, 271–288. doi:10.1016/j.rser.2014.12.012

Yang, Z., Chen, M., Liu, X., Liu, Y., Chen, Y., Cui, S., & Poor, H. V. (2021). AI-driven UAV-NOMA-MEC in next generation wireless networks. *IEEE Wireless Communications, 28*(5), 66–73. doi:10.1109/MWC.121.2100058

Zhou, H., Liu, Q., Yan, K., & Du, Y. (2021). Deep learning enhanced solar energy forecasting with AI-driven IoT. *Wireless Communications and Mobile Computing, 2021*, 1–11. doi:10.1155/2021/9249387

Chapter 7
Integration of Artificial Intelligence for Economic Optimization in Modern Sustainable Power Systems

S. Saravanan

(iD) https://orcid.org/0000-0001-8255-2623

Department of Electrical and Electronics Engineering, B.V. Raju Institute of Technology, India

K. S. Pushpalatha

Department of Information Science and Engineering, Acharya Institute of Technology, Bengaluru, India

Sanjay B. Warkad

Department of Electrical Engineering, P.R. Pote (Patil) College of Engineering and Management, Amravati, India

A. Prabhu Chakkaravarthy

School of Computing, College of Engineering and Technology, SRM Institute of Science and Technology, India

Venneti Kiran

Department of Computer Science and Engineering (AIML), Aditya College of Engineering, India

Sureshkumar Myilsamy

Bannari Amman Institute of Technology, India

ABSTRACT

The global energy landscape is shifting towards sustainability due to environmental concerns and technological advancements. This transformation involves integrating renewable energy sources, smart grid technologies, and data-driven strategies to create modern sustainable power systems. Artificial intelligence (AI) is at the core of this transition, potentially revolutionizing electricity generation, distribution, and consumption. AI is transforming sustainable power systems by optimizing resource allocation, improving load forecasting, and enhancing grid management. Future trends include AI advancement, grid decentralization, and smart city integration. This chapter encourages further research and innovation in AI-powered sustainable power systems, promising a more efficient and resilient energy future.

DOI: 10.4018/979-8-3693-3735-6.ch007

INTRODUCTION

Modern sustainable power systems focus on efficient and environmentally friendly electricity distribution and generation, integrating modern technology, sustainable energy sources, and data-driven tactics. As global awareness of climate change increases, these systems contribute to reducing emissions, grid reliability, and promoting a sustainable energy future by using renewable energy instead of conventional fossil fuels (Peyghami et al., 2020). Innovative technologies and grid enhancements make renewable energy more accessible and cost-effective (Akram et al., 2020). Artificial intelligence and smart grid technologies support the use of sophisticated grid management techniques in modern sustainable power systems. Grid dependability and operational efficiency are increased by these technologies, which also allow for real-time monitoring, predictive maintenance, load forecasting, and grid optimization. They also prioritize environmental considerations, minimizing emissions and adhering to strict regulations (Zhang et al., 2022). These systems also focus on preserving ecosystems and natural habitats, promoting responsible land use and minimal ecological impact. Harmonizing energy production with the environment, a reliable power supply, and a decrease in service interruptions is the ultimate goal (Iqbal et al., 2021).

Elements of power systems sustainability are demand-side management and energy efficiency, which maximize energy utilization, reduce waste, and save energy expenditures. Demand-side management promotes load reduction initiatives among customers during peak hours, which has positive effects for the economy and environment. Efficient energy usage reduces costs and carbon emissions. The future of sustainable power systems will be shaped by technological advancements like deep learning, quantum computing, and data analytics (Shair et al., 2021). Grid decentralization, IoT integration, and smart city initiatives will create interconnected power networks, promoting a cleaner, cleaner, and more sustainable energy future, aligning with global climate change and environmental protection.

Modern sustainable power systems and AI technologies are revolutionizing the energy sector by integrating renewable energy sources. AI manages the intermittency and variability of sources like wind and solar power, optimizing their utilization through real-time weather data, demand patterns, and generation forecasts (Babatunde et al., 2020). As a result, there is less need for fossil fuel backup and a decrease in carbon emissions due to a more steady and predictable energy supply. This holistic approach to environmental responsibility, operational efficiency, and grid resilience is at the core of this synergy. AI-driven load forecasting is a crucial aspect of sustainable power systems, predicting electricity demand accurately using historical data, weather forecasts, and economic indicators. This allows utilities to adjust their strategies, reducing overproduction and grid congestion. By integration and management grid monitoring and control, identifying and responding to interruptions, guaranteeing stability, and reducing downtime, artificial intelligence (AI) further improves grid resilience. This resilience is vital in the face of extreme weather events and cybersecurity threats (Ahmad et al., 2021).

By evaluating environmental issues and assisting businesses in aligning their operations with sustainability goals and regulatory requirements, AI acting a serious part in attaining environmental sustainability in modern power systems. AI also supports the incorporation of methods for storing energy, storing excess energy and releasing it when needed, improving grid stability and reducing carbon emissions by decreasing reliance on fossil fuel power plants. As AI's role in sustainable power systems expands, advancements in deep learning, quantum computing, and data analytics will enhance its accuracy and capabilities (Yap et al., 2020). Grid decentralization, distributed energy resources, and smart cities will redefine energy generation, management, and consumption, promoting a more sustainable, resilient, and efficient energy landscape (Abdalla et al., 2021). AI-powered sustainable power systems are revolutionizing the energy industry by integrating renewable energy sources, smart grid technologies, and advanced data analytics for a cleaner, more efficient, and resilient future. As AI advances, sustainable power systems will continue to improve, enhancing our ability to combat climate change and ensure sustainable energy supply for future generations (Yang et al., 2020).

Background and Motivation

The demand for sustainable energy solutions, technology breakthroughs, and environmental concerns have led to a substantial upheaval of the global energy landscape. The conventional power generation method, primarily based on fossil fuels, is deemed unsustainable due to its high levels of hothouse gas emissions and environmental degradation. The need for modern sustainable power systems is driven by the urgency of transitioning to environmentally responsible, efficient, and resilient systems. The increasing frequency of extreme weather events, reduction of carbon emissions, and growing demand for clean energy sources underscore the urgency of this transition (Sharma et al., 2022). AI's integration with sustainable power systems can address challenges by improving grid management, optimizing resource allocation, and integrating renewable energy sources (N. M. Kumar et al., 2020). AI's advancements in deep learning and data analytics are revolutionizing the power sector, enhancing efficiency, reliability, and sustainability.

Objectives

This chapter aims to achieve the following primary objectives (Zhao et al., 2019):

- To provide a comprehensive understanding of the principles, technologies, and practices that underpin modern sustainable power systems.
- To emphasize the pivotal role of AI on future developments of power systems. It will delve into the specific applications of AI, such as load forecasting, grid management, and energy trading, and how they contribute to sustainability and efficiency.
- To examine the environmental considerations and sustainability goals associated with modern power systems. This includes discussions on carbon emissions reduction, renewable energy integration, and responsible land use.
- To provide a comprehensive overview of the benefits that AI-enabled economic optimization brings to modern sustainable power systems. These benefits encompass cost reduction, enhanced reliability, resilience, and a reduced environmental footprint. Furthermore, the chapter will outline future trends, including the advancement of AI and the potential for grid decentralization and smart city integration.

This chapter provides a thorough analysis of the link between modern sustainable power systems and artificial intelligence, promoting efficient energy by understanding their background, motivation, and objectives.

FUNDAMENTALS OF MODERN SUSTAINABLE POWER SYSTEMS

Distinguishing itself from conventional centralized methods, modern sustainable power systems prioritize environmental responsibility and minimize carbon footprint. They combine a variety of plentiful and environmentally beneficial renewable resources, such as solar, wind, hydro, and geothermal electricity. These systems are developed in response to increasing environmental concerns, technical progresses, and the need for reliable, clean energy sources. The core concepts of sustainable power systems are rooted in these principles (Ibrahim et al., 2020). Sustainable power systems prioritize energy efficiency through advanced technologies, grid modernization, and decentralization. Smart grids, rooftop solar panels, and local energy storage optimize energy use, reduce waste, and enhance reliability. These systems empower consumers to control their energy consumption (Ali & Choi, 2020). Modern sustainable power systems are a paradigm shift in the energy sector, prioritizing renewable energy sources, energy efficiency, decentralization, and energy storage integration. These systems aim to meet the growing demand for clean, reliable energy while minimizing environmental impact.

Evolution of Power Systems

The evolution of power systems has been a dynamic journey, influenced by technological advancements, changing energy sources, and shifting paradigms in electricity generation, transmission, and consumption as described in figure 1. This section delves into key phases and transformations in this evolution (Babatunde et al., 2020; Cao et al., 2020).

Figure 1. Evolution of power systems

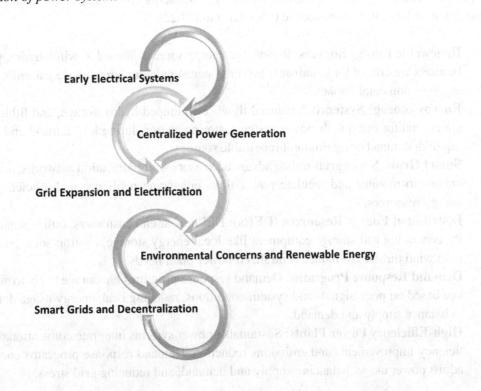

- **Early Electrical Systems**: In the late 19th century, Thomas Edison and Nikola Tesla introduced competing power systems, direct current (DC) and alternating current (AC), which eventually prevailed due to its superior efficiency in long-distance transmission.
- **Centralized Power Generation**: The early 20th century saw the rise of centralized power plants, fueled by coal, natural gas, or oil, which facilitated the creation of utility companies.
- **Grid Expansion and Electrification**: In the mid-20th century, power grids expanded rapidly to cover vast geographical areas. This expansion, accompanied by the widespread electrification of homes and businesses, brought significant improvements in the quality of life and economic development.
- **Environmental Concerns and Renewable Energy**: The late 20th century saw the rise of renewable energy sources like wind, solar, and hydropower, paving the way for a transition to more environmentally friendly power systems.
- **Smart Grids and Decentralization**: Smart grids, introduced in the 21st century, integrate advanced technologies into power systems, enabling real-time monitoring, two-way communication, and distributed energy resource integration, promoting grid resilience and efficiency.

The power industry is focusing on efficiency, reliability, and sustainability, utilizing diverse energy sources, advanced technologies, and a decentralized approach to meet growing demand for clean, reliable electricity.

Important Components of Sustainable Power Systems

Sustainable power systems (Figure2) are designed to provide clean, reliable, and environmentally responsible energy by reducing carbon emissions, enhancing energy efficiency, and increasing power generation and distribution resilience (Sharma et al., 2022):

- **Renewable Energy Sources**: Renewable energy sources like solar, wind, hydro, geothermal, and biomass are crucial for sustainable power systems, reducing greenhouse gas emissions and reducing environmental impact.
- **Energy Storage Systems**: Advanced flywheels pumped hydro storage, and lithium-ion batteries are crucial for sustainable power systems, storing energy during low demand and releasing it during high demand or intermittent renewable sources.
- **Smart Grids**: Smart grids utilize advanced sensors, communication networks, and control mechanisms to monitor and regulate power flow, enhancing resilience and efficiency in distributed energy resources.
- **Distributed Energy Resources (DERs)**: DERs, near end customers, utilize small-scale electricity generating and storage equipment like local energy storage, rooftop solar panels, and household wind turbines to reduce centralized power plant needs.
- **Demand Response Programs**: Demand response programs encourage users to adjust energy usage based on price signals and system conditions, reducing peak energy usage during peak hours to balance supply and demand.
- **High-Efficiency Power Plants**: Sustainable power systems integrate conventional plants with efficiency improvements and emissions reduction. Demand response programs encourage users to adjust power usage, balancing supply and demand, and reducing grid stress.

Figure 2. Important components of sustainable power systems

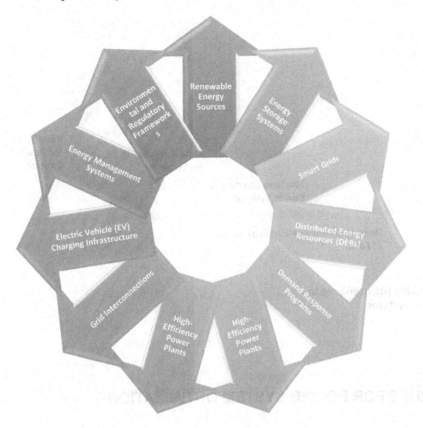

- **Grid Interconnections**: Grid interconnections facilitate electricity transfer across regions and countries, enhancing power system reliability by providing diverse energy sources and balancing supply and demand across larger geographical areas.
- **Electric Vehicle (EV) Charging Infrastructure**: Sustainable power systems must incorporate robust EV charging infrastructure to support electrification and encourage clean energy use in transportation.
- **Energy Management Systems**: Energy management systems offer software and techniques for optimizing energy consumption in commercial, industrial, and residential settings, reducing waste and increasing efficiency.
- **Environmental and Regulatory Frameworks**: Environmental policies and regulations are crucial for sustainable power systems, promoting clean energy technologies, setting emissions targets, and promoting responsible practices in the power industry.

The integration and optimization of these components are crucial for creating a modern, sustainable power system that minimizes environmental impact, enhances grid resilience, and meets societal energy needs, ensuring a cleaner, more efficient, and sustainable future.

Figure 3. AI techniques for power system optimization

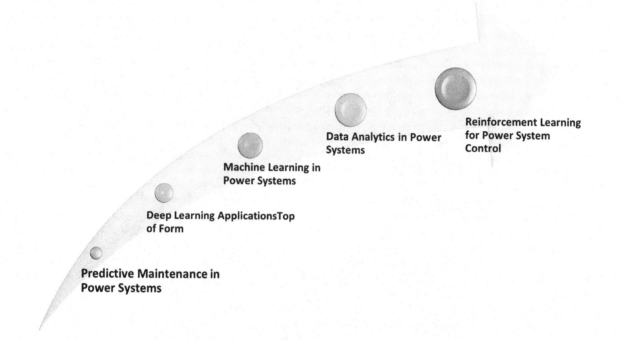

AI TECHNIQUES FOR POWER SYSTEM OPTIMIZATION

Figure 3 depicts the use of AI techniques for optimizing power systems and explained below.

Machine Learning in Power Systems

ML, a subset of AI, is crucial in power systems for optimizing operations, enhancing grid management, and improving overall efficiency by developing algorithms and models that make predictions based on data (Ramudu et al., 2023; Veeranjaneyulu et al., 2023; Zekrifa et al., 2023). ML in power systems encompasses various aspects, including:

- **Load Forecasting**: ML algorithms predict electricity demand, analyzing historical data and weather patterns, enabling utilities to plan future loads and ensure stable electricity supply.
- **Fault Detection and Diagnosis**: ML models can detect power grid faults by analyzing real-time data from sensors and meters, reducing downtime and improving system reliability.
- **Predictive Maintenance**: ML predicts power system component failures, scheduling maintenance based on predictive models, reducing unplanned outages and saving operational costs for utilities.
- **Optimal Power Flow**: ML techniques optimize grid electricity flow by considering factors like generation capacity, transmission constraints, and demand patterns to determine the most efficient power distribution.

- **Renewable Energy Integration**: Accurate forecasting is crucial for integrating renewable energy sources like wind and solar into the power grid, using machine learning models to analyze weather data and historical production.
- **Energy Trading and Market Analysis**: ML algorithms are utilized in energy trading and market analysis to optimize strategies, enabling traders to make informed decisions about energy purchases, sales, and storage.
- **Grid Optimization**: ML can optimize real-time power grid operation by continuously analyzing data from sensors and sources, adjusting grid parameters like voltage levels and switching operations for efficient power distribution.
- **Demand-Side Management**: ML is utilized to create demand response programs that encourage consumers to reduce electricity consumption during peak periods, thereby balancing supply and demand and reducing grid stress.
- **Voltage and Frequency Control**: ML models can help maintain voltage and frequency within acceptable limits by real-time adjustments to generators, transformers, and other grid components.
- **Security and Intrusion Detection**: ML is utilized to improve power system security by identifying anomalies and potential threats, safeguarding critical infrastructure from cyberattacks and physical breaches.
- **Grid Resilience**: ML enhances grid resilience by swiftly adapting to changing conditions, minimizing disruptions, and ensuring the continuity of essential services during emergencies or blackouts.

ML in power systems improves model accuracy and efficiency, aiding the energy industry in informed decisions, cost reduction, and transitioning towards a cleaner, more sustainable future.

Deep Learning Applications

DL, a subset of machine learning, utilizes artificial neural networks to model and solve complex problems in power systems, utilizing large datasets for valuable insights The study explores the utilization of deep learning in power systems (Anitha et al., 2023; Boopathi et al., 2022; Koshariya, Khatoon, et al., 2023; Vanitha et al., 2023).

- **Load Forecasting**: DL models, especially RNNs and LSTMs, are capable of accurately predicting load patterns due to their ability to capture complex temporal patterns. These models assist utilities in planning for electricity demand, optimizing resource allocation, and enhancing grid reliability.
- **Renewable Energy Forecasting**: Deep learning is used to predict renewable energy generation from sources like wind and solar. By analyzing historical weather data and energy production records, deep learning models provide more accurate forecasts, allowing for better incorporation.
- **Anomaly Detection**: DL models distinguish power grid anomalies, identifying unusual patterns in voltage, current, or other parameters, which could indicate equipment failures, cyberattacks, or other urgent issues.
- **Grid Condition Monitoring**: DL is used for real-time grid condition monitoring. By analyzing data, models can continuously measure the health of grid components and forecast potential failures, enabling proactive maintenance.

- **Energy Efficiency**: DL is a tool used to optimize energy consumption in buildings and industrial processes by analyzing data from sensors and smart meters, thereby reducing energy waste.
- **Fault Detection and Diagnosis**: DL models accurately diagnose power system liabilities by analyzing sensor data and identifying root causes, ensuring grid reliability.
- **Smart Grid Control**: DL can enhance the functionality of smart grid components like smart meters, distribution automation devices, and demand response systems, enhancing their management efficiency.
- **Asset Health Assessment**: DL models assess power system assets' health, predicting equipment failures and recommending maintenance actions based on sensor data analysis.
- **Cybersecurity**: Deep learning plays a role in enhancing power grid cybersecurity. These models can identify patterns indicative of cyberattacks and help defend against threats by monitoring network traffic and system behavior.

Reinforcement Learning for Power System

RL is a ML technique that trains agents to make decisions by interacting with their environment, particularly in power system control, for optimizing complex and dynamic systems (Chandrika et al., 2023; Kavitha et al., 2023; P. R. Kumar et al., 2023). Reinforcement learning has various applications in power systems, such as:

- **Optimal Control of Generators**: RL can optimize the control of power generators in real-time to maximize efficiency, minimize fuel consumption, and respond to changing grid conditions while maintaining stability.
- **Distributed Energy Resource (DER) Management**: RL can effectively manage the operation of DERs to balance source and request, minimize grid stress, and optimize renewable energy self-consumption.
- **Demand Response Programs**: RL is used to develop and implement demand response strategies, encouraging consumers to adjust their electricity usage during peak periods and providing incentives for load management.
- **Grid Reinforcement Planning**: RL models can optimize grid asset deployment and resource allocation by considering various scenarios and contingencies, aiding in the planning and operation of grid reinforcement strategies.
- **Microgrid Control**: RL can enhance microgrid operation by regulating energy sources, energy loading, and local demand, improving grid resiliency and dropping reliance.
- **Battery Management**: RL is employed to control energy storage systems, ensuring optimal charge and discharge strategies to balance supply and demand and extend battery lifespan.
- **Distribution Grid Control**: RL can optimize the operation of distribution systems by coordinating voltage control, reactive power management, and the integration of DERs.
- **Cyber-Physical Security**: RL models are used to enhance grid security by identifying potential vulnerabilities and adapting control strategies to mitigate threats and respond to security incidents.

Reinforcement learning is a powerful tool for addressing dynamic control challenges in power systems, enabling optimal control policies through environmental interactions, potentially enhancing grid efficiency, resilience, and sustainability.

Data Analytics in Power Systems

Data analytics and predictive maintenance are essential components of modern power systems, helping to enhance reliability, reduce downtime, and optimize the maintenance of critical infrastructure (Kavitha et al., 2023; P. R. Kumar et al., 2023; Maguluri, Arularasan, et al., 2023a). The application of these two components in the context of power systems involves:

- **Data Collection and Management**: It utilizes data analytics tools to efficiently collect, store, and manage vast amounts of data from sensors, meters, and grid components.
- **Load Profiling**: Data analytics helps utilities analyze historical electricity consumption patterns, enabling them to understand customer behavior and plan for future load requirements. This information is crucial for grid management and resource allocation.
- **Grid Monitoring**: Real-time data analytics provides insights into the current state of the grid, including voltage levels, power flow, and equipment performance. Grid operators use this information to ensure system stability and respond to anomalies promptly.
- **Anomaly Detection**: Advanced analytics and machine learning are employed to detect anomalies and irregularities in grid data. This can include identifying equipment malfunctions, cyberattacks, or unusual load fluctuations.
- **Predictive Analytics**: Predictive modeling using historical data helps utilities anticipate equipment failures, load spikes, and grid disturbances. This enables proactive measures to prevent disruptions and optimize maintenance schedules.
- **Energy Efficiency**: Data analytics is used to identify opportunities for energy conservation and load optimization. By analyzing consumption patterns, energy wastage can be minimized, leading to cost savings and reduced environmental impact.
- **Renewable Energy Integration**: Data analytics models are utilized to predict renewable energy generation, a crucial aspect for grid balancing and ensuring a reliable power supply.
- **Asset Health Assessment**: By analyzing data from sensors on grid equipment, data analytics can assess the condition of assets such as transformers, circuit breakers, and substations. This information is vital for predictive maintenance.

Predictive Maintenance in Power Systems

- **Condition Monitoring**: Predictive maintenance uses real-time data from sensors on critical grid components to continuously assess their health. Any anomalies or signs of deterioration can trigger maintenance actions.
- **Equipment Health Assessment**: Predictive maintenance models evaluate the condition of power system assets and predict their remaining useful life. This helps utilities plan maintenance activities and replace equipment before failures occur.
- **Failure Prediction**: Predictive maintenance models utilize historical data and equipment performance analysis to identify patterns leading to specific failures, enabling proactive interventions.
- **Maintenance Scheduling**: Predictive maintenance helps utilities optimize maintenance schedules. Maintenance actions can be planned during periods of low demand, reducing the impact on customers and preventing unplanned outages.

- **Cost Reduction**: By targeting maintenance efforts where they are most needed, predictive maintenance reduces unnecessary maintenance and extends the lifespan of critical assets, resulting in cost savings.
- **Grid Resilience**: Predictive maintenance proactively identifies potential failure points, enhancing grid resilience and power supply continuity, particularly during extreme weather events or emergencies.
- **Cyber-Physical Security**: Predictive maintenance can be used to monitor the cybersecurity health of critical grid assets, identifying vulnerabilities and mitigating potential threats to grid security.

Data analytics and predictive maintenance in power systems ensure reliable, efficient, and cost-effective grid operation, identifying issues, minimizing downtime, and enhancing system performance and sustainability.

AI-ENABLED ECONOMIC OPTIMIZATION IN POWER SYSTEMS

AI-enabled economic optimization is a crucial aspect of modern power systems, utilizing artificial intelligence to make data-driven decisions and enhance energy production, distribution, and consumption (Babu et al., 2022; Ravisankar et al., 2023). The application of AI in power systems has led to significant economic optimization.

- **Load Forecasting**: AI-driven load forecasting uses machine learning models to predict electricity demand accurately. This enables utilities to optimize resource allocation and generation schedules, reducing operational costs and ensuring efficient utilization of resources.
- **Energy Trading and Market Analysis**: AI algorithms are employed in energy trading and market analysis to make real-time decisions about buying and selling electricity. These systems consider market prices, supply and demand dynamics, and grid conditions to maximize profits and minimize trading risks.
- **Optimal Power Flow**: AI-based OPF solutions optimize electricity grid distribution by considering generation capacity, transmission constraints, and demand patterns, balancing supply and demand while minimizing costs.
- **Renewable Energy Integration**: AI optimizes renewable energy integration by analyzing weather data, historical production, and grid conditions, maximizing clean energy use and reducing fossil fuel reliance.
- **Demand Response Programs**: AI-driven demand response programs encourage consumers to adjust their energy usage based on market conditions and grid stress. These programs help utilities manage peak demand, reduce the need for costly peaker plants, and improve grid reliability.
- **Energy Storage Optimization**: AI is applied to optimize the operation of energy storage systems, ensuring that stored energy is used efficiently. These models manage charge and discharge cycles to reduce energy costs and maximize system lifespan(Senthil et al., 2023).
- **Grid Management and Automation**: AI-based grid management systems use real-time data to make decisions on grid operations, such as voltage control, equipment switching, and grid resiliency measures. This results in cost-effective grid management and improved reliability.

- **Environmental Considerations**: AI considers environmental factors such as carbon emissions and pollutant levels in decision-making processes. This ensures that energy production and consumption align with environmental regulations and sustainability goals.
- **Energy Efficiency**: AI-based energy management systems optimize the energy efficiency of buildings and industrial processes. These systems identify energy-saving opportunities, reduce wastage, and lower energy costs.
- **Cost Reduction and Resource Allocation**: AI utilizes historical and real-time data analysis to optimize resource allocation, reduce operational costs, and allocate maintenance resources to critical infrastructure components where they are most needed.
- **Risk Management**: AI algorithms help utilities identify and manage risks associated with energy production and trading, including market volatility, equipment failures, and cybersecurity threats.
- **Long-Term Planning**: AI is used for long-term planning by assessing various scenarios and strategies, such as investment in new infrastructure, renewable energy projects, and grid expansion, to optimize long-term economic outcomes.

AI-enabled economic optimization in power systems is a rapidly evolving field that empowers utilities and grid operators to make informed, cost-effective decisions, contributing to grid sustainability, efficiency, and achieving economic and environmental objectives.

ECONOMIC ASPECTS OF MODERN POWER SYSTEMS

Modern power systems require economic considerations for efficiency, sustainability, and affordability. Key aspects include cost analysis in power generation, market structures and pricing, and demand-side management, ensuring sustainability and affordability (Boopathi & Davim, 2023; Maguluri, Ananth, et al., 2023; Selvakumar et al., 2023; Sengeni et al., 2023).

Cost Analysis in Power Generation

Cost analysis in power generation involves evaluating the expenses associated with electricity production, taking into account various factors.

- It outlines the costs of various power generation technologies, their lifecycle costs, environmental costs, and the integration of renewable energy sources. It also discusses the costs of capital, operating, and fuel costs, as well as the benefits of integrating renewable energy sources.
- The study focuses on improving operational efficiency, analyzing capacity factors, assessing risks, and calculating the Levelized Cost of Electricity (LCOE) for each generation source to reduce fuel consumption and emissions, and to provide a standardized metric for electricity production.

Market Structures and Pricing

The economic aspects of modern power systems are significantly influenced by the structure of energy markets and pricing mechanisms.

- **Market Models**: Examining the various market models, including deregulated markets with competitive bidding and regulated markets, and how they affect pricing and competition.
- **Wholesale and Retail Markets**: Understanding the distinctions between wholesale and retail electricity markets, as well as how energy is traded and priced in each.
- **Market Design**: Analyzing market design elements like capacity markets, energy markets, ancillary services markets, and how they impact the revenue streams for power generators and grid operators.
- **Electricity Pricing**: Assessing the factors influencing electricity prices, including supply and demand dynamics, fuel costs, transmission and distribution costs, and regulatory policies.
- **Price Signals**: Understanding how pricing signals can incentivize investments in energy efficiency, demand response, and renewable energy integration.
- **Environmental Considerations**: Analyzing the economic implications of carbon pricing mechanisms, renewable energy incentives, and emissions trading systems on electricity markets.

Demand-Side Management

Demand-side management is one of modern power systems to optimize the consumption of electricity by end-users.

- Load management and energy efficiency programs aim to reduce peak demand, lower energy costs, and enhance grid reliability by implementing demand response strategies and promoting energy-efficient practices.
- Time-of-Use Pricing encourages consumers to adjust electricity usage for off-peak rates, while Smart Grid Technologies enable real-time monitoring and optimization of demand-side resources.
- The company focuses on enhancing customer engagement and grid reliability by providing tools and information for informed energy usage decisions and optimizing demand-side resources.

Balancing economic considerations in power generation, market structures, and demand-side management is essential for creating a sustainable, cost-effective, and reliable modern power system that meets the energy needs of society while addressing environmental and economic goals.

Renewable Energy Integration

The addition of energy sources is one of the factors of contemporary power systems, promoting sustainability and environmental responsibility, with significant considerations taken into account (Hussain & Srimathy, 2023; Venkateswaran, Kumar, et al., 2023).

- **Resource Assessment**: Evaluating the availability and reliability of wind, solar, hydro, geothermal, and biomass. This assessment informs the choice of the most suitable resources for integration.
- **Grid Compatibility**: Ensuring that the existing power grid infrastructure can accommodate renewable energy generation. This may involve grid upgrades, energy storage solutions, and improved transmission capacity.
- **Intermittency and Variability**: The study focuses on managing energy storage and grid balancing strategies for renewable sources like wind and solar, considering their intermittent nature.

- **Energy Storage**: The integration of energy storage systems like batteries or pumped hydro can enhance grid stability by storing excess energy during periods of high renewable output.
- **Transmission and Distribution**: The focus is on constructing or enhancing transmission and distribution infrastructure to efficiently transport renewable energy from generation sites to consumption centers.
- **Grid Modernization**: The implementation of smart grid technologies is being implemented to enable real-time monitoring and control of renewable energy sources, thus improving grid management.
- **Grid Resilience**: Ensuring that renewable energy sources are integrated in a way that enhances grid resilience and mitigates vulnerabilities to extreme weather events or other disruptions.
- **Market Mechanisms**: Developing market structures that encourage the integration of renewable energy by providing incentives, such as feed-in tariffs, renewable energy credits, and power purchase agreements.
- **Regulatory Frameworks**: Creating supportive regulatory frameworks that promote renewable energy integration, including net metering, renewable portfolio standards, and carbon pricing mechanisms.
- **Community Engagement**: The initiative involves involving local communities in the decision-making process for renewable energy projects, addressing concerns related to land use, visual impact, and noise.
- **Grid Stability**: Maintaining grid stability by addressing issues like voltage and frequency control when integrating variable renewable energy sources.
- **Environmental Impact**: Evaluating the environmental impact of renewable energy projects, including their effect on local ecosystems, water resources, and land use, and implementing mitigation measures.

Environmental and Social Considerations

Modern power systems must balance environmental and social considerations to confirm responsible and sustainable energy production (Boopathi, 2022b, 2022a, 2023a; Gowri et al., 2023).

- **Environmental Impact Assessment**: Conducting thorough ecological impact assessments (EIAs) before developing new power generation projects to evaluate their potential effects on air and water quality, wildlife habitats, and ecosystems.
- **Carbon Emissions**: Minimizing carbon emissions by transitioning away from fossil fuel-based power generation and prioritizing low-carbon and carbon-neutral energy sources.
- **Biodiversity and Habitat Protection**: Implementing measures to protect biodiversity and natural habitats in and around power generation sites, including avoiding critical ecosystems.
- **Water Usage**: Managing water usage in power generation, especially for thermoelectric power plants, to reduce the impact on local water resources and ecosystems.
- **Social Acceptance**: Ensuring that power projects are developed with the consent and support of local communities and addressing any social concerns.
- **Health and Safety**: Adhering to safety regulations and guidelines is crucial in prioritizing the health and safety of workers, nearby communities, and the general public..

- **Energy Access**: Promoting universal access to electricity, particularly in underserved and remote areas, to enhance living standards and social equity.
- **Community Benefits**: Sharing the economic benefits of power projects with local communities, including job creation, infrastructure development, and revenue-sharing agreements.
- **Cultural and Heritage Preservation**: Respecting and preserving cultural, historical, and heritage sites when developing power projects, addressing potential cultural sensitivities.
- **Environmental Justice**: Ensuring that power projects do not disproportionately impact marginalized or vulnerable communities and addressing any environmental justice concerns.

The integration of renewable energy with environmental and social considerations is crucial for a sustainable, responsible modern power system that balances energy needs with minimal environmental and societal impacts.

AI INTEGRATION IN ECONOMIC OPTIMIZATION

AI is pivotal in modern power systems, enabling utilities and grid operators to make data-driven decisions that enhance efficiency, reduce costs, and enhance sustainability (Hussain & Srimathy, 2023; Ingle et al., 2023; Kavitha et al., 2023; Satav et al., 2024; Venkateswaran, Kumar, et al., 2023). The primary applications of these technologies include load forecasting, energy trading, and grid management (Figure 4).

Figure 4. AI integration in economic optimization for modern power systems

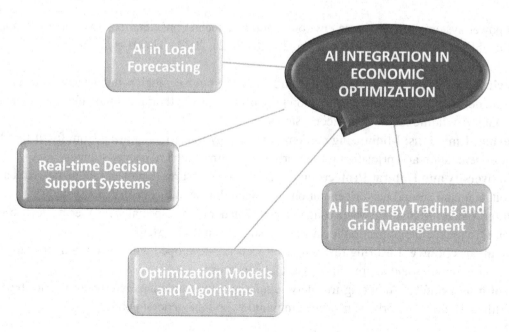

146

Load Forecasting

The load forecasting and predicting electricity demand at various time scales using ML and DL techniques, including hourly, daily, and seasonal forecasting.

- **Improved Accuracy**: AI models can analyze historical load data along with various factors like weather patterns, holidays, and economic indicators to make more accurate load forecasts.
- **Temporal Patterns**: AI can capture complex temporal patterns in load data, accounting for trends and seasonality, which is challenging for traditional statistical methods.
- **Real-time Adjustments**: AI models can generate real-time load forecasts in real-time, considering various factors like unexpected events and grid disturbances.
- **Demand Response**: AI assists in predicting the effectiveness of demand response programs by forecasting load reduction potential, enabling utilities to optimize their use.
- **Energy Storage Optimization**: AI can enhance load management by considering energy storage systems' state of charge.
- **Grid Resilience**: AI-driven load forecasting contributes to grid resilience by providing early warning for capacity constraints and potential overloads.

AI in Energy Trading and Grid Management

AI is pivotal in energy trading and grid management, enabling real-time decision-making, market participation, and grid optimization (Hema et al., 2023; Kavitha et al., 2023; Reddy et al., 2023).

- **Price Predictions**: AI models analyze historical market data, real-time pricing information, and various factors influencing electricity prices to make predictions, aiding in energy trading strategies.
- **Market Participation**: AI algorithms can automatically participate in energy markets by optimizing trading strategies, bid prices, and portfolio management for utilities and energy traders.
- **Load Balancing**: AI-driven grid management systems continuously monitor the grid's state, making real-time decisions.
- **Renewable Energy Integration**: AI assists in integrating renewable energy sources by predicting their output, ensuring efficient use and maximizing revenue from renewable generation.
- **Grid Optimization**: AI models make grid optimization decisions, such as voltage control, switching operations, and reactive power management, to enhance grid efficiency and reliability.
- **Demand Response**: AI is utilized in demand response programs to automatically advise consumers to reduce or adjust their energy consumption during peak periods, based on market conditions and grid constraints.
- **Grid Security**: AI enhances grid security by real-timely identifying anomalies, threats, and vulnerabilities, enabling swift responses to cyberattacks or system breaches.
- **Grid Resilience**: AI-driven grid management systems respond to grid disturbances and adapt to changing conditions, improving grid resilience during emergencies and unexpected events.
- **Optimal Dispatch**: AI is used for optimal dispatch of generators and energy storage assets to minimize costs and reduce emissions while ensuring grid reliability.

AI integration in economic optimization is a crucial tool for power systems to operate efficiently, adapt to changing conditions, and align with economic and sustainability objectives, ensuring a clean, reliable, and cost-effective energy future.

Optimization Models and Algorithms

Optimization models and algorithms are crucial in power systems for informed decision-making and maximizing efficiency, reliability, and economic performance, being applied across various power system areas (Agrawal et al., 2023, 2024; Satav et al., 2024).

- **Optimal Power Flow (OPF)**: OPF is used to optimize the allocation of electricity generation resources, minimize costs, and meet grid constraints while maintaining system stability. OPF models utilize mathematical optimization algorithms like linear programming or mixed-integer programming.
- **Unit Commitment**: Unit commitment models determine the optimal scheduling of power generation units, considering start-up costs, ramp rates, and operating constraints. Metaheuristic algorithms like genetic algorithms and simulated annealing are often used to solve unit commitment problems.
- **Transmission Expansion Planning**: Optimization models aid in planning the expansion and reinforcement of transmission networks to accommodate energy incorporation and ensure reliable power delivery.
- **Energy Storage Optimization**: Models and algorithms optimize the operation of energy storage systems, deciding when to charge and discharge energy to reduce costs, balance supply and demand, and enhance grid stability.
- **Load Shedding and Restoration**: Decision support systems use optimization to determine load shedding strategies during grid emergencies, ensuring critical loads are maintained while minimizing disruptions.
- **Market Bidding and Pricing**: Utilities and energy traders use optimization models to determine optimal bidding strategies in energy markets. This ensures competitive pricing and efficient use of resources.
- **Demand Response Optimization**: Optimization algorithms help utilities and consumers implement demand response programs by determining the optimal load reduction strategies during peak periods, considering customer preferences and market conditions.
- **Distribution System Optimization**: These models optimize the distribution of electricity at the lower voltage levels of the grid. They consider factors such as voltage regulation, loss minimization, and optimal placement of distributed energy resources.
- **Renewable Energy Integration**: Optimization algorithms are used to integrate renewable energy sources efficiently, taking into account their intermittent nature and grid constraints. This includes the optimal scheduling of generation and storage resources.
- **Supply Chain Management**: Optimization models help utilities manage their supply chains efficiently, from fuel procurement for power plants to spare parts inventory management for maintenance.

Decision Support Systems

Decision support systems (DSS) are crucial for power systems to make decisions in dynamic environments, using real-time data and predictive models to ensure grid reliability and resilience (Ugandar et al., 2023).

- **Grid Monitoring**: Real-time DSS systems collect data from sensors and meters to monitor grid conditions, providing immediate insights into voltage levels, line currents, and equipment health.
- **Event Detection**: DSS use algorithms to detect and identify grid events, such as equipment failures, line faults, or cyberattacks. Early event detection is critical for prompt response and minimizing disruptions.
- **Load Forecasting**: Real-time DSS incorporate load forecasting models to predict electricity demand, enabling grid operators to proactively manage supply and demand imbalances.
- **Grid Control**: DSS support grid control by making real-time decisions on voltage control, generator dispatch, and network reconfiguration to optimize grid performance and stability.
- **Cybersecurity**: Real-time DSS include cybersecurity features, such as intrusion detection systems, to identify and respond to cyber threats, helping protect critical grid infrastructure.
- **Outage Management**: DSS assist in outage management by identifying affected areas, predicting outage durations, and coordinating restoration efforts to minimize service disruptions.
- **Emergency Response**: DSS provide guidance for emergency response, ensuring that grid operators follow pre-defined procedures and protocols during grid disturbances or extreme events.
- **Visualizations and Alerts**: Real-time DSS present information through interactive visualizations and generate alerts to grid operators, allowing them to quickly understand the situation and make informed decisions.

Real-time decision support systems are crucial for grid reliability, ensuring swift responses to contingencies, optimizing operation in dynamic environments, enhancing service quality, and minimizing downtime.

CHALLENGES AND CONSIDERATIONS

Modern power systems face numerous challenges for efficient, reliable, and sustainable energy production and distribution, including data privacy and security, regulatory frameworks, technology integration challenges, and human resources and training issues (Nishanth et al., 2023; Samikannu et al., 2022; Satav et al., 2024; Vanitha et al., 2023; Venkateswaran, Vidhya, et al., 2023).

Data Privacy and Security

Modern power systems require data privacy and security due to the vast amount of sensitive information they collect and manage (Karthik et al., 2023; Kumar Reddy R. et al., 2023; Maguluri, Arularasan, et al., 2023b; Srinivas et al., 2023).

- **Data Protection**: Ensuring that customer data, operational data, and grid data are adequately protected against unauthorized access, breaches, and cyberattacks.
- **Compliance**: The text focuses on the importance of ensuring data management practices are compliant with legal requirements, such as the GDPR in the European Union.
- **Cybersecurity**: Protecting critical infrastructure from cyber threats, including vulnerabilities in smart grid technologies and the potential for large-scale disruptions caused by cyberattacks.
- **Data Sharing**: Balancing the need for data sharing among stakeholders for grid management, research, and innovation with the privacy and security concerns of individuals and organizations.
- **Secure Communication**: Ensuring that data communication within the grid, including between sensors, control systems, and data centers, is secure and encrypted to prevent interception or tampering.
- **Incident Response**: Establishing incident response and recovery plans to address data breaches or cyber incidents promptly and minimize their impact on grid operations.

Regulatory and Policy Frameworks

The regulatory and policy landscape significantly impacts the operation and evolution of modern power systems, presenting challenges and considerations (Ingle et al., 2023; Saravanan et al., 2022).

- **Market Structures**: The regulatory frameworks are being adjusted to accommodate the changing market structures, incorporating renewable energy, distributed energy resources, and demand response.
- **Environmental Regulations**: Navigating and complying with environmental regulations, such as carbon emissions limits and renewable energy mandates.
- **Grid Modernization**: Developing policies that support grid modernization efforts, including smart grid technologies and advanced metering infrastructure (AMI).
- **Cybersecurity Regulations**: Ensuring that regulations are in place to enforce cybersecurity standards for critical infrastructure and protect against cyber threats.
- **Interconnection Standards**: Establishing interconnection standards for distributed energy resources and microgrids to facilitate their integration into the grid.
- **Grid Resilience**: The focus is on improving the power grid's resilience to extreme weather events and emergencies by implementing policies to enhance its resilience.

Technology Integration Challenges

Modernizing power systems and integrating new technologies present various challenges and considerations (Boopathi, 2023c, 2023b; Koshariya, Kalaiyarasi, et al., 2023; Syamala et al., 2023). The integration of renewable energy, distributed generation, and electric vehicles into the physical grid requires compatibility, cyber-physical integration, and scalability, with data management and investment sources considered.

Human Resources and Training

The power industry necessitates a proficient workforce to efficiently operate and maintain modern power systems (Boopathi et al., 2023; Domakonda et al., 2022; Samikannu et al., 2022).

- **Workforce Transition**: Addressing the transition to a modern power system, including reskilling and upskilling the existing workforce to operate new technologies and systems.
- **Recruitment and Retention**: Attracting and retaining talent in a competitive job market, especially in technical fields related to power systems.
- **Training Programs**: Developing training programs and certifications for workers to become proficient in the operation and maintenance of modern power systems.
- **Knowledge Transfer**: Facilitating knowledge transfer from experienced workers to the next generation of power industry professionals.
- **Diversity and Inclusion**: Promoting diversity and inclusion in the power industry to ensure a broader range of perspectives and ideas for addressing challenges and innovation.

Addressing challenges is crucial for power system transformation into efficient, reliable, and sustainable networks that meet societal energy needs while achieving environmental and economic goals.

CONCLUSION

One revolutionary way to tackle the problems facing the energy sector is to incorporate artificial intelligence (AI) into power systems. AI-driven solutions provide sustainability, resilience, cost savings, and increased efficiency. The electricity sector is going through a paradigm transition, where success now largely depends on enhanced automation.

AI-powered economic optimization can significantly decrease costs and enhance operational efficiency for utilities and grid operators. By using AI for load forecasting, resource allocation, and energy trading, utilities can reduce operational expenses and improve asset management through predictive maintenance. AI-driven grid management systems ensure real-time decision-making, balancing supply and demand, and reducing service disruption risks. This increased grid resilience is crucial for uninterrupted power supply, especially in the face of weather events and cyber threats. Energy storage systems also provide backup power during outages, further enhancing reliability.

AI addition in power systems is crucial for environmental sustainability, as it encourages renewable energy sources, reduces fossil fuel dependency, and decreases carbon emissions. This aligns operations with environmental regulations and sustainability goals, promoting cleaner energy production and consumption. Future prospects for AI-enabled economic optimization in power systems are promising, with advancements in deep learning, quantum computing, and data analytics. AI-facilitated grid decentralization will lead to distributed energy resources and microgrids, enabling a more dynamic power system. Smart city initiatives and the Internet of Things will provide a holistic view of energy data, facilitating more precise decision-making.

AI is revolutionizing the power industry, offering a cost-effective, reliable, and sustainable energy solution. Its economic optimization allows power systems to meet societal demands while addressing environmental and economic objectives.

ABBREVIATIONS

AI - Artificial Intelligence

OPF - Optimal Power Flow
IoT - Internet of Things
GDPR - General Data Protection Regulation
AMI - Advanced Metering Infrastructure

REFERENCES

Abdalla, A. N., Nazir, M. S., Tao, H., Cao, S., Ji, R., Jiang, M., & Yao, L. (2021). Integration of energy storage system and renewable energy sources based on artificial intelligence: An overview. *Journal of Energy Storage*, *40*, 102811. doi:10.1016/j.est.2021.102811

Agrawal, A. V., Magulur, L. P., Priya, S. G., Kaur, A., Singh, G., & Boopathi, S. (2023). Smart Precision Agriculture Using IoT and WSN. In *Handbook of Research on Data Science and Cybersecurity Innovations in Industry 4.0 Technologies* (pp. 524–541). IGI Global. doi:10.4018/978-1-6684-8145-5.ch026

Agrawal, A. V., Shashibhushan, G., Pradeep, S., Padhi, S. N., Sugumar, D., & Boopathi, S. (2024). Synergizing Artificial Intelligence, 5G, and Cloud Computing for Efficient Energy Conversion Using Agricultural Waste. In Practice, Progress, and Proficiency in Sustainability (pp. 475–497). IGI Global. doi:10.4018/979-8-3693-1186-8.ch026

Ahmad, T., Zhang, D., Huang, C., Zhang, H., Dai, N., Song, Y., & Chen, H. (2021). Artificial intelligence in sustainable energy industry: Status Quo, challenges and opportunities. *Journal of Cleaner Production*, *289*, 125834. doi:10.1016/j.jclepro.2021.125834

Akram, U., Nadarajah, M., Shah, R., & Milano, F. (2020). A review on rapid responsive energy storage technologies for frequency regulation in modern power systems. *Renewable & Sustainable Energy Reviews*, *120*, 109626. doi:10.1016/j.rser.2019.109626

Ali, S. S., & Choi, B. J. (2020). State-of-the-art artificial intelligence techniques for distributed smart grids: A review. *Electronics (Basel)*, *9*(6), 1030. doi:10.3390/electronics9061030

Anitha, C., Komala, C., Vivekanand, C. V., Lalitha, S., & Boopathi, S. (2023). Artificial Intelligence driven security model for Internet of Medical Things (IoMT). *IEEE Explore*, 1–7.

Babatunde, O. M., Munda, J. L., & Hamam, Y. (2020). Power system flexibility: A review. *Energy Reports*, *6*, 101–106. doi:10.1016/j.egyr.2019.11.048

Babu, B. S., Kamalakannan, J., Meenatchi, N., Karthik, S., & Boopathi, S. (2022). Economic impacts and reliability evaluation of battery by adopting Electric Vehicle. *IEEE Explore*, 1–6.

Boopathi, S. (2022a). An investigation on gas emission concentration and relative emission rate of the near-dry wire-cut electrical discharge machining process. *Environmental Science and Pollution Research International*, *29*(57), 86237–86246. doi:10.1007/s11356-021-17658-1 PMID:34837614

Boopathi, S. (2022b). Cryogenically treated and untreated stainless steel grade 317 in sustainable wire electrical discharge machining process: A comparative study. *Environmental Science and Pollution Research*, 1–10. Springer.

Boopathi, S. (2023a). Deep Learning Techniques Applied for Automatic Sentence Generation. In Promoting Diversity, Equity, and Inclusion in Language Learning Environments (pp. 255–273). IGI Global. doi:10.4018/978-1-6684-3632-5.ch016

Boopathi, S. (2023b). Internet of Things-Integrated Remote Patient Monitoring System: Healthcare Application. In *Dynamics of Swarm Intelligence Health Analysis for the Next Generation* (pp. 137–161). IGI Global. doi:10.4018/978-1-6684-6894-4.ch008

Boopathi, S. (2023c). Securing Healthcare Systems Integrated With IoT: Fundamentals, Applications, and Future Trends. In Dynamics of Swarm Intelligence Health Analysis for the Next Generation (pp. 186–209). IGI Global.

Boopathi, S., & Davim, J. P. (2023). *Sustainable Utilization of Nanoparticles and Nanofluids in Engineering Applications*. IGI Global. doi:10.4018/978-1-6684-9135-5

Boopathi, S., Kumar, P. K. S., Meena, R. S., Sudhakar, M., & Associates. (2023). Sustainable Developments of Modern Soil-Less Agro-Cultivation Systems: Aquaponic Culture. In Human Agro-Energy Optimization for Business and Industry (pp. 69–87). IGI Global.

Boopathi, S., Sureshkumar, M., & Sathiskumar, S. (2022). Parametric Optimization of LPG Refrigeration System Using Artificial Bee Colony Algorithm. *International Conference on Recent Advances in Mechanical Engineering Research and Development*, (pp. 97–105). IEEE.

Cao, D., Hu, W., Zhao, J., Zhang, G., Zhang, B., Liu, Z., Chen, Z., & Blaabjerg, F. (2020). Reinforcement learning and its applications in modern power and energy systems: A review. *Journal of Modern Power Systems and Clean Energy*, 8(6), 1029–1042. doi:10.35833/MPCE.2020.000552

Chandrika, V., Sivakumar, A., Krishnan, T. S., Pradeep, J., Manikandan, S., & Boopathi, S. (2023). Theoretical Study on Power Distribution Systems for Electric Vehicles. In *Intelligent Engineering Applications and Applied Sciences for Sustainability* (pp. 1–19). IGI Global. doi:10.4018/979-8-3693-0044-2.ch001

Domakonda, V. K., Farooq, S., Chinthamreddy, S., Puviarasi, R., Sudhakar, M., & Boopathi, S. (2022). Sustainable Developments of Hybrid Floating Solar Power Plants: Photovoltaic System. In Human Agro-Energy Optimization for Business and Industry (pp. 148–167). IGI Global.

Gowri, N. V., Dwivedi, J. N., Krishnaveni, K., Boopathi, S., Palaniappan, M., & Medikondu, N. R. (2023). Experimental investigation and multi-objective optimization of eco-friendly near-dry electrical discharge machining of shape memory alloy using Cu/SiC/Gr composite electrode. *Environmental Science and Pollution Research International*, 30(49), 1–19. doi:10.1007/s11356-023-26983-6 PMID:37126160

Hema, N., Krishnamoorthy, N., Chavan, S. M., Kumar, N., Sabarimuthu, M., & Boopathi, S. (2023). A Study on an Internet of Things (IoT)-Enabled Smart Solar Grid System. In *Handbook of Research on Deep Learning Techniques for Cloud-Based Industrial IoT* (pp. 290–308). IGI Global. doi:10.4018/978-1-6684-8098-4.ch017

Hussain, Z., & Srimathy, G. (2023). *IoT and AI Integration for Enhanced Efficiency and Sustainability*.

Ibrahim, M. S., Dong, W., & Yang, Q. (2020). Machine learning driven smart electric power systems: Current trends and new perspectives. *Applied Energy*, 272, 115237. doi:10.1016/j.apenergy.2020.115237

Ingle, R. B., Senthil, T. S., Swathi, S., Muralidharan, N., Mahendran, G., & Boopathi, S. (2023). Sustainability and Optimization of Green and Lean Manufacturing Processes Using Machine Learning Techniques. In IGI Global. doi:10.4018/978-1-6684-8238-4.ch012

Iqbal, A., Malik, H., Riyaz, A., Abdellah, K., & Bayhan, S. (2021). Renewable Power for Sustainable Growth: *Proceedings of International Conference on Renewal Power (ICRP 2020)*. Springer.

Karthik, S. A., Hemalatha, R., Aruna, R., Deivakani, M., Reddy, R. V. K., & Boopathi, S. (2023). Study on Healthcare Security System-Integrated Internet of Things (IoT). In M. K. Habib (Ed.), (pp. 342–362). Advances in Systems Analysis, Software Engineering, and High Performance Computing. IGI Global. doi:10.4018/978-1-6684-7684-0.ch013

Kavitha, C. R., Varalatchoumy, M., Mithuna, H. R., Bharathi, K., Geethalakshmi, N. M., & Boopathi, S. (2023). Energy Monitoring and Control in the Smart Grid: Integrated Intelligent IoT and ANFIS. In M. Arshad (Ed.), (pp. 290–316). Advances in Bioinformatics and Biomedical Engineering. IGI Global. doi:10.4018/978-1-6684-6577-6.ch014

Koshariya, A. K., Kalaiyarasi, D., Jovith, A. A., Sivakami, T., Hasan, D. S., & Boopathi, S. (2023). AI-Enabled IoT and WSN-Integrated Smart Agriculture System. In *Artificial Intelligence Tools and Technologies for Smart Farming and Agriculture Practices* (pp. 200–218). IGI Global. doi:10.4018/978-1-6684-8516-3.ch011

Koshariya, A. K., Khatoon, S., Marathe, A. M., Suba, G. M., Baral, D., & Boopathi, S. (2023). Agricultural Waste Management Systems Using Artificial Intelligence Techniques. In *AI-Enabled Social Robotics in Human Care Services* (pp. 236–258). IGI Global. doi:10.4018/978-1-6684-8171-4.ch009

Kumar, N. M., Chand, A. A., Malvoni, M., Prasad, K. A., Mamun, K. A., Islam, F., & Chopra, S. S. (2020). Distributed energy resources and the application of AI, IoT, and blockchain in smart grids. *Energies*, 13(21), 5739. doi:10.3390/en13215739

Kumar, P. R., Meenakshi, S., Shalini, S., Devi, S. R., & Boopathi, S. (2023). Soil Quality Prediction in Context Learning Approaches Using Deep Learning and Blockchain for Smart Agriculture. In R. Kumar, A. B. Abdul Hamid, & N. I. Binti Ya'akub (Eds.), (pp. 1–26). Advances in Computational Intelligence and Robotics. IGI Global., doi:10.4018/978-1-6684-9151-5.ch001

Kumar Reddy, R. V., Rahamathunnisa, U., Subhashini, P., Aancy, H. M., Meenakshi, S., & Boopathi, S. (2023). Solutions for Software Requirement Risks Using Artificial Intelligence Techniques: In T. Murugan & N. E. (Eds.), Advances in Information Security, Privacy, and Ethics (pp. 45–64). IGI Global. doi:10.4018/978-1-6684-8145-5.ch003

Maguluri, L. P., Ananth, J., Hariram, S., Geetha, C., Bhaskar, A., & Boopathi, S. (2023). Smart Vehicle-Emissions Monitoring System Using Internet of Things (IoT). In Handbook of Research on Safe Disposal Methods of Municipal Solid Wastes for a Sustainable Environment (pp. 191–211). IGI Global.

Maguluri, L. P., Arularasan, A. N., & Boopathi, S. (2023a). Assessing Security Concerns for AI-Based Drones in Smart Cities. In R. Kumar, A. B. Abdul Hamid, & N. I. Binti Ya'akub (Eds.), (pp. 27–47). Advances in Computational Intelligence and Robotics. IGI Global.., doi:10.4018/978-1-6684-9151-5.ch002

Maguluri, L. P., Arularasan, A. N., & Boopathi, S. (2023b). Assessing Security Concerns for AI-Based Drones in Smart Cities. In R. Kumar, A. B. Abdul Hamid, & N. I. Binti Ya'akub (Eds.), (pp. 27–47). Advances in Computational Intelligence and Robotics. IGI Global.., doi:10.4018/978-1-6684-9151-5.ch002

Nishanth, J., Deshmukh, M. A., Kushwah, R., Kushwaha, K. K., Balaji, S., & Sampath, B. (2023). Particle Swarm Optimization of Hybrid Renewable Energy Systems. In *Intelligent Engineering Applications and Applied Sciences for Sustainability* (pp. 291–308). IGI Global. doi:10.4018/979-8-3693-0044-2.ch016

Peyghami, S., Palensky, P., & Blaabjerg, F. (2020). An overview on the reliability of modern power electronic based power systems. *IEEE Open Journal of Power Electronics*, *1*, 34–50. doi:10.1109/OJPEL.2020.2973926

Ramudu, K., Mohan, V. M., Jyothirmai, D., Prasad, D., Agrawal, R., & Boopathi, S. (2023). Machine Learning and Artificial Intelligence in Disease Prediction: Applications, Challenges, Limitations, Case Studies, and Future Directions. In Contemporary Applications of Data Fusion for Advanced Healthcare Informatics (pp. 297–318). IGI Global.

Ravisankar, A., Sampath, B., & Asif, M. M. (2023). Economic Studies on Automobile Management: Working Capital and Investment Analysis. In Multidisciplinary Approaches to Organizational Governance During Health Crises (pp. 169–198). IGI Global.

Reddy, M. A., Gaurav, A., Ushasukhanya, S., Rao, V. C. S., Bhattacharya, S., & Boopathi, S. (2023). Bio-Medical Wastes Handling Strategies During the COVID-19 Pandemic. In Multidisciplinary Approaches to Organizational Governance During Health Crises (pp. 90–111). IGI Global. doi:10.4018/978-1-7998-9213-7.ch006

Samikannu, R., Koshariya, A. K., Poornima, E., Ramesh, S., Kumar, A., & Boopathi, S. (2022). Sustainable Development in Modern Aquaponics Cultivation Systems Using IoT Technologies. In *Human Agro-Energy Optimization for Business and Industry* (pp. 105–127). IGI Global.

Saravanan, A., Venkatasubramanian, R., Khare, R., Surakasi, R., Boopathi, S., Ray, S., & Sudhakar, M. (2022). POLICY TRENDS OF RENEWABLE ENERGY AND NON. *Renewable Energy*.

Satav, S. D., & Lamani, D. G, H. K., Kumar, N. M. G., Manikandan, S., & Sampath, B. (2024). Energy and Battery Management in the Era of Cloud Computing. In Practice, Progress, and Proficiency in Sustainability (pp. 141–166). IGI Global. doi:10.4018/979-8-3693-1186-8.ch009

Selvakumar, S., Shankar, R., Ranjit, P., Bhattacharya, S., Gupta, A. S. G., & Boopathi, S. (2023). E-Waste Recovery and Utilization Processes for Mobile Phone Waste. In *Handbook of Research on Safe Disposal Methods of Municipal Solid Wastes for a Sustainable Environment* (pp. 222–240). IGI Global. doi:10.4018/978-1-6684-8117-2.ch016

Sengeni, D., Padmapriya, G., Imambi, S. S., Suganthi, D., Suri, A., & Boopathi, S. (2023). Biomedical Waste Handling Method Using Artificial Intelligence Techniques. In *Handbook of Research on Safe Disposal Methods of Municipal Solid Wastes for a Sustainable Environment* (pp. 306–323). IGI Global. doi:10.4018/978-1-6684-8117-2.ch022

Senthil, T. S., Ohmsakthi Vel, R., Puviyarasan, M., Babu, S. R., Surakasi, R., & Sampath, B. (2023). Industrial Robot-Integrated Fused Deposition Modelling for the 3D Printing Process. In R. Keshavamurthy, V. Tambrallimath, & J. P. Davim (Eds.), (pp. 188–210). Advances in Chemical and Materials Engineering. IGI Global., doi:10.4018/978-1-6684-6009-2.ch011

Shair, J., Li, H., Hu, J., & Xie, X. (2021). Power system stability issues, classifications and research prospects in the context of high-penetration of renewables and power electronics. *Renewable & Sustainable Energy Reviews, 145*, 111111. doi:10.1016/j.rser.2021.111111

Sharma, P., Said, Z., Kumar, A., Nizetic, S., Pandey, A., Hoang, A. T., Huang, Z., Afzal, A., Li, C., Le, A. T., Nguyen, X. P., & Tran, V. D. (2022). Recent advances in machine learning research for nanofluid-based heat transfer in renewable energy system. *Energy & Fuels, 36*(13), 6626–6658. doi:10.1021/acs.energyfuels.2c01006

Srinivas, B., Maguluri, L. P., Naidu, K. V., Reddy, L. C. S., Deivakani, M., & Boopathi, S. (2023). Architecture and Framework for Interfacing Cloud-Enabled Robots: In T. Murugan & N. E. (Eds.), Advances in Information Security, Privacy, and Ethics (pp. 542–560). IGI Global. doi:10.4018/978-1-6684-8145-5.ch027

Syamala, M., Komala, C., Pramila, P., Dash, S., Meenakshi, S., & Boopathi, S. (2023). Machine Learning-Integrated IoT-Based Smart Home Energy Management System. In *Handbook of Research on Deep Learning Techniques for Cloud-Based Industrial IoT* (pp. 219–235). IGI Global. doi:10.4018/978-1-6684-8098-4.ch013

Ugandar, R. E., Rahamathunnisa, U., Sajithra, S., Christiana, M. B. V., Palai, B. K., & Boopathi, S. (2023). Hospital Waste Management Using Internet of Things and Deep Learning: Enhanced Efficiency and Sustainability. In M. Arshad (Ed.), (pp. 317–343). Advances in Bioinformatics and Biomedical Engineering. IGI Global. doi:10.4018/978-1-6684-6577-6.ch015

Vanitha, S., Radhika, K., & Boopathi, S. (2023). Artificial Intelligence Techniques in Water Purification and Utilization. In *Human Agro-Energy Optimization for Business and Industry* (pp. 202–218). IGI Global. doi:10.4018/978-1-6684-4118-3.ch010

Veeranjaneyulu, R., Boopathi, S., Kumari, R. K., Vidyarthi, A., Isaac, J. S., & Jaiganesh, V. (2023). Air Quality Improvement and Optimisation Using Machine Learning Technique. *IEEE- Explore*, 1–6.

Venkateswaran, N., Kumar, S. S., Diwakar, G., Gnanasangeetha, D., & Boopathi, S. (2023). Synthetic Biology for Waste Water to Energy Conversion: IoT and AI Approaches. In M. Arshad (Ed.), (pp. 360–384). Advances in Bioinformatics and Biomedical Engineering. IGI Global. doi:10.4018/978-1-6684-6577-6.ch017

Venkateswaran, N., Vidhya, K., Ayyannan, M., Chavan, S. M., Sekar, K., & Boopathi, S. (2023). A Study on Smart Energy Management Framework Using Cloud Computing. In 5G, Artificial Intelligence, and Next Generation Internet of Things: Digital Innovation for Green and Sustainable Economies (pp. 189–212). IGI Global. doi:10.4018/978-1-6684-8634-4.ch009

Yang, T., Zhao, L., Li, W., & Zomaya, A. Y. (2020). Reinforcement learning in sustainable energy and electric systems: A survey. *Annual Reviews in Control, 49*, 145–163. doi:10.1016/j.arcontrol.2020.03.001

Yap, K. Y., Sarimuthu, C. R., & Lim, J. M.-Y. (2020). Artificial intelligence based MPPT techniques for solar power system: A review. *Journal of Modern Power Systems and Clean Energy*, 8(6), 1043–1059. doi:10.35833/MPCE.2020.000159

Zekrifa, D. M. S., Kulkarni, M., Bhagyalakshmi, A., Devireddy, N., Gupta, S., & Boopathi, S. (2023). Integrating Machine Learning and AI for Improved Hydrological Modeling and Water Resource Management. In *Artificial Intelligence Applications in Water Treatment and Water Resource Management* (pp. 46–70). IGI Global. doi:10.4018/978-1-6684-6791-6.ch003

Zhang, Y., Shi, X., Zhang, H., Cao, Y., & Terzija, V. (2022). Review on deep learning applications in frequency analysis and control of modern power system. *International Journal of Electrical Power & Energy Systems*, 136, 107744. doi:10.1016/j.ijepes.2021.107744

Zhao, Y., Li, T., Zhang, X., & Zhang, C. (2019). Artificial intelligence-based fault detection and diagnosis methods for building energy systems: Advantages, challenges and the future. *Renewable & Sustainable Energy Reviews*, 109, 85–101. doi:10.1016/j.rser.2019.04.021

Chapter 10
Adaptive Intelligence in Microgrid Systems:
Harnessing Machine Learning for Efficiency

S. Saravanan

https://orcid.org/0000-0001-8255-2623

Department of Electrical and Electronics Engineering, B.V. Raju Institute of Technology, India

N. M. G. Kumar

https://orcid.org/0000-0003-1494-5737

Department of Electrical and Electronics Engineering, Sree Vidyanikethan Engineering College, Mohan Babu University, India

Putchakayala Yanna Reddy

Department of Electrical & Electronics Engineering, Bharath Institute of Engineering and Technology, India

R. Ramya Sri

Department of English, Kongu Engineering College, India

M. Ramesh

Department of Aerospace Engineering, SNS College of Technology, Coimbatore, India

B. Sampath

Mechanical Engineering, Mythayammal Engineering College (Autonomous), India

ABSTRACT

Microgrid systems, with diverse energy sources and decentralized control, are revolutionizing energy management. However, integrating renewable energy sources and consumer demands poses challenges to grid stability. Machine learning (ML) is a key tool for optimizing energy management and adaptive control in microgrids, enabling accurate load forecasting, renewable energy output prediction, and efficient resource utilization. This abstract discusses the potential of ML in revolutionizing microgrid systems by enabling adaptive control mechanisms, predictive maintenance, early fault detection, and proactive scheduling. ML also addresses cybersecurity concerns, providing sophisticated solutions for intrusion detection and secure data management. This approach optimizes operations, drives innovation, and ensures resilient, cost-effective, and sustainable energy infrastructures.

DOI: 10.4018/979-8-3693-3735-6.ch010

INTRODUCTION

Microgrid systems are a new approach to traditional centralized energy distribution, offering a localized, adaptable system for power generation, distribution, and consumption. These systems consist of interconnected energy sources and loads, which can operate independently or in conjunction with the main grid, providing increased flexibility and resilience. They address challenges like aging infrastructure, climate change, and increased demand, making them a promising alternative (Chandraratne et al., 2020).

Microgrid systems utilize various energy sources like solar panels and wind turbines, along with energy storage systems. Advanced control systems and intelligent algorithms ensure efficient resource utilization. Microgrids offer improved energy reliability, especially in remote areas, and enable the integration of renewable energy sources, contributing to environmental sustainability by reducing reliance on fossil fuels and lowering carbon emissions. Microgrid systems offer a decentralized, reliable, and autonomous power supply, benefiting communities, businesses, and critical facilities (Barra et al., 2020). The evolution of microgrid technology is driven by advancements in control systems, energy storage, and smart technologies like IoT and machine learning. This evolution optimizes microgrid performance, enhances energy efficiency, and ensures seamless integration with the broader electrical infrastructure. Understanding and harnessing the potential of microgrid systems is crucial for shaping the future of energy distribution and consumption, as the world seeks more sustainable energy solutions (Kaur et al., 2021).

Microgrids are a revolutionary energy distribution system that offer localized, self-sufficient networks capable of generating, managing, and distributing electricity. They are smaller versions of traditional power grids, but possess unique characteristics that differentiate them from centralized systems. Microgrids are adaptable, resilient, and integrate diverse energy sources. They can operate connected to the main grid or autonomously, serving various purposes like supporting remote communities and optimizing energy use in urban environments. Their modularity allows for scalability, allowing them to be tailored to specific needs and expanded as demand increases (Khaleel, 2023).

Microgrids are energy systems that use renewable energy sources like solar panels, wind turbines, and biomass or hydroelectric. They are integrated with energy storage systems like batteries or flywheels to ensure continuous power supply even when renewable sources fluctuate. Advanced control and monitoring systems, using algorithms, real-time data analytics, and smart sensors, optimize energy flow, balance supply and demand, and ensure grid stability. These technologies also enable predictive maintenance, enhancing reliability and mitigating potential issues (Behera & Dev Choudhury, 2021).

Microgrids are crucial for grid resilience and energy management due to their decentralized approach to power generation and distribution. They reduce vulnerabilities to outages and enhance system reliability. As the energy landscape evolves, microgrids' adaptability, sustainability, and reliability make them essential for future energy infrastructure. Efficiency is a cornerstone in microgrid operations, optimizing resource utilization, enhancing cost-effectiveness, and ensuring power supply reliability (Leonori et al., 2020). This focus on efficiency encompasses energy generation, distribution, and system management. Microgrid efficiency is a system that optimizes energy resources by combining renewable sources like solar, wind, and hydro with conventional ones. This ensures clean and sustainable power, reducing dependency on non-renewable resources and minimizing environmental impact. Energy storage technologies, like batteries or flywheels, store excess energy during peak production periods, which can be used during high demand or less productive periods, ensuring a stable power supply. This approach significantly contributes to microgrid efficiency (Ramesh et al., 2021).

Microgrid operations are efficient due to advanced control systems and intelligent algorithms that manage energy flow and distribution. This ensures optimal resource utilization, minimizes transmission losses, and enhances the microgrid's effectiveness. Demand response mechanisms also contribute to a balanced load profile and grid stability. These mechanisms allow adjustments in energy consumption based on real-time pricing or grid conditions, encouraging consumers to use electricity during off-peak hours or reduce usage during high demand periods. Efficiency in microgrid operations is crucial for ensuring reliability, resilience, and sustainability of energy supply (Karimi et al., 2021). Continuous efforts to improve efficiency standards can maximize benefits and contribute to a more robust and adaptive energy infrastructure.

Background

The chapter discusses the role of machine learning in microgrid systems, highlighting its significant impact on efficiency and resilience. It highlights how machine learning optimizes energy usage, adapts to dynamic conditions, and improves microgrid operations. It also highlights the ability of machine learning to facilitate accurate load forecasting, ensuring optimal energy distribution within microgrids and minimizing waste. Additionally, machine learning enhances the integration of renewable energy sources, enabling adaptive control mechanisms that improve grid resilience and stability in uncertain and disturbance-affected conditions.

The chapter explores the benefits of machine learning-driven predictive maintenance strategies in microgrid operations, highlighting their cost-efficiency and sustainability. It also addresses challenges like data security, scalability, and interoperability, emphasizing the need for improved cybersecurity measures, standardized protocols, and continuous advancements in machine learning techniques to address these issues and revolutionize maintenance practices in microgrids.

The chapter highlights the transformative impact of machine learning on the energy industry, reshaping energy management, promoting renewable energy integration, and contributing to sustainable energy ecosystems. It emphasizes the need for ongoing advancements and a proactive approach to address challenges, ensuring the integration of machine learning in microgrid systems propels the industry towards a more sustainable future.

FOUNDATIONS OF MACHINE LEARNING IN ENERGY

Machine Learning Applications in Energy Systems

- **Predictive Maintenance:** Machine learning models analyze data from sensors and equipment to predict potential failures or maintenance needs in energy infrastructure. This proactive approach helps in scheduling maintenance before breakdowns occur, minimizing downtime, and optimizing operational efficiency (Khokhar & Parmar, 2022).
- **Load Forecasting:** ML algorithms analyze historical data to forecast energy demand accurately. These forecasts aid utilities in planning and optimizing power generation, ensuring sufficient supply to meet demand, and preventing under or overproduction.
- **Optimization of Energy Distribution:** Machine learning optimizes the distribution of energy within the grid. Algorithms manage transmission lines, adjust voltage levels, and route energy

flows efficiently, reducing losses and enhancing grid reliability (Boopathi et al., 2023; Nishanth et al., 2023; Sampath et al., 2022).

- **Energy Consumption Analytics:** ML algorithms analyze consumer data to understand patterns in energy usage. This insight helps in designing personalized energy-saving recommendations for consumers and implementing demand-side management strategies (Kavitha et al., 2023a; Satav et al., 2024; Venkateswaran, Vidhya, Ayyannan, et al., 2023).
- **Renewable Energy Integration:** Machine learning assists in integrating renewable energy sources into the grid by forecasting solar or wind energy generation. This aids in better grid balancing and management when dealing with intermittent renewable sources (Nishanth et al., 2023).
- **Fault Detection and Diagnostics:** ML models detect anomalies in the grid, identifying faults or abnormalities in real-time. This enables swift responses to potential issues, minimizing downtime and ensuring grid stability.
- **Energy Trading and Market Forecasting:** Machine learning is employed in predicting energy market prices and trends. This aids in strategic decision-making for energy trading, optimizing revenue generation, and managing risks associated with market fluctuations.
- **Grid Optimization and Control:** ML algorithms enable adaptive and self-learning control systems for the grid. These systems can adjust parameters in real-time, optimizing grid operations and ensuring efficient energy delivery (Hema et al., 2023a; Kavitha et al., 2023a).
- **Resource Allocation and Asset Management:** Machine learning helps in optimal resource allocation, such as deploying energy storage systems or allocating resources to different parts of the grid based on demand patterns. It also assists in asset management by predicting equipment lifespan and identifying the need for replacements.

These fundamental applications demonstrate the versatility of machine learning in enhancing efficiency, reliability, and sustainability across various facets of energy systems.

Role of Machine Learning in Microgrid Optimization

Machine learning plays a crucial role in optimizing microgrid systems, contributing significantly to their efficiency, reliability, and overall performance (Barra et al., 2020; Kaur et al., 2021).

- **Energy Forecasting and Demand Prediction:** Machine learning models analyze historical data to forecast energy generation from renewable sources like solar or wind, as well as predict consumer demand. Accurate predictions aid in scheduling operations, optimizing energy storage, and balancing supply and demand within the microgrid.
- **Dynamic Control and Energy Management:** ML algorithms enable real-time monitoring and control of energy flow within the microgrid. These systems adjust parameters, such as battery charging/discharging rates or switching between energy sources, to maximize efficiency and minimize costs based on current conditions.
- **Optimal Resource Allocation:** Machine learning assists in determining the most efficient use of resources within the microgrid. This includes deciding when to use renewable sources, energy storage systems, or even when to draw from or supply energy to the main grid for economic or reliability reasons.

- **Fault Detection and Diagnostics:** ML models continuously analyze data from sensors within the microgrid, detecting anomalies or faults. Early detection allows for swift responses to mitigate potential issues, minimizing downtime, and ensuring grid stability.
- **Adaptive Learning for Improved Efficiency:** Machine learning algorithms can adapt and learn from changing conditions and operational data. Over time, these systems become more accurate and efficient in optimizing microgrid operations, adjusting strategies based on performance feedback (Ramudu et al., 2023; Syamala, C. R., et al., 2023; Venkateswaran, Vidhya, Naik, et al., 2023).
- **Demand Response and Consumer Behavior Analysis:** ML helps in understanding consumer behavior patterns regarding energy usage. This insight enables the implementation of demand response programs, encouraging consumers to adjust consumption during peak hours or in response to price signals, contributing to overall grid stability.
- **Integration of New Technologies:** Machine learning aids in integrating emerging technologies into microgrid systems, such as incorporating AI-driven predictive maintenance for equipment or utilizing advanced control systems for optimal energy distribution.

Overall, machine learning's adaptive and data-driven capabilities are instrumental in continually improving the efficiency, resilience, and sustainability of microgrid operations by optimizing decision-making and system control in response to dynamic and complex conditions.

DATA ACQUISITION AND PREPROCESSING

Data acquisition and preprocessing are fundamental stages in harnessing the potential of data for optimizing microgrid operations. These processes lay the groundwork for effective analysis, modeling, and decision-making within the microgrid system (Dhanya et al., 2023a; Pramila et al., 2023).

The first step involves identifying and gathering data from various sources within the microgrid. This encompasses information from smart meters, sensors monitoring energy generation and consumption, weather data for renewable energy sources, and grid status data. Ensuring the quality and consistency of data from these diverse sources is critical. Advanced metering infrastructure and IoT devices play a pivotal role in continuous data collection, providing real-time insights into the performance and status of the microgrid components (Rebecca et al., 2023).

Once collected, the data undergoes preprocessing, which involves cleaning, formatting, and transforming raw data into a usable format for analysis. This step includes handling missing values, removing outliers, and standardizing data to ensure uniformity. Additionally, data normalization and scaling techniques are employed to bring data within consistent ranges, facilitating accurate analysis and model training. Preprocessing also involves timestamp alignment and synchronization of disparate datasets to enable meaningful correlations and insights across various parameters (Dhanya et al., 2023b).

Feature engineering involves creating new features or extracting meaningful information from existing data to enhance its predictive power. Techniques like dimensionality reduction, where irrelevant or redundant features are eliminated, and feature scaling, which brings features to a similar scale, are applied. Feature selection identifies the most relevant variables that significantly impact the performance of models. This process streamlines the dataset, focusing on the most informative features, reducing

Figure 1. Sources of data in microgrid systems

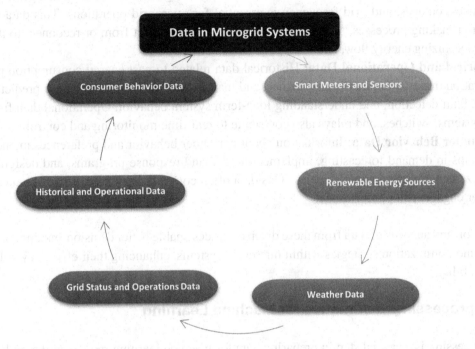

computational complexity, and enhancing the accuracy of predictive models used in optimizing microgrid operations (Hema et al., 2023b; Kavitha et al., 2023b).

Sources of Data in Microgrid Systems

Microgrid systems utilize data from various sources to monitor, analyze, and optimize energy generation, distribution, and consumption, with primary sources being (Moharm, 2019; Portalo et al., 2021; Tan et al., 2020). Figure 1 depicts the various sources of data in microgrid systems.

- **Smart Meters and Sensors:** Smart meters installed at different points within the microgrid measure energy consumption in real-time. These meters provide granular data on electricity usage for individual households, buildings, or specific equipment. Additionally, sensors deployed throughout the microgrid monitor parameters such as voltage, current, frequency, and temperature, offering insights into system performance and health.
- **Renewable Energy Sources:** Data from renewable sources like solar panels, wind turbines, or hydroelectric systems are crucial for understanding the fluctuating nature of renewable energy generation. This data includes solar irradiance, wind speed, hydroelectric flow rates, and other environmental variables that affect energy production.
- **Weather Data:** Weather conditions significantly impact the performance of renewable energy sources. Meteorological data, such as sunlight intensity, wind speed, temperature, and humidity, is collected from weather stations or online sources to predict energy generation and optimize microgrid operations.

- **Grid Status and Operations Data:** Information regarding the state of the main grid, grid disturbances, outages, and grid frequency is essential for microgrid operations. This data helps in decision-making processes, such as deciding when to disconnect from or reconnect to the main grid, optimizing energy flow, and managing grid stability.
- **Historical and Operational Data:** Historical data related to past energy consumption patterns, generation trends, equipment performance, and maintenance logs are valuable for predictive analytics, fault detection, and understanding long-term system behavior. Operational data from control systems, switches, and relays also contribute to real-time monitoring and control.
- **Consumer Behavior Data:** Information about consumer behavior and preferences in energy usage helps in demand forecasting, implementing demand response programs, and designing personalized energy efficiency strategies. This data often comes from surveys, smart home devices, or user engagement platforms.

Integration and analysis of data from these diverse sources enable better decision-making, predictive modeling, and optimization strategies within microgrid systems, enhancing their efficiency, reliability, and sustainability.

Data Preprocessing Techniques for Machine Learning

Data preprocessing is a crucial step in preparing data for machine learning models, utilizing key techniques such as (Portalo et al., 2021):

- **Handling Missing Values:** Dealing with missing data is crucial. Techniques include imputation, where missing values are filled in using statistical measures (like mean, median, or mode), or deletion of rows or columns with missing values if appropriate.
- **Data Cleaning:** This involves removing or correcting noisy data, outliers, or irrelevant information that could negatively impact model performance. Outliers might be corrected or removed based on domain knowledge.
- **Normalization and Standardization:** Normalization scales numeric features within a specific range (commonly between 0 and 1) to ensure uniformity, while standardization rescales data to have a mean of 0 and a standard deviation of 1. These techniques prevent features with larger scales from dominating the model training.
- **Encoding Categorical Variables:** Categorical variables (like 'red,' 'blue,' 'green' in a 'color' column) are encoded into numerical values. One-hot encoding and label encoding are common methods used for this purpose.
- **Feature Scaling:** Scaling features to a similar range helps algorithms converge faster during model training. Techniques like Min-Max scaling or Z-score normalization are employed depending on the nature of the data.
- **Handling Imbalanced Data:** In cases where one class significantly outweighs the others in a classification problem, techniques like oversampling, under-sampling, or using synthetic data generation methods like SMOTE (Synthetic Minority Over-sampling Technique) can balance the dataset.

- **Feature Engineering:** Creating new features or transforming existing ones to improve model performance. Techniques include binning, polynomial features, or extracting meaningful information from date/time variables.
- **Dimensionality Reduction:** Techniques like Principal Component Analysis (PCA) or t-Distributed Stochastic Neighbor Embedding (t-SNE) are used to reduce the number of features while preserving important information. This helps in dealing with high-dimensional data and reducing computational complexity.
- **Splitting Data into Training and Validation Sets:** Data is split into training and validation sets to train the model on one portion and evaluate its performance on unseen data. Common splits include 70-30 or 80-20 for training and validation, respectively.

MACHINE LEARNING ALGORITHMS FOR MICROGRID OPTIMIZATION

Machine learning algorithms are crucial in optimizing microgrid operations by utilizing data-driven insights to improve efficiency, reliability, and sustainability. Linear Regression is a useful algorithm for modeling relationships between dependent and independent variables, predicting energy demand or generation based on historical data and weather conditions. Decision Trees are useful for determining optimal decisions based on specific microgrid conditions, such as when to switch between energy sources or storage systems (Boopathi & Kanike, 2023; Syamala, Komala, et al., 2023; Veeranjaneyulu, Boopathi, Kumari, et al., 2023; Veeranjaneyulu, Boopathi, Narasimharao, et al., 2023).

Random Forest, Support Vector Machines (SVMs), and neural networks are three deep learning techniques used in microgrid optimization. Random Forests are decision trees that handle large datasets and complex relationships, effectively predicting energy generation and fault detection. SVMs are used for load forecasting and fault detection, while neural networks, including artificial neural networks, convolutional neural networks, and recurrent neural networks, capture complex patterns and dependencies in microgrid data for load forecasting, predictive maintenance, and control strategies (Syamala, Komala, et al., 2023).

Clustering algorithms, reinforcement learning, genetic algorithms, and ensemble learning are used to optimize microgrid operations. Clustering algorithms group similar elements, identifying patterns for efficient energy distribution. Reinforcement learning learns optimal control policies through interaction with the microgrid environment, adapting to changing conditions. Genetic algorithms determine the most efficient configuration or scheduling of energy resources. Ensemble learning techniques like gradient boosting or AdaBoost combine multiple models for improved prediction accuracy. The choice of machine learning algorithm for microgrid optimization depends on the specific use case, data nature, and desired outcome, often involving a combination of algorithms or ensemble methods for optimal performance enhancement.

Supervised Learning Techniques for Microgrid Efficiency

Supervised learning techniques in machine learning involve training models on labeled data, allowing algorithms to make predictions or decisions based on input-output pairs, particularly in improving microgrid efficiency (Maheswari et al., 2023; Zekrifa et al., 2023).

- **Regression Analysis:** Regression models, such as Linear Regression, Polynomial Regression, or Support Vector Regression (SVR), are used for predicting continuous variables. In microgrids, regression can forecast energy demand or predict energy generation from renewable sources based on historical data, weather conditions, and other relevant factors.

- **Time Series Forecasting:** Techniques like Autoregressive Integrated Moving Average (ARIMA), Exponential Smoothing Methods, or Long Short-Term Memory (LSTM) networks are valuable for predicting future energy consumption patterns within the microgrid. These models capture temporal dependencies and seasonality in energy data.

- **Decision Trees and Random Forest:** Decision trees and their ensemble, Random Forests, can assist in making decisions regarding optimal energy flow paths or control strategies within the microgrid. They can aid in classifying various operational conditions or predicting faults based on sensor data.

- **Support Vector Machines (SVM):** SVMs are effective for classification tasks or regression problems in microgrid systems. They can be utilized in fault detection and classification of grid anomalies based on data from sensors and smart meters (Kumar et al., 2023; Ugandar et al., 2023; Venkateswaran, Vidhya, Naik, et al., 2023).

- **Gradient Boosting and XGBoost:** Gradient boosting algorithms, like Gradient Boosting Machines (GBM) or XGBoost, are powerful for improving model accuracy and efficiency. They excel in tasks such as load forecasting, optimizing energy distribution, or predicting renewable energy output.

- **Neural Networks:** Supervised learning with neural networks, including Multi-Layer Perceptrons (MLPs) or Recurrent Neural Networks (RNNs), can model complex relationships within microgrid data. They are beneficial for forecasting, fault detection, and optimizing energy management strategies(Satav et al., 2024).

- **Ensemble Learning Techniques:** Ensemble methods, such as AdaBoost or Bagging, combine multiple base models to improve predictive accuracy. These techniques can be applied for load forecasting or energy generation prediction within the microgrid.

The choice of supervised learning technique in microgrid contexts depends on the task, data type, and model accuracy. Combining techniques or ensemble methods can enhance microgrid efficiency optimization.

Unsupervised Learning Approaches in Microgrid Control

Unsupervised learning techniques in machine learning extract patterns or structures from unlabeled data, particularly useful in microgrid control and optimization (Boopathi & Kanike, 2023; Maheswari et al., 2023).

- **Clustering Algorithms:** Clustering techniques like k-means clustering, hierarchical clustering, or DBSCAN (Density-Based Spatial Clustering of Applications with Noise) can be used to group similar elements or devices within the microgrid based on energy consumption patterns, operational characteristics, or other features. This grouping aids in understanding system behavior and optimizing energy distribution strategies.

- **Anomaly Detection:** Unsupervised learning helps in identifying anomalies or irregularities within the microgrid system. Techniques such as Isolation Forest, One-Class SVM, or Autoencoders can detect unusual behaviors in energy consumption, generation, or grid parameters, signaling potential faults or abnormalities for further investigation.
- **Dimensionality Reduction:** Methods like Principal Component Analysis (PCA) or t-Distributed Stochastic Neighbor Embedding (t-SNE) can reduce the dimensionality of data while preserving essential information. These techniques help in visualizing high-dimensional data, identifying correlations, or reducing computational complexity in microgrid analytics.
- **Association Rule Mining:** Unsupervised learning can reveal associations or relationships between different components or events within the microgrid. Apriori algorithm or FP-growth algorithm can discover patterns in energy consumption or device interactions, aiding in decision-making for optimal energy management.
- **Self-Organizing Maps (SOM):** SOMs are neural network-based algorithms that facilitate visualization and clustering of high-dimensional data. In microgrids, SOMs can organize and represent spatial relationships between devices or energy sources, aiding in understanding system structure and optimizing control strategies.
- **Density Estimation:** Gaussian Mixture Models (GMM) or Kernel Density Estimation (KDE) techniques estimate the probability density function of the data. These methods can assist in understanding the distribution of energy consumption or generation patterns, enabling better resource allocation or load balancing.

Unsupervised learning techniques help identify hidden patterns in microgrid data, providing operators with insights into system behavior, optimizing energy distribution, and enhancing control and efficiency.

PREDICTIVE MAINTENANCE AND FAULT DETECTION

Utilizing Machine Learning for Predictive Maintenance

Machine learning is being utilized for predictive maintenance in microgrid systems, using data-driven methods to predict equipment failures or maintenance needs(Divya et al., 2023; Yu et al., 2019). Figure 2 illustrates the use of machine learning for predictive maintenance.

- **Data Collection:** Gather data from sensors, smart meters, SCADA systems, and historical maintenance records. This data includes parameters like voltage, current, temperature, vibration, and other relevant metrics that indicate equipment health.
- **Feature Engineering:** Extract meaningful features from the collected data, such as statistical measures, trends, or patterns in sensor readings. This step involves cleaning, normalizing, and transforming raw data to make it suitable for predictive modeling.
- **Model Selection:** Choose appropriate machine learning models for predictive maintenance, such as:
  - **Regression Models:** Utilize models like Linear Regression, Decision Trees, or Random Forests to predict the remaining useful life of equipment based on degradation patterns observed in historical data.

Figure 2. Utilizing machine learning for predictive maintenance

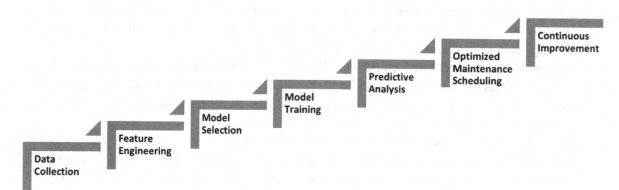

- ○ **Survival Analysis:** Techniques like Cox Proportional Hazards Model or Kaplan-Meier Estimator predict the probability of failure at a given time, considering time-to-failure scenarios.
- ○ **Deep Learning:** Neural networks, especially Recurrent Neural Networks (RNNs) or Long Short-Term Memory (LSTM) networks, can capture complex temporal patterns in sensor data for predictive maintenance.
- **Model Training:** Train the chosen model(s) on historical data to learn patterns indicative of equipment failure or degradation.
- **Predictive Analysis:** Apply the trained model to real-time or streaming data from the microgrid to predict potential equipment failures. Algorithms continuously monitor incoming sensor data, generating alerts or maintenance schedules when deviations or patterns indicating potential failures are detected.
- **Optimized Maintenance Scheduling:** Use predictive maintenance insights to schedule maintenance proactively, minimizing downtime, optimizing resource allocation, and extending the lifespan of critical equipment within the microgrid.
- **Continuous Improvement:** Continuously update and refine the predictive maintenance model by retraining it with new data and incorporating feedback to enhance accuracy and reliability.

Machine learning-based predictive maintenance can enhance microgrid system reliability and performance by reducing operational costs, minimizing unplanned downtime, and promoting proactive maintenance strategies.

Fault Detection and Diagnosis in Microgrid Components

The implementation of machine learning techniques in fault detection and diagnosis in microgrid components aims to identify and address system abnormalities or malfunctions (Badihi et al., 2019).

- **Data Collection:** Gather data from sensors, smart meters, and monitoring devices across the microgrid components, including inverters, batteries, renewable energy sources, transformers, switches, and other critical equipment. This data typically includes voltage, current, temperature, frequency, and other relevant parameters.

- **Feature Selection and Engineering:** Extract relevant features from the collected data, such as statistical measures, spectral analysis results, harmonic distortions, or transient behavior patterns. These features help in identifying normal operation and deviations that indicate faults or anomalies.

- **Unsupervised Learning for Anomaly Detection:** Utilize unsupervised learning techniques like clustering, density estimation, or autoencoders to identify anomalies or outliers within the data. These methods help in detecting deviations from normal behavior without needing labeled data.

- **Supervised Learning for Fault Classification:** Train supervised learning models such as decision trees, support vector machines (SVM), or neural networks on labeled data to classify different fault types once anomalies are detected. This step involves creating a labeled dataset with instances of known fault types for training the model.

- **Fault Diagnostics and Localization:** Combine fault classification outputs with system topology and engineering knowledge to diagnose the root cause and localize faults within the microgrid components. This step may involve rule-based systems, knowledge graphs, or additional machine learning models tailored for diagnostic purposes.

- **Real-Time Monitoring and Response:** Implement the fault detection and diagnosis system in real-time to continuously monitor the microgrid. Set up alerts or automated responses when faults are detected to initiate corrective actions, such as isolation of faulty components, reconfiguration of energy flow, or dispatching maintenance teams.

- **Continuous Improvement and Adaptation:** Regularly update and refine the fault detection models by incorporating new data and feedback from maintenance actions or system updates. This iterative process improves the accuracy and adaptability of the fault detection and diagnosis system.

Machine learning and data-driven approaches can improve fault detection and diagnosis in microgrids, enhancing system reliability, reducing downtime, and proactively addressing issues for efficient operation.

ADAPTIVE CONTROL AND ENERGY MANAGEMENT

Adaptive Intelligence for Real-Time Control in Microgrids

Adaptive control and energy management in microgrids utilize intelligent systems to dynamically adjust to changing conditions, optimize energy usage, and maintain grid stability, using real-time control methods (Badihi et al., 2019).

- **Data Acquisition and Monitoring:** Gather real-time data from sensors, smart meters, and monitoring devices across the microgrid. This data includes information on energy generation, consumption, grid voltage, frequency, weather conditions, and other relevant parameters.

- **Adaptive Control Algorithms:** Implement adaptive control algorithms that can dynamically adjust to varying conditions. Techniques like Model Predictive Control (MPC), Fuzzy Logic Control, or Reinforcement Learning-based control systems can be used to optimize energy flow and maintain grid stability in response to changing demand, weather fluctuations, or equipment malfunctions.

- **Machine Learning for Adaptive Intelligence:** Employ machine learning models that continuously learn and adapt to real-time data. Reinforcement learning algorithms, in particular, enable systems to learn optimal control policies by interacting with the microgrid environment, making decisions, and receiving feedback on their actions.
- **Dynamic Energy Optimization:** Develop algorithms that optimize energy dispatch, storage, and distribution within the microgrid. These systems continuously analyze real-time data to dynamically allocate energy resources, manage demand-response programs, and balance supply and demand to minimize costs and maximize efficiency.
- **Predictive Analytics for Load Forecasting:** Utilize machine learning models for accurate load forecasting. Predictive analytics help in anticipating future energy demands, enabling proactive energy management strategies and efficient resource allocation.
- **Real-Time Decision Making:** Implement intelligent decision-making systems that can respond swiftly to changing conditions. These systems use AI-based algorithms to make decisions on energy routing, load shedding, or grid reconfiguration to ensure stability and reliability.
- **Resilience and Adaptability:** Design the control system to be resilient to uncertainties and adaptable to unforeseen events. The adaptive intelligence should allow for quick adaptation to varying energy sources, grid disturbances, or changes in consumer behavior.
- **Continuous Learning and Optimization:** Continuously update and refine the control algorithms based on real-time feedback and evolving grid conditions. This adaptive learning loop ensures that the system continuously improves its performance and adapts to changing microgrid dynamics.

By integrating adaptive intelligence and machine learning into control strategies, microgrid operators can create dynamic systems capable of efficient, reliable, and adaptable energy management in real-time, contributing to grid stability and optimized operations.

Energy Management Strategies using Machine Learning

Machine learning-based energy management strategies are utilized to enhance energy usage, distribution, and efficiency in microgrids through data-driven approaches (Jadidi, Badihi, Yu, et al., 2020; Jadidi, Badihi, & Zhang, 2020). Figure 3 depicts energy management strategies utilizing machine learning.

- **Load Forecasting:** Machine learning models predict future energy demand based on historical consumption patterns, weather forecasts, and other relevant factors. Accurate load forecasting helps in planning resource allocation and optimizing energy generation and distribution.
- **Demand Response Optimization:** Utilize machine learning to analyze consumer behavior and preferences in energy usage. Implement demand response programs that incentivize consumers to adjust their energy consumption based on real-time pricing, grid conditions, or personalized recommendations.
- **Renewable Energy Integration:** Machine learning algorithms forecast the output of renewable energy sources like solar or wind. This information is used to optimize the integration of renewable energy into the microgrid, balancing supply and demand efficiently.

Figure 3. Energy management strategies using machine learning

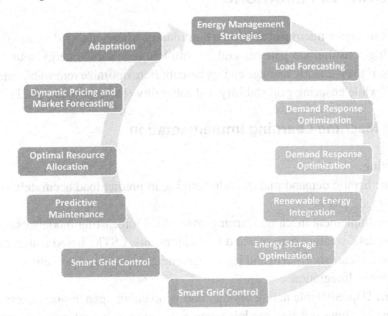

- **Energy Storage Optimization:** Optimize the utilization of energy storage systems, such as batteries or flywheels, using predictive analytics. Machine learning algorithms determine the best times for storage charging and discharging to ensure maximum efficiency and grid stability.

- **Smart Grid Control:** Implement intelligent control systems that adapt to changing conditions in real-time. Machine learning-based control algorithms manage energy flow, grid voltage, and frequency, optimizing overall grid performance.

- **Predictive Maintenance:** Use machine learning to predict equipment failures or maintenance needs. Proactively scheduling maintenance based on predictive analytics minimizes downtime and optimizes the lifespan of critical components.

- **Optimal Resource Allocation:** Machine learning helps in dynamically allocating energy resources based on current demand, cost, and availability. This includes deciding when to switch between energy sources, minimizing costs, and reducing reliance on non-renewable resources.

- **Dynamic Pricing and Market Forecasting:** Machine learning models predict energy market prices and trends. This information assists in strategic decision-making for energy trading, optimizing revenue generation, and managing risks associated with market fluctuations.

- **Adaptation:** Regularly update and refine energy management models based on real-time data and feedback. Adaptive learning ensures that the system continuously improves its performance and adapts to changing energy dynamics.

By leveraging machine learning for energy management, microgrid operators can optimize energy usage, reduce costs, enhance grid reliability, and facilitate the integration of renewable energy sources, contributing to a more efficient and sustainable energy infrastructure.

CASE STUDIES AND APPLICATIONS

A utility company operates a microgrid serving a community where energy demand fluctuates significantly due to varying consumption patterns and intermittent renewable energy sources like solar and wind. The challenge is to efficiently manage energy distribution, optimize renewable energy integration, and minimize costs while ensuring grid stability and reliability (Hossain et al., 2021).

Solution Using Machine Learning Implementation

- Load Forecasting
  - **Problem:** Erratic demand makes it challenging to predict load accurately (Her et al., 2021; Hossain et al., 2019).
  - **Solution:** Implement machine learning-based load forecasting models using historical consumption data, weather patterns, and calendar events. LSTM-based neural networks are employed to forecast short-term and long-term energy demand accurately.
- Renewable Energy Integration
  - **Problem:** Unpredictable nature of solar and wind energy generation affects grid stability.
  - **Solution:** Machine learning models using regression or neural networks predict renewable energy generation based on weather forecasts, historical data, and sensor inputs. These predictions aid in optimal integration and management of renewable sources, minimizing reliance on non-renewable backup sources.
- Optimal Energy Storage Management
  - **Problem:** Inefficient use of energy storage systems leads to underutilization or overloading.
  - **Solution:** Employ reinforcement learning algorithms to control energy storage systems. These algorithms learn optimal charging/discharging strategies based on real-time data, grid conditions, and historical patterns, ensuring efficient utilization while maintaining grid stability.
- Demand Response and Smart Grid Control
  - **Problem:** Lack of adaptive control systems to manage dynamic grid conditions.
  - **Solution:** Implement machine learning-based control systems that adjust energy flow, voltage levels, and grid configurations in real-time. Reinforcement learning models continuously learn and adapt to changing conditions, optimizing grid control strategies for stability and efficiency.
- Predictive Maintenance
  - **Problem:** Reactive maintenance leads to increased downtime and maintenance costs.
  - **Solution:** Deploy predictive maintenance models using machine learning algorithms that analyze sensor data to predict equipment failures or maintenance needs. This proactive approach schedules maintenance before failures occur, reducing downtime and optimizing equipment lifespan.
- Outcomes
  - **Improved Efficiency:** Accurate load forecasting and renewable energy integration optimize energy distribution and utilization.
  - **Cost Reduction:** Efficient energy storage management and predictive maintenance reduce operational costs.

Figure 4. Emerging trends in machine learning for microgrid systems

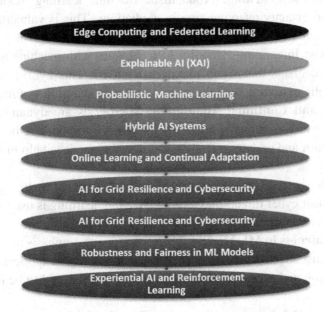

- ○ **Enhanced Reliability:** Adaptive control systems ensure grid stability and reliability under varying conditions.
- ○ **Transition to Sustainable Energy:** Better integration of renewable sources promotes a greener energy infrastructure.

The case study showcases how machine learning is effectively applied in microgrids to enhance energy management, thereby promoting a more sustainable and reliable energy system.

FUTURE DIRECTIONS AND CHALLENGES

Emerging Trends in Machine Learning for Microgrid Systems

Figure 4 depicts the latest trends in machine learning for microgrid systems.

- **Edge Computing and Federated Learning:** With the rise of edge computing, machine learning models are being deployed directly on edge devices within the microgrid. Federated learning techniques allow these decentralized devices to collaboratively learn models while preserving data privacy and security (Chandrasekaran et al., 2020).
- **Explainable AI (XAI) for Decision Support:** As machine learning models become more complex, XAI techniques are gaining importance in microgrid systems. XAI helps in interpreting and explaining model decisions, providing insights into how and why certain actions or predictions are made, crucial for decision-making in energy management.

- **Probabilistic Machine Learning:** Probabilistic machine learning models, such as Bayesian methods, enable uncertainty quantification in predictions. This is valuable in microgrids where uncertainties exist due to fluctuating renewable energy sources or dynamic grid conditions.
- **Hybrid AI Systems:** Integrating symbolic AI techniques with machine learning models allows for reasoning and decision-making in microgrid systems. This hybrid approach combines the strengths of symbolic reasoning with the pattern recognition abilities of machine learning.
- **Online Learning and Continual Adaptation:** Microgrids are dynamic systems, and online learning techniques that continually adapt to changing conditions are becoming more relevant. Models that can learn and update in real-time as new data arrives help in maintaining efficiency and reliability.
- **AI for Grid Resilience and Cybersecurity:** AI-powered systems are being developed to enhance grid resilience against cyber threats and attacks. Machine learning is used for anomaly detection, intrusion detection, and response in securing microgrid systems.
- **Robustness and Fairness in ML Models:** There's a growing emphasis on developing ML models that are robust against adversarial attacks and biases. In microgrid applications, ensuring fairness in energy distribution and access is essential, and ML models need to be designed with fairness considerations.
- **Experiential AI and Reinforcement Learning:** Experiential AI systems learn from interactions with the microgrid environment, similar to how humans learn from experience. Reinforcement learning techniques, especially model-based RL, are increasingly applied for adaptive control and energy management in microgrids.

These emerging trends in machine learning for microgrid systems indicate a shift toward more adaptive, secure, and explainable AI models. They aim to address the complexities and uncertainties inherent in microgrid operations while optimizing efficiency, resilience, and sustainability.

Addressing Challenges and Potential Solutions

Microgrid systems face challenges that can be addressed by utilizing technology advancements, smart algorithms, and innovative approaches to address common issues (Duchesne et al., 2020; Ibrahim et al., 2020).

- Integration of Renewable Energy Sources
  - **Challenge:** Intermittency and variability of renewable sources impact grid stability.
  - **Potential Solution:** Advanced forecasting models using machine learning predict renewable energy generation, enabling better integration and management within the microgrid.
- Grid Resilience and Reliability
  - **Challenge:** Grid disturbances, equipment failures, or cyber threats affect grid resilience.
  - **Potential Solution:** Implement AI-based adaptive control systems that respond dynamically to grid conditions, ensuring resilience and reliability. Additionally, predictive maintenance reduces downtime and enhances system reliability.
- Optimal Energy Management
  - **Challenge:** Balancing supply and demand for efficient energy management.

- ○ **Potential Solution:** Utilize machine learning for load forecasting, demand response, and intelligent control algorithms to optimize energy distribution, storage, and usage.
- Data Security and Privacy
  - ○ **Challenge:** Protecting sensitive grid data from cyber threats and ensuring consumer privacy.
  - ○ **Potential Solution:** Implement robust cybersecurity measures leveraging AI for intrusion detection, anomaly detection, and encryption techniques while complying with privacy regulations.
- Scalability and Interoperability
  - ○ **Challenge:** Integrating diverse components and scaling microgrid systems.
  - ○ **Potential Solution:** Develop standardized communication protocols and interoperable systems. AI-driven solutions facilitate adaptive scalability by learning from different grid configurations.
- Cost Efficiency and Financial Viability
  - ○ **Challenge:** Balancing costs with the implementation of advanced technologies.
  - ○ **Potential Solution:** Use AI-powered optimization models for cost-effective resource allocation, predictive maintenance to reduce operational costs, and strategic energy trading based on market forecasts.
- Skill Gap and Knowledge Transfer
  - ○ **Challenge:** Shortage of skilled personnel for managing complex microgrid systems.
  - ○ **Potential Solution:** Provide training programs and knowledge-sharing platforms for professionals to understand and utilize AI and advanced technologies in microgrid operations effectively.

The integration of advanced technologies like machine learning, AI, and data analytics in microgrids can enhance operational efficiency, resilience, and promote a sustainable and efficient energy future.

CONCLUSION

The chapter provides important insights and implications on the integration of machine learning in microgrid systems. Machine learning is a vital tool in microgrid systems, improving efficiency and resilience by analyzing data, predicting patterns, optimizing energy usage, and adapting to changing conditions. It enables accurate load forecasting, optimal energy distribution, and effective integration of renewable energy sources. This integration of adaptive control mechanisms contributes to grid resilience and stability, ensuring robust operations despite uncertainties and disturbances. The chapter discusses the benefits of machine learning-driven predictive maintenance strategies in microgrid operations, highlighting their cost-efficiency and sustainability. It also discusses challenges such as data security, scalability, and interoperability, suggesting the need for improved cybersecurity measures, standardized protocols, and continuous advancements in machine learning techniques to tackle these evolving issues in microgrid systems. The chapter concludes by highlighting the transformative impact of machine learning on the energy industry, highlighting its role in reshaping energy management, promoting renewable energy integration, and fostering sustainable energy ecosystems.

ABBREVIATIONS

MLSMG - Machine Learning in Microgrid Systems
RENEML - Renewable Energy and Machine Learning
ACML - Adaptive Control with Machine Learning
PDMML - Predictive Maintenance and Machine Learning
GRSML - Grid Resilience and Machine Learning
CESML - Cost Efficiency with Machine Learning
CSMML - Cybersecurity in Microgrid Systems with Machine Learning

REFERENCES

Badihi, H., Jadidi, S., Zhang, Y., Su, C.-Y., & Xie, W.-F. (2019). AI-driven intelligent fault detection and diagnosis in a hybrid AC/DC microgrid. *2019 1st International Conference on Industrial Artificial Intelligence (IAI)*, 1–6.

Barra, P. H. A., Coury, D. V., & Fernandes, R. A. S. (2020). A survey on adaptive protection of microgrids and distribution systems with distributed generators. *Renewable & Sustainable Energy Reviews*, *118*, 109524. doi:10.1016/j.rser.2019.109524

Behera, S., & Dev Choudhury, N. B. (2021). A systematic review of energy management system based on various adaptive controllers with optimization algorithm on a smart microgrid. *International Transactions on Electrical Energy Systems*, *31*(12), e13132. doi:10.1002/2050-7038.13132

Boopathi, S., & Kanike, U. K. (2023). Applications of Artificial Intelligent and Machine Learning Techniques in Image Processing. In *Handbook of Research on Thrust Technologies' Effect on Image Processing* (pp. 151–173). IGI Global. doi:10.4018/978-1-6684-8618-4.ch010

Boopathi, S., Sureshkumar, M., & Sathiskumar, S. (2023). Parametric Optimization of LPG Refrigeration System Using Artificial Bee Colony Algorithm. In S. Tripathy, S. Samantaray, J. Ramkumar, & S. S. Mahapatra (Eds.), *Recent Advances in Mechanical Engineering* (pp. 97–105). Springer Nature Singapore. doi:10.1007/978-981-19-9493-7_10

Chandraratne, C., Naayagi Ramasamy, T., Logenthiran, T., & Panda, G. (2020). Adaptive protection for microgrid with distributed energy resources. *Electronics (Basel)*, *9*(11), 1959. doi:10.3390/electronics9111959

Chandrasekaran, K., Kandasamy, P., & Ramanathan, S. (2020). Deep learning and reinforcement learning approach on microgrid. *International Transactions on Electrical Energy Systems*, *30*(10), e12531. doi:10.1002/2050-7038.12531

Dhanya, D., Kumar, S. S., Thilagavathy, A., Prasad, D., & Boopathi, S. (2023a). Data Analytics and Artificial Intelligence in the Circular Economy: Case Studies. In Intelligent Engineering Applications and Applied Sciences for Sustainability (pp. 40–58). IGI Global.

Dhanya, D., Kumar, S. S., Thilagavathy, A., Prasad, D. V. S. S. S. V., & Boopathi, S. (2023b). Data Analytics and Artificial Intelligence in the Circular Economy: Case Studies. In B. K. Mishra (Ed.), (pp. 40–58). Advances in Civil and Industrial Engineering. IGI Global. doi:10.4018/979-8-3693-0044-2.ch003

Divya, D., Marath, B., & Santosh Kumar, M. (2023). Review of fault detection techniques for predictive maintenance. *Journal of Quality in Maintenance Engineering, 29*(2), 420–441. doi:10.1108/JQME-10-2020-0107

Duchesne, L., Karangelos, E., & Wehenkel, L. (2020). Recent developments in machine learning for energy systems reliability management. *Proceedings of the IEEE, 108*(9), 1656–1676. doi:10.1109/JPROC.2020.2988715

Hema, N., Krishnamoorthy, N., Chavan, S. M., Kumar, N., Sabarimuthu, M., & Boopathi, S. (2023b). A Study on an Internet of Things (IoT)-Enabled Smart Solar Grid System. In *Handbook of Research on Deep Learning Techniques for Cloud-Based Industrial IoT* (pp. 290–308). IGI Global. doi:10.4018/978-1-6684-8098-4.ch017

Hema, N., Krishnamoorthy, N., Chavan, S. M., Kumar, N. M. G., Sabarimuthu, M., & Boopathi, S. (2023a). A Study on an Internet of Things (IoT)-Enabled Smart Solar Grid System. In P. Swarnalatha & S. Prabu (Eds.), (pp. 290–308). Advances in Computational Intelligence and Robotics. IGI Global. doi:10.4018/978-1-6684-8098-4.ch017

Her, C., Sambor, D. J., Whitney, E., & Wies, R. (2021). Novel wind resource assessment and demand flexibility analysis for community resilience: A remote microgrid case study. *Renewable Energy, 179,* 1472–1486. doi:10.1016/j.renene.2021.07.099

Hossain, M. A., Chakrabortty, R. K., Ryan, M. J., & Pota, H. R. (2021). Energy management of community energy storage in grid-connected microgrid under uncertain real-time prices. *Sustainable Cities and Society, 66,* 102658. doi:10.1016/j.scs.2020.102658

Hossain, M. A., Pota, H. R., Squartini, S., Zaman, F., & Guerrero, J. M. (2019). Energy scheduling of community microgrid with battery cost using particle swarm optimisation. *Applied Energy, 254,* 113723. doi:10.1016/j.apenergy.2019.113723

Ibrahim, M. S., Dong, W., & Yang, Q. (2020). Machine learning driven smart electric power systems: Current trends and new perspectives. *Applied Energy, 272,* 115237. doi:10.1016/j.apenergy.2020.115237

Jadidi, S., Badihi, H., Yu, Z., & Zhang, Y. (2020). Fault detection and diagnosis in power electronic converters at microgrid level based on filter bank approach. *2020 IEEE 3rd International Conference on Renewable Energy and Power Engineering (REPE),* (pp. 39–44). IEEE.

Jadidi, S., Badihi, H., & Zhang, Y. (2020). Fault diagnosis in microgrids with integration of solar photovoltaic systems: A review. *IFAC-PapersOnLine, 53*(2), 12091–12096. doi:10.1016/j.ifacol.2020.12.763

Karimi, H., Beheshti, M. T., Ramezani, A., & Zareipour, H. (2021). Intelligent control of islanded AC microgrids based on adaptive neuro-fuzzy inference system. *International Journal of Electrical Power & Energy Systems, 133,* 107161. doi:10.1016/j.ijepes.2021.107161

Kaur, G., Prakash, A., & Rao, K. U. (2021). A critical review of Microgrid adaptive protection techniques with distributed generation. *Renewable Energy Focus*, *39*, 99–109. doi:10.1016/j.ref.2021.07.005

Kavitha, C. R., Varalatchoumy, M., Mithuna, H. R., Bharathi, K., Geethalakshmi, N. M., & Boopathi, S. (2023a). Energy Monitoring and Control in the Smart Grid: Integrated Intelligent IoT and ANFIS. In M. Arshad (Ed.), (pp. 290–316). Advances in Bioinformatics and Biomedical Engineering. IGI Global. doi:10.4018/978-1-6684-6577-6.ch014

Kavitha, C. R., Varalatchoumy, M., Mithuna, H. R., Bharathi, K., Geethalakshmi, N. M., & Boopathi, S. (2023b). Energy Monitoring and Control in the Smart Grid: Integrated Intelligent IoT and ANFIS. In M. Arshad (Ed.), (pp. 290–316). Advances in Bioinformatics and Biomedical Engineering. IGI Global. doi:10.4018/978-1-6684-6577-6.ch014

Khaleel, M. (2023). Intelligent Control Techniques for Microgrid Systems. *Brilliance: Research of Artificial Intelligence*, *3*(1), 56–67. doi:10.47709/brilliance.v3i1.2192

Khokhar, B., & Parmar, K. S. (2022). A novel adaptive intelligent MPC scheme for frequency stabilization of a microgrid considering SoC control of EVs. *Applied Energy*, *309*, 118423. doi:10.1016/j.apenergy.2021.118423

Kumar, P. R., Meenakshi, S., Shalini, S., Devi, S. R., & Boopathi, S. (2023). Soil Quality Prediction in Context Learning Approaches Using Deep Learning and Blockchain for Smart Agriculture. In R. Kumar, A. B. Abdul Hamid, & N. I. Binti Ya'akub (Eds.), (pp. 1–26). Advances in Computational Intelligence and Robotics. IGI Global. doi:10.4018/978-1-6684-9151-5.ch001

Leonori, S., Martino, A., Mascioli, F. M. F., & Rizzi, A. (2020). Microgrid energy management systems design by computational intelligence techniques. *Applied Energy*, *277*, 115524. doi:10.1016/j.apenergy.2020.115524

Maheswari, B. U., Imambi, S. S., Hasan, D., Meenakshi, S., Pratheep, V., & Boopathi, S. (2023). Internet of Things and Machine Learning-Integrated Smart Robotics. In Global Perspectives on Robotics and Autonomous Systems: Development and Applications (pp. 240–258). IGI Global. doi:10.4018/978-1-6684-7791-5.ch010

Moharm, K. (2019). State of the art in big data applications in microgrid: A review. *Advanced Engineering Informatics*, *42*, 100945. doi:10.1016/j.aei.2019.100945

Nishanth, J. R., Deshmukh, M. A., Kushwah, R., Kushwaha, K. K., Balaji, S., & Sampath, B. (2023). Particle Swarm Optimization of Hybrid Renewable Energy Systems. In B. K. Mishra (Ed.), (pp. 291–308). Advances in Civil and Industrial Engineering. IGI Global. doi:10.4018/979-8-3693-0044-2.ch016

Portalo, J. M., González, I., & Calderón, A. J. (2021). Monitoring system for tracking a PV generator in an experimental smart microgrid: An open-source solution. *Sustainability (Basel)*, *13*(15), 8182. doi:10.3390/su13158182

Pramila, P., Amudha, S., Saravanan, T., Sankar, S. R., Poongothai, E., & Boopathi, S. (2023). Design and Development of Robots for Medical Assistance: An Architectural Approach. In Contemporary Applications of Data Fusion for Advanced Healthcare Informatics (pp. 260–282). IGI Global.

Ramesh, M., Yadav, A. K., & Pathak, P. K. (2021). Intelligent adaptive LFC via power flow management of integrated standalone micro-grid system. *ISA Transactions*, *112*, 234–250. doi:10.1016/j.isatra.2020.12.002 PMID:33303227

Ramudu, K., Mohan, V. M., Jyothirmai, D., Prasad, D. V. S. S. S. V., Agrawal, R., & Boopathi, S. (2023). Machine Learning and Artificial Intelligence in Disease Prediction: Applications, Challenges, Limitations, Case Studies, and Future Directions. In G. S. Karthick & S. Karupusamy (Eds.), (pp. 297–318). Advances in Healthcare Information Systems and Administration. IGI Global. doi:10.4018/978-1-6684-8913-0.ch013

Rebecca, B., Kumar, K. P. M., Padmini, S., Srivastava, B. K., Halder, S., & Boopathi, S. (2023). Convergence of Data Science-AI-Green Chemistry-Affordable Medicine: Transforming Drug Discovery. In B. B. Gupta & F. Colace (Eds.), (pp. 348–373). Advances in Computational Intelligence and Robotics. IGI Global. doi:10.4018/978-1-6684-9999-3.ch014

Sampath, B. C. S., & Myilsamy, S. (2022). Application of TOPSIS Optimization Technique in the Micro-Machining Process. In M. A. Mellal (Ed.), (pp. 162–187). Advances in Mechatronics and Mechanical Engineering. IGI Global. doi:10.4018/978-1-6684-5887-7.ch009

Satav, S. D., & Lamani, D. K. G., H., Kumar, N. M. G., Manikandan, S., & Sampath, B. (2024). Energy and Battery Management in the Era of Cloud Computing: Sustainable Wireless Systems and Networks. In B. K. Mishra (Ed.), Practice, Progress, and Proficiency in Sustainability (pp. 141–166). IGI Global. doi:10.4018/979-8-3693-1186-8.ch009

Syamala, M. C. R., K., Pramila, P. V., Dash, S., Meenakshi, S., & Boopathi, S. (2023). Machine Learning-Integrated IoT-Based Smart Home Energy Management System: In P. Swarnalatha & S. Prabu (Eds.), Advances in Computational Intelligence and Robotics (pp. 219–235). IGI Global. doi:10.4018/978-1-6684-8098-4.ch013

Syamala, M., Komala, C., Pramila, P., Dash, S., Meenakshi, S., & Boopathi, S. (2023). Machine Learning-Integrated IoT-Based Smart Home Energy Management System. In *Handbook of Research on Deep Learning Techniques for Cloud-Based Industrial IoT* (pp. 219–235). IGI Global. doi:10.4018/978-1-6684-8098-4.ch013

Tan, S., Wu, Y., Xie, P., Guerrero, J. M., Vasquez, J. C., & Abusorrah, A. (2020). New challenges in the design of microgrid systems: Communication networks, cyberattacks, and resilience. *IEEE Electrification Magazine*, *8*(4), 98–106. doi:10.1109/MELE.2020.3026496

Ugandar, R. E., Rahamathunnisa, U., Sajithra, S., Christiana, M. B. V., Palai, B. K., & Boopathi, S. (2023). Hospital Waste Management Using Internet of Things and Deep Learning: Enhanced Efficiency and Sustainability. In M. Arshad (Ed.), (pp. 317–343). Advances in Bioinformatics and Biomedical Engineering. IGI Global., doi:10.4018/978-1-6684-6577-6.ch015

Veeranjaneyulu, R., Boopathi, S., Kumari, R. K., Vidyarthi, A., Isaac, J. S., & Jaiganesh, V. (2023). Air Quality Improvement and Optimisation Using Machine Learning Technique. *IEEE- Explore*, (pp. 1–6). IEEE.

Veeranjaneyulu, R., Boopathi, S., Narasimharao, J., Gupta, K. K., Reddy, R. V. K., & Ambika, R. (2023). Identification of Heart Diseases using Novel Machine Learning Method. *IEEE- Explore*, (pp. 1–6). IEEE.

Venkateswaran, N., Vidhya, K., Ayyannan, M., Chavan, S. M., Sekar, K., & Boopathi, S. (2023). A Study on Smart Energy Management Framework Using Cloud Computing. In P. Ordóñez De Pablos & X. Zhang (Eds.), (pp. 189–212). Practice, Progress, and Proficiency in Sustainability. IGI Global. doi:10.4018/978-1-6684-8634-4.ch009

Venkateswaran, N., Vidhya, R., Naik, D. A., Michael Raj, T. F., Munjal, N., & Boopathi, S. (2023). Study on Sentence and Question Formation Using Deep Learning Techniques. In O. Dastane, A. Aman, & N. S. Bin Mohd Satar (Eds.), (pp. 252–273). Advances in Business Strategy and Competitive Advantage. IGI Global. doi:10.4018/978-1-6684-6782-4.ch015

Yu, W., Dillon, T., Mostafa, F., Rahayu, W., & Liu, Y. (2019). A global manufacturing big data ecosystem for fault detection in predictive maintenance. *IEEE Transactions on Industrial Informatics*, *16*(1), 183–192. doi:10.1109/TII.2019.2915846

Zekrifa, D. M. S., Kulkarni, M., Bhagyalakshmi, A., Devireddy, N., Gupta, S., & Boopathi, S. (2023). Integrating Machine Learning and AI for Improved Hydrological Modeling and Water Resource Management. In *Artificial Intelligence Applications in Water Treatment and Water Resource Management* (pp. 46–70). IGI Global. doi:10.4018/978-1-6684-6791-6.ch003

Chapter 11
Review on Artificial Intelligence–Based Sustainable System for Autonomous Vehicles

Sundaram Sankar Ganesh

ⓘ https://orcid.org/0000-0001-7687-4562

KPR Institute of Engineering and Technology, India

K. Bala Murugan

KPR Institute of Engineering and Technology, India

G. Pandiya Rajan

KPR Institute of Engineering and Technology, India

M. Muthukumar

ⓘ https://orcid.org/0009-0003-7643-0197

KPR Institute of Engineering and Technology, India

Alagar Karthick

ⓘ https://orcid.org/0000-0002-0670-5138

KPR Institute of Engineering and Technology, India

ABSTRACT

Artificial intelligence (AI) has played a pivotal role in developing autonomous vehicles and revolutionizing the automotive industry. AI is at the forefront of the autonomous vehicle revolution, with the potential to transform transportation, enhance safety, and reduce environmental impact. While there are challenges and concerns to overcome, ongoing research, technological advancements, and regulatory efforts are helping to pave the way for a future where autonomous vehicles are an integral part of our transportation landscape. The successful integration of AI in autonomous vehicles will depend on a collaborative effort between the automotive industry, regulators, and the public to ensure the safe and responsible deployment of this transformative technology.

DOI: 10.4018/979-8-3693-3735-6.ch011

INTRODUCTION

In the late 19th century, with the advent of electricity brought about by the Second Industrial Revolution, the first electric cars came into being. Because of the advantages it offered over gasoline-powered cars in terms of noise reduction, passenger comfort, and ease of operation, electric vehicles were a popular choice for motor vehicle propulsion in the early 20th century. However, concerns about having insufficient Energy stored in the batteries used at the time prevented widespread use of these vehicles. In the late 90s, hybrid electric cars, which combine internal combustion engines with electric motors, started to gain popularity. Mass manufacturing of plug-in hybrid electric vehicles did not begin until the late 2000s, and battery electric cars were not commercially viable choices for buyers until the 2010s.

Electric motors are used as the primary propulsion source in these vehicles rather than an afterthought. Vehicle sales increased in the 2010s thanks to government incentives implemented in 1990 in Norway and later in the 2000s in other big markets, such as the US and EU. The electric vehicle industry is anticipated to significantly increase due to rising public interest, awareness, and structural incentives, such as those included in the green recovery efforts after the COVID-19 epidemic. As a result of quarantines imposed during the COVID-19 epidemic, less greenhouse gases were released by diesel and gasoline automobiles. Policies about large electric cars are among the many things the International Energy Agency has called on nations to undertake to achieve climate targets. In 2022, electric vehicles accounted for 14% of all new vehicle sales, up from 9% in 2020 and 5% in 2021. The percentage of sales of electric cars may rise from 1% in 2016 to over 35% by 2030. The worldwide electric vehicle industry was valued at $280 billion in July 2022 and was projected to reach $1 trillion by 2026. American, European, and Chinese markets are anticipated to account for a significant portion of this expansion. Electric two-wheelers and three-wheelers are expected to increase in emerging nations, according to a literature study in 2020. However, four-wheeled electric vehicles seem economically unlikely to have such growth. More than 20% of all EVs are two- or three-wheelers, making them the most electrified road transport category now and in the future.

The fuel and technology used to generate Energy impact the carbon footprint and other emissions of electric cars. A battery, flywheel, or supercapacitors may all be used to store the power in the vehicle. Most vehicles' ability comes from a limited number of sources, primarily fossil fuels that aren't replenishable when it comes to internal combustion engines. Electric cars benefit greatly from regenerative braking, which returns the kinetic Energy usually wasted as heat during friction braking to the onboard battery as electricity. Figure 1 shows the overall layout of the Electric vehicle.

BATTERY PERFORMANCE PREDICTION ALGORITHM

While several computational approaches may be used to estimate stability, this paper's Free Energy Perturbation method explicitly uses a rigorous molecular dynamics simulation methodology to account for solvent effects and sample conformational dynamics. (Scarabelli et al., 2022)The model's prediction performance is greatly enhanced using appropriate training data, as AE is kept below 2.5% and MRE approaches 0.7%. The model's superiority in SOC prediction is shown by comparing simulation results and the determination of the ideal hyperparameters. Accurate state of charge prediction for real-world battery systems using a novel dual-dropout-based neural network(R. Li et al., 2022). Compared to battery systems, this one has clear benefits. An exponential forgetting-like time-weighting strategy

Figure 1. Shows the overall layout of the electric vehicle

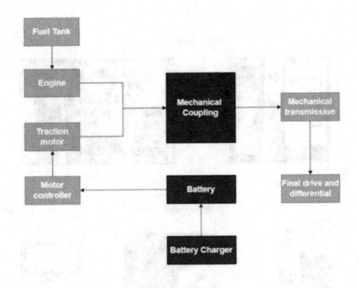

is used to exponentially reduce the impact of earlier data points on the regression analysis (Verbrugge et al., 2005). Using empirical mode decomposition (EMD) to reduce local fluctuation, It can be accurately predict SOH in a unified framework that includes one-step, multistep, and long-term forecasts. Then, the use the decoupled residual SOH series as the training set (Cai et al., 2022). Online optimum scheduling for building energy management is an everyday use of reinforcement learning. Regarding renewable energy sources, machine learning is often used in solar and wind power systems to predict solar irradiance, wind resources, PV power, intelligent control, fault diagnostics, and maximum power point tracking (MPPT) (Y. Zhou, 2022).

This peak demand happens during a set period of around three hours, and the LHTS integration was made to decrease it (Pignata et al., 2023; Y. Wang & Chen, 2020),(Shibl et al., 2023),(Khalid & Savkin, 2010; Shu et al., 2020), (Lin et al., 2021)A dot-line is drawn with the initial and final points preserved, and every two data points are skipped to represent the experimental data curves. Extensive analysis is conducted on the three model states corresponding to battery charging, discharging, and resting principles. A genetic algorithm (GA) is used to identify the parameters of the OCV estimating equations, which are driven by detailed mathematical formulas (B. Chen et al., 2020). To create a new three-dimensional vector that can be used as an input matrix for the filtered current and voltage, an optimal sliding balancing window is built for the measured current filtering. Battery charging capacity drops dramatically with time, according to long-term discharge capacity decay rate findings. (S. Wang, Takyi-Aninakwa, Jin, et al., 2022). Crewless aerial vehicle (UAV) battery management systems and lithium-ion battery condition monitoring are studied subjects into generation mechanisms and preventative measuring techniques. Predicting its endurance is essential to managing the power lithium battery pack's Energy and safety (S. L. Wang et al., 2019). In conjunction with the updating process's convergence, the optimal iterative state initialization is done by utilizing the uncertainty covariance matrix of the three previous time points. The dynamic properties of lithium-ion batteries are described by the two RC circuits of the standard SO-ECM, which consider battery polarisation (S. Wang, Takyi-Aninakwa, Fan, et al., 2022). The CKF technique generates the model's output data, which is then utilized to train SVM. In the meantime, the

Figure 2. Construction of test platforms
(Song et al., 2021)

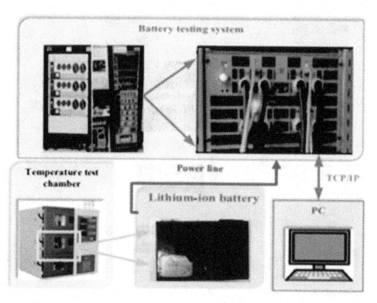

initial SOC is adjusted using the model's output data to obtain a more precise estimate of SOC. Less data is required for prediction once the SVM-CKF method is implemented (Song et al., 2021). Figure 2 shows the Construction of test platforms

Conventional approaches use ideas from system identification and optimization to forecast fixed attributes. The proposed method may be integrated with current system identification techniques to predict the voltage at the battery terminals (Narayanan & Thangavel, 2023). Results showing a linear relationship between precursor charge state and ionization energies agree with theoretical forecasts and molecular capacitors' scaling law. The low signal-to-noise ratio and the considerable background caused by interference with other ions made it impossible to reliably determine the AE for C6~ (Matt et al., 1997). The necessary battery energy is defined as a minimal State of Charge (SoC) using a Backwards Vehicle Model and a Speed Profile Prediction method. Before feeding the BVM with an energetically comparable driving profile, the Speed Profile Prediction is run (Brunelli et al., 2023).

We first use the recurrent neural network to train the mode identification neural network to achieve an online co-state estimation application based on the optimum findings of dynamic programming (Wu et al., 2020). This model was purpose-built for use in energy management simulation tools in smart grids. Accordingly, this model balances ease of use, precision, and generalizability. This model was custom-built to facilitate intelligent grid modelling, prediction, and control (Homan et al., 2019), (Nejad et al., 2016). The Naive model, the most popular approach, oversimplifies the process by assuming that every peptide bond would be cut with an equal probability, resulting in fragments with charges lower than the precursor ion. Basophiles improve identification rates by decreasing the occurrence of false positive matches and enhancing the accuracy of predictions (D. Wang et al., 2013). Two distinct batteries were chosen as examples to demonstrate the process of developing and verifying battery algorithms.

A systematic verification approach was designed to ensure that different battery algorithms' SOC and power capability estimations are accurate (He et al., 2010). The development and enhancement of

Figure 3. Shows the Electrical Testing Platform for Battery Packs
(Yoon et al., 2019)

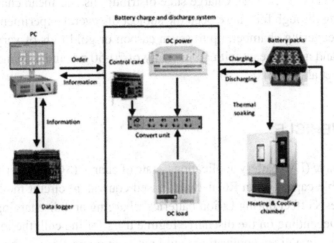

EIS-based prediction systems are impeded by the lack of clarity on the mechanical link between charging curves and impedance spectrum (Guo et al., 2023). The battery management system faces more demands and obstacles due to this (BMS). Regarding electric vehicles, BMS is the key that unlocks the door to the industrialization of EVs. Consequently, research into developing high-performance BMSs for lithium-ion batteries has exploded recently (Karthick et al., 2022),(Stroe et al., 2023).

This study's focus on series-parallel hybrid electric vehicles (SP-HEVs) reveals a correlation between the battery's health and the charging and discharging states under various vehicle operating models (Shah & Kurchania, 2023). The space industry's certification requirements and international certifications for shock, sine vibration, and random vibration form the basis of the environmental test. Two different battery packs, one rectangular and one cube, are made and tested in different environments to see how different shapes affect performance. Figure 3 shows the electrical testing platform for battery packs. One method for reducing the complexity of models is to use a profile approximation to solve the problem of solid phase diffusion within electrode particles. This allows for more efficient computing and may be used onboard for state estimation (Bharathraj et al., 2023). The state of charge (SOC) is analogous to the battery's fuel indicator. To prevent the battery from dying, it is crucial to have information on the remaining capacity of the storm. According to the USABC, a specific regime called the dynamic stress test (DST) is used to get training data from the battery (Álvarez Antón et al., 2023). Several operational factors are used to study the association. The results of the correlation study show that the relationship between the cell voltage and the anode potential is almost linear. The relation is slightly perturbed when changing the operating parameters (current, temperature, and the starting state of charge) (Aitken et al., 1986),(Armenta, 1989)(Alzieu & Smimite, 1995), (Zoerr et al., 2023). Despite its desirable convenience benefits, fast charging has the potential to harm batteries severely and potentially pose safety risks.

The suggested method uses the dynamic information stored inside the battery to balance the two opposing goals and guarantee the fulfilment of all health-related requirements, in contrast to conventional charging techniques like CV (Zou et al., 2017). Using a one-state hysteresis formulation, the hysteresis phenomena are modelled independently of the cell voltage. The physical model is used to determine the other voltage contributions. With a root-mean-square error (RMSE) of less than 18 mV, the extended model is proven to match closely observed voltage after being simplified using the discrete-time realization approach (Gao et al., 2022). Charge state distributions and mean charge states of energetic heavy ions after passing through foils have been the subject of several experimental reports throughout the last decade. In a scattering chamber, up to seven carbon or gold foils of varying thicknesses were placed in the middle, and an ion beam with an intensity of 20-50 nA and a diameter of less than 2 mm was injected (Ishihara et al., 1982).

AUTONOMOUS VEHICLE

Regarding electric vehicle (EV) battery applications, state of charge (SOC) is a critical metric representing the remaining usable capacity. An RC network-based equivalent circuit model was constructed to account for the hysteresis effect during Li-ion batteries' charging and discharging switching processes (Y. Ma et al., 2016). Depending on the discharge regime used for the cell, the total capacity produced during the cell's lifetime in a given application could not even come close to the stored power. Several factors come into play here, including the load resistance amount, the discharge duration, the recovery period between shots, and the terminal voltage on the load after the discharge (Barton & Mitchell, 1989). Based on the premise that batteries have a relative maximum absolute capacity (rather than the 20-hour standards given by manufacturers) and that the pace at which this capacity is removed depends on the precise discharge circumstances, the suggested technique captures these needs (Hausmann & Depcik, 2013). The LIBs and ultra-capacitors are put through their paces using the Neware BTS-8000, a programmable DC power supply and electrical load function generator made by Shenzhen Neware Technology Co., LTD. A personal computer is used to record and store data (Y. Wang et al., 2017a). Regarding electric vehicles, the dynamic driving model considers several factors, including how traffic affects speed and power consumption and how different vehicle states, including driving, parking, and charging, affect the road network (Liu et al., 2022).

A solar-thermal system was designed utilizing this innovative structure. It has a flat plate collector for collecting sunlight, an organic Rankine cycle for power production, and a thermal energy storage tank that links the two loops together (Osorio et al., 2022). All the unique properties of the ALEPH data,

Figure 4. Berkeley autonomous vehicle
(Karimi Pour et al., 2021)

including the cross-section value, the diet mass difference, the dijet charge content, the lack of bottom quarks in the end state, and the associated formation of a left and proper selection, may be explained by this (Carenaatb et al., 1997). As the system grows, finding a self-consistent solution to the Konhn-Sham equations becomes more difficult. To provide a physically well-grounded local description, the innovative method of synaptic weight symmetrization used by the RE-ANN ensures consistency in spatial rotations (Mitran & Nemnes, 2021)—figure 4. Berkeley Autonomous vehicle, this work presents a novel health-aware control method for autonomous racing cars, which aims to maximize the RUL of the battery while simultaneously controlling the vehicle to its operating limitations and maintaining a predetermined course. A Linear Parameter Varying (LPV) model is developed to address the vehicle's non-linear behaviour (Karimi Pour et al., 2021), (J. Hu et al., 2022).

This research focuses on predicting the terminal voltage and state-of-charge (SoC) of Lithium-ion batteries under two different kinds of dynamic loads. Exploring the potential application of a direct multistep forecasting technique in conjunction with Machine Learning is the main topic of inquiry. In addition, discharge profiles acquired at various C-rates comprise a feature bank (Dineva et al., 2021). Different unexpected noises will be encountered while using lithium-ion batteries. Due to the limitations of Extended Kalman Filtering (EKF) in this context, the authors of this study use a dynamic technique to integrate two algorithms—a combination of EKF and Bayesian regularised back propagation neural networks—to enhance the processing speed and accuracy (Y. Hu et al., 2021), (Zazoum, 2023) . Energy management and optimization are becoming crucial to overcome the future development bottleneck of wearable exoskeleton robots (J. Li & Chen, 2023).

BATTERY TEST BED

When data is scarce in the energy and mass domains, it is not easy to use theoretical models to accurately forecast the distributions of charge states. Therefore, direct measurement of charge state fractions is often required (Marshall et al., 2023). Most current methods use theoretical values to set or identify the model parameters offline without adaptation. The EKF is used for the identification of parameters. To further reduce convergence time and tiny variation, the starting values are determined using the RLS and offline identification methodologies [58].

Figure 5. Testbed for the battery packs
(Y. Wang et al., 2017b)

Because of this, measuring charge state fractions directly is often required. Here, Thedescribe an innovative approach to imaging the charge-dispersed beam following a set of magnetic dipoles using a scintillation screen and a CMOS camera. The model parameters in most current methods are either set to theoretical values or found offline without adaptation. The EKF is used to identify parameters. Also, the starting values are determined using RLS and offline identification techniques, which have reduced convergence time and minimal variance (Y. Wang et al., 2017b). With its higher energy density, possibility for reduced material prices, and safety record, the lithium-sulfur battery presents a promising alternative to the well-established lithium-ion battery. It is possible to deploy 500-600Wh/kg in the following years. Both the cathode and anode sides of the cell have their descriptions of the first impacts of the volume change (Brieske et al., 2023).

REAL-TIME PREDICTION OF BATTERY

There is still a long way to go before off-road HEVs have an adequate power management system. In response to this problem, this research suggests a method for off-road HEVs to use predictive power coordinated control using online learning. In one-step-ahead predictive control, the coordinated regulation of EGS speed and power is defined, considering the dynamic and postponed reaction (R. Chen et al., 2022). A BMS is necessary for an electric vehicle's battery pack to balance the cells' energy capacity and estimate the state of charge (SOC). SOC forecasting is a significant worry in the study field, and it is still difficult to answer correctly (Nefraoui et al., 2023). We need EV battery inspection technologies that can be used in real time for various applications to lower average lifespan costs and ensure safe battery usage.

Consequently, this work analyses possible application approaches and real application situations (Yi et al., 2023) before discussing the methodologies for predicting the state of charge (SOC), state of health (SOH), and remaining life of electric vehicle (EV) batteries. Figure 6 Battery real-time applications [69We define the voltage-based state of charge to represent the maximum possible remaining capacity of the storm and to minimize the impacts of polarization. There are three primary points that this study

Figure 6. Battery Real-Time Applications
(Z. Chen et al., 2019)

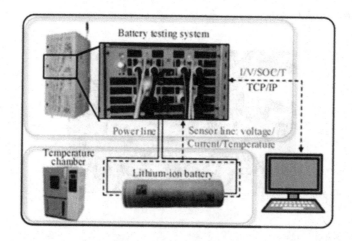

brings forward. The LIB's dynamic performance might be firstly modelled using an OCV-based state space equation (Z. Chen et al., 2019)

The sample cells are chosen using two readily accessible variables, typical voltage and ohmic resistance, considering the state of charge and voltage restrictions for power prediction. The experimental findings confirm this selection procedure's practicality and resilience (Z. Zhou et al., 2019). This study aims to improve prediction accuracy, decrease training time, and apply it to online estimation, which is why it focuses on optimizing the training data set using basic quantifiable data sets. The results show that a dataset produced at random might be a good choice for training purposes (J. Li et al., 2021).

CONCLUSION

The review highlights the diversity of approaches employed in battery performance prediction, from empirical models to machine learning algorithms and physics-based simulations. This diversity reflects the complexity of battery systems and the need for versatile prediction methods. Real-time monitoring and adaptive prediction strategies are crucial for dynamic applications, such as electric vehicles and renewable energy storage. The review emphasizes the need for continuous data acquisition to enhance the accuracy and reliability of predictions in real-world scenarios. Understanding of battery behaviour. These technologies have the potential to address existing challenges in performance prediction.

REFERENCES

Aitken, T. W., Groome, A. E., Joy, T., & Myring, W. J. (1986). Design And Operation Of A Cryogenically Pumped Gas Stripper In The Terminal Of The Daresbury Tandem. In Nuclear Instruments and Methods in Physics Research, 244.

Álvarez Antón, J. C., García-Nieto, P. J., García-Gonzalo, E., González Vega, M., & Blanco Viejo, C. (2023). Data-driven state-of-charge prediction of a storage cell using ABC/GBRT, ABC/MLP and LASSO machine learning techniques. *Journal of Computational and Applied Mathematics, 433*, 115305. Advance online publication. doi:10.1016/j.cam.2023.115305

Alzieu, J., & Smimite, H. (1995). Development of an onboard charge and discharge management system for electric-vehicle batteries. In Journal of Power Sources, 53.

Armenta, C. (1989). Determination Of The State-Of-Charge In Lead-Acid Batteries By Means Of A Reference Cell. In Journal of Power Sources, 27.

Barton, R. T., & Mitchell, P. J. (1989). Estimation of the residual capacity of maintenance-free lead-acid batteries part 1. Identification of a parameter for the prediction of state-of-charge. In Journal of Power Sources, 27.

Bharathraj, S., Adiga, S. P., Mayya, K. S., Song, T. W., & Kim, J. H. (2023). Considering solid phase diffusion penetration depth to improve profile approximations: Towards accurate State estimations in lithium-ion batteries at low characteristic diffusion lengths. *Journal of Power Sources, 554*, 232325. doi:10.1016/j.jpowsour.2022.232325

Brieske, D. M., Warnecke, A., & Sauer, D. U. (2023). Modeling the volumetric expansion of the lithium-sulfur battery considering charge and discharge profiles. *Energy Storage Materials, 55,* 289–300. doi:10.1016/j.ensm.2022.11.053

Brunelli, L., Capancioni, A., Canè, S., Cecchini, G., Perazzo, A., Brusa, A., & Cavina, N. (2023). A predictive control strategy based on A-ECMS to handle Zero-Emission Zones: Performance assessment and testing using an HiL equipped with vehicular connectivity. *Applied Energy, 340,* 121008. doi:10.1016/j.apenergy.2023.121008

Cai, L., Lin, J., & Liao, X. (2022). A data-driven method for state of health prediction of lithium-ion batteries in a unified framework. *Journal of Energy Storage, 51,* 104371. doi:10.1016/j.est.2022.104371

Carenaatb, M. Giudicea,', G. F., Lolaa, S., & Wagner, C. E. M. (1997). c&A. __ * __ l!fiB Four-jet signal at LEP2 and supersymmetry. In Physics Letters B, 395.

Chen, B., Jiang, H., Sun, H., Yu, M., Yang, J., Li, H., Wang, Y., Chen, L., & Pan, C. (2020). A new gas–liquid dynamics model towards robust state of charge estimation of lithium-ion batteries. *Journal of Energy Storage, 29,* 101343. doi:10.1016/j.est.2020.101343

Chen, R., Yang, C., Ma, Y., Wang, W., Wang, M., & Du, X. (2022). Online learning predictive power coordinated control strategy for off-road hybrid electric vehicles considering the dynamic response of engine generator set. *Applied Energy, 323,* 119592. doi:10.1016/j.apenergy.2022.119592

Chen, Z., Sun, H., Dong, G., Wei, J., & Wu, J. (2019). Particle filter-based state-of-charge estimation and remaining-dischargeable-time prediction method for lithium-ion batteries. *Journal of Power Sources, 414,* 158–166. doi:10.1016/j.jpowsour.2019.01.012

Dineva, A., Csomós, B., Kocsis Sz, S., & Vajda, I. (2021). Investigation of the performance of direct forecasting strategy using machine learning in State-of-Charge prediction of Li-ion batteries exposed to dynamic loads. *Journal of Energy Storage, 36,* 102351. doi:10.1016/j.est.2021.102351

Gao, Y., Plett, G. L., Fan, G., & Zhang, X. (2022). Enhanced state-of-charge estimation of LiFePO4 batteries using an augmented physics-based model. *Journal of Power Sources, 544,* 231889. Advance online publication. doi:10.1016/j.jpowsour.2022.231889

Guo, J., Che, Y., Pedersen, K., & Stroe, D. I. (2023). Battery impedance spectrum prediction from partial charging voltage curve by machine learning. *Journal of Energy Chemistry, 79,* 211–221. doi:10.1016/j.jechem.2023.01.004

Hausmann, A., & Depcik, C. (2013). Expanding the Peukert equation for battery capacity modeling through inclusion of a temperature dependency. *Journal of Power Sources, 235,* 148–158. doi:10.1016/j.jpowsour.2013.01.174

He, Y., Liu, W., & Koch, B. J. (2010). Battery algorithm verification and development using hardware-in-the-loop testing. *Journal of Power Sources, 195*(9), 2969–2974. doi:10.1016/j.jpowsour.2009.11.036

Homan, B., ten Kortenaar, M. V., Hurink, J. L., & Smit, G. J. M. (2019). A realistic model for battery state of charge prediction in energy management simulation tools. *Energy, 171,* 205–217. doi:10.1016/j.energy.2018.12.134

Hu, J., Wang, Z., Du, H., & Zou, L. (2022). Hierarchical energy management strategy for fuel cell/ultracapacitor/battery hybrid vehicle with life balance control. *Energy Conversion and Management, 272*, 116383. Advance online publication. doi:10.1016/j.enconman.2022.116383

Hu, Y., Zhang, Y., Wang, S., Xu, W., Fan, Y., & Liu, Y. (2021). Joint Dynamic Strategy of Bayesian Regularized Back Propagation Neural Network with Strong Robustness - Extended Kalman Filtering for the Battery State-of-Charge Prediction. *International Journal of Electrochemical Science, 16*(11), 1–15. doi:10.20964/2021.11.07

Ishihara, T., Shima, K., Kimura, T., Ishii, S., Momoi, T., Yamaguchi, H., Umetani, K., Moriyama, M., Yamanouchi, M., & Mikumo, T. (1982). Equilibrium Charge State Distributions Of Fast Si And Ci Ions In Carbon And Gold Foils. In Nuclear Instruments and Methods, 204.

Karimi Pour, F., Theilliol, D., Puig, V., & Cembrano, G. (2021). Health-aware control design based on remaining useful life estimation for autonomous racing vehicle. *ISA Transactions, 113*, 196–209. doi:10.1016/j.isatra.2020.03.032 PMID:32451079

Karthick, A., Mohanavel, V., Chinnaiyan, V. K., Karpagam, J., Baranilingesan, I., & Rajkumar, S. (2022). State of charge prediction of battery management system for electric vehicles. In Active Electrical Distribution Network: Issues, Solution Techniques, and Applications (pp. 163–180). Elsevier. doi:10.1016/B978-0-323-85169-5.00012-5

Khalid, M., & Savkin, A. V. (2010). A model predictive control approach to the problem of wind power smoothing with controlled battery storage. *Renewable Energy, 35*(7), 1520–1526. doi:10.1016/j.renene.2009.11.030

Li, J., & Chen, C. (2023). Machine learning-based energy harvesting for wearable exoskeleton robots. *Sustainable Energy Technologies and Assessments, 57*, 103122. doi:10.1016/j.seta.2023.103122

Li, J., Ziehm, W., Kimball, J., Landers, R., & Park, J. (2021). Physical-based training data collection approach for data-driven lithium-ion battery state-of-charge prediction. *Energy and AI, 5*, 100094. doi:10.1016/j.egyai.2021.100094

Li, R., Wang, H., Dai, H., Hong, J., Tong, G., & Chen, X. (2022). Accurate state of charge prediction for real-world battery systems using a novel dual-dropout-based neural network. *Energy, 250*, 123853. doi:10.1016/j.energy.2022.123853

Lin, X., Wu, J., & Wei, Y. (2021). An ensemble learning velocity prediction-based energy management strategy for a plug-in hybrid electric vehicle considering driving pattern adaptive reference SOC. *Energy, 234*, 121308. doi:10.1016/j.energy.2021.121308

Liu, Y., Liu, W., Gao, S., Wang, Y., & Shi, Q. (2022). Fast charging demand forecasting based on the intelligent sensing system of dynamic vehicle under EVs-traffic-distribution coupling. *Energy Reports, 8*, 1218–1226. doi:10.1016/j.egyr.2022.02.261

Ma, Y., Li, B., Xie, Y., & Chen, H. (2016). Estimating the State of Charge of Lithium-ion Battery based on Sliding Mode Observer. *IFAC-PapersOnLine, 49*(11), 54–61. doi:10.1016/j.ifacol.2016.08.009

Marshall, C., Meisel, Z., Montes, F., Wagner, L., Hermansen, K., Garg, R., Chipps, K. A., Tsintari, P., Dimitrakopoulos, N., Berg, G. P. A., Brune, C., Couder, M., Greife, U., Schatz, H., & Smith, M. S. (2023). Measurement of charge state distributions using a scintillation screen. *Nuclear Instruments & Methods in Physics Research. Section A, Accelerators, Spectrometers, Detectors and Associated Equipment, 1056*, 168661. doi:10.1016/j.nima.2023.168661

Matt, S., Echt, O., Wisrgistter, R., Grill, V., Scheier, P., Lifshitz, C., & Miirk, T. D. (1997). Appearance and ionization energies of multiply-charged C70 parent ions produced by electron impact ionization. In Chemical Physics Letters, 264. doi:10.1016/S0009-2614(96)01303-6

Mitran, T. L., & Nemnes, G. A. (2021). Ground state charge density prediction in C-BN nanoflakes using rotation equivariant feature-free artificial neural networks. *Carbon, 174*, 276–283. doi:10.1016/j.carbon.2020.12.048

Narayanan, S. S. S., & Thangavel, S. (2023). A novel static model prediction method based on machine learning for Li-ion batteries operated at different temperatures. *Journal of Energy Storage, 61*, 106789. doi:10.1016/j.est.2023.106789

Nefraoui, A., Kandoussi, K., Louzazni, M., Boutahar, A., Elotmani, R., & Daya, A. (2023). Optimal battery state of charge parameter estimation and forecasting using non-linear autoregressive exogenous. *Materials Science for Energy Technologies, 6*, 522–532. doi:10.1016/j.mset.2023.05.003

Nejad, S., Gladwin, D. T., & Stone, D. A. (2016). A systematic review of lumped-parameter equivalent circuit models for real-time estimation of lithium-ion battery states. In *Journal of Power Sources* (Vol. 316, pp. 183–196). Elsevier B.V., doi:10.1016/j.jpowsour.2016.03.042

Osorio, J. D., Wang, Z., Karniadakis, G., Cai, S., Chryssostomidis, C., Panwar, M., & Hovsapian, R. (2022). Forecasting solar-thermal systems performance under transient operation using a data-driven machine learning approach based on the deep operator network architecture. *Energy Conversion and Management, 252*, 115063. doi:10.1016/j.enconman.2021.115063

Pignata, A., Minuto, F. D., Lanzini, A., & Papurello, D. (2023). A feasibility study of a tube bundle exchanger with phase change materials: A case study. *Journal of Building Engineering, 78*, 107622. doi:10.1016/j.jobe.2023.107622

Scarabelli, G., Oloo, E. O., Maier, J. K. X., & Rodriguez-Granillo, A. (2022). Accurate Prediction of Protein Thermodynamic Stability Changes upon Residue Mutation using Free Energy Perturbation. *Journal of Molecular Biology, 434*(2), 167375. doi:10.1016/j.jmb.2021.167375 PMID:34826524

Shah, N. L., & Kurchania, A. K. (2023). Comparative analysis of predictive models for SOC estimation in EV under different running conditions. *E-Prime - Advances in Electrical Engineering. Electronics and Energy, 5*, 100207. doi:10.1016/j.prime.2023.100207

Shibl, M. M., Ismail, L. S., & Massoud, A. M. (2023). A machine learning-based battery management system for state-of-charge prediction and state-of-health estimation for unmanned aerial vehicles. *Journal of Energy Storage, 66*, 107380. doi:10.1016/j.est.2023.107380

Shu, X., Li, G., Shen, J., Lei, Z., Chen, Z., & Liu, Y. (2020). An adaptive multi-state estimation algorithm for lithium-ion batteries incorporating temperature compensation. *Energy*, *207*, 118262. doi:10.1016/j.energy.2020.118262

Song, Q., Wang, S., Xu, W., Shao, Y., & Fernandez, C. (2021). A Novel Joint Support Vector Machine - Cubature Kalman Filtering Method for Adaptive State of Charge Prediction of Lithium-Ion Batteries. *International Journal of Electrochemical Science*, *16*(8), 1–15. doi:10.20964/2021.08.26

Stroe, D. I., Qi, J., Chen, L., Wang, S., Wang, Y., Fan, Y., & Liu, Y. (2023). State of charge estimation strategy based on fractional-order model. In *State Estimation Strategies in Lithium-ion Battery Management Systems* (pp. 191–206). Elsevier. doi:10.1016/B978-0-443-16160-5.00005-6

Verbrugge, M. W., Liu, P., & Soukiazian, S. (2005). Activated-carbon electric-double-layer capacitors: Electrochemical characterization and adaptive algorithm implementation. *Journal of Power Sources*, *141*(2), 369–385. doi:10.1016/j.jpowsour.2004.09.034

Wang, D., Dasari, S., Chambers, M. C., Holman, J. D., Chen, K., Liebler, D. C., Orton, D. J., Purvine, S. O., Monroe, M. E., Chung, C. Y., Rose, K. L., & Tabb, D. L. (2013). Basophile: Accurate Fragment Charge State Prediction Improves Peptide Identification Rates. *Genomics, Proteomics & Bioinformatics*, *11*(2), 86–95. doi:10.1016/j.gpb.2012.11.004 PMID:23499924

Wang, S., Takyi-Aninakwa, P., Fan, Y., Yu, C., Jin, S., Fernandez, C., & Stroe, D. I. (2022). A novel feedback correction-adaptive Kalman filtering method for the whole-life-cycle state of charge and closed-circuit voltage prediction of lithium-ion batteries based on the second-order electrical equivalent circuit model. *International Journal of Electrical Power & Energy Systems*, *139*, 108020. doi:10.1016/j.ijepes.2022.108020

Wang, S., Takyi-Aninakwa, P., Jin, S., Yu, C., Fernandez, C., & Stroe, D. I. (2022). An improved feedforward-long short-term memory modeling method for the whole-life-cycle state of charge prediction of lithium-ion batteries considering current-voltage-temperature variation. *Energy*, *254*, 124224. doi:10.1016/j.energy.2022.124224

Wang, S. L., Tang, W., Fernandez, C., Yu, C. M., Zou, C. Y., & Zhang, X. Q. (2019). A novel endurance prediction method of series connected lithium-ion batteries based on the voltage change rate and iterative calculation. *Journal of Cleaner Production*, *210*, 43–54. doi:10.1016/j.jclepro.2018.10.349

Wang, Y., & Chen, Z. (2020). A framework for state-of-charge and remaining discharge time prediction using unscented particle filter. *Applied Energy*, *260*, 114324. doi:10.1016/j.apenergy.2019.114324

Wang, Y., Liu, C., Pan, R., & Chen, Z. (2017a). Experimental data of lithium-ion battery and ultracapacitor under DST and UDDS profiles at room temperature. *Data in Brief*, *12*, 161–163. doi:10.1016/j.dib.2017.01.019 PMID:28459088

Wang, Y., Liu, C., Pan, R., & Chen, Z. (2017b). Modeling and state-of-charge prediction of lithium-ion battery and ultracapacitor hybrids with a co-estimator. *Energy*, *121*, 739–750. doi:10.1016/j.energy.2017.01.044

Wu, Y., Zhang, Y., Li, G., Shen, J., Chen, Z., & Liu, Y. (2020). A predictive energy management strategy for multi-mode plug-in hybrid electric vehicles based on multi neural networks. *Energy*, *208*, 118366. Advance online publication. doi:10.1016/j.energy.2020.118366

Yi, Y., Zhou, Y., Su, H., Fang, C., Wang, H., Feng, D., & Li, H. (2023). Overview of EV battery testing and evaluation of EES systems located in EV charging station with PV. *Energy Reports*, *9*, 134–144. doi:10.1016/j.egyr.2023.04.075

Yoon, C. O., Lee, P. Y., Jang, M., Yoo, K., & Kim, J. (2019). Comparison of internal parameters varied by environmental tests between high-power series/parallel battery packs with different shapes. *Journal of Industrial and Engineering Chemistry*, *71*, 260–269. doi:10.1016/j.jiec.2018.11.034

Zazoum, B. (2023). Lithium-ion battery state of charge prediction based on machine learning approach. *Energy Reports*, *9*, 1152–1158. doi:10.1016/j.egyr.2023.03.091

Zhou, Y. (2022). Advances of machine learning in multi-energy district communities– mechanisms, applications and perspectives. In *Energy and AI* (Vol. 10). Elsevier B.V., doi:10.1016/j.egyai.2022.100187

Zhou, Z., Kang, Y., Shang, Y., Cui, N., Zhang, C., & Duan, B. (2019). Peak power prediction for series-connected LiNCM battery pack based on representative cells. *Journal of Cleaner Production*, *230*, 1061–1073. doi:10.1016/j.jclepro.2019.05.144

Zoerr, C., Sturm, J. J., Solchenbach, S., Erhard, S. V., & Latz, A. (2023). Electrochemical polarization-based fast charging of lithium-ion batteries in embedded systems. *Journal of Energy Storage*, *72*, 108234. doi:10.1016/j.est.2023.108234

Zou, C., Hu, X., Wei, Z., & Tang, X. (2017). Electrothermal dynamics-conscious lithium-ion battery cell-level charging management via state-monitored predictive control. *Energy*, *141*, 250–259. doi:10.1016/j.energy.2017.09.048

Chapter 12
AI–IoT–Enabled Biomass–to–Biofuel Conversion:
Advancements in Synthetic Processes and Bio–Remediation

Smriti Khare

iD https://orcid.org/0000-0001-6701-1992

Department of Environmental Science, Amity School of Applied Sciences, Amity University, India

N. Ahalya

Department of Biotechnology, M.S. Ramaiah Institute of Technology, India

Kode Jaya Prakash

iD https://orcid.org/0000-0002-4947-7745

Department of Mechanical Engineering, VNR Vignana Jyothi Institute of Engineering and Technology, India

P. Booma Devi

Department of Aeronautical Engineering, Sathyabama Institute of Science and Technology, India

Sumanta Bhattacharya

Department of Textile Technology, Maulana Abul Kalam Azad University of Technology, India

ABSTRACT

The integration of AI and IoT is revolutionizing the energy sector by improving efficiency and minimizing downtime in power plants. IoT sensors and smart grid technologies enhance transmission efficiency by real-time monitoring and power flow optimization. However, challenges like initial investment costs and skill requirements persist. Ethical considerations, data privacy, and equitable access are crucial for fully harnessing the potential of AI and IoT in the energy sector. This chapter synthesizes successful case studies, lessons learned, and future trends, emphasizing the pivotal role of AI and IoT in fostering innovation, optimizing energy systems, and driving the industry towards a cleaner, more sustainable energy landscape.

DOI: 10.4018/979-8-3693-3735-6.ch012

INTRODUCTION

Biomass-to-biofuel conversion is a key area in sustainable energy, offering a sustainable alternative to traditional fossil fuels. The integration of artificial intelligence (AI) in this field holds transformative potential, enhancing efficiency, production, and sustainable energy practices. This convergence is crucial in the pursuit of renewable energy sources to combat climate change and reduce non-renewable resource reliance (L. Kumar & Bharadvaja, 2020). At its core, biomass-to-biofuel conversion involves the transformation of organic materials such as agricultural residues, wood waste, or algae into biofuels like ethanol, biodiesel, or biogas. This conversion, traditionally a complex and resource-intensive process, has seen remarkable advancements due to the integration of AI technologies. AI brings forth a suite of tools—from machine learning algorithms to predictive analytics that optimize every stage of biofuel production, from feedstock selection to refining methodologies (Maharana et al., 2023).

The impact of AI-driven insights in this realm extends far beyond mere process optimization. It encompasses a broader spectrum, enabling a deeper understanding of biomass characteristics, refining techniques, and environmental impacts. By leveraging AI's capabilities to analyze vast datasets and model intricate processes, researchers and industry experts gain invaluable insights crucial for scaling up biofuel production sustainably. One of the fundamental challenges in biomass-to-biofuel conversion lies in the variability of feedstock sources and their composition (Raj et al., 2023). AI tackles this challenge head-on, offering solutions for efficient feedstock selection, blending, and preprocessing. Through data-driven algorithms and predictive models, AI optimizes the choice of feedstock mixtures and preprocessing methods, ensuring higher yields and cost-effectiveness while reducing waste and resource consumption (Goswami et al., 2022).

The integration of Synthetic Processes, Bio-Remediation, and IoT (Internet of Things) is transforming sustainability and environmental conservation. IoT's network of interconnected devices and sensors has revolutionized data collection, monitoring, and analysis in real-time. In synthetic processes, IoT sensors capture real-time information about chemical reactions, material properties, and production parameters, which is fed into AI-driven systems for precise control and optimization. In bio-remediation, IoT devices monitor environmental factors like soil composition, water quality, and air pollutants, enabling better understanding and management of ecological remediation efforts. This transformative force offers unprecedented insights and control over environmental processes (Reddy et al., 2023; Venkateswaran, Kumar, et al., 2023).

IoT plays a crucial role in interconnectedness and automation, enabling a holistic approach to synthetic processes and bio-remediation. It facilitates the deployment of autonomous drones or sensors in bio-remediation practices, offering efficient solutions for identifying and treating contaminated areas. IoT-powered automation streamlines manufacturing workflows, ensuring precision and consistency while minimizing waste (Sengeni et al., 2023; Venkateswaran, Kumar, et al., 2023). The integration of IoT, synthetic processes, and bio-remediation is transforming our approach to environmental sustainability. IoT's real-time data insights enable scientists, engineers, and environmentalists to develop efficient, eco-friendly solutions, fostering a new era of innovation and preserving and restoring our planet's ecosystems.

AI-driven insights significantly improve biofuel production efficiency and environmental sustainability by fine-tuning operational parameters, optimizing energy consumption, and minimizing waste. This not only streamlines the production process but also contributes to the creation of eco-friendly biofuels, reducing carbon emissions and promoting a greener energy landscape. The amalgamation of AI and biomass-to-biofuel conversion is not merely a technological convergence but a paradigm shift in sustainable energy practices. This synthesis holds the promise of revolutionizing the biofuel industry, propelling it towards greater efficiency, scalability, and environmental responsibility. In this chapter, we delve into the myriad ways AI-driven insights are transforming biomass-to-biofuel conversion, highlighting its significance in shaping the future of sustainable energy (Nunes, 2023). AI's role in biomass-to-biofuel conversion is multifaceted, extending from the initial stages of feedstock identification and characterization to the intricate processes of biochemical conversion and refining. Within the realms of feedstock management, AI algorithms process diverse data sets, encompassing geographical, climatic, and agronomic information to identify optimal sources for biomass. These algorithms assess factors such as yield potential, sustainability, and logistical considerations, empowering biofuel producers to make informed decisions regarding the selection and sourcing of raw materials (Loy et al., 2023).

The complexity of biofuel production necessitates precise control and optimization. AI's predictive modeling and process optimization enable the creation of real-time closed-loop control systems, minimizing energy consumption, enhancing conversion efficiency, and mitigating process deviations, resulting in higher-quality biofuel outputs. Another critical facet of AI integration in biomass-to-biofuel conversion lies in its capacity for predictive analytics and anomaly detection. By analyzing vast amounts of operational data, AI models can forecast potential bottlenecks, equipment failures, or deviations in the production process. This proactive approach enables preemptive maintenance and process adjustments, minimizing downtime and optimizing productivity (Tahir et al., 2023).

AI-enabled life cycle assessments (LCA) are crucial for understanding the environmental impact of biofuel production, identifying areas for improvement, optimizing resource utilization, and ensuring the production remains environmentally sustainable and economically viable, thereby contributing to the overall sustainability of biofuel production. The integration of AI-driven insights in biomass-to-biofuel conversion isn't confined to laboratory settings; it extends to large-scale industrial applications. Through adaptive learning and continuous improvement, AI systems evolve, fine-tuning their models based on real-time data feedback from operational biofuel production facilities. This iterative learning process enhances efficiency and scalability, fostering a more agile and responsive biofuel industry (Goswami et al., 2022).

Ultimately, the synergy between AI and biomass-to-biofuel conversion represents a significant stride towards sustainable energy practices. It not only catalyzes advancements in biofuel production but also contributes to a more resilient and environmentally conscious energy landscape. In exploring the convergence of these fields, we uncover a trajectory that holds immense promise in meeting global energy demands while mitigating the environmental impact of conventional fuel sources.

ADVANCEMENTS IN SYNTHETIC PROCESSES AND BIO-REMEDIATION

Advancements in synthetic processes and bio-remediation have been remarkable in recent years. In synthetic processes, the development of new materials, chemicals, and compounds through advanced techniques like synthetic biology, nanotechnology, and chemical engineering has seen significant progress (Dutta et al., 2022).

Synthetic Processes

- **Synthetic Biology**: This field focuses on designing and constructing new biological parts, devices, and systems, often using genetic engineering and manipulation of biological organisms. It has applications in medicine, agriculture, energy production, and environmental remediation.
- **Nanotechnology**: Advancements in nanomaterials have led to the creation of innovative products with unique properties. These materials find applications in various industries, including medicine, electronics, and manufacturing.
- **Chemical Engineering**: Novel chemical processes are being developed to create more efficient and environmentally friendly methods of producing chemicals and materials. Green chemistry principles are increasingly incorporated to minimize waste and reduce the environmental impact of chemical manufacturing.

Bio-Remediation

Bio-remediation involves using biological organisms to degrade or eliminate contaminants in soil, water, and air (Singh et al., 2020). Advancements in this field have been capable as:

- **Microbial Biodegradation**: Harnessing the power of naturally occurring microorganisms to break down pollutants such as oil spills, industrial chemicals, and pesticides is a key area of research. Genetically engineered microbes are being developed to target specific contaminants more effectively.
- **Phytoremediation**: Plants are utilized to absorb, detoxify, or accumulate contaminants from the environment. Researchers are exploring ways to enhance the ability of certain plants to remediate soil and water contaminated with heavy metals, organic pollutants, and radioactive elements.
- **Bioremediation Techniques**: Scientists are developing innovative bioremediation techniques involving bioaugmentation (adding specialized microbes), bioventing (using microorganisms to degrade contaminants in soil), and biofiltration (using living material to capture and biologically degrade pollutants from air or water).

These advancements in synthetic processes and bio-remediation offer promising solutions for various environmental challenges, from pollution cleanup to sustainable production methods. However, ethical considerations, regulatory frameworks, and careful risk assessment remain crucial in deploying these technologies responsibly.

BIOMASS CHARACTERIZATION THROUGH AI AND IOT SENSORS

The use of Artificial Intelligence (AI) and Internet of Things (IoT) sensors (Anitha et al., 2023; Boopathi et al., 2022; Koshariya et al., 2023; Venkateswaran, Vidhya, et al., 2023) in biomass characterization is a significant advancement in understanding the properties of biomass resources as shown in Figure 1.

Figure 1. Artificial intelligence (AI) and internet of things (IoT) sensors in biomass characterization

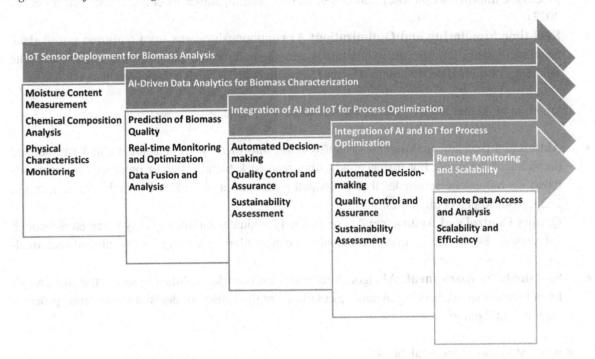

IoT Sensor Deployment for Biomass Analysis

- **Moisture Content Measurement:** IoT sensors equipped with humidity and temperature sensors quantify moisture content in biomass. This data is crucial for assessing biomass quality and determining its suitability for conversion processes.
- **Chemical Composition Analysis:** Spectroscopic IoT sensors or probes can assess the chemical composition of biomass. These sensors utilize techniques like near-infrared spectroscopy (NIRS) or Raman spectroscopy to determine biomass constituents like cellulose, hemicellulose, lignin, and moisture content.
- **Physical Characteristics Monitoring:** IoT sensors track physical parameters such as particle size, density, and bulk density, providing insights into the handling and processing characteristics of biomass feedstock.

AI-Driven Data Analytics for Biomass Characterization

- **Data Fusion and Analysis:** AI algorithms process data collected from IoT sensors, combining information on moisture, composition, and physical properties. Machine learning models are trained to identify patterns and correlations within this multi-dimensional dataset (Agrawal et al., 2024; Hussain & Srimathy, 2023).
- **Prediction of Biomass Quality:** AI models trained on large datasets can predict biomass quality based on sensor data. These predictions assist in selecting optimal biomass for various conversion

processes, improving efficiency and output quality (Hanumanthakari et al., 2023; Sengeni et al., 2023).

- **Real-time Monitoring and Optimization:** AI continuously learns from real-time sensor data, enabling dynamic adjustments in biomass processing parameters. This optimization enhances the efficiency of biofuel production processes (Sampath et al., 2023; Veeranjaneyulu et al., 2023).

Integration of AI and IoT for Process Optimization

- **Automated Decision-making:** AI-driven systems, informed by IoT sensor data, make automated decisions in real time. For instance, adjusting feedstock blend ratios or preprocessing methods based on immediate sensor feedback (Boopathi, Kumar, et al., 2023; Boopathi, Pandey, et al., 2023; Domakonda et al., 2022).
- **Quality Control and Assurance:** AI continuously evaluates biomass quality parameters through IoT sensors. Deviations from desired standards trigger alerts, ensuring consistent feedstock quality for biofuel conversion.
- **Sustainability Assessment:** AI algorithms analyze sensor-derived data to assess the sustainability of biomass sources, aiding in making environmentally conscious decisions in biomass procurement and utilization.

Remote Monitoring and Scalability

- **Remote Data Access and Analysis:** IoT-enabled systems facilitate remote monitoring and analysis, allowing for centralized data collection and analysis across multiple biomass processing facilities (Boopathi, 2023; P. R. Kumar et al., 2023; Subha et al., 2023).
- **Scalability and Efficiency:** The combination of AI and IoT enables scalable biomass characterization processes, ensuring consistent and efficient analysis of diverse biomass sources at various scales of operation.

The integration of AI and IoT in biomass characterization enhances efficiency, decision-making, and sustainability in biomass-to-biofuel conversion processes.

AI AND IOT IN ENERGY GENERATION

Smart Grids: Enhancing Energy Generation with AI and IoT

Smart grids represent an evolution in the traditional power grid infrastructure, incorporating advanced communication and control technologies. At their core, these grids leverage IoT sensors, real-time data analytics, and AI-driven algorithms to optimize energy generation, transmission, and distribution (Hema et al., 2023; Kavitha et al., 2023).

AI Optimization Techniques for Energy Generation

AI enhances energy generation efficiency in smart grids through machine learning algorithms, predictive analytics, and optimization models, enabling facilities to operate at peak performance levels while minimizing operational costs and environmental impact (Hussain & Srimathy, 2023; Ingle et al., 2023).

Predictive Maintenance and Fault Detection

AI's predictive maintenance capabilities are evident in power generation equipment. By analyzing data from IoT sensors, AI algorithms can anticipate potential faults or failures, minimizing downtime and reducing maintenance costs, thereby ensuring the continuous and efficient operation of energy generation assets.

Dynamic Load Balancing and Demand Response

AI-powered systems, in conjunction with IoT devices, enable dynamic load balancing by analyzing real-time data on energy demand and supply. These systems adjust energy generation and distribution in response to fluctuations in demand, optimizing the grid's performance and ensuring a reliable energy supply.

Optimizing Renewable Energy Integration

The variability of renewable energy sources like solar and wind poses challenges to grid stability. AI algorithms forecast renewable energy generation patterns based on weather data collected through IoT sensors. This foresight enables grid operators to efficiently integrate and manage fluctuating renewable energy outputs within the grid, ensuring a reliable and stable power supply (Boopathi et al., 2022; Boopathi & Sivakumar, 2013; Nishanth et al., 2023; Saravanan et al., 2022).

Enhanced Decision-Making with Data Analytics

AI and IoT technologies are transforming energy generation by enabling data-driven decision-making. AI algorithms analyze vast data from IoT devices, providing insights for grid operators to make informed decisions about generation, transmission, and distribution strategies. This combination of AI and IoT technologies in smart grids leads to more efficient, resilient, and sustainable energy practices.

IoT Sensors for Generation Monitoring

IoT sensors are essential for monitoring energy generation, providing real-time data to optimize operations and ensure efficient performance (P. R. Kumar et al., 2023; Ugandar et al., 2023; Venkateswaran, Kumar, et al., 2023). Common types include smart meters, sensors for power generation, and smart sensors for monitoring as shown in Figure 2.

Figure 2. Various types of IoT sensors for energy generation monitoring process

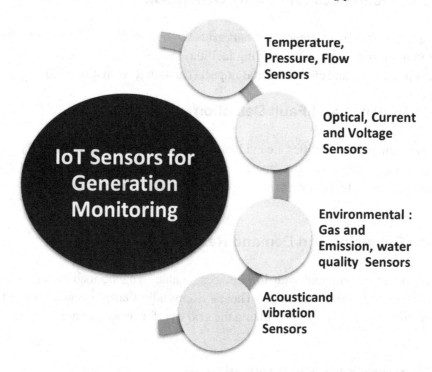

- **Vibration Sensors:** These sensors monitor the vibrations in rotating equipment like turbines and generators. Anomalies in vibration patterns can indicate potential faults or mechanical issues, enabling predictive maintenance.

- **Temperature Sensors:** Monitoring temperatures within different components of generation machinery helps detect overheating or abnormal temperature fluctuations that could signify operational inefficiencies or potential breakdowns.
- **Pressure Sensors:** These sensors measure pressure variations within systems such as boilers, pipelines, or hydraulic systems. Deviations from normal pressure levels could indicate leaks or inefficiencies.
- **Flow Meters:** Flow meters track the flow rates of liquids or gases within pipelines, aiding in monitoring fuel or coolant flow through various parts of the generation process. Deviations might signal leaks or blockages.
- **Current and Voltage Sensors:** These sensors monitor electrical parameters like current and voltage in generators or transmission lines. They help ensure proper functioning and detect irregularities that could lead to power quality issues or equipment failure.
- **Environmental Sensors:** Sensors for environmental factors like humidity, air quality, and atmospheric conditions provide data on the ambient conditions affecting the generation facility, aiding in optimizing operations and equipment performance.

- **Optical Sensors:** Optical sensors measure light, allowing for monitoring flame or combustion quality in combustion-based generation systems. They ensure efficient and safe combustion processes.
- **Gas and Emission Sensors:** These sensors detect and measure the presence of gases and emissions like CO_2, NO_x, or SO_x. Monitoring emissions aids in compliance with environmental regulations and optimizing combustion processes.
- **Water Quality Sensors:** In hydroelectric or steam-driven generation systems, water quality sensors measure parameters such as pH, conductivity, or turbidity, ensuring the quality and efficiency of water used in the process.
- **Acoustic Sensors:** Acoustic sensors monitor sound frequencies emitted by equipment to detect anomalies or irregularities in machinery operation. They aid in predictive maintenance by identifying unusual noise patterns.

IoT sensors in energy generation facilities improve efficiency, reduce downtime, and optimize power generation performance by enabling real-time data collection, analysis, and predictive maintenance.

AI-Driven Predictive Maintenance in Power Plants

AI-driven predictive maintenance in power plants has revolutionized maintenance strategies, providing a proactive approach to equipment upkeep and enhancing operational efficiency (Maguluri, Arularasan, et al., 2023).

- **Data-Driven Insights:** AI systems gather and analyze vast amounts of data from various sensors, including temperature, pressure, vibration, and operational parameters, to establish equipment behavior patterns.
- **Anomaly Detection:** Using machine learning algorithms, AI models identify deviations from normal equipment behavior. These anomalies might indicate potential failures or deterioration in machinery performance.
- **Predictive Analytics:** AI algorithms forecast equipment health based on historical data and real-time sensor inputs. By predicting potential breakdowns or malfunctions, maintenance activities can be scheduled before issues arise.
- **Condition Monitoring:** Continuous monitoring of machinery conditions through IoT sensors allows AI systems to assess the health and performance of critical components, providing insights into wear and tear, fatigue, or potential failures.
- **Proactive Maintenance Scheduling:** AI-driven predictive maintenance enables plant operators to schedule maintenance activities during planned downtime or off-peak periods, minimizing disruptions to power generation.
- **Failure Probability Assessment:** AI models calculate the probability of equipment failure based on the analysis of historical failure data and current operating conditions, allowing for prioritization of maintenance tasks.
- **Cost Optimization:** By predicting equipment failures and scheduling maintenance activities accordingly, power plants can optimize costs by reducing unplanned downtime, minimizing repair costs, and extending equipment lifespan.

- **Integration of Decision Support Systems:** AI-driven predictive maintenance systems provide actionable insights and recommendations to maintenance teams, aiding in decision-making regarding repair, replacement, or optimization strategies.
- **Continuous Improvement:** AI systems continuously learn and improve their predictive capabilities through feedback loops, adjusting algorithms based on new data and performance feedback, thus enhancing their accuracy over time.
- **Safety Enhancement:** Predictive maintenance helps in preventing catastrophic equipment failures, enhancing overall plant safety by minimizing the risks associated with unexpected breakdowns.

TRANSMISSION OPTIMIZATION WITH AI AND IOT

Intelligent Transmission Line Monitoring and Management

The integration of Artificial Intelligence (AI) and the Internet of Things (IoT) in Intelligent Transmission Line Monitoring and Management significantly enhances the efficiency, reliability, and safety of power transmission networks (Hussain & Srimathy, 2023; Koshariya et al., 2023; Maguluri, Ananth, et al., 2023; Samikannu et al., 2022; Venkateswaran, Kumar, et al., 2023).

- **Remote Sensing and Monitoring:** IoT sensors installed along transmission lines collect real-time data on parameters like temperature, humidity, voltage, and current. These sensors enable remote monitoring of the health and performance of transmission infrastructure.
- **Predictive Maintenance:** AI algorithms analyze data from IoT sensors to predict potential faults or failures in transmission lines. Early detection allows for proactive maintenance, minimizing downtime and preventing costly outages.
- **Fault Detection and Localization:** AI-driven analytics swiftly detect and locate faults on transmission lines by analyzing data patterns from IoT sensors. This capability enables rapid response and precise fault isolation, reducing outage durations.
- **Dynamic Line Rating:** IoT sensors, combined with weather data and AI algorithms, enable dynamic line rating. This technology assesses real-time weather conditions to optimize the transmission capacity of lines, maximizing efficiency without compromising safety.
- **Condition-Based Monitoring:** Continuous data collection from IoT sensors allows for condition-based monitoring. AI systems analyze this data to assess the health of transmission assets, facilitating targeted maintenance interventions as needed.
- **Load Balancing and Optimization:** AI-driven systems, utilizing IoT data, optimize load balancing across transmission lines. By dynamically redistributing power flow, they prevent overloads and improve network stability.
- **Grid Resilience and Self-Healing Networks:** AI algorithms help in designing self-healing networks. In case of a disruption or fault, these systems autonomously reroute power, enhancing grid resilience and minimizing the impact of outages.
- **Predictive Analytics for Asset Management:** AI analyzes historical and real-time data to predict asset performance and aging trends. This facilitates informed decisions regarding asset maintenance, repair, or replacement, optimizing asset life cycles.

- **Enhanced Grid Security:** AI-driven systems monitor data anomalies that might indicate cybersecurity threats or attempted breaches on the transmission network, bolstering grid security measures.
- **Adaptive Learning and Continuous Improvement:** AI systems continuously learn from new data patterns, adapting their models to improve accuracy and efficiency in transmission line management over time.

AI and IoT technologies enable Intelligent Transmission Line Monitoring and Management, enabling grid operators to make data-driven decisions, optimize transmission infrastructure, and ensure a more resilient, reliable, and efficient power grid.

AI-Based Fault Detection and Prediction in Transmission Systems

AI-based fault detection and prediction in transmission systems is a revolutionary method to improve power grid reliability and efficiency. This approach uses AI algorithms and machine learning techniques to identify and anticipate faults in transmission systems(Maguluri, Arularasan, et al., 2023).

- **Data-driven Fault Detection:** AI algorithms analyze data collected from sensors installed along transmission lines to detect anomalies. These anomalies may include deviations in voltage, current, temperature, or other parameters, indicating potential faults or abnormalities.
- **Pattern Recognition and Anomaly Detection:** AI models utilize pattern recognition to distinguish normal operating patterns from irregularities or faults in the transmission system. By learning from historical data, these systems identify unusual patterns that may precede faults.
- **Predictive Maintenance Models:** Machine learning algorithms forecast potential faults by learning from historical data patterns and correlating them with precursor signals. These predictive models help in scheduling maintenance before faults escalate into critical issues.
- **Real-time Monitoring and Analysis:** AI systems continuously monitor real-time data from sensors along transmission lines, instantly analyzing and interpreting this data to identify potential fault signatures or early warning signs.
- **Early Fault Identification and Localization:** AI-driven fault detection algorithms swiftly identify the location and type of potential faults along transmission lines. This rapid identification aids in quick response and targeted interventions, minimizing downtime.
- **Probabilistic Risk Assessment:** AI models calculate the probability of specific fault occurrences based on historical data, environmental conditions, and system behavior. This probabilistic assessment assists in prioritizing mitigation strategies.
- **Adaptive and Self-learning Systems:** AI-powered fault detection systems continuously adapt and refine their algorithms based on new data. They self-learn from each identified fault scenario, improving accuracy and reducing false alarms over time.
- **Integration of Predictive Analytics:** By integrating predictive analytics, AI systems forecast potential fault scenarios based on evolving system conditions, enabling proactive measures to prevent failures before they happen.
- **Improved Grid Resilience:** Early fault detection and prediction contribute to grid resilience by enabling faster response times and reducing the impact of faults, ultimately ensuring a more robust and reliable power transmission network.

- **Optimization of Maintenance Activities:** AI-based fault prediction allows utilities to optimize maintenance schedules by focusing efforts on areas with higher likelihoods of faults. This approach minimizes unnecessary maintenance while addressing critical issues promptly.

AI-based fault detection and prediction in transmission systems improve power grid reliability, minimize downtime, optimize maintenance strategies, and ensure efficient network operation, thereby enhancing reliability.

IoT-Enabled Grid Resilience and Security Measures

IoT-enabled grid resilience and security measures are crucial in enhancing power infrastructure's protection against potential threats and disruptions (P. R. Kumar et al., 2023; Maguluri, Arularasan, et al., 2023).

- **Real-time Monitoring and Surveillance:** IoT devices deployed across the grid continuously monitor various parameters, including voltage, current, temperature, and network traffic. Real-time data collection enables swift detection of anomalies or suspicious activities.
- **Intrusion Detection Systems (IDS):** IoT sensors coupled with AI-driven analytics form the backbone of IDS. These systems detect unusual patterns or behaviors in the grid, indicating potential cyber threats or physical intrusions.
- **Predictive Threat Analysis:** IoT data, when analyzed with AI algorithms, helps in predictive threat analysis. Patterns identified from historical data aid in forecasting potential security breaches or vulnerabilities, allowing preemptive action.
- **Adaptive Response Mechanisms:** IoT-enabled systems, combined with AI, facilitate adaptive response mechanisms. Upon detecting anomalies or threats, these systems autonomously trigger responses such as isolating compromised segments or rerouting power flows to minimize disruptions.
- **Cybersecurity Measures:** IoT devices include security protocols to safeguard against cyber threats. Encryption, authentication, and access controls are implemented to protect data integrity and prevent unauthorized access to critical grid components.
- **Resilient Communication Networks:** IoT devices establish robust communication networks, often leveraging multiple communication protocols or technologies. Redundant communication paths ensure continuous data transmission, even in the face of network disruptions.
- **Self-healing Networks:** IoT-enabled grid systems, in conjunction with AI, are designed to create self-healing networks. In the event of a disruption or attack, these systems autonomously reroute power flows or isolate affected areas to restore functionality.
- **Threat Intelligence and Collaboration:** IoT devices contribute to a broader threat intelligence network by sharing data and insights across utilities or agencies. Collaborative efforts based on shared information bolster the overall security posture of the grid.
- **Vulnerability Assessments and Patch Management:** IoT devices facilitate continuous vulnerability assessments and patch management. Regular updates and patches are deployed to address identified vulnerabilities and strengthen grid security.
- **Disaster Recovery Planning:** IoT-driven insights aid in disaster recovery planning. By analyzing historical data and real-time information, utilities can better prepare for and respond to potential grid disruptions caused by natural disasters or other emergencies.

Figure 3. AI and IoT: Enhancing energy consumption efficiency

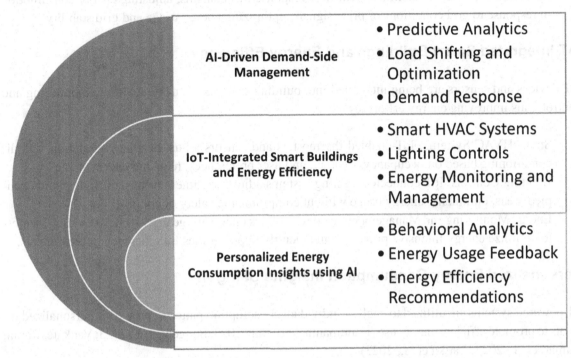

IoT-enabled grid resilience and security measures, backed by advanced analytics and adaptive technologies, are crucial for safeguarding critical infrastructure, ensuring power supply continuity, and minimizing disruption impact, thereby fortifying the grid against evolving threats.

ENHANCING ENERGY CONSUMPTION EFFICIENCY

The study explores the use of Artificial Intelligence (AI) and the Internet of Things (IoT) to improve energy consumption efficiency in three areas as shown in Figure 3.

AI-Driven Demand-Side Management

AI is instrumental in enhancing energy consumption by analyzing patterns, predicting demand changes, and implementing efficient energy management strategies (Agrawal et al., 2024; B et al., 2024; Doma-konda et al., 2022; Nishanth et al., 2023; Satav et al., 2024).

- Predictive Analytics: AI algorithms forecast energy demand patterns based on historical data, weather forecasts, and other factors, allowing utilities to adjust supply accordingly.
- Load Shifting and Optimization: AI enables shifting non-urgent energy usage to off-peak hours by automating devices or systems to consume energy when rates are lower or when renewable sources are more available.

- Demand Response: AI facilitates real-time response mechanisms, adjusting energy consumption in response to grid conditions or price signals, optimizing energy usage and grid stability.

IoT-Integrated Smart Buildings and Energy Efficiency

IoT devices and sensors are being integrated into building systems to enable real-time monitoring and control, thus improving energy efficiency.

- Smart HVAC Systems: IoT-enabled thermostats and sensors adjust heating, ventilation, and air conditioning based on occupancy, weather, and user preferences, reducing energy waste.
- Lighting Controls: IoT-connected lighting systems adjust brightness or turn off lights in unoccupied areas, optimizing energy usage without compromising safety or comfort.
- Energy Monitoring and Management: IoT devices track energy usage in real time, providing insights to optimize energy-intensive processes and identify opportunities for efficiency improvements.

Personalized Energy Consumption Insights Using AI

AI-powered systems are utilized to analyze individual consumption patterns, providing personalized insights to promote efficient energy use (Hanumanthakari et al., 2023; Sengeni et al., 2023; Venkateswaran, Kumar, et al., 2023; Yatika et al., 2023).

- Behavioral Analytics: AI algorithms analyze user behavior, preferences, and historical energy consumption data to offer personalized recommendations for reducing energy waste.
- Energy Usage Feedback: AI-driven applications provide real-time feedback and actionable insights to consumers, empowering them to make informed decisions to optimize their energy consumption.
- Energy Efficiency Recommendations: AI systems suggest personalized energy-saving tips or upgrades, encouraging users to adopt energy-efficient practices or technologies tailored to their specific needs.

These initiatives enhance energy efficiency and sustainability by optimizing usage, reducing waste, and empowering users to make informed choices for energy conservation.

IOT MONITORING AND CONTROL IN BIOFUEL CONVERSION PLANTS

IoT systems in biofuel conversion plants use interconnected sensors and devices to collect real-time data, enabling remote control and management, contributing significantly to the efficient monitoring and control of these plants (Kavitha et al., 2023; Maguluri, Ananth, et al., 2023; Venkateswaran, Kumar, et al., 2023).

- Real-time Data Collection:

- **Sensor Networks:** IoT sensors are deployed throughout the biofuel conversion plant to monitor critical parameters. These sensors measure variables like temperature, pressure, flow rates, pH levels, moisture content, and other relevant process parameters.
- **Data Aggregation:** IoT devices collect data from various sensors and transmit it to a centralized system for real-time monitoring and analysis.

- Remote Monitoring and Control:
  - **Cloud-Based Platforms:** IoT systems leverage cloud-based platforms to enable remote monitoring and control of biofuel conversion processes. This facilitates access to real-time data and plant operations from anywhere with an internet connection.
  - **Remote Alerts and Notifications:** IoT systems trigger alerts and notifications for abnormal conditions or deviations from set parameters. This enables quick response and intervention to prevent potential issues or optimize processes remotely.

- Process Optimization:
  - **Predictive Maintenance:** IoT sensors continuously monitor equipment health, detecting early signs of malfunction or wear. Predictive maintenance strategies can be employed based on this data to prevent breakdowns and optimize equipment performance.
  - **Optimization Algorithms:** IoT data feeds into optimization algorithms and machine learning models. These models analyze the data to identify process inefficiencies or bottlenecks, allowing for process optimization and improved production efficiency.

- Energy Management:
  - **Energy Monitoring:** IoT sensors track energy consumption across the plant, identifying areas of high energy usage or potential energy-saving opportunities. This data aids in implementing energy-efficient practices.
  - **Demand-Side Management:** IoT systems help manage energy demand by optimizing the timing of energy-intensive processes based on grid conditions or energy pricing.

- Safety and Compliance:
  - **Environmental and Safety Monitoring:** IoT sensors monitor environmental conditions and safety parameters within the plant, ensuring compliance with safety standards and environmental regulations.
  - **Quality Control:** IoT systems aid in maintaining product quality by monitoring and controlling various process parameters to ensure consistency in biofuel production.

- Integration with Control Systems:

  SCADA Integration: IoT systems can integrate with Supervisory Control and Data Acquisition (SCADA) systems, allowing for centralized control and visualization of plant operations.

IoT-enabled monitoring systems in biofuel conversion plants improve operational efficiency, reduce downtime, optimize resource utilization, and enhance safety. This integration of technologies provides valuable insights into complex biofuel production processes, contributing to sustainable and efficient practices.

ALGORITHM: AI-IOT ENABLED BIOMASS-TO-BIOFUEL CONVERSION

This section provides an explanation of the AI-IoT-enabled biomass-to-biofuel conversion(Bedi et al., 2022).

Inputs:

Biomass Data: Information about the biomass type, quantity, composition, and quality.

IoT Sensor Data: Real-time data from IoT sensors monitoring factors like temperature, moisture, pH levels, and nutrient content.

AI Model: Trained AI model for biomass analysis, process optimization, and decision-making.

Synthetic Process Parameters: Parameters for biofuel conversion processes.

Bio-Remediation Techniques: Strategies for managing waste and enhancing biofuel yield.

Steps:

Data Acquisition and Monitoring: Collect real-time data from IoT sensors embedded in the biomass processing facilities. This includes monitoring environmental conditions and biomass characteristics.

Data Preprocessing: Clean and preprocess the acquired data to remove noise, outliers, or inconsistencies.

AI Analysis and Optimization: Utilize the AI model to analyze biomass data and predict optimal parameters for biofuel conversion. This involves:

    i.   Analyzing biomass composition to determine the most efficient conversion process.

    ii.   Optimizing parameters for the synthetic processes involved in biofuel production based on the AI recommendations.

    iii.   Predicting maintenance schedules and potential issues through predictive analytics to avoid downtime.

Synthetic Processes for Biofuel Conversion: Implement the recommended synthetic processes for biofuel conversion based on the AI-driven optimization. This may involve fermentation, pyrolysis, or other techniques tailored to the specific biomass type and composition.

Bio-Remediation Integration: Incorporate bio-remediation techniques for managing waste generated during the conversion process. This could involve the use of microorganisms to break down by-products or pollutants.

Continuous Monitoring and Feedback Loop: Continuously monitor the conversion process using IoT sensors to ensure parameters are within optimal ranges. Collect real-time feedback data to update the AI model and improve its predictive capabilities for future conversions.

Output Generation: Generate biofuel yield data and quality metrics based on the conversion process. Assess the effectiveness of bio-remediation techniques in managing waste and enhancing biofuel yield.

Evaluation and Optimization: Evaluate the overall efficiency, yield, and environmental impact of the conversion process. Use this information to refine the AI model, adjust synthetic processes, and improve bio-remediation strategies for subsequent iterations.

Output: The algorithm provides biofuel yield data, quality assessment, optimization recommendations, and insights for improvement. Implementation involves detailed specifications, AI models,

Figure 4. Sustainability and future developments

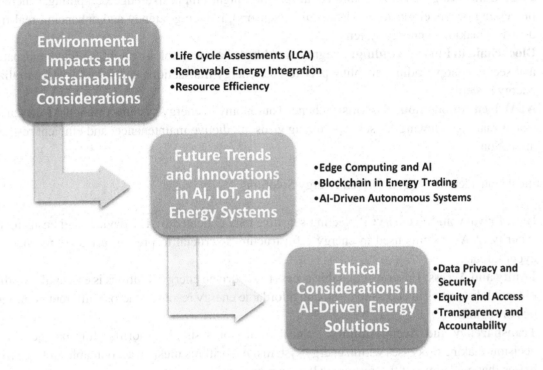

IoT devices, synthetic processes, and bio-remediation techniques tailored to biomass and biofuel production facility characteristics.

SUSTAINABILITY AND FUTURE PERSPECTIVES

The Figure 4 depicts the potential for sustainability and future developments.

Environmental Impacts and Sustainability Considerations

- **Life Cycle Assessments (LCA):** Sustainability assessments using LCAs evaluate the environmental impacts of AI, IoT, and energy systems across their life cycle. This includes analyzing the sourcing of materials, manufacturing, operation, and end-of-life disposal or recycling of technology components.
- **Renewable Energy Integration:** The increased adoption of renewable energy sources, facilitated by AI and IoT, contributes significantly to sustainability by reducing reliance on fossil fuels and mitigating greenhouse gas emissions.
- **Resource Efficiency:** AI and IoT-driven optimization in energy systems focus on resource efficiency, minimizing waste generation, and maximizing the utilization of renewable resources, contributing to a more sustainable energy landscape.

Future Trends and Innovations in AI, IoT, and Energy Systems

- **Edge Computing and AI:** Future advancements might emphasize edge computing, where AI processing occurs closer to the data source (sensors), reducing latency and enhancing real-time decision-making in energy systems.
- **Blockchain in Energy Trading:** Integration of blockchain technology might facilitate transparent and secure energy trading, enabling peer-to-peer energy transactions and fostering decentralized energy systems.
- **AI-Driven Autonomous Systems:** Enhanced autonomy in energy systems through AI-driven decision-making, allowing for self-optimizing grids, predictive maintenance, and efficient resource allocation.

Ethical Considerations in AI-Driven Energy Solutions

- **Data Privacy and Security:** Protecting sensitive data collected by IoT devices and ensuring the security of AI systems used in energy infrastructure is crucial to prevent privacy breaches and cyber threats.
- **Equity and Access:** Ensuring equitable access to AI-driven energy solutions is essential, avoiding potential disparities in access to clean and affordable energy resources across different communities or regions.
- **Transparency and Accountability:** Ethical frameworks should prioritize transparency in AI decision-making processes within energy systems. Algorithms must be accountable and free from biases that could impact fair resource allocation or access.

The integration of AI and IoT in energy systems necessitates a focus on sustainability, anticipating future trends, and addressing ethical considerations for responsible, efficient, and equitable energy solutions.

CASE STUDIES AND REAL-WORLD APPLICATIONS

Case Study One: Successful Implementation of AI in Energy Generation

A utility company is incorporating AI into its power generation facilities to improve efficiency and reliability, specifically by optimizing operations in a combined-cycle gas turbine power plant (Nunes, 2023; Raj et al., 2023).

Implementation: AI monitors and predicts turbine performance using sensor data, weather forecasts, and historical patterns. Machine learning models adjust operating parameters for optimal energy production, considering demand and fuel costs. AI-based maintenance strategies reduce downtime.

Outcome: Optimized operations led to a 10% increase in energy output, 15% reduction in maintenance costs through predictive maintenance, and reduced greenhouse gas emissions.

Case Study Two: IoT Implementation for Improved Transmission Efficiency

A utility company is utilizing IoT sensors and smart grid technologies to improve transmission efficiency and reliability (Goswami et al., 2022; Lim et al., 2023).

Implementation: IoT sensors on transmission lines monitor real-time data on voltage, current, and weather conditions, enabling smart grid analytics to optimize power flow and facilitate quick response to faults or disruptions. *Outcome:* Optimized power flow control reduced transmission losses by 20%, improved grid reliability by 30%, and increased fault response time by 40%, minimizing service disruptions.

Case Study Three: Biomass-to-Biofuel Conversion: Lessons Learned

The biofuel production plant utilizes AI and IoT for optimizing biomass-to-biofuel conversion processes (Hussain & Srimathy, 2023; Venkateswaran, Kumar, et al., 2023).

- *Lessons Learned:* AI analysis aids in feedstock selection, optimizing biofuel yield. IoT sensors enable real-time monitoring, enhancing efficiency. Lessons learned include optimizing enzyme usage and fermentation agents, minimizing resource consumption.
- *Outcomes and Challenges:* The biofuel yield was enhanced by 25% through optimized processes and feedstock selection, despite initial investment costs and the need for skilled personnel.

Case studies showcase the advantages and challenges of incorporating AI in energy generation, utilizing IoT for transmission efficiency, and learning from biomass-to-biofuel conversion processes.

LIMITATIONS OR CHALLENGES ASSOCIATED WITH THE AI-IOT TECHNOLOGIES

AI and IoT technology have the potential to revolutionize industries, but they also present challenges and limitations in their integration and deployment (Ingle et al., 2023; Ramudu et al., 2023).

- **Security Concerns**: IoT devices often lack robust security measures, making them vulnerable to cyberattacks. When combined with AI, these vulnerabilities can lead to more sophisticated attacks. Ensuring the security of both the devices and the AI algorithms becomes critical to prevent data breaches, unauthorized access, and potential manipulation of AI-powered systems.
- **Data Privacy**: The interconnected nature of IoT devices generates vast amounts of data. AI algorithms rely on this data for analysis and decision-making. However, ensuring data privacy and complying with regulations (like GDPR) can be challenging, especially when dealing with sensitive information collected by IoT sensors.
- **Interoperability Issues**: IoT devices come from various manufacturers and often use different protocols and standards. Integrating these devices and ensuring they can communicate seamlessly with AI systems poses a significant challenge. Standardization efforts are ongoing but are yet to achieve widespread adoption.
- **Scalability and Complexity**: As the number of IoT devices increases, managing and maintaining a large-scale AI-IoT ecosystem becomes complex. Scaling AI algorithms to handle vast amounts of data from diverse sources while maintaining efficiency and accuracy is a considerable challenge.
- **Power Consumption and Efficiency**: Many IoT devices operate on limited battery power. Implementing AI algorithms on these devices can strain their resources, impacting their efficiency

and lifespan. Optimizing AI models for energy efficiency without compromising performance is an ongoing challenge.

- **Ethical and Social Implications**: AI-IoT systems can lead to automation, impacting jobs and societal norms. There are ethical concerns surrounding AI decision-making, especially when it comes to critical areas like healthcare or autonomous vehicles. Ensuring fairness, transparency, and accountability in AI decision-making processes remains a challenge.
- **Reliability and Trustworthiness**: AI-driven decisions in IoT systems must be accurate and reliable. Ensuring the trustworthiness of AI models in real-time applications is crucial, as errors or biases can have significant consequences, especially in critical systems like healthcare or autonomous vehicles.

The full potential of AI-IoT technologies can only be realized through a balance between innovation and regulation, security, privacy, and ethical considerations, necessitating collaboration among technology developers, policymakers, and stakeholders.

CONCLUSION

In conclusion, the integration of AI and IoT technologies in the energy sector, coupled with advancements in synthetic processes and bio-remediation, marks a transformative shift towards sustainability. The energy sector must prioritize innovation, cost optimization, ethical considerations, and data privacy measures to achieve a cleaner, more efficient, and environmentally conscious future, ensuring a cleaner, more efficient, and sustainable future.

The integration of Artificial Intelligence and the Internet of Things in the energy sector has significantly enhanced processes, efficiency, and sustainability. AI-driven optimization in power generation facilities has led to improved operational efficiency, predictive maintenance, and increased energy output. IoT sensors and smart grid technologies have improved transmission efficiency, real-time monitoring, power flow optimization, and grid reliability. The use of AI and IoT in biomass-to-biofuel conversion has increased biofuel yield, but initial investment costs and skill requirements remain challenges.

- Innovative synthetic processes have facilitated the development of cleaner energy sources and materials, reducing the ecological footprint of energy production. Bio-remediation techniques have aided in cleaning up pollution from energy-related activities, contributing to environmental conservation.
- Advancements in synthetic processes have optimized resource utilization, making energy production more efficient. Similarly, bio-remediation has provided eco-friendly solutions for waste management, enhancing sustainability in energy-related industries.
- Combining AI-IoT systems with synthetic processes and bio-remediation techniques opens avenues for synergistic solutions. These collaborations can further enhance energy efficiency, minimize waste generation, and facilitate a circular economy approach within the energy sector.
- The integration of AI and IoT in biomass-to-biofuel conversion has boosted biofuel yield, offering a renewable energy alternative. Despite initial investment costs and skill requirements, these advancements pave the way for a more sustainable and diversified energy mix.

- Ongoing research and development in both AI-IoT and synthetic biology/bio-remediation domains are crucial. Collaboration between these fields could lead to breakthroughs, driving more sustainable practices and solutions in energy production and environmental conservation.
- Case studies demonstrate the transformative impact of AI and IoT in the energy sector, leading to sustainable, efficient, and resilient systems. With continued innovation, focus on initial costs, data privacy, and ethical considerations, the industry will transition to sustainable energy practices.

In conclusion, the integration of AI and IoT technologies in the energy sector, coupled with advancements in synthetic processes and bio-remediation, marks a transformative shift towards sustainability. To further this progress, continued innovation, cost optimization, ethical considerations, and data privacy measures must remain at the forefront of development efforts. This collective focus will propel the energy sector towards a cleaner, more efficient, and environmentally conscious future.Top of Form

ABBREVIATIONS

- AI: Artificial Intelligence
- IoT: Internet of Things
- LCA: Life Cycle Assessment
- SCADA: Supervisory Control and Data Acquisition
- NIRS: Near-Infrared Spectroscopy

REFERENCES

Agrawal, A. V., Shashibhushan, G., Pradeep, S., Padhi, S. N., Sugumar, D., & Boopathi, S. (2024). Synergizing Artificial Intelligence, 5G, and Cloud Computing for Efficient Energy Conversion Using Agricultural Waste. In Practice, Progress, and Proficiency in Sustainability (pp. 475–497). IGI Global. doi:10.4018/979-8-3693-1186-8.ch026

Anitha, C., Komala, C., Vivekanand, C. V., Lalitha, S., & Boopathi, S. (2023). Artificial Intelligence driven security model for Internet of Medical Things (IoMT). *IEEE Explore*, 1–7.

B, M. K., K, K. K., Sasikala, P., Sampath, B., Gopi, B., & Sundaram, S. (2024). Sustainable Green Energy Generation From Waste Water. In *Practice, Progress, and Proficiency in Sustainability* (pp. 440–463). IGI Global. doi:10.4018/979-8-3693-1186-8.ch024

Bedi, P., Goyal, S., Rajawat, A. S., Shaw, R. N., & Ghosh, A. (2022). Application of AI/IoT for smart renewable energy management in smart cities. *AI and IoT for Smart City Applications*, 115–138.

Boopathi, S. (2023). Internet of Things-Integrated Remote Patient Monitoring System: Healthcare Application. In *Dynamics of Swarm Intelligence Health Analysis for the Next Generation* (pp. 137–161). IGI Global. doi:10.4018/978-1-6684-6894-4.ch008

Boopathi, S., Kumar, P. K. S., Meena, R. S., Sudhakar, M., & Associates. (2023). Sustainable Developments of Modern Soil-Less Agro-Cultivation Systems: Aquaponic Culture. In Human Agro-Energy Optimization for Business and Industry (pp. 69–87). IGI Global.

Boopathi, S., Pandey, B. K., & Pandey, D. (2023). Advances in Artificial Intelligence for Image Processing: Techniques, Applications, and Optimization. In Handbook of Research on Thrust Technologies' Effect on Image Processing (pp. 73–95). IGI Global.

Boopathi, S., & Sivakumar, K. (2013). Experimental investigation and parameter optimization of near-dry wire-cut electrical discharge machining using multi-objective evolutionary algorithm. *International Journal of Advanced Manufacturing Technology, 67*(9–12), 2639–2655. doi:10.1007/s00170-012-4680-4

Boopathi, S., Sureshkumar, M., & Sathiskumar, S. (2022). Parametric Optimization of LPG Refrigeration System Using Artificial Bee Colony Algorithm. *International Conference on Recent Advances in Mechanical Engineering Research and Development*, (pp. 97–105). IEEE.

Domakonda, V. K., Farooq, S., Chinthamreddy, S., Puviarasi, R., Sudhakar, M., & Boopathi, S. (2022). Sustainable Developments of Hybrid Floating Solar Power Plants: Photovoltaic System. In Human Agro-Energy Optimization for Business and Industry (pp. 148–167). IGI Global.

Dutta, N., Usman, M., Ashraf, M. A., Luo, G., & Zhang, S. (2022). A critical review of recent advances in the bio-remediation of chlorinated substances by microbial dechlorinators. *Chemical Engineering Journal Advances, 12*, 100359. doi:10.1016/j.ceja.2022.100359

Goswami, R. K., Agrawal, K., Upadhyaya, H. M., Gupta, V. K., & Verma, P. (2022). Microalgae conversion to alternative energy, operating environment and economic footprint: An influential approach towards energy conversion, and management. *Energy Conversion and Management, 269*, 116118. doi:10.1016/j.enconman.2022.116118

Hanumanthakari, S., Gift, M. M., Kanimozhi, K., Bhavani, M. D., Bamane, K. D., & Boopathi, S. (2023). Biomining Method to Extract Metal Components Using Computer-Printed Circuit Board E-Waste. In *Handbook of Research on Safe Disposal Methods of Municipal Solid Wastes for a Sustainable Environment* (pp. 123–141). IGI Global. doi:10.4018/978-1-6684-8117-2.ch010

Hema, N., Krishnamoorthy, N., Chavan, S. M., Kumar, N., Sabarimuthu, M., & Boopathi, S. (2023). A Study on an Internet of Things (IoT)-Enabled Smart Solar Grid System. In *Handbook of Research on Deep Learning Techniques for Cloud-Based Industrial IoT* (pp. 290–308). IGI Global. doi:10.4018/978-1-6684-8098-4.ch017

Hussain, Z., & Srimathy, G. (2023). *IoT and AI Integration for Enhanced Efficiency and Sustainability*.

Ingle, R. B., Senthil, T. S., Swathi, S., Muralidharan, N., Mahendran, G., & Boopathi, S. (2023). Sustainability and Optimization of Green and Lean Manufacturing Processes Using Machine Learning Techniques. In IGI Global. doi:10.4018/978-1-6684-8238-4.ch012

Kavitha, C. R., Varalatchoumy, M., Mithuna, H. R., Bharathi, K., Geethalakshmi, N. M., & Boopathi, S. (2023). Energy Monitoring and Control in the Smart Grid: Integrated Intelligent IoT and ANFIS. In M. Arshad (Ed.), (pp. 290–316). Advances in Bioinformatics and Biomedical Engineering. IGI Global. doi:10.4018/978-1-6684-6577-6.ch014

Koshariya, A. K., Kalaiyarasi, D., Jovith, A. A., Sivakami, T., Hasan, D. S., & Boopathi, S. (2023). AI-Enabled IoT and WSN-Integrated Smart Agriculture System. In *Artificial Intelligence Tools and Technologies for Smart Farming and Agriculture Practices* (pp. 200–218). IGI Global. doi:10.4018/978-1-6684-8516-3.ch011

Kumar, L., & Bharadvaja, N. (2020). A review on microalgae biofuel and biorefinery: Challenges and way forward. *Energy Sources. Part A, Recovery, Utilization, and Environmental Effects*, 1–24. doi:10.1080/15567036.2020.1836084

Kumar, P. R., Meenakshi, S., Shalini, S., Devi, S. R., & Boopathi, S. (2023). Soil Quality Prediction in Context Learning Approaches Using Deep Learning and Blockchain for Smart Agriculture. In R. Kumar, A. B. Abdul Hamid, & N. I. Binti Ya'akub (Eds.), (pp. 1–26). Advances in Computational Intelligence and Robotics. IGI Global. doi:10.4018/978-1-6684-9151-5.ch001

Lim, H. Y., Rashidi, N. A., Othman, M. F. H., Ismail, I. S., Saadon, S. Z. A. H., Chin, B. L. F., Yusup, S., & Rahman, M. N. (2023). Recent advancement in thermochemical conversion of biomass to biofuel. *Biofuels*, 1–18. doi:10.1080/17597269.2023.2261788

Loy, A. C. M., Kong, K. G. H., Lim, J. Y., & How, B. S. (2023). Frontier of digitalization in Biomass-to-X supply chain: Opportunity or threats? *Journal of Bioresources and Bioproducts*.

Maguluri, L. P., Ananth, J., Hariram, S., Geetha, C., Bhaskar, A., & Boopathi, S. (2023). Smart Vehicle-Emissions Monitoring System Using Internet of Things (IoT). In Handbook of Research on Safe Disposal Methods of Municipal Solid Wastes for a Sustainable Environment (pp. 191–211). IGI Global.

Maguluri, L. P., Arularasan, A. N., & Boopathi, S. (2023). Assessing Security Concerns for AI-Based Drones in Smart Cities. In R. Kumar, A. B. Abdul Hamid, & N. I. Binti Ya'akub (Eds.), (pp. 27–47). Advances in Computational Intelligence and Robotics. IGI Global. doi:10.4018/978-1-6684-9151-5.ch002

Maharana, D., Kommadath, R., & Kotecha, P. (2023). An innovative approach to the supply-chain network optimization of biorefineries using metaheuristic techniques. *Engineering Optimization*, 55(8), 1278–1295. doi:10.1080/0305215X.2022.2080204

Nishanth, J., Deshmukh, M. A., Kushwah, R., Kushwaha, K. K., Balaji, S., & Sampath, B. (2023). Particle Swarm Optimization of Hybrid Renewable Energy Systems. In *Intelligent Engineering Applications and Applied Sciences for Sustainability* (pp. 291–308). IGI Global. doi:10.4018/979-8-3693-0044-2.ch016

Nunes, L. J. (2023). Exploring the present and future of biomass recovery units: Technological innovation, policy incentives and economic challenges. *Biofuels*, ●●●, 1–13.

Raj, S., Sajith, A., Sreenikethanam, A., Vadlamani, S., Satheesh, A., Ganguly, A., Rajesh Banu, J., Varjani, S., Gugulothu, P., & Bajhaiya, A. K. (2023). Renewable biofuels from microalgae: Technical advances, limitations and economics. *Environmental Technology Reviews*, 12(1), 18–36. doi:10.1080/21622515.2023.2167126

Ramudu, K., Mohan, V. M., Jyothirmai, D., Prasad, D., Agrawal, R., & Boopathi, S. (2023). Machine Learning and Artificial Intelligence in Disease Prediction: Applications, Challenges, Limitations, Case Studies, and Future Directions. In Contemporary Applications of Data Fusion for Advanced Healthcare Informatics (pp. 297–318). IGI Global.

Reddy, M. A., Gaurav, A., Ushasukhanya, S., Rao, V. C. S., Bhattacharya, S., & Boopathi, S. (2023). Bio-Medical Wastes Handling Strategies During the COVID-19 Pandemic. In Multidisciplinary Approaches to Organizational Governance During Health Crises (pp. 90–111). IGI Global. doi:10.4018/978-1-7998-9213-7.ch006

Samikannu, R., Koshariya, A. K., Poornima, E., Ramesh, S., Kumar, A., & Boopathi, S. (2022). Sustainable Development in Modern Aquaponics Cultivation Systems Using IoT Technologies. In *Human Agro-Energy Optimization for Business and Industry* (pp. 105–127). IGI Global.

Sampath, B., Sasikumar, C., & Myilsamy, S. (2023). Application of TOPSIS Optimization Technique in the Micro-Machining Process. In IGI:Trends, Paradigms, and Advances in Mechatronics Engineering (pp. 162–187). IGI Global.

Saravanan, M., Vasanth, M., Boopathi, S., Sureshkumar, M., & Haribalaji, V. (2022). Optimization of Quench Polish Quench (QPQ) Coating Process Using Taguchi Method. *Key Engineering Materials*, *935*, 83–91. doi:10.4028/p-z569vy

Satav, S. D., & Lamani, D. G, H. K., Kumar, N. M. G., Manikandan, S., & Sampath, B. (2024). Energy and Battery Management in the Era of Cloud Computing. In Practice, Progress, and Proficiency in Sustainability (pp. 141–166). IGI Global. doi:10.4018/979-8-3693-1186-8.ch009

Sengeni, D., Padmapriya, G., Imambi, S. S., Suganthi, D., Suri, A., & Boopathi, S. (2023). Biomedical Waste Handling Method Using Artificial Intelligence Techniques. In *Handbook of Research on Safe Disposal Methods of Municipal Solid Wastes for a Sustainable Environment* (pp. 306–323). IGI Global. doi:10.4018/978-1-6684-8117-2.ch022

Singh, S., Kumar, V., Datta, S., Dhanjal, D. S., Sharma, K., Samuel, J., & Singh, J. (2020). Current advancement and future prospect of biosorbents for bioremediation. *The Science of the Total Environment*, *709*, 135895. doi:10.1016/j.scitotenv.2019.135895 PMID:31884296

Subha, S., Inbamalar, T., Komala, C., Suresh, L. R., Boopathi, S., & Alaskar, K. (2023). A Remote Health Care Monitoring system using internet of medical things (IoMT). *IEEE Explore*, 1–6.

Tahir, F., Arshad, M. Y., Saeed, M. A., & Ali, U. (2023). Integrated process for simulation of gasification and chemical looping hydrogen production using Artificial Neural Network and machine learning validation. *Energy Conversion and Management*, *296*, 117702. doi:10.1016/j.enconman.2023.117702

Ugandar, R. E., Rahamathunnisa, U., Sajithra, S., Christiana, M. B. V., Palai, B. K., & Boopathi, S. (2023). Hospital Waste Management Using Internet of Things and Deep Learning: Enhanced Efficiency and Sustainability. In M. Arshad (Ed.), (pp. 317–343). Advances in Bioinformatics and Biomedical Engineering. IGI Global. doi:10.4018/978-1-6684-6577-6.ch015

Veeranjaneyulu, R., Boopathi, S., Kumari, R. K., Vidyarthi, A., Isaac, J. S., & Jaiganesh, V. (2023). Air Quality Improvement and Optimisation Using Machine Learning Technique. *IEEE- Explore*, 1–6.

Venkateswaran, N., Kumar, S. S., Diwakar, G., Gnanasangeetha, D., & Boopathi, S. (2023). Synthetic Biology for Waste Water to Energy Conversion: IoT and AI Approaches. In M. Arshad (Ed.), (pp. 360–384). Advances in Bioinformatics and Biomedical Engineering. IGI Global. doi:10.4018/978-1-6684-6577-6.ch017

Venkateswaran, N., Vidhya, K., Ayyannan, M., Chavan, S. M., Sekar, K., & Boopathi, S. (2023). A Study on Smart Energy Management Framework Using Cloud Computing. In 5G, Artificial Intelligence, and Next Generation Internet of Things: Digital Innovation for Green and Sustainable Economies (pp. 189–212). IGI Global. doi:10.4018/978-1-6684-8634-4.ch009

Yatika, G., Kumar, S. R., Patel, P. B., Singh, D. P., Rajkamal, M., & Boopathi, S. (2023). Experimental investigation of RHA biochar and tamarind fibre epoxy composite. *Materials Today: Proceedings*. doi:10.1016/j.matpr.2023.02.439

Chapter 17

Estimating Parameters for Implantable Hydroelectric Asynchronous Generators Field Simulations and Modified Standard Measurements Approach

S. Angalaeswari

iD https://orcid.org/0000-0001-9875-9768

Vellore Institute of Technology, Chennai, India

Kaliappan Seeniappan

iD https://orcid.org/0000-0002-5021-8759

KCG College of Technology, Chennai, India

ABSTRACT

This study introduces a novel methodology for estimating the parameters of implantable hydroelectric asynchronous generators, combining field simulations with a modified standard measurements approach. The research focuses on developing a reliable yet straightforward technique for determining the constant parameters essential for the optimal functioning of these generators. The proposed method is tailored specifically to the unique requirements of implantable hydroelectric generators and is primarily based on specialized testing conducted during various stages of the generator's operation cycle. Initially, the approach involves an independent assessment of the stator inductor's field in these generators. Subsequently, all remaining constant parameters are estimated using stationary tests, specifically designed for the unique operational environment of implantable generators. To validate the effectiveness of this methodology, comprehensive field simulations are conducted.

DOI: 10.4018/979-8-3693-3735-6.ch017

INTRODUCTION

Precise techniques of identifying ac system parameters by utilising simple tests and processes are extremely beneficial to the market. Although if originally supplied, equipment settings might change significantly from factory settings over time and should be revised (Kanimozhi et al. 2022). Additionally, required again for evaluation of such technologies is parameter estimation without appropriate precision. Several articles on models and also parameter estimates for switching devices were released (Nagarajan et al. 2022). Its primary and known worldwide computational algorithms are based on impedance testing and fem simulations, accordingly. It still does not exclude several additional indications of importance (Angalaeswari et al. 2022).

Stalemate or active spectral studies generate information in the Fourier transform instead of discrete time. These are modest transmission evaluations that are performed on home appliances. In either event, the device is fed via an adjustable-speed line rather than the commercial network (Darshan et al. 2022). Theorists, particularly with numerical simulation techniques, need the precise inner structure and mechanical mechanics of a device as well as extremely complex and pricey technology. If those same prerequisites are not satisfied, as is entirely plausible, the numerical simulation approach cannot be used (Balamurugan et al. 2023). Our current various reports present a straightforward method for evaluating the parameters of conspicuous or non-salient pole rotating generators (Velmurugan et al. 2023).

The product's strengths persist in the premise that almost all testing is reasonably understood, in stable equilibrium, just at the desired frequency, and in a manner comparable to electrical machines (Suman et al. 2023). A resulting value for achieving recognition is depicted in Figure 1, with test scenarios provided within every stage (Asha et al. 2022). Because the sequencing shows that perhaps the technique is essentially reliant on understanding the magnetic to stator substitute goods, particular attention is made to quantifying this coefficient using three separate basic techniques (Reddy et al. 2023). It should be highlighted that the analysis is entirely in line with random values; irregular magnetising limb features are generally expected to be accessible from the function object or analysed independently as is customary (Josphineleela et al. 2023a). The following is how the article is structured: In this first part, they present a technique for calculating the field/skeleton switching frequency (Santhosh Kumar et al. 2022). A novel technique for stator leaking resistive estimation is presented in the following chapter. The assessment of a field's electrical characteristics is covered in the third part.

DIMENSION OF FIELD TO ARMATURE TURNS PROPORTION

As aforementioned, the current to actuator 10 to 21 factor is crucial to a suggested technique (Mahesha et al. 2022). It has an impact on the entire identifier, starting with the actuator leaking characteristic impedance (Divya et al. 2022). As a result, it's going to be decided. from two different exams: unloading, voltage collapse, and trailing 0 component testing A contrast will indeed be created in order to determine the appropriate value followed by the most comfortable check (Sharma et al. 2022). It is worth noting that the expansive characteristics and score of equipment under consideration are provided. DC testing at high temperatures yields spindle as well as field coil conductance (Seeniappan et al. 2023).

Testing

At a synchronous motor, the quasi-electromagnetic curves are either produced via activating the field circuits and exposing the generator airports or conversely, as shown in Figure 1. Ignoring metal and core loss, it is assumed that the actuator present in Va is a clean magnetising element (Nagarajan et al. 2022). As a result, for a particular stator current, the associated if and Ia are determined, establishing this proportion combined (Thakre et al. 2023). It isn't questionable that any analogous component is along, which is easy to demonstrate using the power constant d-q matrices conversion (Kaushal et al. 2023).

Short Circuit Testing

In contrast to the earlier scenario, just one system is required in which the magneto connections are crimped as well as the field circuits are agitated by a very small concentration of power whenever (Subramanian et al. 2022a). Let us first recall that this test can provide a number. In reality, the full employment formulae of prominent pole power converters in the d-q axis are Ldm as well as Lqm, where Ldm as well as Lqm are both dd as well as q chisel magnetising inductors, and is the rotor stimulation flow (Sendrayaperumal et al. 2021). If the oscillating system R is ignored, Eq. (3) yields iq = 0, indicating that the battery voltage vector flow has v e. With the same premise, Eq. (4) converts the null and void of a world flux across the d-axis, i.e., this same flow amount created by some motor torque is completely balanced either by fluorescence marginal revenue by the primary coil (Sendrayaperumal et al. 2021).

Lagging Zero-Power Factor Load Test

This Pitier technique is extensively used to determine not just the substitute goods as well as the stator leaking resistive Xs to a generation of entirely magnetic loading conditions. That in itself is a visual system that relies both on the renowned Pitier chart (Subramanian et al. 2022a). This technique, including all related operations, was being completely calculated, which is rare, but it will increase the recognition rate. For identifying, the accessible characteristics E(if), a situational constraint that shapes inquiry at completely inductance, as well as the field potential at three phase briefs, equivalent to the same inductance, are required (Selvi et al. 2023). The phase shift graph in Figure 1 is produced without taking into account the armature impact. This graphic demonstrates how charged particles in each formula may be expressed algebraically (Loganathan et al. 2023). Er is just the real or expected inductance produced by a fictitious illumination flow. As a result, the pairing is associated with the constant current shape E. (if). Except for Er as well as fir, all variables just on top of (7) and (8) are given. Because that's what we'll be doing in the next (Darshan et al. 2022).

Comparative Study

That filament impedance is neglected in three different conditions (Josphineleela et al. 2023b). The estimate is carried out in a fairly constant mode, and also the mean of a sought variable as a proportion of present Ia was used in figure 2. Inside this deflection test, Figure 2 shows that psychological inductance is less than each observed change in voltage Va due to the leaking power loss (Natrayan and Kaliappan 2023). As a result, the operating point that produces the same electromotive force as those used in the test procedure is I′ a > Ia, and the corresponding proportion is less than 0. 1923.Readers can make the if Xs is understood, which is not currently the case (Kaliappan et al. 2023a).

However, at this point, it is simple to choose the most efficient method of finding the substitute goods. Because the average scores of 0.1369 matched to = 0.09 acquired with lagged minimal energy component testing are somewhat better than the ones with past trials, it appears to be far more genuine that they will be employed in the life orientation (Kaliappan et al. 2023c). This is possible because Eqs. (6) through (10) translate a true account of electrical occurrences in a device San approximation (Natrayan et al. 2023).

RESULT AND DISCUSSIONS

The investigate goes into the inner dynamics of spindle leak inductance, highlighting its existence and relevance in each intermittent and reasonably constant activities (Ramaswamy et al. 2022b). Despite the intrinsic properties of dampers, spindle leak inductance appears as a vital component, performing a key role in forecasting the asynchronous efficacy of the system (Muralidaran et al. 2023). This reputation underlines the complicated nature of device conduct, noting that spindle leak inductance is a dynamic and impactful problem independent of the operating circumstances (Josphineleela et al. 2023c; Kaliappan et al. 2023b).

Traditionally, the Pitier approach has been a conventional and frequently employed way for determining stator leakage resistive parameter Xs. However, the look at shines light on a notable discovery from investigations done in which difficult conditions the usual dependence on Pitier's feature impedance (Arun et al. 2022; Kaliappan et al. 2023d; Sivakumar et al. 2023). The tests reveals that Pitier's feature impedance is much larger than that of a filament leaking resistor, forcing a review of its suitability for positive circumstances (Balaji et al. 2022; Ramaswamy et al. 2022a; Selvi et al. 2023). This location presents a key aspect, pushing researchers to reconsider mounted approaches within the quest of precise parameter estimation.

The drawbacks of approaches wholly dependent on the Schrödinger equation are placed to the forefront in the take a look at (Natrayan et al. 2022; Balaji et al. 2023; Pragadish et al. 2023). These tactics, however generally wide-spread, have fundamental bad qualities that prevent their efficiency. The investigations reveals those not uncommon drawbacks—they're discovered to be fully invalid within the longitudinal region and imprecise inside the knees of the system (Lakshmaiya 2023). This realisation stimulates a reevaluation of the underlying equations that have long been depended upon, signifying the necessity for potential tactics that may cope with such limits.

Simultaneously, the observe realises the continuous nature of the discourse, noting the relevance of gathering new data and insights (Devarajan and Lakshmaiya 2022; Subramanian et al. 2022b). The examination of a potential practical response is teased, offering a look into the coming section. This expectation underlines the dynamic and developing character of the research, demonstrating that the examination isn't finest recognising issues however actively hunting for options and alternatives to beautify the information and estimate of gadget parameters (Ramachandran et al. 2022; Sai et al. 2023; Ugle et al. 2023).

In the quest of expanding parameter estimation procedures, the research community is urged to recall and analyse existing strategies considerably (Rajagopalan et al. 2022; Lakshmaiya et al. 2023). The discovery about Pitier's distinctive impedance challenges the established quo, demanding a rethink of widely adopted approaches. As in addition information is obtained within the existing conversation

Figure 1. Zoom of the phase voltage of the moments

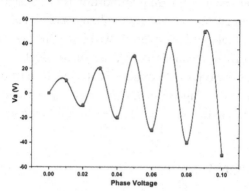

thread, the promise of an other sensible answer in the future phase indicates a proactive and ahead-looking approach to tackling the issues offered via spindle leak inductance. This dynamic and iterative system encompasses the essence of research, whereby every discovery and goal will become a stepping stone in the direction of refining procedures and expanding the collective knowledge of complex structures.

However, the curves only provide the value of Ed without such an explanation; alternatively, Zs as well as Xs may be determined immediately. To solve a problem, the iteration strategy of adjusting Xs by an evolutionary step Xs out of an original amount Xs0, subsequently calculating, was used . The presented method is simple and may be used to match prominent poles or smoother air gaps (Sabarinathan et al. 2022). As previously stated, the technique is limited to the constants of a baseline general framework (Niveditha VR. and Rajakumar PS. 2020). These adjustable components, or magnetising components, are analysed independently as is customary. Its saturation features are now only required during the undertaking as most standing steady tests at duty cycle are power surge tests owing to damping coils (Singh et al. 2017).

As a result, the transmitted voltage must still be restricted. This same unconventional software deals with the founding of grid voltage as well as the vibrational existence of an inverter, which would be secluded but instead automatically activated at its critical frequency (Hemalatha et al. 2020; Anupama et al. 2021). This same power supply is purposefully packed even though commutator tides are still not void throughout this scenario, as well as all d-q circuit boards are procured during the reactor operation, allowing the entire infrastructure to just be verified. A filament flow resistor is acknowledged as an important metric for converters regardless of operating condition. Its Pitier method calculation using equation (8) resulted in Xemena = 36.15. This one is consistent with the study of the experiment, which shows that the Pitier characteristic impedance is greater than that of the genuine leak resistor, notwithstanding the results in growth by the statistical dental procedures.

A repetitive technique described in Section 3 is proposed in the unique method for evaluating Xs. The spindle velocity must always be determined in accordance with Series of Standards 115 A. At duty cycle, electronic equipment having reasonable accuracy is recommended for measuring absorbing energy, rotor as well as outfield circuit voltage, including load winding. The posterior assessment of a corresponding component is assumed for calculating Xs. This would be frequently determined by an unloaded check, a four-test for short circuits, as well as a zero-factor load test. All of the experiments discussed in section II were run, as well as the analysis reveals that perhaps the Pitier number was the most accurate, while

Figure 2. Zoom of the phase voltage of the steady state moments

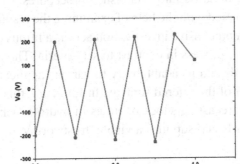

another must first be changed. Its main criterion is that there are conclusions that can be made in Pitier electromagnetic suggesting the proportion.

For the sake of comparability, modifications as per (18) provide 0.0837 as well as 0.097, which are both extremely near to the Pitier number of 0.08. This remainder of the detection phase emphasises the suggested product's versatility and ease of use. All damping plus magnetic circuitry characteristics are calculated in a manner similar to that utilised for inverters during locked rotor testing. Despite the fact that they are calculated by two distinct methods, both the leakage of dark-skinned people as well as the lD are nearly equivalent. This suggests that the alternator is equal to a miniature induction motor drive along with more support for the stated technique. The goal of this study is to build a simple and hence convenient authentication system.

CONCLUSION

This research supplies a distinctive technique for anticipating shock features in hydro-synchronized technology, imparting a new strategy for specific computing. The investigations demonstrates a series of trustworthy methods developed to examine all consistent sections, notably highlighting the relevance of correctly selecting the sector to actuator switching frequency. Notably, the typical graphic strategy outlined within the IEEE-a hundred and fifteen norm, utilised as a comparing instrument in the course of the take a look at, is judged now not simply constrained but furthermore incapable of producing credible estimates.

In a departure from usual conventions, the research independently creates the leak inductor of the electromagnetic transmission line. This departure is vital in confirming intermittent models of an exceptional pole Stacom in both daily and deteriorated overall performance possibilities. The study expands its examination to comprise the same method for analysing capacitance variation throughout the solidification reaction. Utilizing data, the take a look at no longer just confirms intermittent models but additionally supplies insights into the behaviour of a distinctive pole Stacom underneath various settings.

This study resides in its usefulness to telecommuting, making the offered approach clearly acceptable for contemporary applications. The look at attracts similarities among its offered approach and the equipment base favoured model, which operates one damping circuit in every column. This comparative review contextualizes the novel technique within current industry standards, highlighting its flexibility and possible incorporation into established procedures. The dependability of the newly offered approach is carefully verified with the help of examining the results in evaluation to those of the conventional machine. This comprehensive investigation acts as a litmus check, measuring the efficacy and correctness of the innovative process in contrast to set criteria. The examine's focus on validation using actual-international records and its adaptability to various applications underline the pragmatic and forward-searching character of the offered strategy. In summary, this study now just most effective provides a fresh approach for forecasting shock qualities in hydro-synchronized period but moreover questions contemporary standards and supplies a viable, trustworthy alternative for present programs within the area.

REFERENCE

Angalaeswari, S., Jamuna, K., Mohana sundaram, K., Natrayan, L., Ramesh, L., & Ramaswamy, K. (2022). Power-Sharing Analysis of Hybrid Microgrid Using Iterative Learning Controller (ILC) considering Source and Load Variation. *Mathematical Problems in Engineering*, 2022, 1–6. doi:10.1155/2022/7403691

Anupama. (2021). Deep learning with backtracking search optimization based skin lesion diagnosis model. *Computers, Materials & Continua*, 70(1). doi:10.32604/cmc.2022.018396

Arun. (2022). Mechanical, fracture toughness, and Dynamic Mechanical properties of twill weaved bamboo fiber-reinforced Artocarpus heterophyllus seed husk biochar epoxy composite. *Polymer Composites*, 43(11), 8388–8395. doi:10.1002/pc.27010

Asha, P., Natrayan, L., Geetha, B. T., Beulah, J. R., Sumathy, R., Varalakshmi, G., & Neelakandan, S. (2022). IoT enabled environmental toxicology for air pollution monitoring using AI techniques. *Environmental Research*, 205, 112574. doi:10.1016/j.envres.2021.112574 PMID:34919959

Balaji. (2022). Annealed peanut shell biochar as potential reinforcement for aloe vera fiber-epoxy biocomposite: Mechanical, thermal conductivity, and dielectric properties. *Biomass Conversion and Biorefinery*. doi:10.1007/s13399-022-02650-7

Balaji, N., Gurupranes, S. V., Balaguru, S., Jayaraman, P., Natrayan, L., Subbiah, R., & Kaliappan, S. (2023). Mechanical, wear, and drop load impact behavior of Cissus quadrangularis fiber–reinforced moringa gum powder–toughened polyester composite. *Biomass Conversion and Biorefinery*. doi:10.1007/s13399-023-04491-4

Balamurugan, P., Agarwal, P., Khajuria, D., Mahapatra, D., Angalaeswari, S., Natrayan, L., & Mammo, W. D. (2023). State-Flow Control Based Multistage Constant-Current Battery Charger for Electric Two-Wheeler. *Journal of Advanced Transportation*, 2023, 1–11. doi:10.1155/2023/4554582

Darshan, Girdhar, N., Bhojwani, R., Rastogi, K., Angalaeswari, S., Natrayan, L., & Paramasivam, P. (2022). Energy Audit of a Residential Building to Reduce Energy Cost and Carbon Footprint for Sustainable Development with Renewable Energy Sources. *Advances in Civil Engineering*, 2022, 1–10. Advance online publication. doi:10.1155/2022/4400874

Devarajan, & Lakshmaiya, N. (2022). Effective utilization of waste banana peel extracts for generating activated carbon-based adsorbent for emission reduction. *Biomass Conversion and Biorefinery*. Advance online publication. doi:10.1007/s13399-022-03470-5

Divya. (2022). Analysing Analyzing the performance of combined solar photovoltaic power system with phase change material. *Energy Reports*, 8. Advance online publication. doi:10.1016/j.egyr.2022.06.109

Hemalatha. (2020). Analysis of RCC T-beam and prestressed concrete box girder bridges super structure under different span conditions. In *Materials Today*. Proceedings.

Kaliappan, S., Natrayan, L., Kumar, P. V. A., & Raturi, A. (2023b). Mechanical, fatigue, and hydrophobic properties of silane-treated green pea fiber and egg fruit seed powder epoxy composite. *Biomass Conversion and Biorefinery*. doi:10.1007/s13399-023-04534-w

Kaliappan, S., Velumayil, R., & Pravin, P. (2023d). Mechanical, DMA, and fatigue behavior of Vitis vinifera stalk cellulose Bambusa vulgaris fiber epoxy composites. *Polymer Composites*, 44(4), 2115–2121. doi:10.1002/pc.27228

Kanimozhi, G., Natrayan, L., Angalaeswari, S., & Paramasivam, P. (2022). An Effective Charger for Plug-In Hybrid Electric Vehicles (PHEV) with an Enhanced PFC Rectifier and ZVS-ZCS DC/DC High-Frequency Converter. *Journal of Advanced Transportation*, 2022, 1–14. doi:10.1155/2022/7840102

Kaushal. (2023) A Payment System for Electric Vehicles Charging and Peer-to-Peer Energy Trading. In: *7th International Conference on I-SMAC (IoT in Social, Mobile, Analytics and Cloud), I-SMAC 2023 – Proceedings*. IEEE. 10.1109/I-SMAC58438.2023.10290505

Kumar, S. (2022). IoT battery management system in electric vehicle based on LR parameter estimation and ORMeshNet gateway topology. *Sustainable Energy Technologies and Assessments*, 53, 102696. doi:10.1016/j.seta.2022.102696

Lakshmaiya, N. (2023) Investigation on ultraviolet radiation of flow pattern and particles transportation in vanishing raindrops. In: *Proceedings of SPIE*. The International Society for Optical Engineering 10.1117/12.2675556

Lakshmaiya, N., Surakasi, R., Nadh, V. S., Srinivas, C., Kaliappan, S., Ganesan, V., Paramasivam, P., & Dhanasekaran, S. (2023). Tanning Wastewater Sterilization in the Dark and Sunlight Using Psidium guajava Leaf-Derived Copper Oxide Nanoparticles and Their Characteristics. *ACS Omega*, 8(42), 39680–39689. doi:10.1021/acsomega.3c05588 PMID:37901496

Loganathan, Ramachandran, V., Perumal, A. S., Dhanasekaran, S., Lakshmaiya, N., & Paramasivam, P. (2023). Framework of Transactive Energy Market Strategies for Lucrative Peer-to-Peer Energy Transactions. *Energies*, 16(1), 6. doi:10.3390/en16010006

Mahesha, C. R., Rani, G. J., Dattu, V. S. N. C. H., Rao, Y. K. S. S., Madhusudhanan, J., L, N., Sekhar, S. C., & Sathyamurthy, R. (2022). Optimization of transesterification production of biodiesel from Pithecellobium dulce seed oil. *Energy Reports*, 8, 489–497. doi:10.1016/j.egyr.2022.10.228

Muralidaran, Natrayan, L., Kaliappan, S., & Patil, P. P. (2023). Grape stalk cellulose toughened plain weaved bamboo fiber-reinforced epoxy composite: Load bearing and time-dependent behavior. *Biomass Conversion and Biorefinery*. doi:10.1007/s13399-022-03702-8

Nagarajan, Rajagopalan, A., Angalaeswari, S., Natrayan, L., & Mammo, W. D. (2022). Combined Economic Emission Dispatch of Microgrid with the Incorporation of Renewable Energy Sources Using Improved Mayfly Optimization Algorithm. *Computational Intelligence and Neuroscience*, *2022*, 1–22. doi:10.1155/2022/6461690 PMID:35479598

Natrayan. (2023) Control and Monitoring of a Quadcopter in Border Areas Using Embedded System. In: *Proceedings of the 4th International Conference on Smart Electronics and Communication, ICOSEC 2023*. IEEE. 10.1109/ICOSEC58147.2023.10276196

Natrayan, L., & Kaliappan, S. (2023) Mechanical Assessment of Carbon-Luffa Hybrid Composites for Automotive Applications. In: SAE Technical Papers. doi:10.4271/2023-01-5070

Natrayan, L., Kaliappan, S., Sethupathy, B. S., Sekar, S., Patil, P. P., Velmurugan, G., & Tariku Olkeba, T. (2022). Effect of Mechanical Properties on Fibre Addition of Flax and Graphene-Based Bionanocomposites. *International Journal of Chemical Engineering*, *2022*, 1–8. doi:10.1155/2022/5086365

Niveditha, V. R., & Rajakumar, P. S. (2020). Pervasive computing in the context of COVID-19 prediction with AI-based algorithms. *International Journal of Pervasive Computing and Communications*, *16*(5). Advance online publication. doi:10.1108/IJPCC-07-2020-0082

Pragadish, N., Kaliappan, S., Subramanian, M., Natrayan, L., Satish Prakash, K., Subbiah, R., & Kumar, T. C. A. (2023). Optimization of cardanol oil dielectric-activated EDM process parameters in machining of silicon steel. *Biomass Conversion and Biorefinery*, *13*(15), 14087–14096. doi:10.1007/s13399-021-02268-1

Rajagopalan, Nagarajan, K., Montoya, O. D., Dhanasekaran, S., Kareem, I. A., Perumal, A. S., Lakshmaiya, N., & Paramasivam, P. (2022). Multi-Objective Optimal Scheduling of a Microgrid Using Oppositional Gradient-Based Grey Wolf Optimizer. *Energies*, *15*(23), 9024. doi:10.3390/en15239024

Ramachandran, Perumal, A. S., Lakshmaiya, N., Paramasivam, P., & Dhanasekaran, S. (2022). Unified Power Control of Permanent Magnet Synchronous Generator Based Wind Power System with Ancillary Support during Grid Faults. *Energies*, *15*(19), 7385. doi:10.3390/en15197385

Ramaswamy. (2022b). Pear cactus fiber with onion sheath biocarbon nanosheet toughened epoxy composite: Mechanical, thermal, and electrical properties. *Biomass Conversion and Biorefinery*. doi:10.1007/s13399-022-03335-x

Ramaswamy, R., Gurupranes, S. V., Kaliappan, S., Natrayan, L., & Patil, P. P. (2022a). Characterization of prickly pear short fiber and red onion peel biocarbon nanosheets toughened epoxy composites. *Polymer Composites*, *43*(8), 4899–4908. Advance online publication. doi:10.1002/pc.26735

Reddy. (2023) Development of Programmed Autonomous Electric Heavy Vehicle: An Application of IoT. In: *Proceedings of the 2023 2nd International Conference on Electronics and Renewable Systems, ICEARS 2023*. IEEE. 10.1109/ICEARS56392.2023.10085492

Sabarinathan, P., Annamalai, V. E., Vishal, K., Nitin, M. S., Natrayan, L., Veeeman, D., & Mammo, W. D. (2022). Experimental study on removal of phenol formaldehyde resin coating from the abrasive disc and preparation of abrasive disc for polishing application. *Advances in Materials Science and Engineering, 2022*, 1–8. doi:10.1155/2022/6123160

Sai, Venkatesh, S. N., Dhanasekaran, S., Balaji, P. A., Sugumaran, V., Lakshmaiya, N., & Paramasivam, P. (2023). Transfer Learning Based Fault Detection for Suspension System Using Vibrational Analysis and Radar Plots. *Machines, 11*(8), 778. doi:10.3390/machines11080778

Seeniappan. (2023). *Modelling and development of energy systems through cyber physical systems with optimising interconnected with control and sensing parameters.*

Selvi. (2023). Optimization of Solar Panel Orientation for Maximum Energy Efficiency. In: *Proceedings of the 4th International Conference on Smart Electronics and Communication, ICOSEC 2023.* IEEE. 10.1109/ICOSEC58147.2023.10276287

Sendrayaperumal, Mahapatra, S., Parida, S. S., Surana, K., Balamurugan, P., Natrayan, L., & Paramasivam, P. (2021). Energy Auditing for Efficient Planning and Implementation in Commercial and Residential Buildings. *Advances in Civil Engineering, 2021*, 1–10. doi:10.1155/2021/1908568

Sharma, Raffik, R., Chaturvedi, A., Geeitha, S., Akram, P. S., L, N., Mohanavel, V., Sudhakar, M., & Sathyamurthy, R. (2022). Designing and implementing a smart transplanting framework using programmable logic controller and photoelectric sensor. *Energy Reports, 8*, 430–444. doi:10.1016/j.egyr.2022.07.019

Singh. (2017). An experimental investigation on mechanical behaviour of siCp reinforced Al 6061 MMC using squeeze casting process. *International Journal of Mechanical and Production Engineering Research and Development, 7*(6). doi:10.24247/ijmperddec201774

Sivakumar, V., Kaliappan, S., Natrayan, L., & Patil, P. P. (2023). Effects of Silane-Treated High-Content Cellulose Okra Fibre and Tamarind Kernel Powder on Mechanical, Thermal Stability and Water Absorption Behaviour of Epoxy Composites. *Silicon, 15*(10), 4439–4447. doi:10.1007/s12633-023-02370-1

Subramanian, Lakshmaiya, N., Ramasamy, D., & Devarajan, Y. (2022a). Detailed analysis on engine operating in dual fuel mode with different energy fractions of sustainable HHO gas. *Environmental Progress & Sustainable Energy, 41*(5), e13850. doi:10.1002/ep.13850

Subramanian, Solaiyan, E., Sendrayaperumal, A., & Lakshmaiya, N. (2022b). Flexural behaviour of geopolymer concrete beams reinforced with BFRP and GFRP polymer composites. *Advances in Structural Engineering, 25*(5), 954–965. doi:10.1177/13694332211054229

Suman. (2023) IoT based Social Device Network with Cloud Computing Architecture. In: *Proceedings of the 2023 2nd International Conference on Electronics and Renewable Systems, ICEARS 2023.* IEEE. 10.1109/ICEARS56392.2023.10085574

Thakre, Pandhare, A., Malwe, P. D., Gupta, N., Kothare, C., Magade, P. B., Patel, A., Meena, R. S., Veza, I., Natrayan L, & Panchal, H. (2023). Heat transfer and pressure drop analysis of a microchannel heat sink using nanofluids for energy applications. *Kerntechnik, 88*(5), 543–555. doi:10.1515/kern-2023-0034

Ugle, Arulprakasajothi, M., Padmanabhan, S., Devarajan, Y., Lakshmaiya, N., & Subbaiyan, N. (2023). Investigation of heat transport characteristics of titanium dioxide nanofluids with corrugated tube. *Environmental Quality Management, 33*(2), 127–138. doi:10.1002/tqem.21999

Velmurugan, G., Natrayan, L., Chohan, J. S., Vasanthi, P., Angalaeswari, S., Pravin, P., Kaliappan, S., & Arunkumar, D. (2023). Investigation of mechanical and dynamic mechanical analysis of bamboo/ olive tree leaves powder-based hybrid composites under cryogenic conditions. *Biomass Conversion and Biorefinery.* doi:10.1007/s13399-023-04591-1

Chapter 20
Power Consumption Predictive Analysis

Aditi Singh
Vellore Institute of Technology, India

Akhil Kodalapuram
iD https://orcid.org/0000-0001-6655-0089
Vellore Institute of Technology, India

Prathamesh Prabhu
Vellore Institute of Technology, India

Arnava Soni
Vellore Institute of Technology, India

Vijayapriya Ramachandran
iD https://orcid.org/0000-0001-6655-0089
Vellore Institute of Technology, India

ABSTRACT

Initially the project is started by recording power consumption from a laptop at different performance levels using a software known as "HWInfo." Next cleaned this data by filtering the potential variables needed for the analysis and obtained average values for unique utility percentages. The variables to be used for training different models at the time of model choosing were power and temperature for CPU and GPU individually. Upon training different models with the training data, ExtraTreeRegressor was the ML model that gave the least RMSE error, in turn being used for the prediction. Power consumption of VIT university had two variables and one predicting target, them being date, power units, and maximum demand respectively; for which the authors shifted to an ML model called NeuralProphet. This model takes care of any outliers in the data. To deal with the many inconsistencies and unpredicted data, NeuralProphet performs adaptive analysis which is used for training and prediction. After obtaining results from both data, the model is integrated into a web page's front-end.

DOI: 10.4018/979-8-3693-3735-6.ch020

INTRODUCTION

The objective of this project is to provide the user with an estimated cost of the electricity that will be consumed in a particular duration based on previous consumption tracking and analysis; and show the total cost of the actual consumption of electricity in said duration. We are trying to use a device which meters the consumption of Power over a small time period. The device will store the data collected over a period of time on the cloud, this data will be the data set used to teach the machine learning model. The model then will predict the Power consumption of the given device and return the cost as well as the units consumed. We are trying to find the most efficient way to reduce the cost and consumption of Power.

A review of the techniques used to optimize Power consumption and examine the factors influencing the creation of thermal comfort, visual comfort and convenience in air quality and reduce Power consumption is detailed in Salam et al. (2019). The paper provides a broad review of various power consumption optimization techniques, but it does not provide a detailed comparison or evaluation of these techniques. Therefore, it may be challenging to determine which techniques are most effective for a specific application or building type. The study focuses specifically on IoT-based smart building environments, which may not be applicable to other types of buildings or settings. Additionally, the effectiveness of the reviewed techniques may depend on the specific hardware and software used in each building, which may vary across different settings.

A review that is focused on the prediction analytics using Web and Android technology platforms is presented in Kewo et al. (2015). In this case, to predict the Power consumption three regression models such as simple linear regression, KLM a and KLM b are applied. All models can be applied to predict the next period of Power consumption based on the independent variable of X = da y and dependent variables of Y = current, voltage, and power. The paper focuses only on regression analysis as the method for predicting electricity consumption and does not explore other machine-learning algorithms that could potentially provide better results. The stud y uses data from only one site, which ma y limit the generalizability of the results to other sites with different characteristics.

In Fard et al. (2022), the internal Power consumption of the IoT has been studied. The purpose of this paper is to predict the factors affecting Power consumption in buildings by considering machine learning algorithms such as k-nearest neighbors (KNN), Ada boost, random forest and neural network The study evaluates the performance of machine learning algorithms for predicting power consumption and IoT modelling but does not provide a comparison with other traditional prediction methods. The proposed model for IoT in complex networks relies on accurate data collection, which may be challenging to obtain in some real-world settings

The potentials of deep reinforcement learning techniques are investigated in Liu et al (2020). Three commonly-used DRL techniques, i.e. A3C, DDPG and RDPG are utilized for building. Power consumption forecasting. The prediction performances of 6 popular predictive algorithms are studied The study proposes a model based on deep in reinforcement learning techniques, which may require a large amount of data and computational power for training and implementation. The proposed model is evaluated on a limited dataset, which may limit its generalizability to other datasets with different characteristics

Advanced metering infrastructure takes and saves Power usage data as datasets. Understanding Power consumption statistics' data structure and consumer behavior may be laborious. Some research has explored smart home Power usage records and consumer behavior, although the approaches are difficult Prakasha et al. (2022). This research presents a simple method for exploring smart home Power use data. This method works for any numerical smart home dataset. The study analyses the impact of different energy-saving techniques on energy consumption and customer behavior, but does not provide a comparison of the effectiveness of these techniques. The study focuses on smart home energy consumption and customer behavior and may not be directly applicable to other settings such as commercial or industrial buildings.

DATA ANALYSIS

Desktop data is accumulated and analysed to find the relation between the percentage usage and the temperature of the CPU/GPU with the Wattage Consumption or Power consumed of the CPU/GPU. The plots of Percentage CPU usage vs Average core SOC power, Percentage CPU usage vs Average CPU Temperature, Maximum and Minimum values of power usage, Maximum and Minimum values of Temperature, Percentage GPU usage vs Average GPU Temperature, Percentage GPU usage vs Average GPU Power, Maximum and Minimum values of GPU Temperature, Maximum and Minimum values of GPU Power are depicted in Figure 1, 2, 3, 4, 5, 6, 7 and 8, respectively.

The total power that the desktop draws will be the addition of the drawn power of the CPU and the GPU. The range of the power drawn by the CPU/GPU during different percentage usage and temperatures helps us find a more accurate final value of this specific system. This is because the Machine Learning Algorithm can use all these data values and provide an accurate final value of power drawn. For such an estimate calculation the Regression model is best suited.

Figure 1. Percentage CPU usage vs. average core SOC power

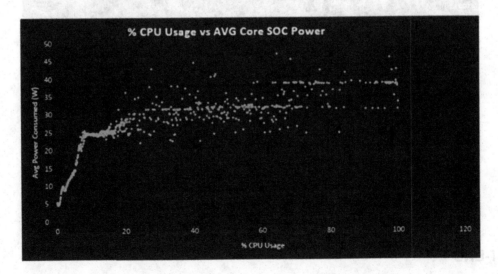

Figure 2. Percentage CPU usage vs. average CPU Temperature

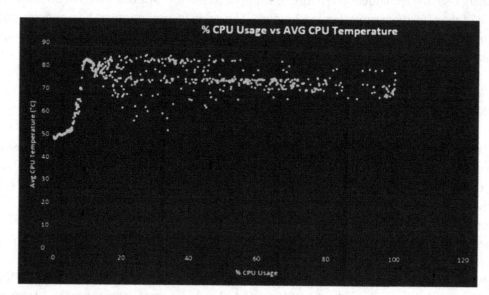

Figure 3. Maximum and minimum values of Power usage

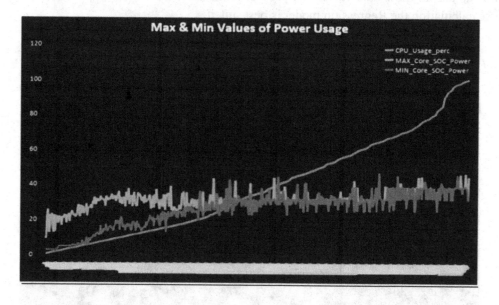

FRONT-END

Website for Energy Consumption of Desktop

In this website, the CPU wattage and GPU wattage at a particular CPU usage, CPU Temperature, GPU usage and GPU temperature are predicted and are respectively depicted in Figure 9. In this a simple background is employed and using CSS the UI/UX is styled accordingly. Consequently, four input boxes

Figure 4. Maximum and minimum values of Temperature

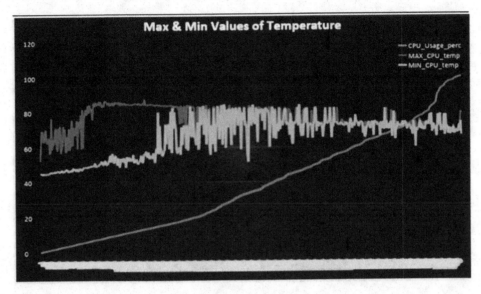

Figure 5. Percentage GPU usage vs. average GPU temperature

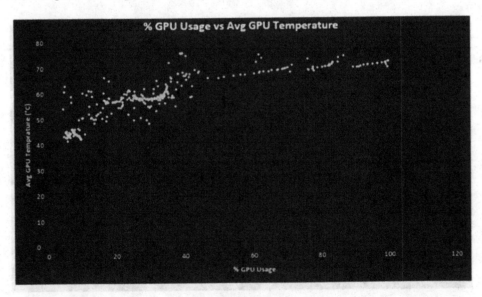

are created with a particular id. These values are sent to the backend using jQuery and API keys. When the Submit button is pressed, it must post the output in output boxes. For this three output boxes named CPU Wattage and GPU Wattage and Total Power are created. These output boxes also have a Unique Id. The result from the backend on CPU Wattage and GPU Wattage are posted using #POST, id and jQuery.

Total Power = CPU Wattage + GPU Wattage

Figure 6. Percentage GPU usage vs. average GPU Power

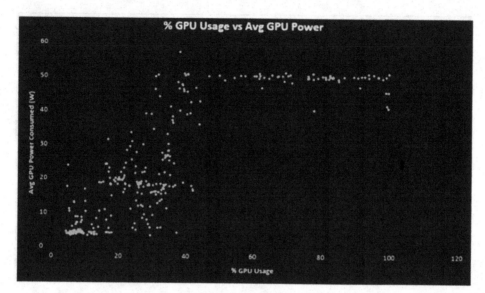

Figure 7. Maximum and Minimum values of GPU Temperature

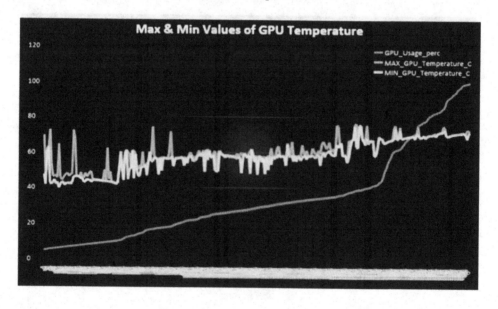

Website for Energy Consumption Prediction

In this website, the consumption of EB units are [predicted and then according to that the energy bill as shown in Figure 10. The Eb units according to the year and month are also predicted. Subsequently, two drop-down menus are created; one for a year and one for month. A submit button and two output boxes one for EB units consumed (the one to be predicted) and one for Energy Bill. The dropdown menu is designed such that you can only predict future values. The previous year or month is not a valid entry. For Example, if the current year is 2023 and the current month is April then in the month

Figure 8. Maximum and minimum values of GPU power

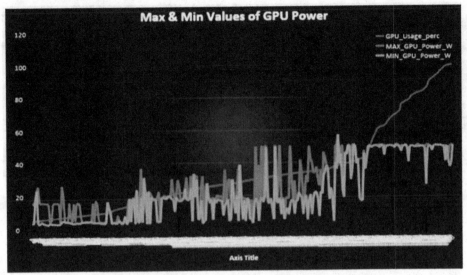

Figure 9. Website for energy consumption of desktop

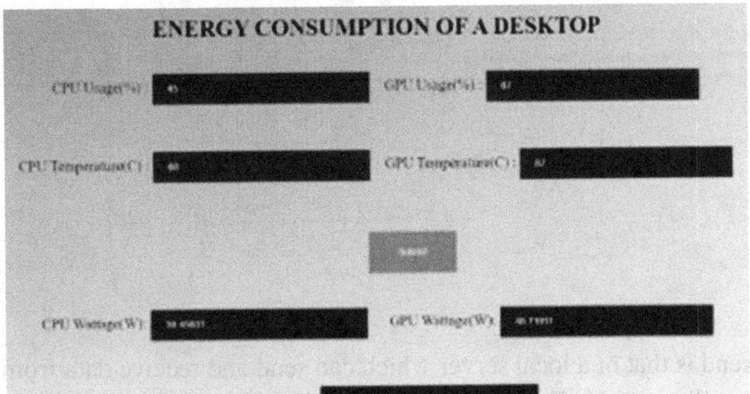

dropdown menu for the year 2023 the month will start from May but for the year 2024, it will start from January. It is made using a simple event lister that if the entered year is the current year, then the month drop-down menu option will start one month greater than the current year i.e. (if (year=current year) {month=current_month+1;}) Here also jQuery is employed and API to transfer the data i.e. (year, month) to backend getting the EB units as output. Posting it in EB units Consumed output box using #POST comment and calculate the Energy bill according to Eb units consumed. For example, if per unit consumed charge is 5 rupees/unit then:

Energy bill = Eb units consumed *5 Rupees. Vijayapriya, R et al. (2023) and Arul S L et al. (2023)

Figure 10. Website for energy consumption prediction

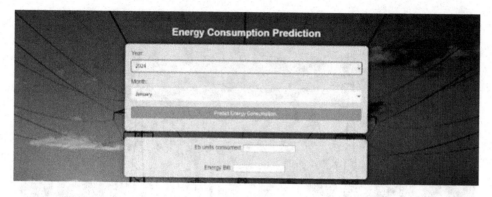

Figure 11. Back-End

BACK-END

The job of the backend is that of a local server which can send and receive data from the ML model as well as the frontend as illustrated in Figure 11. A server is created using the services of Fast Api. FastAPI is a modern, fast (high-performance), web framework for building APIs with Python 3.7+ based on standard Python-type hints. A FastAPI is employed to integrate the front-end, backend and ML models. The use of FastAPI is to receive the data posted from the front-end and run the ML model on the given data (parameters), and then resend the predicted values back to the front-end. Here JQuery is used to send the post requests to the server.

MACHINE LEARNING

The machine learning part of this project is split into 2 parts:

- Device Energy Consumption prediction
- VIT Energy consumption prediction.

Two different models were used to perform the energy consumption prediction. The details of these models will be elaborated below:

Device Energy Consumption Prediction Model Used: ExtraTreesRegressor

For device energy consumption prediction, initially an analysis is carried out to identify which model would be the best fit for predicting the energy consumption of a device. For device energy consumption prediction, fore mostly CPU Power and GPU Power are needed to be predicted. Factors used to predict CPU Power were Total CPU Usage [%] & CPU SOC [°C]. Factors used to predict GPU Power were GPU Memory Usage [%] & GPU Temperature [°C]. A Python code is used, which took in a list of models and ran those models on our data to see which model gave the highest score and the lowest error. The metrics used for this were the R2 score and RMSE Error. In simple terms, R2 score is a statistical measure which measures how well the regression line actually fits the model. RMSE shows how far predictions fall from measured true values using Euclidean Distance. The results of the code are illustrated in Figure 12, Figure 13, Figure 14 and Figure 15.

It is clear that in both cases ExtraTrees Regressor gave better scores for both CPU and GPU Power prediction. Hence ExtraTreesRegressor model was chosen. Extra Trees Regressor (ExtraTreesRegressor) is a type of ensemble learning algorithm used for regression tasks in machine learning. It belongs to the family of decision tree algorithms and is an extension of the Random Forest algorithm. In Extra Trees Regressor, multiple decision trees are created, and each tree is trained on a random subset of the training data and a random subset of the features. The tree splits are chosen randomly instead of using the best split, as is the case in traditional decision trees. This introduces more randomness into the algorithm, which helps to reduce overfitting and improve generalization performance. The "Extra" in Extra Trees Regressor refers to the fact that in this algorithm, the decision trees are built using a larger number of randomly selected features than is typically used in a Random Forest. This further increases the diversity of the trees and helps to reduce the variance of the model. Once the trees are constructed, predictions are made by averaging the predictions of all the trees. The final result is an average of the predictions of all the trees, weighted by the number of samples that each tree represents. Extra Trees Regressor is a powerful algorithm for regression tasks, particularly when dealing with high-dimensional data or noisy data. It is relatively fast and easy to use, as it requires minimal hyperparameter tuning compared to other models, and it can handle missing values in the data.

Figure 12. R2 Score results for CPU analysis: Numerical representation

	Name	Train_Time	Train_R2_Score	Test_R2_Score	Test_RMSE_Score
0	Lasso	0.009570	0.000000	0.000000	1.000000
1	Ridge	0.003032	0.886579	0.889791	0.331978
2	KNeighborsRegressor	0.052885	0.978312	0.966198	0.183852
3	SVR	9.861810	0.968067	0.964033	0.189649
4	RandomForest	2.915440	0.987462	0.964788	0.187649
5	ExtraTreeRegressor	1.691958	0.989461	0.965974	0.184463
6	GradientBoostingClassifier	0.837427	0.974425	0.968674	0.176992
7	MLPRegressor	3.992093	0.965462	0.962621	0.193337

Figure 13. R2 Score results for CPU analysis: Graphical representation

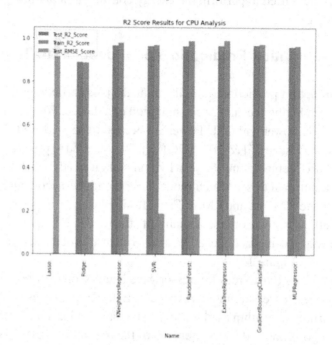

Figure 14. R2 Score results for GPU analysis: Numerical representation

	Name	Train_Time	Train_R2_Score	Test_R2_Score	Test_RMSE_Score
0	Lasso	0.005740	0.000000	0.000000	1.000000
1	Ridge	0.003972	0.980826	0.980243	0.140561
2	KNeighborsRegressor	0.460534	0.994403	0.992765	0.085059
3	SVR	2.560857	0.989928	0.989515	0.102397
4	RandomForest	1.089817	0.996479	0.992379	0.087300
5	ExtraTreeRegressor	0.833205	0.997031	0.993489	0.080691
6	GradientBoostingClassifier	0.729580	0.994228	0.989856	0.100716
7	MLPRegressor	2.269091	0.989859	0.989660	0.101687

VIT ENERGY CONSUMPTION PREDICTION

This differs from device energy prediction because that was a regression problem while this is primarily a time series problem. Hence, the input would be only the month and the year for which the EB units to be predicted Vijayapriya et al. (2022) and Arun et al. (2022). There were a couple of models which could've been implemented. One of them was Autoregression. This is a type of time series analysis model used to predict future values of a time-dependant variable based on its own past values. Basically it uses past values of a variable to predict its future values. The idea behind autoregression is that the past values of a variable can provide insights into its behavior and trends over time. By analyzing these patterns and relationships in the historical data, an autoregression model can learn to make predictions about future values of the variable. The visualization of VIT energy consumption data is depicted in Figure 16.

Figure 15. R2 Score results for GPU analysis: Graphical representation

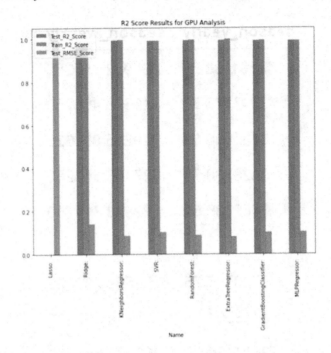

Figure 16. VIT energy consumption

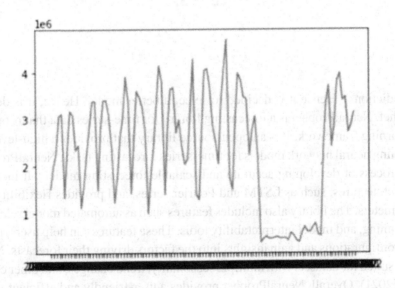

This is a simple graph which plots EB units against time. As it can be observed there is a period where EB units was very less. This happened because of COVID-19. During the COVID-19 period, VIT consumed much less EB units. Because of this anomaly in the data, models like auto regression couldn't be applied. This is because the model would be inaccurate and those data points needed to be dropped (which would result in very few data for training and testing) or normalizes that data (which would lead

Figure 17. Predicted values

season_yearly	season_weekly
3501188.75	360932.656250
3547176.75	319626.468750
3589950.50	29235.890625
3628738.75	-337486.656250
3662730.50	-322828.718750
...	...
3730797.75	-81558.203125
3718004.25	32078.632812
3694480.25	360932.656250
3489275.75	-322828.718750
3535952.00	-81558.203125

to inaccurate predictions). Hence it is decided to select a better model. Hence, it is decided to implement neuralprophet. NeuralProphet is a forecasting library for time-series data that is built on top of the PyTorch deep learning framework. It is an open-source library that provides a high-level API for easily building and training neural network models for time-series forecasting tasks. NeuralProphet is designed to simplify the process of developing accurate and scalable forecasting models. It includes a range of built-in model architectures, such as LSTM and Fourier series, and provides flexibility in configuring model hyperparameters. The library also includes features such as automated model selection, automatic hyperparameter tuning, and model interpretability tools. These features can help users quickly iterate on different model configurations and gain insights into the factors driving their forecasts. NeuralProphet is particularly well-suited for datasets with multiple seasonality, trend changes, and other complex patterns Keytingan et al. (2021). Overall, NeuralProphet provides a user-friendly and efficient way to build and train neural network models for time-series forecasting tasks. This made NeuralProphet the ideal model for our task. On training the model, the predictions obtained is given in Figure 17.

I is evident that NeuralProphet not only gave us the yearly EB units prediction, it also gave the weekly EB units consumption. This is what makes it such a useful model for a time series task.

CONCLUSION

It is clear from the results that both the models are trained and tested successfully. Upon training different models with the training data, it is evident that for the prediction of CPU usage ExtraTreeRegressor ML model produce least RMSE error. Similarly, for predicting the power consumption of VIT university ML model called NeuralProphet is employed. As the model takes care of any outliers in the data, the model yielded accurate result even the data set includes almost zero power consumption during Covid period. After obtaining results from both data, the model is integrated into a web page's front-end for easy visualization and access.

REFERENCES

Arun, S.L, Bingi, K, Vijaya Priya, R, Jacob Raglend, I, & Hanumantha Rao, B. (2023). Novel Architecture for Transactive Energy Management Systems with Various Market Clearing Strategies. *Mathematical Problems in Engineering*.

Arun S L R. (2022). Framework of Transactive Energy Market Strategies for Lucrative Peer-to-Peer Energy Transactions. *Energies*.

Fard, R. H., & Hosseini, S. "Machine Learning algorithms for prediction of Power consumption and IoT modeling in complex networks. Microprocessors and Microsystems", 2022December*Wireless Personal Communications 121* 10 doi:10.1007/s11277-021-08879-1

Kewo, A., Munir, R., & Lapu, A. K. (2015). *IntelligEnSia based electricity consumption prediction analytics using regression method.* Paper presented at 5th International Conference on Electrical Engineering and Informatics, Bali, Indonesia. 10.1109/ICEEI.2015.7352556

Liu, T., Xu, C., Chen, H., & Li, Z. (2020). Study on deep Reinforcement learning techniques for building Power consumption forecasting, Power & Buildings. *Energy and Buildings, 208.* doi:10.1016/j.enbuild.2019.109675

Mel Keytingan, M. (2021). Energy consumption prediction by using machine learning for smart building: Case study in Malaysia. *Developments in the Built Environment, 5.* doi:10.1016/j.dibe.2020.100037

Salam Shah, A., Nasir, H., Fayaz, M., Lajis, A., & Shah, A. (2019). A review on Power consumption optimization techniques in IoT-based smart building environments. *Information (Basel), 10*(3), 108. doi:10.3390/info10030108

Vijayapriya, R. (2022). IoT Based Energy Management System for Net-Zero Energy Building Operation. *Conference Paper, IEEE Delhi Section Conference (DELCON).* IEEE.

Vijayapriya, R., Umamageswari, A, Bhat, R, Dass, R, & Manikandan, N (2023). *Web-based data manipulation to improve the accessibility of factory data using big data analytics: An industry 4.0 approach.* Data Fabric Architectures: Web-Driven Applications.

Section 2

AI and ML Applications to Renewable Energy Applications

Chapter 8
Optimizing Urban Sustainability:
Reinforcement Learning–Driven Energy–Efficient Ubiquitous Robots for Smart Cities

Dharani Jaganathan

iD https://orcid.org/0000-0003-1781-7618

Computer Science and Engineering, KPR Institute of Engineering and Technology, Coimbatore, India

Vishnu Kumar Kaliappan

Computer Science and Engineering, KPR Institute of Engineering and Technology, Coimbatore, India

ABSTRACT

In the rapidly evolving landscape of smart cities, the deployment of ubiquitous robots holds immense potential for enhancing various aspects of urban living. However, the widespread integration of these robots into smart city infrastructures necessitates a careful consideration of energy efficiency to ensure sustainable and long-term operation. By leveraging advanced algorithms, these robots can adapt their behaviors and decision-making processes, leading to reduced energy consumption and increased operational sustainability. This chapter explores the application of reinforcement learning techniques to optimize the energy efficiency of ubiquitous robots operating in smart cities and also investigates various implementation methods of reinforcement learning in the context of smart cities, focusing on enhancing the energy efficiency of ubiquitous robots like search and rescue robots and contributing to the overall development of energy-conscious urban environments.

INTRODUCTION

Smart cities leverage advanced technology and data-driven solutions to enhance urban living, optimize resource utilization, and improve the overall quality of life for residents, fostering sustainable development and innovation. Ubiquitous robots, also known as service robots, are designed to operate autonomously in various environments and perform tasks to assist humans. They can be found in a wide range

DOI: 10.4018/979-8-3693-3735-6.ch008

of applications like Industrial Robots, Domestic Robots, Social Robots, Entertainment Robots, Medical Robots, Search and Rescue Robots, Agricultural Robots, Delivery Robots, Research and Exploration Robots. Educational Robots. These ubiquitous robots, powered by advanced technologies like artificial intelligence, sensors, and automation, are becoming increasingly prevalent in our daily lives, contributing to various industries and improving overall efficiency and convenience.

In the early days of search and rescue efforts, human responders grappled with the limitations of traditional tools and manual methods in confronting the challenges posed by collapsed buildings, natural disasters, and other crisis scenarios. The realization of the potential for robotic assistance marked the inception of a transformative journey. The initial phase witnessed the deployment of remote-controlled robots, allowing human operators to navigate hazardous environments from a safe distance. However, the reliance on direct human control limited the efficacy of these early robotic systems, especially in situations requiring swift and autonomous decision-making. The evolution gained momentum with the shift towards autonomous operation, a pivotal development that liberated robots from constant human guidance. This newfound autonomy significantly improved their ability to navigate unpredictable terrains, assess situations independently, and execute tasks with precision. The integration of artificial intelligence, particularly machine learning algorithms, further elevated the perceptual and decision-making capabilities of search and rescue robots.However, the energy constraints of these robots pose a challenge, as prolonged operation is crucial for successful missions. The motivation behind this research stems from the need to develop intelligent, adaptive algorithms that enable search and rescue robots to navigate complex terrains, identify survivors, and perform various tasks while conserving energy. Traditional rule-based approaches may not be sufficient for handling the dynamic and unpredictable nature of disaster-stricken environments. Reinforcement learning, with its ability to learn from interactions with the environment, offers a promising solution to address these challenges. By training robots to make energy-efficient decisions autonomously, they can operate for extended periods, increasing their chances of completing missions and aiding in disaster response efforts. Reinforcement learning, a machine learning paradigm based on trial and error, and agent-based navigation, which relies on autonomous decision-making, come together to create intelligent robotic systems capable of learning from their interactions with the environment and dynamically adjusting their behaviors. This integration enables robots to adapt to unpredictable terrains, avoid obstacles, and optimize their paths based on real-time data, showcasing the synergy between learning algorithms and autonomous decision-making processes.

This study aims to investigate the potential of reinforcement learning techniques, including deep reinforcement learning, to improve the energy efficiency of search and rescue robots.

OVERVIEW OF AI APPLICATIONS IN ROBOTICS: REINFORCEMENT LEARNING AND AGENT-BASED NAVIGATION

Artificial Intelligence (AI) has revolutionized the field of robotics, enabling machines to perform tasks that traditionally required human-like intelligence. Two key areas of AI application in robotics are reinforcement learning and agent-based navigation.

Reinforcement learning (RL) stands as a pivotal subset of machine learning, empowering robotic systems to navigate complex environments and tasks. In the area of robotics, the essence of RL lies in enabling agents to make decisions, learn from actions, and maximize rewards within their surroundings. The primary objective is to imbue robots with the ability to acquire, enhance, and replicate tasks, adapting seamlessly to constantly changing conditions through exploration and self-guided learning. Through iterative trial and error, robots gain the capability to manipulate objects, walk, fly, or perform diverse physical tasks, honing their skills over time. RL is also instrumental in task automation, allowing robots to automate decision-making processes such as grasping objects, assembly, or navigation in dynamic environments. By learning from their interactions, robots continually refine their performance, ensuring efficiency and accuracy in task execution. Furthermore, RL plays a vital role in optimizing energy efficiency, where algorithms are utilized to streamline robot movements, conserving energy and enabling prolonged operational durations. This aspect is particularly crucial in applications like search and rescue, where sustained robot functionality can be a matter of life-saving significance. Through RL, robots not only navigate their environments but also evolve, adapt, and excel in their designated tasks, marking a significant advancement in the field of robotics (Kalakrishnan et al., 2011; Kohl & Stone, 2004).

In Amsterdam, "Billy" robots have transformed waste management by reducing manual labor, improving operational efficiency, and contributing to cleaner streets. Meanwhile, in Singapore, the "WasteShark" robot operates in waterborne environments, offering sustainable waste collection solutions. Dubai utilizes robotic window cleaners to enhance energy efficiency in building maintenance, optimizing energy consumption. Globally, RL is applied to smart grids, ensuring sustainable power management and the seamless integration of renewable energy sources. In Boston, "Spot" robots are deployed for automated infrastructure inspection, enhancing public safety. Additionally, fire detection drones in various locations enable faster response times to fire outbreaks, further enhancing public safety measures. Sustainable deliveries witness the deployment of delivery robots on college campuses, reducing traffic congestion and lowering emissions. In urban areas, autonomous last-mile vans leverage RL for optimized delivery routes and increased fuel efficiency. These real-world examples underscore the diverse applications of RL-driven ubiquitous robots in fostering smart city sustainability, from waste management to public safety and sustainable deliveries.

Agent-based navigation has emerged as a transformative technology in the realm of autonomous systems, particularly in the context of search and rescue operations. (Murphy Robin R.and Tadokoro, 2008) highlight its application in robots designed for disaster-stricken areas, equipped with sensors and cameras. These robots navigate through hazardous terrains, such as collapsed buildings, employing agents collaborating to explore diverse areas, avoid obstacles, and communicate with each other, optimizing the search process. Drones, as demonstrated by (Tahir et al., 2019),(Yoshimoto et al., 2018), and (Arnold et al., 2018), take advantage of agent-based navigation to form swarms for efficient search and rescue missions in vast and inaccessible areas. Each drone functions as an autonomous agent, adapting search patterns based on real-time information and avoiding collisions.

In disaster zones, humanoid robots equipped with agent-based navigation capabilities, as discussed by (Bouyarmane et al., 2012), play a crucial role. These robots navigate through cluttered environments, providing aid to survivors and performing tasks that demand human-like dexterity and decision-making. In hazardous environments, such as those contaminated by chemicals or radiation, agent-based navigation facilitates the coordination of multi-robot teams (Ristic et al., 2017), ensuring comprehensive coverage without endangering human responders.

Underwater search and rescue operations benefit from agent-based navigation in robots equipped with sonar and cameras ((Murphy et al., 2008). Autonomous underwater vehicles (AUVs) collaborate as agents to explore large underwater areas, aiding in the location of missing persons or submerged wreckage (Cai et al., 2023).

In agriculture, agent-based navigation finds application in robots navigating dynamic fields, adapting actions based on local sensory inputs to optimize planting, irrigation, and harvesting processes (Hameed, 2014). Boston Dynamics' Spot robots exemplify this adaptability in construction sites, autonomously navigating through changing structures, debris, and equipment (Koval et al., 2022). These robots analyze surroundings, avoid obstacles, and adjust routes in real-time, showcasing the versatility of agent-based navigation across diverse industries. In summary, these examples underscore how agent-based navigation enhances adaptability, collaboration, and efficiency in autonomous systems, revolutionizing various sectors .

Significance of Reinforcement Learning and Agent-Based Navigation in Robotics

The significance of reinforcement learning (RL) and agent-based navigation in robotics is profound and multifaceted. Firstly, these techniques provide robots with adaptability, a vital trait for real-world applications in unpredictable environments such as disaster-stricken areas or unstructured terrains. The ability to adapt to changing conditions ensures that robots can operate effectively and make valuable contributions in dynamic and challenging scenarios. Secondly, RL and agent-based navigation contribute to efficiency by optimizing robot actions. Energy-efficient movements not only reduce costs but also enable robots to function for longer durations, which is particularly vital in applications where sustained operation is paramount. Moreover, efficient navigation ensures timely responses, especially in critical situations where quick decision-making and action are essential. Additionally, agent-based navigation encourages collaboration between robots, enabling them to work together in teams to accomplish tasks efficiently. This collaborative approach not only enhances the overall effectiveness of robotic systems but also opens up new avenues for tackling complex challenges that require coordinated efforts. By allowing robots to learn from their experiences, adapt to new situations, and refine their decision-making processes, RL ensures that robots become increasingly proficient over time. This iterative learning process equips robots with the capability to handle diverse tasks and challenges, making them valuable assets in various real-world applications (Lee & Yusuf, 2022).

Ethical Considerations, Data Privacy, and the Social Acceptance of Autonomous Robots in Emergency Scenarios

The ethical considerations and societal impacts of AI-driven search and rescue robotics present complex challenges that demand thoughtful solutions. Privacy concerns arising from data capture during operations can be mitigated through stringent anonymization protocols and sensors designed to respect individuals' privacy rights. Addressing biases in decision-making requires regular audits, diverse datasets, and continuous oversight, ensuring fairness, especially for vulnerable populations. Establishing accountability and transparency frameworks, along with explainable AI models, fosters trust and understanding of AI-driven robots' behavior. Ethical human-robot interaction protocols, coupled with comprehensive training for both responders and the public, promote empathy and mutual respect. Societal impacts, including job displacement and overreliance on technology, necessitate a balanced approach where robots enhance human capabilities, not replace them entirely. Equitable access to technology should be ensured, addressing disparities through community engagement. Additionally, mitigating the environmental impact involves designing recyclable and energy-efficient robotic systems and implementing responsible disposal practices. Collaboration between technologists, ethicists, policymakers, and communities, coupled with open dialogue and transparency, is essential to navigate these ethical and societal challenges responsibly, ensuring the positive integration of AI-driven search and rescue robotics into society.

Statement of the Problem: Energy Efficiency Challenges in Search and Rescue Robots

Search and rescue robots play a crucial role in disaster-stricken areas, providing critical support to human responders. However, these robots face significant energy efficiency challenges that hinder their effectiveness and operational sustainability (Cheng et al., 2011). The following issues listing the pressing concerns related to energy efficiency in search and rescue robots (Ariizumi & Matsuno, 2017; Chitikena et al., 2023; Doroftei et al., 2014).

Limited Adaptability to Dynamic Environments

Challenge: Real-world disaster environments are dynamic and unpredictable, often deviating from controlled testing conditions.

Limitation: Robots may struggle to adapt to rapidly changing conditions, hindering their effectiveness in unpredictable disaster scenarios.

Integration with Existing Infrastructure

Challenge: Seamless integration with other emergency response systems and infrastructure.

Limitation: Lack of standardized communication protocols may impede interoperability with existing systems, slowing down response times.

Human-Robot Collaboration

Challenge: Establishing effective collaboration between search and rescue robots and human responders.

Limitation: Limited understanding of how humans and robots can efficiently work together in high-stress, dynamic environments.

Scalability and Cost Constraints

Challenge: Scaling up deployment to cover larger disaster-stricken areas.

Limitation: High development and deployment costs may limit widespread adoption, especially in regions with limited financial resources.

Operational Endurance

Challenge: Prolonged operations in harsh conditions strain robot components and systems.

Limitation: Robots may experience wear and tear, reducing their operational effectiveness over extended missions.

Communication Reliability

Challenge: Maintaining robust communication links in challenging environments.

Limitation: Signal interference, loss of connectivity, or limited bandwidth can hinder real-time communication between robots and human operators.

Ethical and Legal Considerations

Challenge: Addressing ethical concerns related to privacy, consent, and the potential impact on human lives.

Limitation: Lack of clear legal frameworks for the deployment of robotic technologies in emergency situations.

Limited Autonomy in Unstructured Environments

Challenge: Operating autonomously in unstructured and cluttered disaster environments.

Limitation: Robots may struggle to navigate through debris, hindering their ability to locate and assist survivors effectively.

Maintenance and Repair Challenges

Challenge: Ensuring ongoing maintenance and repair of robotic systems in the field.

Limitation: Limited accessibility to damaged robots in disaster-stricken areas may hinder prompt repairs.

Public Perception and Acceptance

Challenge: Gaining public trust and acceptance of robotic technologies in emergency response.

Limitation: Fear or skepticism from the public may impact the successful integration of robots into search and rescue operations.

Addressing these multifaceted challenges is paramount. Innovative solutions that optimize energy consumption without compromising the robots' performance are crucial, ensuring their successful deployment in critical and life-saving missions.

Fundamentals of AI in Search and Rescue Robotics

In the context of search and rescue robotics, understanding the fundamental AI concepts is crucial for developing intelligent and effective robotic systems. Below are explanations of basic AI concepts relevant to search and rescue robotics:

1. **Artificial Intelligence (AI):** AI refers to the simulation of human intelligence in machines that are programmed to think and learn like humans. In search and rescue robotics, AI enables robots to perceive their environment, make decisions based on that perception, and take actions to accomplish tasks without direct human intervention (Couceiro, 2017a).

2. **Machine Learning (ML):** Machine learning is a subset of AI (Kühl et al., 2022) that involves algorithms and statistical models that enable robots to improve their performance on a specific task over time as they are exposed to more data. ML techniques are used in search and rescue robotics for tasks such as object recognition, pattern detection, and decision-making (Soori et al., 2023).

3. **Deep Learning:** Deep learning is a specialized field of machine learning that involves artificial neural networks with multiple layers (deep neural networks). It is particularly effective for tasks that require processing large amounts of complex data, such as image and speech recognition . Deep learning techniques are valuable in search and rescue robotics for analyzing visual data from sensors and cameras (Sarker, 2021).

4. **Computer Vision:** Computer vision is an interdisciplinary field that enables computers to gain high-level understanding from digital images or videos (Lundgren et al., 2022). In search and rescue robotics, computer vision techniques are used for tasks like object detection, facial recognition, and terrain mapping, allowing robots to perceive and interpret visual information from their surroundings (Lopez-Fuentes et al., 2018).

5. **Natural Language Processing (NLP):** NLP is a branch of AI that focuses on the interaction between humans and computers through natural language (Khurana et al., 2023). While not as commonly used in search and rescue robotics as in other AI applications, NLP can be valuable for human-robot communication, allowing robots to understand and respond to spoken or written commands from rescue teams or survivors (Younis et al., 2023).

6. **Sensor Fusion:** Sensor fusion involves integrating data from multiple sensors to obtain a more accurate and comprehensive understanding of the environment (Alatise & Hancke, 2020). In search and rescue robotics, sensor fusion techniques combine data from various sensors such as cameras, LIDAR, thermal imaging, and GPS, enhancing the robot's perception capabilities and decision-making processes (Tzafestas, 2013).

7. **Reinforcement Learning:** Reinforcement learning is a type of machine learning where an agent learns to make sequences of decisions by interacting with an environment to achieve a goal (Sivamayil et al., 2023). In search and rescue robotics, reinforcement learning can be used to optimize robot behaviors, such as navigation, path planning, and interaction with objects, based on feedback received from the environment (Shriyanti Kulkarni et al., n.d.).

Understanding these fundamental AI concepts is essential for researchers and engineers working on search and rescue robotics. By leveraging these concepts, developers can create intelligent and adaptive robotic systems capable of effectively and autonomously operating in dynamic and challenging environments.

Overview of Reinforcement Learning Algorithms: Emphasizing Their Importance in Autonomous Decision-Making

Reinforcement learning (RL) is a machine learning paradigm where an agent learns to make sequences of decisions by interacting with an environment to achieve a goal

Figure 1. Basics reinforcement learning

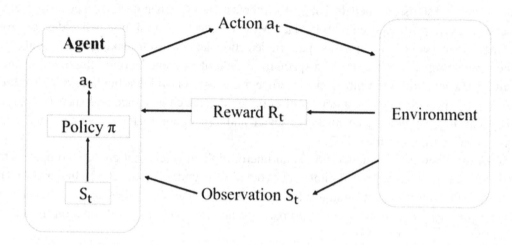

Key Components of Reinforcement Learning

1. **Agent (A):** The learner or decision-maker that interacts with the environment. The agent observes the environment, selects actions, and receives rewards.
2. **Environment (E):** The external system with which the agent interacts. It responds to the agent's actions, presenting new states and rewards.
3. **State (S):** A representation of the current situation or configuration of the environment. It provides the necessary information for the agent to make decisions.
4. **Action (A):** The decision or move made by the agent at a particular state. Actions lead to state transitions.
5. **Reward (R):** A numerical signal indicating the immediate benefit or cost of an agent's action in a specific state. The agent's goal is to maximize the cumulative reward over time.
6. **Policy (π):** A strategy or a mapping from states to actions, defining the agent's behavior. The policy guides the agent's decision-making process.
7. **Value Function (V(s)):** The expected cumulative reward an agent can obtain from a specific state following a particular policy. It represents the long-term desirability of states.
8. **Q-Value Function (Q(s, a)):** Similar to the value function, but considers both the state and the action. It represents the expected cumulative reward of taking action a in state s and following a particular policy thereafter.

Reinforcement learning agents are comprised of three essential components: the model for the environment, policy function, and value function. The environment model represents the agent's anticipation of the upcoming state of the environment and the associated reward. This model consists of two key parts: the state model and the reward model explained in Figure 1. The state (s)model characterizes the probability distribution of the next state when the current state and action are provided. In essence, it predicts how the environment will evolve based on the agent's actions and the current state.

$$\mathcal{P}_{ss'}^{a} = P\left[S_{t+1} = s' \mid S_{t} = s, A_{t} = a\right]$$

The reward model is the reward expectation when current state and action are given

$$\mathcal{R}_{s}^{a} = E\left[R_{t+1} \mid S_{t} = s, A_{t} = a\right]$$

The policy serves as the behavioral blueprint for an agent, dictating its actions in specific states. Essentially, it is a function that maps states to corresponding actions. Policies can be categorized into two main types: deterministic policy and stochastic policy.

A deterministic policy provides a single action for a given state, offering a clear and direct mapping between states and actions. On the other hand, a stochastic policy generates a probability distribution over actions for a given state. Instead of specifying a single action, it provides the likelihood of each possible action, allowing for a more varied and probabilistic decision-making process.

$a = \pi(s)$

$\pi(a|s) = P[At\_a \mid St\_s]$

The value function is a function that predicts how much reward the state and action will return later when following the policy. That is, it is the weighted sum of all rewards to be received after taking the corresponding state and action. At this time, the discounting factor is used to indicate preference for the reward to be received before the reward to be received later.

Policy Evaluation: $v_{\pi}(s) = E_{\pi}[R_{t+1} + \gamma R_{t+2} + \gamma^{2} R_{t+3} + \ldots \mid S_{t} = s]$

$v_{\pi}(s) = E\pi_{[G \mid}t \mid St = {}_{s}]$

When an action is given, the value function is called an action value function and is expressed as q(s,a). It takes action and state as inputs.

$q_{\pi}(s,a) = E_{\pi}[G_{t} \mid S_{t} = s, A_{t} = a]$

DEEP Q-NETWORK

DQN is a type of reinforcement learning that relies on a value function without employing a direct policy. In simpler terms, DQN uses a neural network to learn the value of different actions and selects the action with the highest value implicitly. It's based on the Q-Learning algorithm, utilizing the Bellman Optimal Equation to train the optimal action-value function. The Q-values are updated by minimizing the difference between the estimated value in the current state and the estimated value in the next state.

$$q_*\left(s_t,a_t\right)=E_{s'}\left[R_{t+1}+\gamma\max_{a'}q_*\left(s_{t+1},a'\right)\right]$$

Where $q\left(s,a\right)\leftarrow q\left(s,a\right)+\alpha\left(R+\gamma\max_{a'}q\left(s',a'\right)-q\left(s,a\right)\right)$

The q function using a neural network parameter θ can be expressed as follows.

$q_{\theta(}s,a)$

A loss function of the network is defined as the square of the difference between the target and $q\theta(s,a)$.

$$L(\theta)=E\left[\left(R+\gamma\max_{a'}q_\theta\left(s',a'\right)-q_\theta\left(s,a\right)\right)^2\right]$$

The parameters are updated in the direction of reducing the loss through gradient descent.

$$\theta'=\theta+\alpha(R+\gamma\max_{a'}q_\theta\left(s',a'\right)-q_\theta\left(s,a\right)\nabla_\theta q_\theta\left(s,a\right)$$

Actor-Critic

The actor-critic algorithm is a methodology that uses a policy function and a value function together. The actor selects an action based on the policy, and the critic evaluates the policy based on the value function. In the agent's learning process, the policy and value networks are learned, respectively. The critic network is updated to learn the value of the current policy function. The actor-network is trained in such a way that, through evaluation of the value function, if the result is good, it is reinforced and if it is not good, it is weakened.

Some modifications are made to improve performance of the Actor Critic Algorithm. This is the Q- Actor Critic algorithm.

Actor Critic algorithm.

$$\nabla_\theta J\left(\theta\right)=\bullet\ _{\pi_\theta}[\nabla_\theta log\pi_\theta\left(s,a\right)*Q_{\pi_\theta}\left(s,a\right)$$

In the Advantage Actor Critic algorithm, an advantage function $A\pi\theta(s,a)$ is used to take a specific action compared to the average, general action at the given state

$$A_\lambda\left(s,a\right)=Q_\lambda\left(s,a\right)-V_\lambda\left(s,a\right)$$

Figure 2. Deep Q network (DQN)

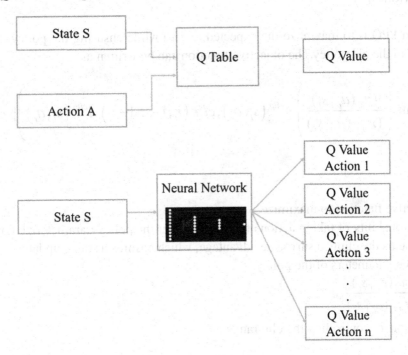

Proximal Policy Optimization (PPO)

PPO is a state-of-the-art policy optimization algorithm that strikes a balance between stability and sample efficiency.

Figure 3. Actor critic network

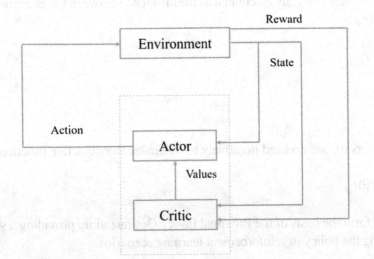

Objective Function

The objective in PPO is to maximize the expected reward while ensuring the policy does not deviate significantly from the old policy. The objective function can be written as

$$J(\theta) = E_t\left[\min\left(\frac{(\pi_\theta(a_t \mid s_t)}{(\pi_{\theta_{old}}(a_t \mid s_t))}\right)A^{clip}(s_t, a_t), clip(r_t, 1-\in, 1+\in)A^{clip}(s_t, a_t)\right]$$

Where:

$J(\theta)$ is the objective function to maximize.
$(\pi\theta_{(a}t\mid_s t)$ is the probability of taking action at $_i$n state st $_u$nder the policy parameterized by θ
$Ac^{lip(}st at_)$ represents the clipped surrogate advantage, which ensures a stable update.
θo_{ld} represents the parameters of the policy.
r_t is the ratio $\dfrac{(\pi_\theta(a_t \mid s_t)}{(\pi_{\theta_{old}}(a_t \mid s_t)}$
$\in$ is a hyperparameter controlling the clip range.

Clipped Surrogate Objective

The clipped surrogate objective ensures that the policy update is within a specified range:

$$L^{CLIP}(\theta) = E_t\left[\min\left(r_t A^{clip}(s_t, a_t), clip(r_t, 1-\in, 1+\in)A^{clip}(s_t, a_t)\right)\right]$$

Advantage Calculation

Advantages ($A^{clip}(s_t a_t)$) are typically calculated as the difference between the estimated value ($V(s_t)$) and the actual rewards(r_t):

$$A^{clip}(s_t a_t) = V(s_t) - r_t$$

Policy Update

The policy parameters (θ) are updated iteratively to maximize the objective function:

$$\theta n_{ew=} \arg\max\theta_L c^{lip(}\theta)$$

These equation form the basis of the Proximal Policy Optimization, providing a stable and effective method for updating the policy in reinforcement learning scenarios.

IMPORTANCE OF AUTONOMOUS DECISION-MAKING

Autonomous decision-making is of paramount importance in the field of autonomous vehicles. Autonomous vehicles need to be able to make split-second decisions to avoid collisions, navigate complex traffic scenarios, and respond to unexpected events on the road. This requires advanced planning and decision-making algorithms that can analyze the current situation, predict future outcomes, and choose the optimal course of action to ensure the safety of the vehicle and its occupants.Efficiency is another important aspect of autonomous decision-making. By making intelligent decisions, autonomous vehicles can optimize their routes, reduce fuel consumption, and minimize travel time. This can lead to improved traffic flow, reduced congestion, and a more efficient transportation system as a whole .(Sana et al., 2023) Moreover, autonomous decision-making plays a crucial role in enabling autonomous vehicles to adapt to complex and unconventional scenarios. These scenarios may include navigating through crowded urban environments, handling adverse weather conditions, or encountering unexpected obstacles on the road. By making informed decisions, autonomous vehicles can effectively handle these challenging situations and ensure a smooth and reliable driving experience (C. Liu et al., 2017) (Couceiro, 2017b; Dosovitskiy et al., 2017; Kamran et al., 2020; Moreau et al., 2019)

Through advanced planning and decision-making algorithms, autonomous vehicles can make intelligent choices that prioritize safety, efficiency, and adaptability (J. Chen et al., 2019; Guan et al., 2023; Huang et al., 2022; Jamgochian et al., 2023; Lillicrap et al., 2016)

1. **Adaptability:** RL algorithms allow robots to adapt to changing environments and unforeseen situations. Through continuous learning, robots can update their policies and strategies, ensuring adaptability in dynamic search and rescue scenarios.
2. **Optimization:** RL algorithms optimize decision-making processes by learning from experiences. Robots learn to choose actions that maximize long-term rewards, ensuring they make informed and strategic decisions in complex and unpredictable environments.
3. **Autonomy:** RL enables robots to operate autonomously, reducing the need for constant human intervention. Autonomous decision-making is vital in search and rescue missions where real-time responses can be the difference between life and death.
4. **Efficiency:** RL algorithms optimize actions based on learned knowledge, making robots more energy-efficient and ensuring they utilize resources effectively, crucial for prolonged operation in search and rescue operations.
5. **Learning from Experience:** RL algorithms enable robots to learn from their interactions with the environment. As robots gather more data and experience, their decision-making capabilities improve, leading to more efficient and effective search and rescue efforts.

ROLE OF AGENT-BASED NAVIGATION IN CHALLENGING ENVIRONMENTS

Agent-based navigation is an approach in robotics where autonomous entities, known as agents, operate independently and interact with their environment to achieve specific goals (Y. Chen & Xiao, 2023). This methodology is particularly valuable in dynamic and unpredictable environments. These robots can navigate disaster-stricken areas, such as collapsed buildings or hazardous terrains, to locate and rescue survivors, such as those encountered in search and rescue missions. Here's an introduction to agent-based navigation and its significant role in such challenging settings:

1. **Autonomous Agents:** In agent-based navigation, robots are equipped with intelligence to act autonomously. Each robot functions as an independent agent capable of sensing its surroundings, making decisions, and executing actions. These agents can work collaboratively or individually, adapting their behavior based on the dynamically changing environment (Cipi & Cico, 2011).
2. **Local Perception and Decision-Making:** Agents rely on local perception, using sensors like cameras, LIDAR, and proximity sensors to understand their immediate surroundings. Through real-time data processing, agents make decisions on navigation, obstacle avoidance, and task execution without relying on a central control system. This local decision-making ability is essential in environments where conditions change rapidly (Kim et al., 2015).
3. **Collaborative Decision-Making:** Multiple agents can collaborate and communicate with each other. Through inter-agent communication, agents share information, coordinate movements, and optimize their paths. Collaboration enables them to cover a larger area efficiently, explore unknown regions, and adapt collectively to unforeseen obstacles or hazards (Kim et al., 2015).
4. **Adaptability and Flexibility:** Agent-based navigation systems are highly adaptable. Agents can modify their paths and behaviors on-the-fly based on new information or changing mission objectives. This adaptability is crucial in search and rescue operations, where the environment can change suddenly due to collapsing structures, shifting debris, or changing weather conditions (J. Liu et al., 2007).
5. **Distributed Intelligence:** Unlike centralized systems, where a single controller dictates all actions, agent-based navigation distributes intelligence across multiple agents. Each agent processes information independently, leading to decentralized decision-making. This distributed intelligence enhances fault tolerance, as the system can continue functioning even if individual agents fail or are compromised (Chitikena et al., 2023).
6. **Exploration and Coverage:** Agent-based systems excel in exploration tasks. Agents can explore unknown or hazardous areas collectively, mapping the environment and identifying potential locations of interest, such as survivors or structural weaknesses. This exploration capability is vital in search and rescue missions to locate victims and assess the extent of damage (Biggie et al., 2023).
7. **Scalability:** Agent-based navigation systems can be easily scaled by adding or removing agents based on the complexity of the mission or the size of the search area. This scalability allows search and rescue teams to deploy a suitable number of robots tailored to the specific requirements of the operation (Biggie et al., 2023).

ENERGY-EFFICIENT ROBOTICS: CHALLENGES AND SOLUTIONS

Addressing the challenges related to energy consumption is essential for developing robots that can operate efficiently, especially in environments where resources may be limited. Here, this chapter delves into the challenges faced in achieving energy-efficient robotics and explore potential solutions to mitigate these issues:

In-Depth Analysis of Energy Consumption Challenges Faced by Search and Rescue Robots

The energy consumption challenges faced by search and rescue robots are multifaceted and demand innovative solutions to ensure their effectiveness in life-or-death situations. High-power sensors and actuators, essential for accurate data collection and precise movements, pose a significant hurdle. To address this, (Paredes et al., 2017) research focuses on low-power sensor technologies, efficient actuation mechanisms, and intelligent algorithms that activate sensors only when necessary. Limited battery life, especially in areas where recharging is impractical, restricts operational time. Solutions involve the development of advanced energy-dense batteries, exploration of alternative power sources like solar or fuel cells, and the implementation of energy-aware algorithms for efficient power management (Alwateer et al., 2019). Modular designs allowing quick battery replacement further extend operational duration. Navigating challenging terrains and obstacles consumes substantial energy, leading to the implementation of energy-efficient path planning algorithms and machine learning techniques, such as reinforcement learning, for optimized robot movements. Real-time communication and high-resolution data transmission demand significant power, impacting battery life. Energy-efficient communication protocols, data compression algorithms, and prioritization methods are explored, alongside the utilization of edge computing for local data processing (Lyu et al., 2023). Quick decision-making under dynamic conditions requires computational power, prompting the employment of lightweight machine learning models and edge AI solutions for rapid, localized decision-making. Extreme temperatures and adverse environmental conditions affect battery efficiency, necessitating robust, temperature-resistant batteries, thermal management systems, and protective casings (Wu et al., 2018). In multi-robot operations, collaborative algorithms enable energy information sharing and coordinated activities, ensuring balanced energy consumption among robots. This holistic approach (Das et al., 2016), integrating advancements in hardware and software, paves the way for search and rescue robots to operate effectively, ensuring prolonged missions and enhancing their life-saving capabilities in critical situations.

Review of Existing Energy-Efficient Solutions and Their Limitations

- *Low-Power Components* (Kim Jong-Hyukand Sukkarieh, 2006) Manufacturers have developed energy-efficient sensors, processors, and actuators specifically designed for robotic applications. These components consume less power, extending the operational time of search and rescue robots
- *Optimized Algorithms* (Riazi et al., 2016): Researchers have devised algorithms for efficient path planning, obstacle avoidance, and sensor data processing. These algorithms minimize unnecessary movements and reduce computational overhead, conserving energy during robot operations
- *Hybrid Power Sources* (Wu et al., 2019): Some robots employ hybrid power systems, combining traditional batteries with alternative sources like solar panels or fuel cells. These systems provide continuous power, especially in outdoor environments where solar energy can be harnessed .
- *Energy-Aware Navigation* Advanced navigation algorithms consider energy constraints when planning robot movements. Energy-aware navigation optimizes paths, avoiding energy-intensive routes and ensuring the robot conserves energy while navigating through challenging terrains (Wallace et al., 2019).

- ***Distributed Computing*** (Y. Zeng et al., 2016): Utilizing edge computing, robots process data locally instead of relying on centralized servers. By reducing the need for continuous high-bandwidth communication, distributed computing solutions conserve energy during data transmission

Limitations of Existing Solutions

Limited Autonomy

While energy-efficient components and algorithms extend operational time, search and rescue robots still face limitations in terms of autonomy. Prolonged missions, especially in remote or hazardous areas, require further advancements to achieve true long-term autonomy.

Scalability Issues

Some solutions are effective for individual robots but face challenges when applied to multi-robot systems. Coordinating energy usage and optimizing tasks across multiple robots in real-time remains a complex problem, especially in dynamic and unpredictable environments.

Cost Constraints

Energy-efficient components and hybrid power systems can be costly, limiting their widespread adoption, especially in resource-constrained environments. Balancing performance and cost-effectiveness is a challenge in the development and deployment of energy-efficient robots.

Environmental Constraints

Solar-powered robots heavily rely on sunlight, making them less effective during nighttime or in environments with limited sunlight. Additionally, fuel cells require specific refueling infrastructure, which may not be readily available in certain disaster-stricken areas.

Maintenance and Durability

Energy-efficient solutions often require careful maintenance and may be more sensitive to environmental conditions. Ensuring the durability and robustness of these solutions, especially in harsh terrains, remains a challenge.

Human-Technology Interaction

Robots in search and rescue scenarios often require human interaction, leading to continuous sensor and communication usage. Balancing energy efficiency with the need for timely communication and human control presents a complex challenge.

While existing energy-efficient solutions have made significant strides in extending the operational time and capabilities of search and rescue robots, there are still limitations to overcome.

Integration Challenges and Potential Strategies for Effective Implementation

The integration of innovative AI-driven approaches in search and rescue robotics is of utmost importance, addressing critical challenges posed by energy limitations in dynamic and unpredictable environments (J. Zeng et al., 2019). Machine learning algorithms, particularly reinforcement learning, play a pivotal role in enabling robots to make intelligent, adaptive, and efficient decisions in real-time. These algorithms optimize robot movements, conserving energy without compromising mission effectiveness in complex and dynamic disaster scenarios. Furthermore, AI algorithms analyze real-time data to optimize resource allocation, ensuring judicious usage of limited energy resources to prolong operational time (Lygouras et al., 2019). Intelligent sensor management is achieved through AI-driven algorithms that selectively process sensor data, activating sensors only when specific conditions, such as detecting survivors, are met. Predictive maintenance models, based on machine learning, enable robots to undergo preventive maintenance by predicting component failures before they occur, ensuring efficient operation and avoiding sudden energy-draining malfunctions (Sepasgozar et al., 2020). Collaborative AI algorithms (Singh, Kurukuru, et al., 2023) facilitate energy-related data sharing among multi-robot systems, allowing robots to distribute tasks based on available energy resources, ensuring balanced energy consumption within the team. Real-time adaptability is achieved through AI algorithms and reinforcement learning models coupled with edge computing (Akhloufi et al., 2021), enabling robots to make context-aware decisions and ensuring energy-efficient responses to rapidly changing environmental conditions. As AI technology continues to evolve, its integration with robotics holds the promise of further enhancing the energy efficiency of search and rescue robots, enabling them to operate effectively and save lives in the most challenging and energy-constrained environments.

REINFORCEMENT LEARNING IN SEARCH AND RESCUE ROBOTICS

Reinforcement learning (RL) techniques have become instrumental in enhancing the energy efficiency of search and rescue robots, as demonstrated by studies such as (Yao et al., 2022) and (Lyu et al., 2023). These algorithms address various challenges associated with optimizing robot behaviors for energy conservation. One key aspect is the incorporation of an energy-aware state representation, where RL algorithms, as highlighted by Le et al. (2020), integrate the robot's battery level and power consumption rates into the state representation. This allows the robot to balance mission objectives with energy conservation goals. RL algorithms also contribute to adaptive path planning, optimizing routes by learning energy-efficient paths through dynamic environments, considering factors like terrain complexity and obstacle density (Shan et al., 2020). Dynamic speed and movement control are achieved by RL algorithms continuously adapting the robot's speed and movement style based on real-time sensor data, ensuring optimal energy utilization, as discussed by (Singh, Ren, et al., 2023).

Moreover, RL techniques address challenges related to sensor activation and data processing, teaching robots when to activate sensors and optimizing data processing algorithms to conserve energy effectively. Task prioritization and resource allocation are improved through RL algorithms that dynamically allocate resources based on task urgency and importance, enhancing overall mission efficiency. The capability of RL algorithms, particularly those involving Deep Q-Networks (DQN), to enable robots to learn from past experiences is crucial for refining decision-making processes and continuously improving energy optimization strategies. In the context of collaboration and multi-robot systems, RL algorithms facilitate

Table 1. Case studies and examples showcasing successful implementations of reinforcement learning in real-world scenarios

Case Study	Scenario	Success
AlphaGo (DeepMind): (Silver et al., 2017)	Mastering the game of Go using RL techniques.	Defeated world champion Go players, showcasing the power of RL in solving complex games.
Robotics Manipulation (OpenAI):(OpenAI et al., 2018)	Robot hand manipulation of objects using RL.	Robots learned dexterous manipulation skills, enabling precise tasks in unstructured environments.
Self-Driving Cars (Waymo) (Z. Li et al., 2020)	Autonomous vehicle navigation with RL.	Waymo's self-driving cars logged millions of miles, demonstrating RL's adaptability in complex traffic scenarios.
Game Playing (OpenAI's DOTA 2 AI):(Berner et al., 2019)	AI agents playing DOTA 2 using RL in dynamic environments.	OpenAI Five defeated professional human players, showcasing RL's adaptability in team-based, dynamic settings with incomplete information.
Adaptive Systems (DeepMind's DeepMind Lab) (Beattie et al., 2016)	AI agents learning in a complex virtual world.	Demonstrated advanced problem-solving skills, adaptability, and efficient navigation, highlighting the versatility of RL algorithms.
Supply Chain Optimization (Alibaba):(Jin et al., 2019)	Optimizing parcel sorting systems in warehouses using RL.	Optimized sorting algorithms, significantly improving supply chain efficiency, and reducing costs, and delivery times.
Robotics in Manufacturing (Siemens)(Indri et al., 2019)	Optimizing robotic manufacturing systems with RL.	Dynamically adjusted robotic movements and energy usage, achieving substantial energy savings while maintaining production efficiency.
Adaptive Energy Management in Data Centers (Google)	Optimizing cooling systems in data centers with RL.	Reduced energy consumption for cooling by around 40%, demonstrating RL's potential in energy-efficient operations.
Healthcare Optimization: (Eckardt et al., 2021)	Optimizing cancer treatment plans using RL.	Integrated RL algorithms into treatment planning, adapting therapy based on patient responses. Improved patient outcomes, minimized side effects, and enhanced overall quality of life. Tumor shrunk at a faster rate than anticipated, showcasing personalized treatment's effectiveness.

communication and coordination among robots, ensuring balanced energy consumption and maximizing the coverage area through shared experiences and coordinated actions.

By incorporating RL techniques in these ways, search and rescue robots can adapt, learn, and evolve their behaviors over time. This adaptability is crucial in dynamic and unpredictable disaster environments, where energy-efficient decisions are vital for prolonged missions, ultimately enhancing the robots' effectiveness in saving lives.

These case studies illustrate the diverse applications of reinforcement learning in solving real-world challenges. From gaming and robotics to autonomous vehicles and supply chain management, RL continues to revolutionize various industries by enabling intelligent decision-making and adaptive behavior in complex and dynamic environments.

AGENT-BASED NAVIGATION: ENHANCING ADAPTABILITY AND EFFICIENCY

Agent-Based Navigation systems enhance adaptability and efficiency in various scenarios, particularly in dynamic environments like disaster-stricken areas. Robots operate autonomously, utilizing sensors such as cameras, LIDAR, and proximity sensors to navigate, avoid obstacles, and achieve goals without centralized control. This approach is valuable in search and rescue missions.By employing agent-based navigation, adaptability and efficiency can be significantly enhanced, leading to safer and more optimized movement of entities in various scenariosv(Turek et al., 2006).

Revolutionizing Search and Rescue Robotics in Dynamic Environments

In search and rescue robotics, Agent-Based Navigation stands out for its ability to adapt in real-time to changing disaster scenarios (T. Chen et al., 2019). Multiple agents collaborate, share information, and coordinate movements, optimizing search processes in large-scale operations (Macek et al., 2002). Moreover, these systems improve exploration efficiency by dividing areas into segments and exploring simultaneously, reducing the time needed to cover extensive and complex environments (Wulfmeier et al., 2020). Sensor fusion capabilities enable informed navigation decisions, augmenting effectiveness in search missions (T. Chen et al., 2019) of search missions. Dynamic path planning allows agents to navigate around obstacles and find efficient routes, even in challenging terrains (Iwano et al., 2004). Redundancy and fault tolerance ensure mission continuity, maximizing the chances of successful rescues. Real-time reevaluation and replanning optimize efforts by adapting to changing survivor locations or unexpected obstacles, increasing the likelihood of success (Wan et al., 2021).

In summary, agent-based navigation systems revolutionize search and rescue operations by combining adaptability, collaboration, efficiency, sensor fusion, redundancy, human-robot collaboration, real-time decision-making, and scalability.

How Agents can Enhance Adaptability, Responsiveness, and Energy Efficiency in Search and Rescue Missions

Robotic agents in search and rescue missions enhance adaptability, responsiveness, and energy efficiency through dynamic decision-making. Continuous assessment of surroundings allows agents to adapt behaviors dynamically, altering paths and search patterns. Collaborative adaptation involves communication and knowledge sharing among agents, collectively adjusting strategies for optimal performance (Bannur et al., 2020). The integration of real-time sensor fusion provides agents with a comprehensive understanding of their environment by amalgamating data from various sensors, such as cameras and thermal imaging devices. Real-time sensor fusion provides a comprehensive understanding of the environment, enabling rapid responses to emergencies, survivor detection, and hazard avoidance (Alatise & Hancke, 2020). Processing sensory inputs locally facilitates swift decision-making without external computation, enhancing overall mission effectiveness. Energy efficiency is achieved through energy-aware path planning, dynamic power management, and collaborative energy optimization (Wilson et al., 2004). Dynamic power management allows agents to adjust power usage based on mission urgency and remaining battery life (Li et al., 2014). Collaborative energy optimization balances tasks and energy consumption, enabling efficient allocation of energy-intensive tasks. In summation, the fusion of dynamic decision-making, collaborative adaptation, real-time sensor fusion, rapid local decision-making, energy-aware path planning, dynamic power management, and collaborative energy optimization elevates the efficiency and effectiveness of search and rescue operations.

COMPARATIVE ANALYSIS OF DIFFERENT REINFORCEMENT LEARNING ALGORITHMS CONCERNING THEIR APPLICABILITY IN AGENT-BASED NAVIGATION

This analysis explores the diverse landscape of reinforcement learning (RL) algorithms and their transformative impact on agent-based navigation systems, specifically in dynamic and unpredictable environments. In the realm of multi-robot systems for search and rescue, innovative approaches, such as mapless collaborative navigation based on deep RL, demonstrate significant advancements, notably enhancing classical algorithms like Deep Deterministic Policy Gradient (DDPG) (W. Chen et al., 2019; Sombolestan et al., 2019). The utilization of mobile robots in disaster areas is optimized through energy-efficient path-planning strategies, reducing risks and improving the efficiency of urban search and rescue operations (Le et al., 2020a; Bing et al., 2020; Beomsoo et al., 2021). Additionally, the integration of the Complete Tileset Energy-Aware Coverage Path Planning (CTPP) framework for self-reconfigurable robots showcases the potential for energy-efficient locomotion, including slithering gaits designed through RL algorithms (Le et al., 2020a; Bing et al., 2020). Further contributions include the development of hierarchical exploration techniques using Graph Neural Networks (Zhang et al., 2022), comparative studies on the energy efficiency of robots with soft-body dynamics (G. Li et al., 2023), and exploration optimization approaches for Visual Simultaneous Localization and Mapping (SLAM) applications (Leong, 2023). The survey also delves into the realm of cooperative agents (Hüttenrauch et al., 2017), complete coverage path planning models (Lakshmanan et al., 2020), and automation applications through energy-aware coverage path planning (Le et al., 2020b), offering insights into the varied and impactful applications of RL in robotics. Finally, the integration of an Autonomous Collaborative Search Learning Algorithm (ACSLA) into a Distributed Multi-Agent Collaborative Search System (DMACSS) for Autonomous Underwater Vehicles (AUVs) highlights adaptive learning algorithms for optimal real-time search path planning (Y. Liu et al., 2020). This survey culminates in an exploration of surveillance and search and rescue with distributed sensor networks (Zhang et al., 2006), assessing RL's general applicability for optimal control policy learning (Lee, 2015), and presenting complete autonomous aerial robotic solutions for complex Search and Rescue (SAR) missions (Sampedro et al., 2019). Furthermore, insights into task allocation strategies and collision avoidance algorithms using RL in multi-robot teams further underscore the versatility and effectiveness of RL in addressing challenges across a myriad of robotics domains (Hu et al., 2020; Saeedvand et al., 2020).

SUMMARY

In this fascinating chapter, we immerse ourselves in the dynamic fusion of artificial intelligence and robotics, working in tandem to shape sustainable and eco-friendly cities. Envision a landscape where robots serve as intelligent assistants, endowed with the capacity to learn and fine-tune their actions for optimal energy efficiency, all with the overarching goal of fostering environmentally conscious urban spaces. These robots emerge as dedicated city helpers, driven by a mission to curtail energy consumption and enhance the sustainability of our daily lives. Operating through the elaborate methodology of reinforcement learning, akin to training them to prioritize eco-friendly choices, these robotic companions navigate cityscapes, drawing insights from experiences to minimize energy usage across transportation, resource management, and waste reduction. Importantly, this narrative emphasizes that one need not possess expertise in AI or robotics to grasp the potential transformative impact. While the chapter delves into the technical details behind the scenes, its essence lies in propelling us towards a future where cities stand as beacons of cleanliness, efficiency, and environmental mindfulness.

DISCUSSION ON POTENTIAL FUTURE DEVELOPMENTS

The advancements outlined in reinforcement learning algorithms, hardware technologies, sensor technologies, and communication technologies represent a significant leap forward in the realm of reinforcement learning-driven agent-based navigation for search and rescue missions.

Firstly, hierarchical reinforcement learning models offer robots the ability to learn at multiple levels of abstraction, enabling efficient decision-making in complex environments. This adaptability allows robots to handle tasks of varying complexities, from high-level mission planning to low-level motor control, enhancing overall efficiency in navigating challenging scenarios. Additionally, lifelong learning algorithms ensure that robots accumulate knowledge and adapt over time, continuously improving their navigation strategies even in novel or unforeseen situations. The development of interpretable and explainable reinforcement learning models provides transparency into the decision-making process, fostering trust among stakeholders, enabling better collaboration, and ensuring reliable decision-making in critical missions.

Powerful onboard processors enable robots to execute complex reinforcement learning algorithms locally, enhancing responsiveness and reducing latency in real-time decision-making, vital in time-sensitive search and rescue missions. The integration of multiple sensors, including visual, LIDAR, thermal, and acoustic sensors, provides robots with a comprehensive understanding of their environment. This multi-modal sensing capability enhances situational awareness, allowing robots to adapt to diverse terrains, improve obstacle detection, identify survivors, and navigate complex environments effectively. Moreover, advances in computer vision and object recognition algorithms enable robots to identify and interact with objects in real-time, distinguishing between survivors, obstacles, and artifacts. This capability enables robots to prioritize tasks efficiently, improving mission effectiveness and potentially saving more lives.

In the branch of communication technologies, enhanced communication protocols enable seamless information exchange between robots and other devices in the field. This edge-to-edge communication facilitates collaborative decision-making, enabling robots to share real-time data, coordinate actions, and explore the environment collectively. Furthermore, the development of decentralized communication networks ensures robust communication even in disrupted or congested environments, enabling continuous coordination and information exchange between robots and command centers. This capability is particularly crucial in disaster-stricken areas with compromised infrastructure, ensuring effective communication and collaboration among robotic system

These advancements in algorithms, hardware, sensor technologies, and communication technologies revolutionize reinforcement learning-driven agent-based navigation. They make robots more intelligent, adaptable, and efficient in search and rescue missions. Continued research and innovation in these areas are essential, paving the way for highly effective autonomous systems that ultimately save more lives and enhance disaster response capabilities globally which mentioned in Figure 4.

Figure 4. Road map for future research and development

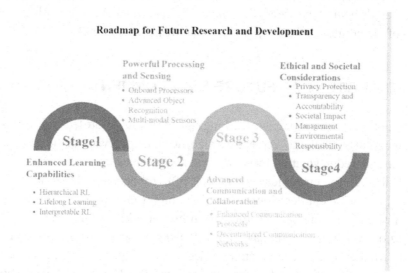

Roadmap for Future Research and Development

Powerful Processing
and Sensing
- Onboard Processors
- Advanced Object Recognition
- Multi-modal Sensors

Ethical and Societal
Considerations
- Privacy Protection
- Transparency and Accountability
- Societal Impact Management
- Environmental Responsibility

Stage1

Stage 2

Stage 3

Stage4

Enhanced Learning
Capabilities
- Hierarchical RL
- Lifelong Learning
- Interpretable RL

Advanced
Communication and
Collaboration
- Enhanced Communication Protocols
- Decentralized Communication Networks

CONCLUSION

In conclusion, this chapter underscores the importance of optimizing energy efficiency in the realm of search and rescue robotics. The application of reinforcement learning techniques has emerged as a powerful tool in achieving this goal, empowering robotic systems to make energy-conscious decisions and adapt to diverse and challenging environments. The integration of reinforcement learning in search and rescue robotics offers a range of key advantages, including adaptive decision-making, optimized navigation, dynamic resource allocation, longer mission durations, and real-time adaptability. These benefits not only enhance the effectiveness of robotic missions but also contribute to sustainability by reducing the environmental impact associated with frequent recharging or battery replacements. The evolution of reinforcement learning-driven agent-based navigation goes beyond technological progress; it symbolizes a commitment to saving lives and building resilient communities. The strides made in this field signify a promising future where search and rescue robots, equipped with intelligent and energy-efficient capabilities, play a pivotal role in addressing critical situations and ensuring the well-being of communities in need.

REFERENCES

Akhloufi, M. A., Couturier, A., & Castro, N. A. (2021). Unmanned aerial vehicles for wildland fires: Sensing, perception, cooperation and assistance. *Drones (Basel)*, 5(1), 15. doi:10.3390/drones5010015

Alatise, M. B., & Hancke, G. P. (2020). A Review on Challenges of Autonomous Mobile Robot and Sensor Fusion Methods. In IEEE Access (Vol. 8). doi:10.1109/ACCESS.2020.2975643

Alwateer, M., Loke, S. W., & Fernando, N. (2019). Enabling drone services: Drone crowdsourcing and drone scripting. *IEEE Access : Practical Innovations, Open Solutions*, 7, 110035–110049. doi:10.1109/ACCESS.2019.2933234

Ariizumi, R., & Matsuno, F. (2017). Dynamic Analysis of Three Snake Robot Gaits. *IEEE Transactions on Robotics, 33*(5), 1075–1087. doi:10.1109/TRO.2017.2704581

Arnold, R. D., Yamaguchi, H., & Tanaka, T. (2018). Search and rescue with autonomous flying robots through behavior-based cooperative intelligence. *Journal of International Humanitarian Action, 3*(1), 18. doi:10.1186/s41018-018-0045-4

Bannur, P., Gujarathi, P., Jain, K., & Kulkarni, A. J. (2020). Application of Swarm Robotic System in a Dynamic Environment using Cohort Intelligence. *Soft Computing Letters, 2*, 100006. doi:10.1016/j.socl.2020.100006

Biggie, H., Rush, E., Riley, D., Ahmad, S., Ohradzansky, M., Harlow, K., Miles, M., Torres, D., McGuire, S., Frew, E., Heckman, C., & Humbert, J. (2023). Flexible Supervised Autonomy for Exploration in Subterranean Environments. *Field Robotics, 3*(1), 125–189. doi:10.55417/fr.2023004

Bouyarmane, K., Vaillant, J., Keith, F., & Kheddar, A. (2012). Exploring humanoid robots locomotion capabilities in virtual disaster response scenarios. *IEEE-RAS International Conference on Humanoid Robots.* IEEE. 10.1109/HUMANOIDS.2012.6651541

Cai, C., Chen, J., Yan, Q., & Liu, F. (2023). A Multi-Robot Coverage Path Planning Method for Maritime Search and Rescue Using Multiple AUVs. *Remote Sensing (Basel), 15*(1), 93. doi:10.3390/rs15010093

Chen, J., Yuan, B., & Tomizuka, M. (2019). Model-free Deep Reinforcement Learning for Urban Autonomous Driving. *2019 IEEE Intelligent Transportation Systems Conference. ITSC, 2019*, 2765–2771. doi:10.1109/ITSC.2019.8917306

ChenT.GuptaS.GuptaA. (2019). Learning Exploration Policies for Navigation. *CoRR, abs/1903.01959.* http://arxiv.org/abs/1903.01959

Chen, Y., & Xiao, J. (2023). *Target Search and Navigation in Heterogeneous Robot Systems with Deep Reinforcement Learning.*

Cheng, F., Liang, J., Tao, Z., & Chen, J. (2011). Functional Materials for Rechargeable Batteries. *Advanced Materials, 23*(15), 1695–1715. doi:10.1002/adma.201003587 PMID:21394791

Chitikena, H., Sanfilippo, F., & Ma, S. (2023). Robotics in Search and Rescue (SAR) Operations: An Ethical and Design Perspective Framework for Response Phase. *Applied Sciences (Basel, Switzerland), 13*(3), 1800. doi:10.3390/app13031800

Cipi, E., & Cico, B. (2011). Simulation of an Agent Based System Behavior in a Dynamic and Unpredicted Environment. [WCSIT]. *World of Computer Science and Information Technology Journal, 1*, 2221–2741.

Couceiro, M. (2017a). An Overview of Swarm Robotics for Search and Rescue Applications. In *Artificial Intelligence.* Concepts, Methodologies, Tools, and Applications., doi:10.4018/978-1-5225-1759-7.ch061

Couceiro, M. (2017b). An Overview of Swarm Robotics for Search and Rescue Applications. In *Artificial Intelligence.* Concepts, Methodologies, Tools, and Applications., doi:10.4018/978-1-5225-1759-7.ch061

Das, P. K., Behera, H. S., & Panigrahi, B. K. (2016). A hybridization of an Improved Particle Swarm optimization and Gravitational Search Algorithm for Multi-Robot Path Planning. *Swarm and Evolutionary Computation, 28*, 14–28. doi:10.1016/j.swevo.2015.10.011

Doroftei, D., Matos, A., & de Cubber, G. (2014). Designing search and rescue robots towards realistic user requirements. *Applied Mechanics and Materials*, *658*, 612–617. Advance online publication. . doi:10.4028/www.scientific.net/AMM.658.612

Dosovitskiy, A., Ros, G., Codevilla, F., Lopez, A., & Koltun, V. (2017). *CARLA: An Open Urban Driving Simulator*.

Guan, J., Chen, G., Huang, J., Li, Z., Xiong, L., Hou, J., & Knoll, A. (2023). A Discrete Soft Actor-Critic Decision-Making Strategy With Sample Filter for Freeway Autonomous Driving. *IEEE Transactions on Vehicular Technology*, *72*(2), 2593–2598. doi:10.1109/TVT.2022.3212996

Hameed, I. A. (2014). Intelligent coverage path planning for agricultural robots and autonomous machines on three-dimensional terrain. *Journal of Intelligent & Robotic Systems*, *74*(3–4), 965–983. doi:10.1007/s10846-013-9834-6

Huang, Z., Wu, J., & Lv, C. (2022). Efficient Deep Reinforcement Learning With Imitative Expert Priors for Autonomous Driving. *IEEE Transactions on Neural Networks and Learning Systems*. doi:10.1109/TNNLS.2022.3142822 PMID:35081030

Iwano, Y., Osuka, K., & Amano, H. (2004). Proposal of a rescue robot system in nuclear-power plants-rescue activity via small vehicle robots. *Proceedings - 2004 IEEE International Conference on Robotics and Biomimetics, IEEE ROBIO 2004*. 10.1109/ROBIO.2004.1521781

Jamgochian, A., Buehrle, E., Fischer, J., & Kochenderfer, M. J. (2023). SHAIL: Safety-Aware Hierarchical Adversarial Imitation Learning for Autonomous Driving in Urban Environments. *Proceedings - IEEE International Conference on Robotics and Automation, 2023-May*. IEEE. 10.1109/ICRA48891.2023.10161449

Kalakrishnan, M., Righetti, L., Pastor, P., & Schaal, S. (2011). Learning force control policies for compliant manipulation. *IEEE International Conference on Intelligent Robots and Systems*. IEEE. 10.1109/IROS.2011.6095096

Kamran, D., Lopez, C. F., Lauer, M., & Stiller, C. (2020). Risk-Aware High-level Decisions for Automated Driving at Occluded Intersections with Reinforcement Learning. *IEEE Intelligent Vehicles Symposium, Proceedings*. IEEE. 10.1109/IV47402.2020.9304606

Khurana, D., Koli, A., Khatter, K., & Singh, S. (2023). Natural language processing: State of the art, current trends and challenges. *Multimedia Tools and Applications*, *82*(3), 3713–3744. doi:10.1007/s11042-022-13428-4 PMID:35855771

Kim, S. W., Liu, W., Ang, M. H., Frazzoli, E., & Rus, D. (2015). The Impact of Cooperative Perception on Decision Making and Planning of Autonomous Vehicles. *IEEE Intelligent Transportation Systems Magazine*, *7*(3), 39–50. Advance online publication. doi:10.1109/MITS.2015.2409883

Kohl, N., & Stone, P. (2004). Policy gradient reinforcement learning for fast quadrupedal locomotion. *Proceedings - IEEE International Conference on Robotics and Automation, 2004*(3). IEEE. 10.1109/ROBOT.2004.1307456

Koval, A., Karlsson, S., & Nikolakopoulos, G. (2022). Experimental evaluation of autonomous map-based Spot navigation in confined environments. *Biomimetic Intelligence and Robotics*, *2*(1), 100035.

Kühl, N., Schemmer, M., Goutier, M., & Satzger, G. (2022). Artificial intelligence and machine learning. *Electronic Markets*, *32*(4), 2235–2244. doi:10.1007/s12525-022-00598-0

Lee, M. F. R., & Yusuf, S. H. (2022). Mobile Robot Navigation Using Deep Reinforcement Learning. *Processes (Basel, Switzerland)*, *10*(12), 2748. doi:10.3390/pr10122748

Li, H., Zhang, G., Ma, R., & You, Z. (2014). Design and Experimental Evaluation on an Advanced Multisource Energy Harvesting System for Wireless Sensor Nodes. *TheScientificWorldJournal*, *671280*, 1–13. doi:10.1155/2014/671280 PMID:25032233

Lillicrap, T. P., Hunt, J. J., Pritzel, A., Heess, N., Erez, T., Tassa, Y., Silver, D., & Wierstra, D. (2016). Continuous control with deep reinforcement learning. *4th International Conference on Learning Representations, ICLR 2016 - Conference Track Proceedings*. IEEE.

Liu, C., Lee, S., Varnhagen, S., & Tseng, H. E. (2017). Path planning for autonomous vehicles using model predictive control. *IEEE Intelligent Vehicles Symposium, Proceedings*. IEEE. 10.1109/IVS.2017.7995716

Liu, J., Wang, Y., Li, B., & Ma, S. (2007). Current research, key performances and future development of search and rescue robots. *Frontiers of Mechanical Engineering in China*, *2*(4), 404–416. doi:10.1007/s11465-007-0070-2

Lopez-Fuentes, L., van de Weijer, J., González-Hidalgo, M., Skinnemoen, H., & Bagdanov, A. D. (2018). Review on computer vision techniques in emergency situations. *Multimedia Tools and Applications*, *77*(13), 17069–17107. doi:10.1007/s11042-017-5276-7

Lundgren, A. V. A., dos Santos, M. A. O., Bezerra, B. L. D., & Bastos-Filho, C. J. A. (2022). Systematic Review of Computer Vision Semantic Analysis in Socially Assistive Robotics. In AI (Switzerland) (Vol. 3, Issue 1). doi:10.3390/ai3010014

Lygouras, E., Santavas, N., Taitzoglou, A., Tarchanidis, K., Mitropoulos, A., & Gasteratos, A. (2019). Unsupervised human detection with an embedded vision system on a fully autonomous UAV for search and rescue operations. *Sensors (Basel)*, *19*(16), 3542. doi:10.3390/s19163542 PMID:31416131

Lyu, M., Zhao, Y., Huang, C., & Huang, H. (2023). Unmanned Aerial Vehicles for Search and Rescue: A Survey. In Remote Sensing, 15(13). doi:10.3390/rs15133266

Macek, K., Petrovic, I., & Peric, N. (2002). A reinforcement learning approach to obstacle avoidance of mobile robots. *7th International Workshop on Advanced Motion Control*, (pp. 462–466). IEEE. 10.1109/AMC.2002.1026964

Moreau, J., Melchior, P., Victor, S., Moze, M., Aioun, F., & Guillemard, F. (2019). Reactive path planning for autonomous vehicle using bézier curve optimization. *IEEE Intelligent Vehicles Symposium, Proceedings, 2019-June*. 10.1109/IVS.2019.8813904

Murphy, A. J., Landamore, M. J., & Birmingham, R. W. (2008). The role of autonomous underwater vehicles for marine search and rescue operations. *Underwater Technology*, *27*(4), 195–205. Advance online publication. doi:10.3723/ut.27.195

Murphy Robin R. & Tadokoro, S. (2008). Search and Rescue Robotics. In O. Siciliano Bruno and Khatib (Ed.), *Springer Handbook of Robotics* (pp. 1151–1173). Springer Berlin Heidelberg. doi:10.1007/978-3-540-30301-5_51

Paredes, J., Saito, C., Abarca, M., & Cuellar, F. (2017). *Study of effects of high-altitude environments on multicopter and fixed-wing UAVs' energy consumption and flight time*. IEEE. doi:10.1109/COA-SE.2017.8256340

Riazi, S., Bengtsson, K., Bischoff, R., Aurnhammer, A., Wigstrom, O., & Lennartson, B. (2016). Energy and peak-power optimization of existing time-optimal robot trajectories. *IEEE International Conference on Automation Science and Engineering, 2016-November*. IEEE. 10.1109/COASE.2016.7743423

Ristic, B., Angley, D., Moran, B., & Palmer, J. L. (2017). Autonomous multi-robot search for a hazardous source in a turbulent environment. *Sensors (Basel)*, *17*(4), 918. doi:10.3390/s17040918 PMID:28430120

Sana, F., Azad, N. L., & Raahemifar, K. (2023). Autonomous Vehicle Decision-Making and Control in Complex and Unconventional Scenarios—A Review. In Machines, 11(7). doi:10.3390/machines11070676

Sarker, I. H. (2021). Deep Learning: A Comprehensive Overview on Techniques, Taxonomy, Applications and Research Directions. In SN Computer Science, 2(6). doi:10.1007/s42979-021-00815-1

Sepasgozar, S., Karimi, R., Farahzadi, L., Moezzi, F., Shirowzhan, S., Ebrahimzadeh, S. M., Hui, F., & Aye, L. (2020). A systematic content review of artificial intelligence and the internet of things applications in smart home. *Applied Sciences (Switzerland)*, *10*(9). doi:10.3390/app10093074

Shan, Y., Zheng, B., Chen, L., Chen, L., & Chen, D. (2020). A Reinforcement Learning-Based Adaptive Path Tracking Approach for Autonomous Driving. *IEEE Transactions on Vehicular Technology*, *69*(10), 10581–10595. doi:10.1109/TVT.2020.3014628

Shriyanti Kulkarni. (n.d.). *Vedashree Chaphekar, Md Moin Uddin Chowdhury, Fatih Erden, & Ismail Guvenc*. UAV Aided Search and Rescue Operation Using Reinforcement Learning.

Singh, R., Kurukuru, V. S. B., & Khan, M. A. (2023). Advanced Power Converters and Learning in Diverse Robotic Innovation: A Review. *Energies*, *16*(20), 7156. doi:10.3390/en16207156

Singh, R., Ren, J., & Lin, X. (2023). A Review of Deep Reinforcement Learning Algorithms for Mobile Robot Path Planning. *Vehicles*, *5*(4), 1423–1451. doi:10.3390/vehicles5040078

Sivamayil, K., Rajasekar, E., Aljafari, B., Nikolovski, S., Vairavasundaram, S., & Vairavasundaram, I. (2023). A Systematic Study on Reinforcement Learning Based Applications. In Energies, 16(3). doi:10.3390/en16031512

Soori, M., Arezoo, B., & Dastres, R. (2023). Artificial intelligence, machine learning and deep learning in advanced robotics, a review. In Cognitive Robotics, 3. doi:10.1016/j.cogr.2023.04.001

Tahir, A., Böling, J., Haghbayan, M. H., Toivonen, H. T., & Plosila, J. (2019). Swarms of Unmanned Aerial Vehicles — A Survey. In Journal of Industrial Information Integration, 16. doi:10.1016/j.jii.2019.100106

Turek, W., Marcjan, R., & Cetnarowicz, K. (2006). Agent-Based Mobile Robots Navigation Framework. *Lecture Notes in Computer Science*, *3993*, 775–782. doi:10.1007/11758532_101

Tzafestas, S. G. (2013). Introduction to Mobile Robot Control. In Introduction to Mobile Robot Control. doi:10.1016/C2013-0-01365-5

Wallace, N., Kong, H., Hill, A., & Sukkarieh, S. (2019). Energy Aware Mission Planning for WMRs on Uneven Terrains. *IFAC-PapersOnLine*, *52*(30). doi:10.1016/j.ifacol.2019.12.513

Wan, Z., Anwar, A., Hsiao, Y. S., Jia, T., Reddi, V. J., & Raychowdhury, A. (2021). Analyzing and Improving Fault Tolerance of Learning-Based Navigation Systems. *Proceedings - Design Automation Conference, 2021-December*. 10.1109/DAC18074.2021.9586116

Wilson, D. G., Robinett, R. D., & Eisler, G. R. (2004). Discrete dynamic programming for optimized path planning of flexible robots. *2004 IEEE/RSJ International Conference on Intelligent Robots and Systems (IROS) (IEEE Cat. No.04CH37566), 3*, (pp. 2918–2923). IEEE. 10.1109/IROS.2004.1389852

Wu, J., Honglun, W., Huang, Y., Su, Z., & Zhang, M. (2018). Energy Management Strategy for Solar-Powered UAV Long-Endurance Target Tracking. *IEEE Transactions on Aerospace and Electronic Systems, PP, 1*. doi:10.1109/TAES.2018.2876738

Wu, J., Wang, H., Huang, Y., Su, Z., & Zhang, M. (2019). Energy Management Strategy for Solar-Powered UAV Long-Endurance Target Tracking. *IEEE Transactions on Aerospace and Electronic Systems, 55*(4), 1878–1891. doi:10.1109/TAES.2018.2876738

WulfmeierM.ByravanA.HertweckT.HigginsI.GuptaA.KulkarniT.ReynoldsM.TeplyashinD.HafnerR. LampeT.RiedmillerM. A. (2020). Representation Matters: Improving Perception and Exploration for Robotics. *CoRR, abs/2011.01758*. https://arxiv.org/abs/2011.01758

Yao, J., Li, X., Zhang, Y., Ji, J., Wang, Y., & Liu, Y. (2022). Path Planning of Unmanned Helicopter in Complex Environment Based on Heuristic Deep Q-Network. *International Journal of Aerospace Engineering, 1360956*, 1–15. doi:10.1155/2022/1360956

Yoshimoto, M., Endo, T., Maeda, R., & Matsuno, F. (2018). Decentralized navigation method for a robotic swarm with nonhomogeneous abilities. *Autonomous Robots, 42*(8), 1583–1599. doi:10.1007/s10514-018-9774-x

Younis, H. A., Ruhaiyem, N. I. R., Ghaban, W., Gazem, N. A., & Nasser, M. (2023). A Systematic Literature Review on the Applications of Robots and Natural Language Processing in Education. In Electronics (Switzerland), 12(13). doi:10.3390/electronics12132864

Zeng, J., Ju, R., Qin, L., Hu, Y., Yin, Q., & Hu, C. (2019). Navigation in unknown dynamic environments based on deep reinforcement learning. *Sensors (Basel), 19*(18), 3837. doi:10.3390/s19183837 PMID:31491927

Zeng, Y., Zhang, R., & Lim, T. J. (2016). Wireless communications with unmanned aerial vehicles: Opportunities and challenges. *IEEE Communications Magazine, 54*(5), 36–42. doi:10.1109/MCOM.2016.7470933

Chapter 9
Recognition of Cyber Physical Systems Through Network Security for Wireless Sensor Networks:
Using Artificial InItelligence in Cyber Physical Systems

S. Selvakanmani
https://orcid.org/0000-0001-6469-6084
RMK Engineering College, India

Seeniappan Kaliappan
https://orcid.org/0000-0002-5021-8759
KCG College of Technology, India

M. Muthukannan
KCG College of Technology, India

Mohammed
SRM Institute of Science and Technology, India

ABSTRACT

This chapter presents a novel approach for recognizing and securing cyber-physical systems (CPS) through the use of artificial intelligence in wireless sensor networks. The increasing use of CPS in various fields has led to a growing need for effective methods of identifying and securing these systems. The proposed approach utilizes artificial intelligence techniques to analyse network traffic and identify patterns that indicate the presence of a CPS. Additionally, the proposed approach uses this information to secure the CPS by implementing appropriate security measures to protect against cyber-attacks. This study highlights the importance of recognizing and securing CPS in wireless sensor networks, and the potential of artificial intelligence to meet this need. It also emphasizes the importance of developing secure and resilient systems in the face of cyber-threats and the need for a holistic security approach for CPS.

DOI: 10.4018/979-8-3693-3735-6.ch009

INTRODUCTION

Cyber-Physical Systems (CPS) represent an intricate fusion of computational and physical components that play a pivotal role across diverse sectors, including but not limited to transportation, healthcare, and industrial automation (Santhosh Kumar et al. 2022). Within the architecture of CPS, Wireless Sensor Networks (WSNs) serve as a fundamental element, facilitating the observation and manipulation of physical processes (Josphineleela et al. 2023a). The amalgamation of WSNs into CPS frameworks, however, introduces a series of security vulnerabilities, chiefly due to the inherent nature of wireless communication and the resource constraints of sensor nodes (Reddy et al. 2023). These vulnerabilities expose the systems to a variety of cyber-attacks. To counteract these security threats, Artificial Intelligence (AI) offers a promising solution by delivering sophisticated mechanisms for the detection and neutralization of potential cyber intrusions (Asha et al. 2022). In our study, we put forth a novel strategy that leverages AI for bolstering the security of CPS, specifically through enhanced safeguarding of WSNs. By integrating machine learning algorithms with established network security protocols, our methodology aims to elevate the efficiency of detecting and countering cyber threats within CPS environments (Suman et al. 2023). This approach not only heightens security measures but also adapts to the evolving landscape of cyber challenges faced by CPS (Darshan et al. 2022).

Recent studies have shown that AI can improve the security of CPS by providing advanced methods for detecting and responding to cyber threats (Loganathan et al. 2023). For example, machine learning algorithms have been used to detect malicious traffic in WSNs by analyzing the behavior of nodes and identifying abnormal patterns (Selvi et al. 2023).

Another approach to enhancing the security of CPS is to use game theory, which allows for the modeling of strategic interactions between attackers and defenders. Game-theoretic techniques have been applied to CPS to study the optimal strategies for defending against cyber attacks, and to design robust and secure control systems (Sendrayaperumal et al. 2021). In addition to AI and game theory, traditional network security methods such as encryption and authentication can also be used to enhance the security of CPS (Subramanian et al. 2022). For example, the use of encryption can protect the confidentiality of communication in WSNs, while authentication can prevent unauthorized access to the network (Kaushal et al. 2023).

Overall, the literature suggests that a combination of AI and traditional network security methods can improve the recognition and response to cyber threats in CPS. Our proposed method aims to build on this by developing a new approach for recognizing CPS through network security for WSNs using AI (Thakre et al. 2023). However, it's worth mentioning that the above is a very broad and generic overview, it would be better to focus on the specific aspect of the proposed method, the problem that it attempts to solve, and its novelty with respect to the existing literature (Nagarajan et al. 2022).

In particular, our proposed method aims to address the problem of accurately recognizing cyber threats in CPS, particularly in WSNs. One of the main challenges in this area is the dynamic and unpredictable nature of cyber attacks, which can evade traditional security methods that rely on predefined rules or signatures (Seeniappan et al. 2023). To address this challenge, our method utilizes machine learning techniques to learn from the behavior of the network and identify abnormal patterns that may indicate a cyber attack. Our method also incorporates traditional network security methods such as encryption and authentication to provide an additional layer of protection (Arockia Dhanraj et al. 2022). One of the key novelties of our proposed method is the use of AI techniques to learn the normal behavior of the network, and then use this knowledge to detect and respond to cyber threats (Sharma et al. 2022). This is different from traditional methods that rely on predefined rules or signatures, which can be easily evaded by attackers (Divya et al. 2022).

Another novelty of our proposed method is the use of a combination of machine learning and traditional network security methods. This approach allows for the identification of cyber threats using both behavioral analysis and predefined security rules, increasing the robustness of the system (Mahesha et al. 2022). In summary, our proposed method for recognizing CPS through network security for WSNs using AI addresses the problem of accurately recognizing cyber threats in CPS. Our approach utilizes machine learning techniques to learn the normal behavior of the network and identify abnormal patterns that may indicate a cyber attack, and also incorporates traditional network security methods for an additional layer of protection. This approach is novel in its use of AI to learn the normal behavior of the network, and the combination of machine learning and traditional network security methods (Vijayaragavan et al. 2022).

CYBERTHREATS OF MANUFACTURING 4.0

Manufacturing 4.0 represents a transformative vision for the industrial sector, driven by the integration of advanced digital technologies to foster a highly interconnected ecosystem. This ecosystem enhances collaboration and cooperation among resources, machinery, and humans, leading to optimized production processes and workflows (Natrayan and Kaliappan 2023). The conceptual framework of this vision is articulated through a cyber-physical systems reference model, which delineates three distinct levels of operation, as illustrated in Figure 1: the consumer system, corporate system, and manufacturing system (Kaliappan et al. 2023a).

The consumer system, or user system level, sits at the forefront of this model. It is primarily concerned with the identification and evaluation of customer demands and preferences (Kaliappan et al. 2023b). This level leverages modern communication tools and platforms, such as smartphones, cellular networks, and applications, to gather and analyze consumer data (Natrayan et al. 2023a). By doing so, it ensures that the production process remains closely aligned with market needs and expectations, enabling companies to respond dynamically to changing consumer trends (Ramaswamy et al. 2022b; Muralidaran et al. 2023).

The corporate system, also referred to as Level 2, serves as the operational backbone of the Manufacturing 4.0 framework. It encompasses a broad range of business-related functions, including but not limited to financial transactions, operations management, and production planning (Josphineleela et al. 2023b; Saravanan et al. 2023). This level integrates traditional information technology systems with advanced data analytics and management tools to streamline corporate operations. It acts as a bridge between the strategic objectives of the organization and the technical capabilities of the manufacturing system, ensuring that business decisions are informed by accurate and timely data (Balaji et al. 2022; Ramaswamy et al. 2022a; Sivakumar et al. 2023). The third level, the manufacturing system, although not detailed in the initial description, can be inferred to focus on the actual production processes. It likely involves the direct application of cyber-physical systems in the factory floor, incorporating robotics, automation technologies, and real-time data monitoring to enhance manufacturing efficiency and flexibility (Lakshmaiya et al. 2023; Selvi et al. 2023).

This level is where the digital and physical elements of Manufacturing 4.0 converge, enabling the creation of smart factories that can adapt their operations in real-time to meet specific production goals and requirements. Together, these three levels create a cohesive and integrated model for Manufacturing 4.0, where data flows seamlessly from consumer insights to corporate strategy and onto the manufactur-

Figure 1. Various levels of Industry 4.0

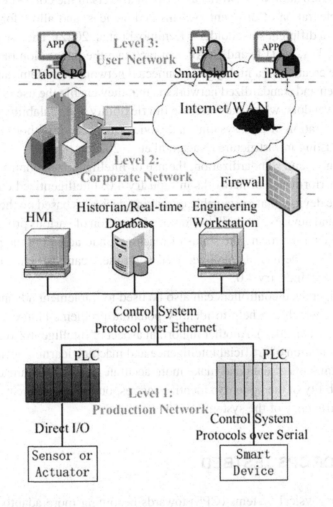

ing floor (Natrayan et al. 2023b). This interconnectedness not only boosts operational efficiency and productivity but also empowers businesses to deliver highly customized and innovative products to the market, reflecting the dynamic interplay between consumer demands, business objectives, and manufacturing capabilities (Arockiasamy et al. 2023).

The production system level, or Level 1, is accountable for monitoring, regulating, detection, and changing corporeal procedures (Balamurugan et al. 2023). This level is additional broken down into three sublayers: the guiding control LAN, control network, and I/O network, which includes devices such as engineering workstations, historians, programmable logic controllers, distributed control systems controllers, remote terminal units, sensors, actuators, and controlled processes (Angalaeswari et al. 2022). Overall, Industry 4.0 aims to create a highly interconnected society where resources, machines, and human collaboration and cooperation are optimized through the use of cyber-physical systems, which are composed of three levels: the consumer network, corporate network, and production network (Nagajothi et al. 2022b). These levels work together to gather and evaluate customer needs, handle business-related operations, and monitor, regulate, sense and change physical processes.

Openness and standardization of networks are critical aspects in the context of Industry 4.0, as they enable the seamless integration of different systems and devices, and allow for the exchange of data and information between different stakeholders (Kanimozhi et al. 2022). The use of open standards and protocols, such as OPC UA, MQTT, and CoAP, can facilitate the integration of different systems and devices, and enable the creation of a highly interconnected network of systems and devices. One of the main advantages of open and standardized networks is that they enable the use of different systems and devices from different vendors, which can increase the flexibility and scalability of the overall system. This allows for the integration of new systems and devices as needed, without the need for extensive modifications to the existing infrastructure (Nagajothi et al. 2022a).

In addition to openness and standardization, the use of Intelligentized controllers can also enhance the functionality and performance of networks in Industry 4.0. Intelligentized controllers, also known as smart controllers, are devices that are capable of making decisions based on the information received from different sensors and devices . This allows for the automation of various processes, such as controlling the flow of materials, monitoring the status of machines, and adjusting the parameters of devices (Lakshmaiya et al. 2023). The use of Intelligentized controllers can also enhance the flexibility and adaptability of networks in Industry 4.0.

Furthermore, Intelligentized controllers can also be used to implement advanced functions such as predictive maintenance, which can help to identify potential problems before they occur and reduce downtime (Subramanian et al. 2022). Another important aspect of Intelligentized controllers is that they can help in the implementation of artificial intelligence and machine learning techniques, which can be used to analyze large amounts of data and make more accurate decisions (Ramachandran et al. 2022). This can improve the ability of the system to identify and respond to cyber threats, and also to optimize the performance and efficiency of the system.

METHODOLOGY OF CPS TESTBED

The evolution of cyber-physical systems (CPS) towards becoming more adaptive, versatile, scalable, diverse, and autonomous presents a double-edged sword. On one hand, these advancements promise significant improvements in efficiency and functionality across various domains. On the other, they increasingly attract the attention of malicious actors intent on exploiting the physical components of these systems (Ugle et al. 2023). This dynamic landscape necessitates sophisticated testing environments, or testbeds, which can simulate real-world scenarios and cyber-attack vectors to ensure the resilience and security of CPS. These testbeds can be broadly categorized into three types: corporeal (physical), software program (simulation-based), and cross (hybrid) testbeds. Corporeal testbeds rely on physical equipment and real network infrastructures to provide an authentic testing environment (Anjankar et al. 2023). While offering a high degree of realism, these testbeds can be costly to set up and maintain, and may not offer the flexibility needed to test against a rapidly evolving array of cyber threats (Rajagopalan et al. 2022). Traditional examples, such as the National SCADA Testbed (NSTB) in the United States, exemplify the strengths and limitations of this approach, offering valuable insights into system vulnerabilities but often at the expense of scalability and adaptability to new technologies (Chennai Viswanathan et al. 2023).

Software program testbeds, on the other hand, utilize simulation techniques to mimic the behavior of cyber-physical systems. This category includes platforms like the Experimental Platform for Internet Contingencies (EPIC), which leverage modeling to construct virtual components of CPS (Seralathan et al. 2023). While these testbeds offer scalability and the ability to rapidly adapt to new scenarios, they sometimes fall short in accurately representing the physical intricacies and real-world unpredictability's of industry-specific environments (Saadh et al. 2024).

Cross testbeds, or hybrid testbeds, represent a convergence of the corporeal and software program methodologies, aiming to balance the realism of physical components with the flexibility and scalability of virtual simulations. The CPS testbed discussed in this context exemplifies this hybrid approach by integrating cloud computation and software-defined network technologies with actual hardware and physical components sourced from real-world environments (Saadh et al. 2023a). This innovative architecture allows for a more nuanced exploration of cyber-physical system vulnerabilities, offering a platform that is both versatile and representative of industry-specific challenges. The hybrid testbed acknowledges the inherent trade-offs between the depth of research achievable and the financial and logistical costs associated with constructing such an advanced testing environment (Saadh et al. 2023b). By leveraging both physical and virtual components, it offers a pragmatic solution that aims to meet the dynamic demands of Industry 4.0, providing a more adaptable and comprehensive framework for testing, analyzing, and improving the security and performance of cyber-physical systems in the face of evolving cyber threats (Seralathan et al. 2023).

ARCHITECTURE OF THE APPROACH

The Cyber-Physical System for Traffic Control and Safety (CPSTCS) is structured into three primary layers as detailed in Figure 2, comprising the application, cloud, and physical resource layers. Central to this architecture, the cloud layer plays a pivotal role by interfacing and processing an array of hardware resources sourced from the physical resource layer (Lakshmaiya et al. 2022). These resources span a comprehensive range, including computing and storage clusters, mobile devices, controllers, and various wireless and traditional networking devices such as Access Points (APs), Air Controllers (ACs), receivers, as well as Layer 2 (L2) and Layer 3 (L3) switches, alongside programmable logic controllers.

To effectively manage and optimize these resources, the cloud layer employs cutting-edge techniques like software-defined networking (SDN) and cloud computing, allowing for the dynamic and efficient digital processing of physical resources. The architecture of the cloud layer is further refined into two sub-layers: the control and datapath layers (Pragadish et al. 2023).

The control layer is essentially composed of OpenFlow controllers and cloud-based controllers. These components leverage virtualization technologies, such as virtual machines and OpenvSwitch, to architect SDN networks, ensuring scalable, flexible, and efficient network management. This layer essentially acts as the brain of the cloud layer, directing traffic and managing network resources through software-based controls (Sabarinathan et al. 2022).

On the other hand, the datapath layer facilitates the cloud layer's interaction with external networks. It utilizes OpenvSwitch to forge a hybrid network environment, seamlessly integrating the internal operations of the CPSTCS with external data flows and network interactions (Niveditha VR. and Rajakumar PS. 2020). This layer is crucial for the exchange of information and ensures that the CPSTCS remains connected and operational within a broader network ecosystem.

Finally, the application layer, also referred to as the claim layer in this context, serves as a user interface for the submission of various requests (Nadh et al. 2021). These requests may include the establishment of topological configurations, host management activities, and traffic monitoring operations. This layer

Figure 2. Architecture of the testbed

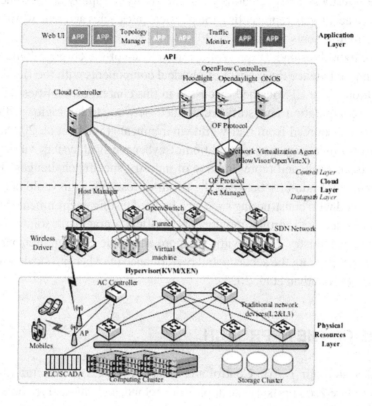

is designed to provide end-users with the ability to tailor the CPSTCS to their specific needs, ensuring that the system can be effectively used for its intended purpose of traffic control and safety (Anupama et al. 2021).

The cloud computing methods employed in the CPSTCS testbed are outlined in Figure 3. The main goal of cloud computing in this testbed is to create a resource pool that combines both physical and virtual resources, allowing for the management of both types of resources. Physical resources include conventional and OpenFlow switches, real routers, physical hosts, and physical controllers, while virtual resources include software simulators, virtual switches, virtual routers, virtual hosts, mathematical models, and various scripts and applications.

The CPSTCS architecture is designed to be adaptive, scalable, diversified, and flexible. It takes into account the dispersion of business traffic and its capacity to manage numerous logical network topology for various application scenarios and fine-grained commercial circulation dispensation are among its key features. Additionally, it complies with Industry 4.0's security and performance standards for cyber-physical systems. Through the use of both physical and virtual resources, the testbed is able to dynamically form a variety of service nodes with the aid of cloud-based functionalities. The "Layer-2-access" and "Layer-3-access" procedures can routinely access actual testbed devices for real resources, while virtualization technologies such as kvm, xen, and virtualbox provide features for virtual resources. These real and virtual service nodes can then be used to form various logical topologies for different scenarios and test cases.

Figure 3. Cloud technology in the testbed

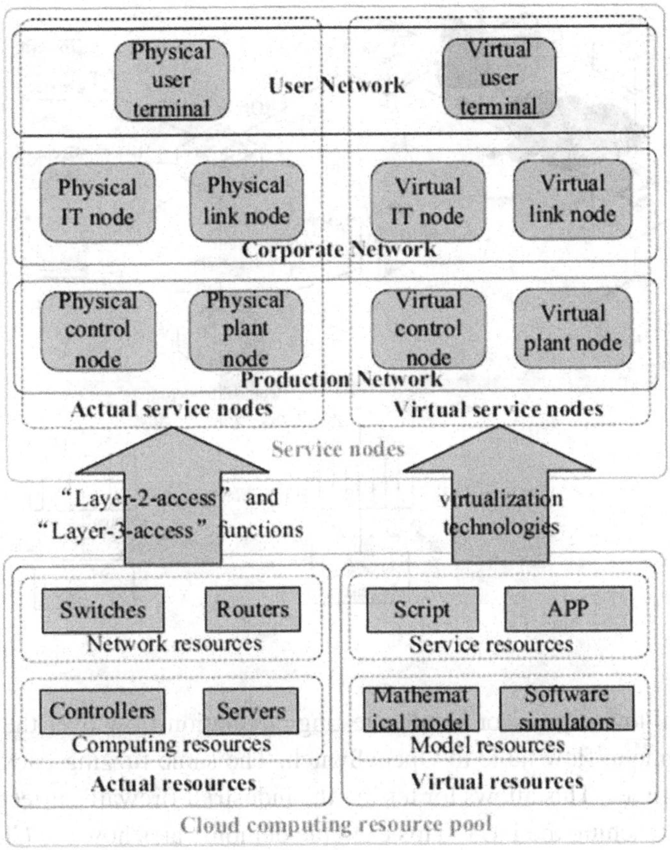

EXPERIMENTAL RESULT AND DISCUSSION

The controllers, such as programmable logic controllers (PLCs), remote terminal units (RTUs), and smart transmitters, that govern process conditions in cyber-physical systems are crucial components that need to be thoroughly tested and secured. The CPSTCS testbed provides a platform for testing and securing these controllers.

Figure 4 illustrates an instance investigational setting on the CPSTCS testbed. As previously mentioned, the CPSTCS testbed allows for commercial traffic of fine-grained processing, which enables quick changes and control of the testing traffic flow across multiple network connections. In the figure 4, the red dotted line represents a connection that employs an industrial firewall, while the green dotted line represents a connection without one. Here are some techniques for conducting experiments on the testbed:

First, the OpenFlow supervisor regulates the challenging traffic flow across the green scattered line by providing a customized flow bench to OpenvSwitch. The tester then creates a fuzzing bundle and runs the facility "Python ModbusTCPTest.py -t Modbus -f modbusTCPfunction code_001.xml -a Y.Y.Y.Y -p 502", where Y.Y.Y.Y is the IP address of the device under test (DUT), the data model is Ox01.xml, the No. 1 purpose code is 001, and the modbus/tcp procedure port is 502.

Figure 4. Testbed example scenario

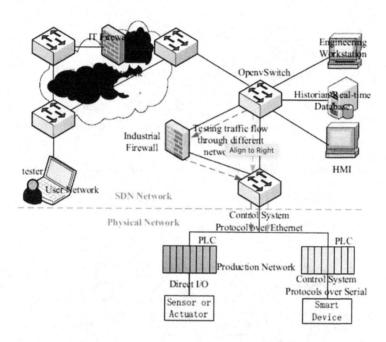

Secondly, the OpenFlow supervisor handles testing circulation flow over the red scattered line by supplying additional unique flow table to OpenvSwitch. The same fuzzing package and command in step 2 is repeated four times. This allows for testing the industrial firewall's effectiveness in preventing malicious traffic from reaching the DUT. This example demonstrates how the CPSTCS testbed can be used to test and secure controllers in cyber-physical systems. It allows for fine-grained control of testing traffic flow, and the ability to test the effectiveness of security measures such as industrial firewalls. The testbed's adaptability and flexibility enable researchers and engineers to test various scenarios and configurations, and to improve the security and performance of cyber-physical systems.

The CPSTCS testbed's comprehensive testing capabilities extend significantly into the domain of controller and end-device security within cyber-physical systems. By accommodating a variety of devices such as Programmable Logic Controllers (PLCs), Remote Terminal Units (RTUs), and Smart Transmitters, it provides a robust platform for probing the security vulnerabilities inherent to these critical components. The inclusion of both physical and virtual resources within the testbed architecture enables the simulation of real-world operational environments, thereby offering a realistic backdrop against which the resilience of controllers can be evaluated. This dual-resource approach not only enriches the testing landscape but also ensures that the testbed can replicate the complexities and nuances of actual cyber-physical systems. Consequently, it offers a more authentic and rigorous testing regime, essential for uncovering subtle vulnerabilities and security loopholes that might not be apparent in less nuanced testing environments.

Moreover, the CPSTCS testbed's support for multiple logical topologies stands out as a vital feature for security research. This capability allows for an exhaustive exploration of potential attack vectors and system configurations, thereby broadening the scope of security assessments. By enabling researchers to experiment with various network setups, the testbed aids in understanding how different cyber-physical system designs can influence vulnerability and security. This, in turn, supports the development of targeted strategies for anomaly detection and cyber-attack prevention, tailored to specific network architectures and operational paradigms. In summary, the CPSTCS testbed offers an unparalleled resource for advancing the security of cyber-physical systems. Its versatile testing environment, capable of accommodating a wide range of controllers and end devices, coupled with the ability to simulate complex, real-world scenarios, positions it as an essential tool for researchers and engineers. Through this testbed, the cyber-physical systems community is better equipped to develop sophisticated security measures, ensuring the integrity and resilience of these critical infrastructures against an ever-evolving landscape of cyber threats.

In essence, the Cyber-Physical System for Traffic Control and Safety (CPSTCS) testbed emerges as a pivotal instrument in the arena of cyber-physical systems' security and performance enhancement. Its engineered capacity for precise manipulation of testing traffic flows, alongside the innovative integration of both physical and virtual resources, positions it as a premier platform for refining the robustness and efficiency of these systems. The testbed's versatility and adaptability are its hallmark features, enabling an exhaustive exploration of myriad scenarios and configurations. This flexibility is crucial for the proactive identification and mitigation of vulnerabilities, fostering the development of advanced protective strategies for controllers and endpoint devices integral to cyber-physical systems.

The CPSTCS testbed's comprehensive testing environment facilitates a deeper understanding of how different components interact within varied logical topologies, thus providing insights into potential security breaches and performance bottlenecks. By simulating real-world conditions and allowing for the evaluation of complex network configurations, the testbed ensures that research and development efforts are both relevant and applicable to actual operational challenges faced by cyber-physical systems.

Therefore, the CPSTCS testbed stands not just as a tool but as a cornerstone for innovation and security in the cyber-physical systems domain. It empowers researchers and engineers to push the boundaries of current security practices, leading to the creation of more secure, resilient, and efficient systems. Through ongoing testing and development facilitated by the CPSTCS testbed, the future of cyber-physical systems looks both promising and secure, with enhanced capabilities to withstand the evolving landscape of cyber threats and technological advancements.

CONCLUSION

The Cyber-Physical System for Traffic Control and Safety (CPSTCS) testbed stands as a critical infrastructure for the experimental validation and security enhancement of control mechanisms within cyber-physical systems. Its design, which facilitates the manipulation of testing traffic flows alongside the integration of both tangible and simulated resources, positions it as a quintessential platform for elevating the security and operational efficacy of cyber-physical systems. The testbed's inherent versatility and adaptability empower researchers to simulate a wide array of scenarios and configurations. This capability is instrumental in forging innovative strategies for safeguarding controllers and terminal devices, which are integral components of cyber-physical systems.

The CPSTCS testbed's comprehensive framework enables the rigorous examination of controller vulnerabilities and security paradigms within cyber-physical environments. It supports the testing of a diverse portfolio of controllers and endpoint devices, including Programmable Logic Controllers (PLCs), Remote Terminal Units (RTUs), and Smart Transmitters, which are pivotal in the functioning of cyber-physical systems. Through this testbed, researchers and engineers are equipped to identify potential security flaws and to conceive robust protective measures against them.

In essence, the CPSTCS testbed serves as an invaluable asset for the cyber-physical systems community, fostering advancements in security and performance. It not only aids in the proactive identification and mitigation of security risks but also catalyzes the development of innovative solutions tailored to the unique challenges of cyber-physical systems. Consequently, the CPSTCS testbed is indispensable for those dedicated to enhancing the resilience and efficiency of cyber-physical systems, thereby contributing significantly to the field's progression.

REFERENCES

Angalaeswari, S., Jamuna, K., Mohana Sundaram, K., Ramesh, L., & Ramaswamy, K. (2022). Power-Sharing Analysis of Hybrid Microgrid Using Iterative Learning Controller (ILC) considering Source and Load Variation. *Mathematical Problems in Engineering*, *2022*, 1–6. doi:10.1155/2022/7403691

Anjankar, P., Lakade, S., Padalkar, A., Nichal, S., Devarajan, Y., Lakshmaiya, N., & Subbaiyan, N. (2023). Experimental investigation on the effect of liquid phase and vapor phase separation over performance of falling film evaporator. *Environmental Quality Management*, *33*(1), 61–69. doi:10.1002/tqem.21952

Anupama. (2021). Deep learning with backtracking search optimization based skin lesion diagnosis model. *Computers, Materials & Continua*, *70*(1). doi:10.32604/cmc.2022.018396

Arockiasamy, Muthukrishnan, M., Iyyadurai, J., Kaliappan, S., Lakshmaiya, N., Djearamane, S., Tey, L.-H., Wong, L. S., Kayarohanam, S., Obaid, S. A., Alfarraj, S., & Sivakumar, S. (2023). Tribological characterization of sponge gourd outer skin fiber-reinforced epoxy composite with Tamarindus indica seed filler addition using the Box-Behnken method. *E-Polymers*, *23*(1), 20230052. doi:10.1515/epoly-2023-0052

Asha, P., Natrayan, L., Geetha, B. T., Beulah, J. R., Sumathy, R., Varalakshmi, G., & Neelakandan, S. (2022). IoT enabled environmental toxicology for air pollution monitoring using AI techniques. *Environmental Research*, *205*, 112574. doi:10.1016/j.envres.2021.112574 PMID:34919959

Balaji. (2022). Annealed peanut shell biochar as potential reinforcement for aloe vera fiber-epoxy biocomposite: Mechanical, thermal conductivity, and dielectric properties. *Biomass Conversion and Biorefinery*. doi:10.1007/s13399-022-02650-7

Balamurugan, P., Agarwal, P., Khajuria, D., Mahapatra, D., Angalaeswari, S., Natrayan, L., & Mammo, W. D. (2023). State-Flow Control Based Multistage Constant-Current Battery Charger for Electric Two-Wheeler. *Journal of Advanced Transportation*, *2023*, 1–11. doi:10.1155/2023/4554582

Darshan, Girdhar, N., Bhojwani, R., Rastogi, K., Angalaeswari, S., Natrayan, L., & Paramasivam, P. (2022). Energy Audit of a Residential Building to Reduce Energy Cost and Carbon Footprint for Sustainable Development with Renewable Energy Sources. *Advances in Civil Engineering*, *2022*, 1–10. doi:10.1155/2022/4400874

Dhanraj, A. (2022). Appraising machine learning classifiers for discriminating rotor condition in 50W–12V operational wind turbine for maximizing wind energy production through feature extraction and selection process. *Frontiers in Energy Research, 10*, 925980. Advance online publication. doi:10.3389/fenrg.2022.925980

Divya. (2022). Analysing Analyzing the performance of combined solar photovoltaic power system with phase change material. *Energy Reports, 8*. Advance online publication. doi:10.1016/j.egyr.2022.06.109

Josphineleela, R., Jyothi, M., Kaviarasu, A., & Sharma, M. (2023a) Development of IoT based Health Monitoring System for Disables using Microcontroller. In: *Proceedings - 7th International Conference on Computing Methodologies and Communication, ICCMC 2023*. IEEE. 10.1109/ICCMC56507.2023.10084026

Josphineleela, R., Kaliappan, S., & Bhatt, U. M. (2023b) Intelligent Virtual Laboratory Development and Implementation using the RASA Framework. In: *Proceedings - 7th International Conference on Computing Methodologies and Communication, ICCMC 2023*. IEEE. 10.1109/ICCMC56507.2023.10083701

Kaliappan, S. (2023a) Checking and Supervisory System for Calculation of Industrial Constraints using Embedded System. In: *Proceedings of the 4th International Conference on Smart Electronics and Communication, ICOSEC 2023*. IEEE. 10.1109/ICOSEC58147.2023.10275952

Kaliappan, S., & Rajput, A. (2023b) Sentiment Analysis of News Headlines Based on Sentiment Lexicon and Deep Learning. In: *Proceedings of the 4th International Conference on Smart Electronics and Communication, ICOSEC 2023*. IEEE. 10.1109/ICOSEC58147.2023.10276102

Kanimozhi, G., Natrayan, L., Angalaeswari, S., & Paramasivam, P. (2022). An Effective Charger for Plug-In Hybrid Electric Vehicles (PHEV) with an Enhanced PFC Rectifier and ZVS-ZCS DC/DC High-Frequency Converter. *Journal of Advanced Transportation, 2022*, 1–14. doi:10.1155/2022/7840102

Kaushal. (2023) A Payment System for Electric Vehicles Charging and Peer-to-Peer Energy Trading. In: *7th International Conference on I-SMAC (IoT in Social, Mobile, Analytics and Cloud), I-SMAC 2023 – Proceedings*. IEEE. 10.1109/I-SMAC58438.2023.10290505

Kumar, S. (2022). IoT battery management system in electric vehicle based on LR parameter estimation and ORMeshNet gateway topology. *Sustainable Energy Technologies and Assessments, 53*, 102696. doi:10.1016/j.seta.2022.102696

Lakshmaiya, N., Ganesan, V., Paramasivam, P., & Dhanasekaran, S. (2022). Influence of Biosynthesized Nanoparticles Addition and Fibre Content on the Mechanical and Moisture Absorption Behaviour of Natural Fibre Composite. *Applied Sciences (Basel, Switzerland), 12*(24), 13030. doi:10.3390/app122413030

Lakshmaiya, N., Surakasi, R., Nadh, V. S., Srinivas, C., Kaliappan, S., Ganesan, V., Paramasivam, P., & Dhanasekaran, S. (2023). Tanning Wastewater Sterilization in the Dark and Sunlight Using Psidium guajava Leaf-Derived Copper Oxide Nanoparticles and Their Characteristics. *ACS Omega, 8*(42), 39680–39689. doi:10.1021/acsomega.3c05588 PMID:37901496

Loganathan, Ramachandran, V., Perumal, A. S., Dhanasekaran, S., Lakshmaiya, N., & Paramasivam, P. (2023). Framework of Transactive Energy Market Strategies for Lucrative Peer-to-Peer Energy Transactions. *Energies, 16*(1), 6. doi:10.3390/en16010006

Mahesha, C. R., Rani, G. J., Dattu, V. S. N. C. H., Rao, Y. K. S. S., Madhusudhanan, J., L, N., Sekhar, S. C., & Sathyamurthy, R. (2022). Optimization of transesterification production of biodiesel from Pithecellobium dulce seed oil. *Energy Reports*, 8, 489–497. doi:10.1016/j.egyr.2022.10.228

Muralidaran, Natrayan, L., Kaliappan, S., & Patil, P. P. (2023). Grape stalk cellulose toughened plain weaved bamboo fiber-reinforced epoxy composite: Load bearing and time-dependent behavior. *Biomass Conversion and Biorefinery*. doi:10.1007/s13399-022-03702-8

Nadh, V. S., Krishna, C., Natrayan, L., Kumar, K. M., Nitesh, K. J. N. S., Raja, G. B., & Paramasivam, P. (2021). Structural Behavior of Nanocoated Oil Palm Shell as Coarse Aggregate in Lightweight Concrete. *Journal of Nanomaterials*, *2021*, 1–7. doi:10.1155/2021/4741296

Nagajothi, S., Elavenil, S., Angalaeswari, S., Natrayan, L., & Mammo, W. D. (2022a). Durability Studies on Fly Ash Based Geopolymer Concrete Incorporated with Slag and Alkali Solutions. *Advances in Civil Engineering*, *2022*, 1–13. Advance online publication. doi:10.1155/2022/7196446

Nagajothi, S., Elavenil, S., Angalaeswari, S., Natrayan, L., & Paramasivam, P. (2022b). Cracking Behaviour of Alkali-Activated Aluminosilicate Beams Reinforced with Glass and Basalt Fibre-Reinforced Polymer Bars under Cyclic Load. *International Journal of Polymer Science*, *2022*, 1–13. Advance online publication. doi:10.1155/2022/6762449

Nagarajan, Rajagopalan, A., Angalaeswari, S., Natrayan, L., & Mammo, W. D. (2022). Combined Economic Emission Dispatch of Microgrid with the Incorporation of Renewable Energy Sources Using Improved Mayfly Optimization Algorithm. *Computational Intelligence and Neuroscience*, *2022*, 1–22. doi:10.1155/2022/6461690 PMID:35479598

Natrayan, L., & Kaliappan, S. (2023) Mechanical Assessment of Carbon-Luffa Hybrid Composites for Automotive Applications. In: SAE Technical Papers. doi:10.4271/2023-01-5070

Natrayan, L., Kaliappan, S., & Pundir, S. (2023a) Control and Monitoring of a Quadcopter in Border Areas Using Embedded System. In: *Proceedings of the 4th International Conference on Smart Electronics and Communication, ICOSEC 2023*. IEEE. 10.1109/ICOSEC58147.2023.10276196

Natrayan, L., Kaliappan, S., Saravanan, A., Vickram, A. S., Pravin, P., Abbas, M., Ahamed Saleel, C., Alwetaishi, M., & Saleem, M. S. M. (2023b). Recyclability and catalytic characteristics of copper oxide nanoparticles derived from bougainvillea plant flower extract for biomedical application. *Green Processing and Synthesis*, *12*(1), 20230030. doi:10.1515/gps-2023-0030

Niveditha, V. R., & Rajakumar, P. S. (2020). Pervasive computing in the context of COVID-19 prediction with AI-based algorithms. *International Journal of Pervasive Computing and Communications*, *16*(5). doi:10.1108/IJPCC-07-2020-0082

Pragadish, N., Kaliappan, S., Subramanian, M., Natrayan, L., Satish Prakash, K., Subbiah, R., & Kumar, T. C. A. (2023). Optimization of cardanol oil dielectric-activated EDM process parameters in machining of silicon steel. *Biomass Conversion and Biorefinery*, *13*(15), 14087–14096. doi:10.1007/s13399-021-02268-1

Rajagopalan, Nagarajan, K., Montoya, O. D., Dhanasekaran, S., Kareem, I. A., Perumal, A. S., Lakshmaiya, N., & Paramasivam, P. (2022). Multi-Objective Optimal Scheduling of a Microgrid Using Oppositional Gradient-Based Grey Wolf Optimizer. *Energies*, *15*(23), 9024. doi:10.3390/en15239024

Ramachandran, Perumal, A. S., Lakshmaiya, N., Paramasivam, P., & Dhanasekaran, S. (2022). Unified Power Control of Permanent Magnet Synchronous Generator Based Wind Power System with Ancillary Support during Grid Faults. *Energies*, *15*(19), 7385. Advance online publication. doi:10.3390/en15197385

Ramaswamy. (2022b). Pear cactus fiber with onion sheath biocarbon nanosheet toughened epoxy composite: Mechanical, thermal, and electrical properties. *Biomass Conversion and Biorefinery*. Advance online publication. doi:10.1007/s13399-022-03335-x

Ramaswamy, R., Gurupranes, S. V., Kaliappan, S., Natrayan, L., & Patil, P. P. (2022a). Characterization of prickly pear short fiber and red onion peel biocarbon nanosheets toughened epoxy composites. *Polymer Composites*, *43*(8), 4899–4908. doi:10.1002/pc.26735

Reddy. (2023) Development of Programmed Autonomous Electric Heavy Vehicle: An Application of IoT. In: *Proceedings of the 2023 2nd International Conference on Electronics and Renewable Systems, ICEARS 2023*. IEEE. 10.1109/ICEARS56392.2023.10085492

Saadh, Almoyad, M. A. A., Arellano, M. T. C., Maaliw, R. R. III, Castillo-Acobo, R. Y., Jalal, S. S., Gandla, K., Obaid, M., Abdulwahed, A. J., Ibrahem, A. A., Sârbu, I., Juyal, A., Lakshmaiya, N., & Akhavan-Sigari, R. (2023a). Long non-coding RNAs: Controversial roles in drug resistance of solid tumors mediated by autophagy. *Cancer Chemotherapy and Pharmacology*, *92*(6), 439–453. doi:10.1007/s00280-023-04582-z PMID:37768333

Saadh, Baher, H., Li, Y., chaitanya, M., Arias-Gonzáles, J. L., Allela, O. Q. B., Mahdi, M. H., Carlos Cotrina-Aliaga, J., Lakshmaiya, N., Ahjel, S., Amin, A. H., Gilmer Rosales Rojas, G., Ameen, F., Ahsan, M., & Akhavan-Sigari, R. (2023b). The bioengineered and multifunctional nanoparticles in pancreatic cancer therapy: Bioresponisive nanostructures, phototherapy and targeted drug delivery. *Environmental Research*, *233*, 116490. doi:10.1016/j.envres.2023.116490 PMID:37354932

Saadh, Rasulova, I., Almoyad, M. A. A., Kiasari, B. A., Ali, R. T., Rasheed, T., Faisal, A., Hussain, F., Jawad, M. J., Hani, T., Sârbu, I., Lakshmaiya, N., & Ciongradi, C. I. (2024). Recent progress and the emerging role of lncRNAs in cancer drug resistance; focusing on signaling pathways. *Pathology, Research and Practice*, *253*, 154999. doi:10.1016/j.prp.2023.154999 PMID:38118218

Sabarinathan, P., Annamalai, V. E., Vishal, K., Nitin, M. S., Natrayan, L., Veeeman, D., & Mammo, W. D. (2022). Experimental study on removal of phenol formaldehyde resin coating from the abrasive disc and preparation of abrasive disc for polishing application. *Advances in Materials Science and Engineering*, *2022*, 1–8. doi:10.1155/2022/6123160

Saravanan, K. G., Kaliappan, S., Natrayan, L., & Patil, P. P. (2023). Effect of cassava tuber nanocellulose and satin weaved bamboo fiber addition on mechanical, wear, hydrophobic, and thermal behavior of unsaturated polyester resin composites. *Biomass Conversion and Biorefinery*. Advance online publication. doi:10.1007/s13399-023-04495-0

Seeniappan. (2023). *Modelling and development of energy systems through cyber physical systems with optimising interconnected with control and sensing parameters.*

Selvi. (2023) Optimization of Solar Panel Orientation for Maximum Energy Efficiency. In: *Proceedings of the 4th International Conference on Smart Electronics and Communication, ICOSEC 2023.* IEEE. 10.1109/ICOSEC58147.2023.10276287

Sendrayaperumal, Mahapatra, S., Parida, S. S., Surana, K., Balamurugan, P., Natrayan, L., & Paramasivam, P. (2021). Energy Auditing for Efficient Planning and Implementation in Commercial and Residential Buildings. *Advances in Civil Engineering, 2021,* 1–10. doi:10.1155/2021/1908568

Seralathan, S., Chenna Reddy, G., Sathish, S., Muthuram, A., Dhanraj, J. A., Lakshmaiya, N., Velmurugan, K., Sirisamphanwong, C., Ngoenmeesri, R., & Sirisamphanwong, C. (2023). Performance and exergy analysis of an inclined solar still with baffle arrangements. *Heliyon, 9*(4), e14807. doi:10.1016/j.heliyon.2023.e14807 PMID:37077675

Sharma, Raffik, R., Chaturvedi, A., Geeitha, S., Akram, P. S., L, N., Mohanavel, V., Sudhakar, M., & Sathyamurthy, R. (2022). Designing and implementing a smart transplanting framework using programmable logic controller and photoelectric sensor. *Energy Reports, 8,* 430–444. doi:10.1016/j.egyr.2022.07.019

Sivakumar, V., Kaliappan, S., Natrayan, L., & Patil, P. P. (2023). Effects of Silane-Treated High-Content Cellulose Okra Fibre and Tamarind Kernel Powder on Mechanical, Thermal Stability and Water Absorption Behaviour of Epoxy Composites. *Silicon, 15*(10), 4439–4447. Advance online publication. doi:10.1007/s12633-023-02370-1

Subramanian, Lakshmaiya, N., Ramasamy, D., & Devarajan, Y. (2022). Detailed analysis on engine operating in dual fuel mode with different energy fractions of sustainable HHO gas. *Environmental Progress & Sustainable Energy, 41*(5), e13850. doi:10.1002/ep.13850

Suman. (2023) IoT based Social Device Network with Cloud Computing Architecture. In: *Proceedings of the 2023 2nd International Conference on Electronics and Renewable Systems, ICEARS 2023.* IEEE. 10.1109/ICEARS56392.2023.10085574

Thakre, Pandhare, A., Malwe, P. D., Gupta, N., Kothare, C., Magade, P. B., Patel, A., Meena, R. S., Veza, I., Natrayan L, & Panchal, H. (2023). Heat transfer and pressure drop analysis of a microchannel heat sink using nanofluids for energy applications. *Kerntechnik, 88*(5), 543–555. doi:10.1515/kern-2023-0034

Ugle, Arulprakasajothi, M., Padmanabhan, S., Devarajan, Y., Lakshmaiya, N., & Subbaiyan, N. (2023). Investigation of heat transport characteristics of titanium dioxide nanofluids with corrugated tube. *Environmental Quality Management, 33*(2), 127–138. doi:10.1002/tqem.21999

Vijayaragavan, Subramanian, B., Sudhakar, S., & Natrayan, L. (2022). Effect of induction on exhaust gas recirculation and hydrogen gas in compression ignition engine with simarouba oil in dual fuel mode. *International Journal of Hydrogen Energy, 47*(88), 37635–37647. doi:10.1016/j.ijhydene.2021.11.201

Viswanathan, C. (2023). Deep Learning for Enhanced Fault Diagnosis of Monoblock Centrifugal Pumps: Spectrogram-Based Analysis. *Machines, 11*(9), 874. doi:10.3390/machines11090874

Section 3
Data Analytics for Energy Management

Chapter 13
Improved Performance of Active Suspension System Using COA Optimized FLC for Full Car With Driver Model

S. Fahira Haseen

College of Engineering Guindy, Anna University, India

P. Lakshmi

College of Engineering Guindy, Anna University, India

ABSTRACT

Suspension in a vehicle is provided primarily to improve the passenger comfort and vehicle handling for the automobiles moving under any road conditions. Because of the non-linear characteristics of the vehicle, fuzzy logic controller (FLC) fed active suspension system is proposed for a full car with driver vehicle model. This controller dynamics are optimized by meta-heuristic optimization algorithm namely big bang–big crunch (BBBC) optimization and coyote optimization algorithm (COA). The passive system dynamics are compared with controller fed and optimized controller fed system under bump and random road inputs. The passive and active model is simulated in MATLAB/Simulink environment. The results are compared based upon root mean square values of head acceleration, body acceleration, pitch acceleration, roll acceleration, and power spectrum density of head acceleration. The results indicate that implementation of COA optimized FLC is effective in improving ride quality and road handling of the vehicle.

INTRODUCTION

Suspension is classified into three types based on how it works: Active Suspension Systems (ASS), Semi-Active Suspension Systems (SASS), and Passive Suspension Systems (PSS). The PSS is the notion that no external force is thrust into a vehicle suspension in order to counteract road disturbances (Tam-

DOI: 10.4018/979-8-3693-3735-6.ch013

boli, J. A., & Joshi, S. G. (1999), Mihai, I., & Andronic, F. (2014)). The principle of using adjustable suspensions, which work on the dynamic changing of the viscosity of the fluid using electro-rheological or magneto-rheological liquids, forms the SASS (Kumar, J., & Bhushan, G. (2023), Ajayi, A. B. et.al., (2023)). The ASS works on the principle of applying an actuator force using controllers, where the controllers are given input from the motion sensors present in the vehicle, which detect road abnormalities (Nguyen, D. N., & Nguyen, T. A. (2023)). Among these suspension systems, ASS plays a major role and has more advantages over the PSS and SASS.

Initially, all the study and implementation of controllers for ASS were tested for 2-Degree-of- Freedom (DoF) vehicle model and then moved to the 4-DoF vehicle model (Ozcan, D. et.al., (2023), Türkay, S., & Akçay, H. (2005), Marzbanrad, J. et.al., (2013)). For real-time implementation of the controller designs for ASS, they must be tested for full car model (Kim, C., & Ro, P. I. (2002)). The controller design for Full car model with a seat that has 8 DOF has been implemented (Yuvapriya, T. et.al., (2023)). But for an accurate analysis, driver dynamics must be included in the study of ASS. Hence, 13 DOF, Full car with Driver (FCD) ASS, is taken as the model for this study (Du, H et.al., (2023)).

The controller designs for ASS are implemented employing Linear Quadratic Regulator (LQR) and Proportional Integral Derivative (PID) controllers (Zhu, S., & He, Y. (2023), Ma, S. et.al., (2023)). However, when these controllers are deployed in systems that have complex dynamics, they tend to perform poorer than other controllers. A model based predictive controller is designed for Electric Vehicle (EV) to control and ensure anti-roll stability and lateral stability (Jing, C. et.al., (2023)). However, the execution time required for MPC is very large compared to other controllers. This is a major drawback of MPC (Ekaputri, C., & Syaichu-Rohman, A. (2013)). Hybrid controllers like Fuzzy-based PID controllers are evaluated for controlling highly non-linear ASS (Kumar, V., & Rana, K. P. S. (2023), Yuvapriya, T., & Lakshmi, P. (2017)). ASS with Fuzzy Logic Controller (FLC) provides enhanced ride comfort (Bingül, Ö., & Yıldız, A. (2023)).

The operation of a controller is currently tuned by employing meta-heuristic algorithms, which play a substantial part in improving the effectiveness of the system. The Genetic Algorithm (GA) and Particle Swarm Optimizer (PSO) techniques are well-known for optimising Fuzzy PID controller (Ji, G., et.al., (2023)) and Fuzzy LQR controller (Abut, T., & Salkim, E. (2023)) for Quarter vehicle ASS. Optimization techniques like Ant Colony Optimization (Manna, S., et.al., (2022)), Artificial Bee Colony Optimization (Abdul Zahra, A. K., & Abdalla, T. Y. (2021)), Differential Evolution based Biogeography based Optimization (Rajagopal, K., & Ponnusamy, L. (2015)), Firefly Optimization (Dif, I., & Dif, N. (2023)), Bat and Grey Wolf Optimization (Yuvapriya, T. et.al., (2022)) etc have been implemented for ASS. Hence application of these techniques has resulted in better results than un-optimized controllers. Big Bang Big Crunch (BBBC) optimization technique is developed based on the natural evolution of planets and the universe (Erol, O. K., & Eksin, I. (2006)). This technique has been implemented to optimize the FLC for 8 DOF vehicle model (Yuvapriya, T. et.al., (2023)). Hence, this evolutionary algorithm can be applied to optimize FLC for the 13 DOF model of ASS. But BBBC has a major drawback of inconsistency in balancing the two phases of the algorithm i.e., the expansion phase and contraction phase. The Coyote Optimization Algorithm (COA) is a technique which has balanced exploration and exploitation rate (Pierezan, J., & Coelho, L. D. S. (2018)). This algorithm is hence tested for optimizing FLC parameters, and its performance in ASS is validated.

The novelty of this work is the implementation of BBBC optimized FLC (B-FLC) and COA optimized FLC (C-FLC) for a four-wheeled vehicle with human driver ASS. This model has 13-DoF. Based on simulation analysis, the optimised controllers are compared to one another.

The main contribution of this paper includes the following:

- The dynamics and mathematical modelling of Full car with driver model which is used in the problem are studied thoroughly using differential equations. These equations are linearized for this study so that, the dynamics of the system can be reduced.
- Implementation of FLC and its components for this problem of ASS are discussed.
- The workings of BBBC and COA techniques is discussed. These algorithms are implemented to optimize the FLC ranges of inputs and outputs for enhanced performance of the controller for the ASS.
- The optimized FLCs, i.e., B-FLC and C-FLC, are implemented for the ASS using Matlab simulation. The results are compiled and presented, which prove that an optimized FLC-fed system performs better than unoptimized FLC and PSS.
- Furthermore, the two optimisation algorithms are compared in terms of ASS performance specifications such as ISO 2631-1 (Duarte, M. L. M. et.al., (2018)) based ride comfort analysis, time domain response of driver head acceleration, vehicle pitch and roll acceleration, Power Spectrum Density (PSD), Vibration Dose Values (VDV), Root Mean Square (RMS) values, and Frequency Weighted RMS (FWRMS) values. These findings indicate that C-FLC outperforms B-FLC.

The paper is divided into the sections listed below. Section 2 is dedicated to vehicle design and modelling. The full car with driver dynamics are studied in this section. Section 3 gives an elaborate detail on the design of FLC and its implementation. The FLC is optimized further using BBBC and COA techniques, and these techniques are discussed in Section 4. Finally, the optimised controllers are simulated for the ASS, and the results are reported and expounded on in Section 5. Section 6 summarises the discussions and outcomes.

MATHEMATICAL MODELLING OF VEHICLE MODEL

This is a four-wheel vehicle structure. This model is considered to have seat suspension and driver (Du, H. et.al., (2013)). The vehicle modelled in the form of a mass, spring, and damper system. The vehicle body, front and rear two-wheel masses is considered as m, m_1 and m_2 respectively and is shown in Fig. [1]. The vehicle body are free for motion along vertical (y), roll (φ) and pitch (θ) direction. Whereas the wheel motion is considered only along vertical direction. The motion of vehicle body in vertical direction is given by y_{sfl}, y_{sfr}, y_{srl} and y_{srr}. The vertical motion of front left and right wheel is given as y_{ufl} and y_{ufr}. Similarly, the rear left and right wheel is y_{url} and y_{urr}. The DOF of full car model is 7 which includes the vehicle body motion, pitch motion, roll motion and four un-sprung mass vertical motion. A seat with frame and cushion is placed in the driver position in the vehicle. The mass of the seat frame is m_3 and cushion is m_4. The displacement of seat frame and cushion is given by y_1 and y_2 vertically. A human model is placed in the seat as a driver sitting in his position. The human model is used from the study obtained in (Wang, Y. et.al.,(2023)). The thighs, lower torso, upper torso and head constitutes the human model. The dynamic equations obtained are given below. Further the movement of these parts are given as y_3, y_4, y_5 and y_6. The mass, stiffness and damping coefficients implemented for the system is given in Table [1].

Figure 1. Vehicle structure and driver body

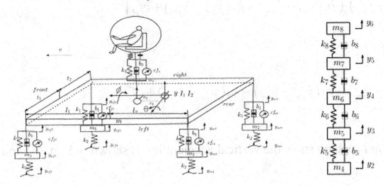

The vehicle body acceleration in vertical direction is given below in Eqn. (1)

$$\ddot{y} = \frac{-1}{m}[k_1(y_{sfl} + y_{sfr} - y_{ufl} - y_{ufr}) + b_1(\dot{y}_{sfl} + \dot{y}_{sfr} - \dot{y}_{ufl} - \dot{y}_{ufr}) + k_1(y_{srl} + y_{srr} - y_{url} - y_{urr})$$
$$+ b_2(\dot{y}_{srl} + \dot{y}_{srr} - \dot{y}_{url} - \dot{y}_{urr}) - k_3(y_1 - y_s) - b_3(\dot{y}_1 - \dot{y}_s) + cf_{fl} + cf_{fr} + cf_{rl} + cf_{rr} - cf_s]$$

(1)

The pitch acceleration experienced by vehicle body due to braking is shown in Eqn. (2)

$$\ddot{\theta} = \frac{1}{I_1}[l_1 k_1(y_{sfl} + y_{sfr} - y_{ufl} - y_{ufr}) + l_1 b_1(\dot{y}_{sfl} + \dot{y}_{sfr} - \dot{y}_{ufl} - \dot{y}_{ufr}) - l_2 k_1(y_{srl} + y_{srr} - y_{url} - y_{urr})$$
$$- l_2 b_2(\dot{y}_{srl} + \dot{y}_{srr} - \dot{y}_{url} - \dot{y}_{urr}) - r_1 k_3(y_1 - y_s) - r_1 b_3(\dot{y}_1 - \dot{y}_s) + l_1 cf_{fl} + l_1 cf_{fr} - l_2 cf_{rl} - l_2 cf_{rr} - r_1 cf_s]$$

(2)

The roll acceleration of the vehicle body due to cornering in a road is given in Eqn. (3)

$$\ddot{\phi} = \frac{1}{I_2}[-t_1 k_1(y_{sfl} - y_{ufl} - y_{sfr} + y_{ufr}) - t_1 b_1(\dot{y}_{sfl} - \dot{y}_{ufl} - \dot{y}_{sfr} + \dot{y}_{ufr}) - t_2 k_1(y_{srl} - y_{url} - y_{srr} + y_{urr})$$
$$- t_2 b_2(\dot{y}_{srl} - \dot{y}_{url} - \dot{y}_{srr} + \dot{y}_{urr}) + r_2 k_3(y_1 - y_s) + r_2 b_3(\dot{y}_1 - \dot{y}_s) - t_1 cf_{fl} + t_1 cf_{fr} - t_2 cf_{rl} + t_2 cf_{rr} + t_2 cf_s]$$

(3)

Eqn. (4)-(7) gives the vertical acceleration of vehicle wheel.

$$\ddot{y}_{ufl} = \frac{1}{m_1}[k_1(y_{sfl} - y_{ufl}) + b_1(\dot{y}_{sfl} - \dot{y}_{ufl}) - k_2(y_{ufl} - y_{rfl}) + cf_{fl}]$$

(4)

$$\ddot{y}_{ufr} = \frac{1}{m_1}[k_1(y_{sfr} - y_{ufr}) + b_1(\dot{y}_{sfr} - \dot{y}_{ufr}) - k_2(y_{ufr} - y_{rfr}) + cf_{fr}]$$

(5)

$$\ddot{y}_{url} = \frac{1}{m_2}[k_1(y_{srl} - y_{url}) + b_2(\dot{y}_{srl} - \dot{y}_{url}) - k_2(y_{url} - y_{rrl}) + cf_{rl}] \tag{6}$$

$$\ddot{y}_{urr} = \frac{1}{m_2}[k_1(y_{srr} - y_{urr}) + b_2(\dot{y}_{srr} - \dot{y}_{urr}) - k_2(y_{urr} - y_{rrr}) + cf_{rr}] \tag{7}$$

The seat frame and cushion show the vertical acceleration is shown in Eqn. (8) and Eqn. (9)

$$\ddot{y}_1 = \frac{1}{m_3}[k_4(y_2 - y_1) + b_4(\dot{y}_2 - \dot{y}_1) - k_s(y_1 - y_s) - b_s(\dot{y}_1 - \dot{y}_s) - cf_s] \tag{8}$$

$$\ddot{y}_2 = \frac{1}{m_4}[-k_4(y_2 - y_1) - b_4(\dot{y}_2 - \dot{y}_1) + k_5(y_3 - y_2) + b_5(\dot{y}_3 - \dot{y}_2)] \tag{9}$$

The Eqn (10) to Eqn. (13) shows the movement of driver body masses in vertical direction.

$$\ddot{y}_3 = \frac{1}{m_5}[-k_5(y_3 - y_2) - b_5(\dot{y}_3 - \dot{y}_2) + k_6(y_4 - y_3) + b_6(\dot{y}_4 - \dot{y}_3)] \tag{10}$$

$$\ddot{y}_4 = \frac{1}{m_6}[-k_6(x_4 - x_3) - b_6(\dot{y}_4 - \dot{y}_3) + k_7(y_5 - y_4) + b_7(\dot{y}_5 - \dot{y}_4)] \tag{11}$$

$$\ddot{y}_5 = \frac{1}{m_7}[-k_7(y_5 - y_4) - b_7(\dot{y}_5 - \dot{x}_4) + k_8(y_6 - y_5) + b_8(\dot{y}_6 - \dot{x}_5)] \tag{12}$$

$$\ddot{y}_6 = \frac{1}{m_8}[-k_8(x_6 - x_5) - b_8(\dot{x}_6 - \dot{x}_5)] \tag{13}$$

where

$$y_s = y + r_2\phi - r_1\theta \tag{14}$$

$$x_{sft} = y + t_1\phi - l_1\theta \tag{15}$$

Table 1. Constant parameters of the system

PARAMETERS	VALUES
Vehicle body mass (m)	1460kg
Front un-sprung masses (m_1)	40kg
Rear un-sprung masses (m_2)	35.5kg
Mass of seat frame (m_3)	15kg
Cushion mass (m_4)	1kg
Driver thighs mass (m_5)	12.78kg
Driver lower body mass (m_6)	8.62kg
Driver upper body mass (m_7)	28.49kg
Driver head mass (m_8)	5.31kg
Suspension stiffness (k_1)	20580N/m
Suspension damping of front (b_1)	946 Ns/m
Tyre stiffness (k_2)	175500N/m
Suspension damping of rear (b_2)	946 Nm/s
Seat stiffness (k_3)	31000N/m
Seat frame damping coefficient (b_3)	830 Ns/m
Cushion stiffness (k_4)	18000N/m
Coefficient of cushion damping (b_4)	200 Ns/m
Thigh stiffness (k_5)	90000N/m
Thigh damping coefficient (b_5)	2064 Ns/m
Lower torso stiffness (k_6)	162800N/m
Lower torso damping coefficient (b_6)	4585 Ns/m
Upper torso stiffness (k_7)	183000N/m
Upper torso damping coefficient (b_7)	4750 Ns/m
Head stiffness (k_8)	310000N/m
Head damping coefficient (b_8)	400 Ns/m
Inertia along pitch angle (I_1)	2460 kgm^2
Inertia along roll angle (I_2)	460 kgm^2
Length of front un-sprung masses from COG (l_1)	1.011 m
Length of left un-sprung masses from COG (t_1)	0.761 m
Distance between driver seat from COG (θ) (r_1)	0.3 m
Length of rear un-sprung masses from COG (l_2)	1.803 m
Length of right un-sprung masses from COG (t_2)	0.755 m
Distance between driver seat from COG (φ) (r_2)	0.25 m

$$x_{sfr} = y - t_1\phi - l_1\theta \tag{16}$$

$$y_{srl} = y + t_2\phi + l_2\theta \tag{17}$$

$$y_{srr} = y - t_2\phi + l_2\theta \tag{18}$$

FUZZY LOGIC CONTROLLER DESIGN

The main purpose of suspension in a vehicle is bringing in passenger comfort and vehicle safety. Fuzzy logic controllers offer the ability to control system uncertainties, which is important in their application in ASS. The fuzzy design implementation is cost-effective when compared to complex controllers such as H-infinity-based state and output feedback controllers (Masoud, U. et.al., (2023), Ab Talib, M. H. (2023)). The FLC has the following segments in its design and implementation, which are discussed below.

Defining the Fuzzy Variables

The Fuzzy Inference System (FIS) system considered for this ASS to control vehicle vibration and vehicle safety is a two-input and single-output FIS. The inputs of this system are considered suspension deflection and suspension velocity. This corresponds to error and derivative of error respectively. The output is considered as the actuator force given as input to the actuator, which in turn acts as resistance to the road disturbance. Triangular membership functions are considered for both inputs and outputs of FLC because of its easy implementation and linear nature. Each input and output are considered to have seven membership functions, namely negative large (β), negative medium (γ), negative small (ε), zero (ζ), positive small (η), positive medium (κ) and positive large (ν) (Yuvapriya, T., & Lakshmi, P. (2017)). The range of the inputs are also pivotsl in fuzzy implementation. The suspension deflection and suspension velocity are taken in the range of [-2.89 2.89] and [-400 400] respectively. The output actuator force is ranged between [-30000 30000]. These ranges play an important role in determining the FLC working. Hence, this parameter can be optimized for obtaining better FLC performance.

Determining Rule Base

The fuzzy rule base provides a logical rationale for the controller's actions. The fuzzy rules used in this system are recorded in the tabular column in Table [2]. The controller works only if the rules are framed with logical justification and are effective in nature. For example, when the suspension velocity is 'β' and the suspension deflection is 'ν', the spring deflection or expansion is very small. This looks like it is closed. Hence, an external force is not necessary for action, the actuator conserves its energy for the future. Similarly, all the firing rules are framed with proper justifications.

Table 2. Fuzzy firing rules

		Suspension Velocity						
		β	γ	ε	ζ	η	κ	ν
Suspension Deflection	ν	ζ	η	κ	ν	ν	ν	ν
	κ	ε	ζ	η	κ	ν	ν	ν
	η	γ	ε	ζ	η	κ	ν	ν
	ζ	β	γ	ε	ζ	η	κ	ν
	ε	β	β	γ	ε	ζ	η	κ
	γ	β	β	β	γ	ε	ζ	η
	β	β	β	β	β	γ	ε	ζ

Scaling Factors

Scaling factors are important in FLC building. The scaling factors are a vital part of the design because they help in amplifying or reducing the inputs and outputs. The faulty scaling factors may sometimes lead to saturation or the rules never being fired. This may lead to un-optimal working of the FLC. The input scaling factors for suspension deflection and suspension velocity are 0.75 and 2 respectively. The output scaling factor is 1. This is obtained through trial and error.

OPTIMIZATION TECHNIQUES

Big Bang Big Crunch (BBBC) Technique

The evolutionary algorithm BBBC has two segments to its approach. The initial segment is called the Big Bang. This is based on the concept of dissipation of energy into nature in randomness. This is followed by convergence to a point by the gravitational force of attraction, which may lead to a global or local optimum point. This segment is called Big Crunch (Erol, O. K., & Eksin, I. (2006)). Nature has always been an inspiration for developing several metaheuristic algorithms such as GA, Simulated Annealing (SA) (Mitra, A. C. et.al., (2016), Shinde, D. et.al., (2018)) etc. But this algorithm has shown improved performance than classical GA and improved GA for many benchmark test functions. For the ASS, the designed FLC is further enhanced by optimizing the range of the inputs and outputs using the BBBC algorithm. The algorithm for conducting this optimisation procedure is as follows:

Algorithm

(Big bang segment)

Step 1: The initial generation's 'n' possibilities are created at random in the search region.
Step 2: This ASS problem's fitness function is generated, and the fitness functions of all candidates are evaluated.

(Big crunch segment)

Step 3: The convergence operator has been employed, and the center of mass is either the best fit individual or the center of mass. The Eqn. (19) is used to determine the center of mass.

$$z^C = \frac{\sum_{j=1}^{n} \frac{1}{f_j} z_i}{\sum_{j=1}^{n} \frac{1}{f_j}} \quad j=1,\dots,D \tag{19}$$

where n is the population size, f_j is the fitness function value of the jth candidate, z_i is the position of the candidate and z^c represents the position of the center of mass.

Step 4: New candidates are calculated around the new point calculated in Step 3 by addition or deletion of a random number whose value decreases as the iterations elapse, which is evaluated as

$$z_{new}^i = z^c + \frac{L_r}{k} \tag{20}$$

where z_{new}^j is the new candidate with upper and lower bounded, L_r is the upper limit of the parameter, r is the random number and k is the iteration step.

Step 5: The fitness function value is calculated for the new candidates and the steps are repeated from step 3. The cycle must be repeated until the stopping criteria is achieved or maximum iterations is finished.

Step 6: Rank all of the 'f' solutions acquired from all iterations in ascending order. The candidate with the lowest fitness value provides the best ideal solution.

Table 3. Optimization parameters of BBBC

Parameters	Values
Population size in numbers (n)	100
Iterations	100
Number of trials	30
Reduction Rate (alpha)	0.01
Dimension	3

Coyote Optimization Technique

The social conditioning and structure of species of coyotes named *Canis latrans* form the basis for the development of COA. This is very much different in its approach when compared to GWO which is developed from the dominance and hierarchical nature of *Canis lupus* (Pierezan, J., & Coelho, L. D. S. (2018)). This approach provides a solution for maintaining a balance between the exploration and exploitation stages of the algorithm (Arfaoui, J. et.al., (2020), Tabak, A., & Duman, S. (2022)). This is a major advantage of this technique over BBBC. This algorithm is employed in enhancing the performance of FLC in ASS and it is compared with the BBBC. The steps involved in this algorithm are formulated below.

Algorithm

Step 1: N_p and N_c represent the number of packs and coyotes in each. The initial coyote population is created, and the social conditions for each coyote are picked at random. This is performed by allocating values in the search space for the cth coyote of the pth pack of the jth dimension at random, as follows:

$$s_{c,j}^{p,t} = (U_j - L_j) * r_j + L_j \tag{21}$$

where U_j and L_j gives the upper and lower limits of the jth decision variable respectively, D is the dimension of the search space and r_j is a real random number generated in the range of [0 1] using uniform probability.

Step 2: The adaption of the coyote in a variety of present societal situations is generated via

$$f_c^{p,t} = f(s_c^{p,t}) \tag{22}$$

Step 3: In some cases, coyotes are ejected from the pack based on the number of coyotes in the pack, with the likelihood given by,

$$P = N_c^2 * 0.005 \tag{23}$$

Step 4: The pack's alpha 'A' coyote is defined as

$$A^{p,t} = \left\{ s_c^{p,t} \middle| \min f\left(s_c^{p,t}\right) \right\} \tag{24}$$

Step 5: The pack's cultural tendency contributes to the pack's survival and the sharing of social conditions with others. This is calculated as follows:

$$C_j^{p,t} = \begin{cases} O_{\frac{(N_c+1)}{2},j}^{p,t} & N_c = odd \\[2ex] \dfrac{O_{\frac{N_c}{2},j}^{p,t} + O_{(\frac{N_c}{2}+1),j}^{p,t}}{2} & otherwise \end{cases} \tag{25}$$

Step 5: The biological events of birth and death have an effect on the age of coyotes. The birth of a new coyote is written as a mix of the social situations of both parents, who are picked at random.

$$P_j^{p,t} = \begin{cases} s_{r_1,j}^{p,t} & rnd_j < P_s \\ s_{r_2,j}^{p,t} & rnd_j \geq P_s + P_a \\ R_j & otherwise \end{cases} \tag{26}$$

Here r_1 and r_2 represents random coyotes in the pack. P_s and P_a represents the association and scatter probability respectively. R_j is a random number inside the limits.

Step 6: The alpha influence (δ_1) and pack influence (δ_2) inside a pack is determined by

$$\delta_1 = A^{p,t} - s_{cr_1}^{p,t} \tag{27}$$

$$\delta_2 = C^{p,t} - s_{cr_2}^{p,t} \tag{28}$$

Step 7: The new social condition of the coyote's is updated as follows

$$s_c^{p,t,new} = r_1\delta_1 + r_2\delta_2 + s_c^{p,t,old} \tag{29}$$

Step 7: Calculate the birth and death of coyotes and the transition between the packs.

Step8: Update the coyote's ages and select the best adapted coyote using

$$new\_f_c^{p,t} = f(new\_s_c^{p,t}) \tag{30}$$

Simulation Results and Discussions

The ASS is tested for bump input and random input. The bump magnitudes are 0.08 m and 0.04 m for front and rear wheel since this helps in testing the effect of controllers in pitch direction. The automobile is travelling at an average rate of 60 km/h. Figure [2] depicts the bump figure input. The ASS's intention is to deliver optimal ride comfort while simultaneously keeping the vehicle's suspension deflection under

Table 4. Optimization parameters of COA

Parameters	Values
No. of Iterations (iter)	100
Population size	100
N_p	20
N_c	5
Dimension	3

control. Table 5 summarises the RMS values of driver head, vehicle body, roll and pitch acceleration. This table shows that reduction driver head acceleration by FLC is about 48%, B-FLC is 65% and C-FLC is 78%. The reduction of RMS of vehicle body acceleration by FLC is 37%, B-FLC is 52% and C-FLC is 60%. FLC shows 34%, B-FLC shows 51% and C-FLC shows 58% RMS reduction in pitch acceleration. Similarly, the FLC, B-FLC and C-FLC gives 30%, 43% and 49% RMS reduction in roll acceleration respectively. Table 6 exhibits the ISO 2631-1 ride comfort analysis based on FWRMS and VDV data. The PSS gives Not Uncomfortable range. Also, C-FLC provides comfort of Not Uncomfortable o range. This is may be due to lower amplitude in bump input.

The simulation results of the system under bump road disturbances are given in Fig. 3- Fig. 7. The time domain representation of driver head acceleration is shown Fig. 3. C-FLC shows maximum reduction when compared to B-FLC and FLC. Similarly, Figs. 4 and 5 depict the pitch and roll accelerations encountered by the vehicle body mass. The FLC, B-FLC and C-FLC exerts maximum of 15 N, 16 N and 18 N respectively as shown in Fig. 6. The analysis of the system in frequency domain is done in Power Spectrum Density Analysis which shows significant reduction of amplitude by C-FLC in the human sensitivity range of noise. Fig. 7 depicts it.

The vehicle speed has been increased from 30 km/h to 120 km/h to test the performance of the designed optimised FLC controller. The % reduction of driver head acceleration by the designed FLC, B-FLC and C-FLC is plotted as a bar graph in Fig. 8. Similarly, the body mass of the driver is varied

Figure 2. Bump road

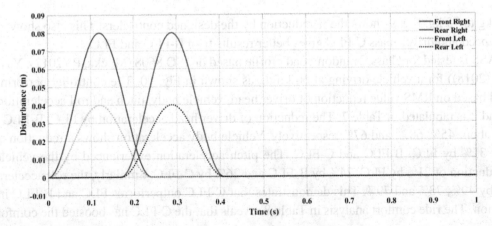

Table 5. Performance of the controllers - bump road

	Driver Head Acceleration (m/s²)	Body Acceleration (m/s²)	Pitch Acceleration (rad/s²)	Roll Acceleration (rad/s²)
PSS	0.5458	0.382	0.3144	0.2673
FLC	0.2798	0.2408	0.206	0.187
B-FLC	0.1906	0.1814	0.1522	0.1521
C-FLC	0.1564	0.1561	0.1334	0.1346

Table 6. Ride Comfort Analysis - Bump Road

	FWRMS of Head Acceleration (m/s²)	VDV oh Head Acceleration (m/s^{1.75})	Ride comfort analysis
PSS	0.1971	0.6499	Not Uncomfortable
FLC	0.0656	0.2226	Not Uncomfortable
B-FLC	0.0428	0.1322	Not Uncomfortable
C-FLC	0.0363	0.1174	Not Uncomfortable

Figure 3. Driver head acceleration under bump road disturbance

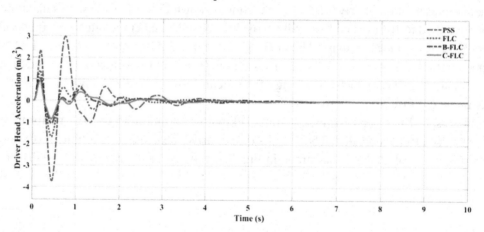

from 35 kg to 75 kg. Fig. 9 shows the % reduction by the designed controllers. This also shows that for varying speed and driver mass C-FLC gives better results than B-FLC and FLC.

The ASS is tested for Class E random road profile based on ISO 8608 (Múčka, P. (2017), Yuvapriya, T. et.al., (2018)) for a vehicle driving at 60 km/h, as shown in Fig. 10. The controller performance is validated based on RMS value reduction of driver head, vehicle body, pitch and roll acceleration of the model and it is tabulated in Table 7. The reduction of driver head acceleration by FLC, B-FLC and C-FLC are about 45%, 60%, and 67% respectively. Vehicle body acceleration shows a reduction of 24%, 27% and 31% by FLC, B-FLC and C-FLC. The pitch acceleration experienced by the vehicle body shows reduction of 21% by FLC, 24% by B-FLC and 26% by C-FLC. Similarly, the roll acceleration is reduced by 49%, 73% and 79%. This demonstrates that C-FLC outperforms FLC and B-FLC in terms of reduction. The ride comfort analysis in Table 8 reveals that the C-FLC has boosted the comfort level

Figure 4. Pitch acceleration experienced by vehicle body mass under bump road disturbance

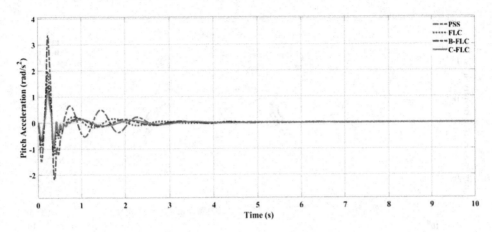

Figure 5. Roll acceleration experienced by vehicle body mass under bump road disturbance

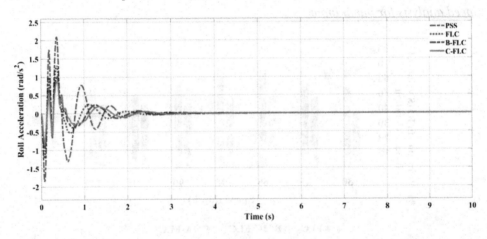

Figure 6. Control force exerted by the designed controllers for bump road disturbance

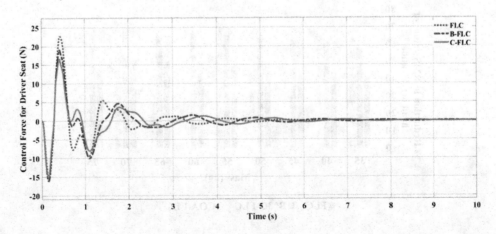

Figure 7. Power spectrum density analysis for the ASS under bump road input

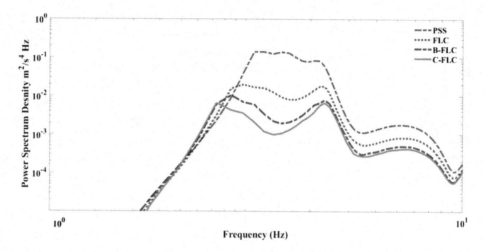

Figure 8. Speed analysis for bump input

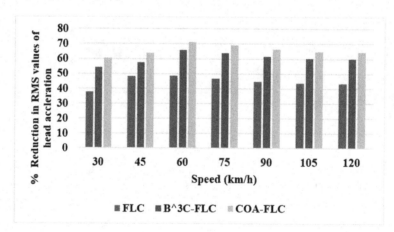

Figure 9. Driver mass analysis for bump input

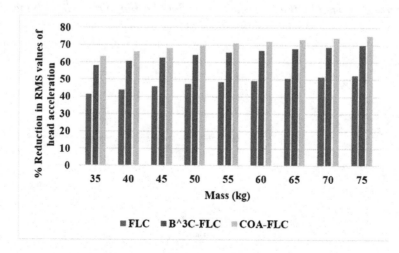

Figure 10. Class E random road disturbance

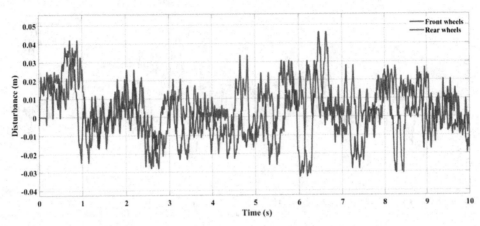

Table 7. Performance of the controllers under random road disturbance

	Driver Head Acceleration (m/s²)*10⁻¹	Sprung Body Acceleration (m/s²) *10⁻¹	Pitch Acceleration (rad/s²) *10⁻¹	Roll Acceleration (rad/s²) *10⁻²
PSS	14.32	12.57	11.01	9.75
FLC	7.907	9.605	8.669	5.01
B-FLC	5.774	9.177	8.42	2.67
C-FLC	4.676	8.702	8.276	2.028

dramatically from Very Uncomfortable to Little Uncomfortable. The VDV also displays the vibration level in the system. This is supported by time domain analysis of driver head acceleration, pitch acceleration and roll acceleration experienced by vehicle body is shown in Fig. 11 to Fig. 13. The designed controllers exert an actuator force which is helpful in vibration suppression. This is shown in Fig. 14. The power spectrum density analysis for the system under random road disturbance is given in Fig. 15. The speed and driver mass analysis for the system under random road disturbance is depicted as a bar graph in Fig. 16 and Fig. 17 respectively.

Table 8. Ride comfort analysis - random road disturbance

	FWRMS of Head Acceleration (m/s²) *10⁻¹	VDV oh Head Acceleration (m/s¹·⁷⁵) *10⁻¹	Ride comfort analysis
PSS	14.36	33.05	Very Uncomfortable
FLC	7.715	18.02	Fairly Uncomfortable
B-FLC	5.359	12.38	Little Uncomfortable
C-FLC	4.227	9.747	Little Uncomfortable

Figure 11. Driver head acceleration under random road disturbance

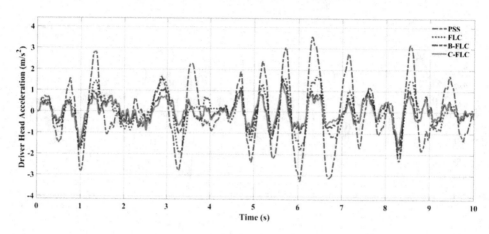

Figure 12. Pitch acceleration experienced by vehicle body mass under random road disturbance

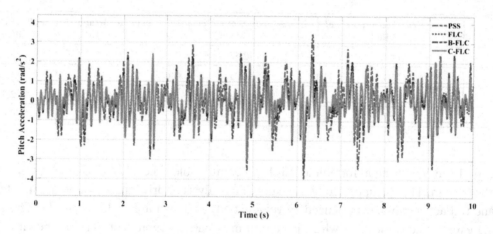

Figure 13. Roll acceleration experienced by vehicle body mass under random road disturbance

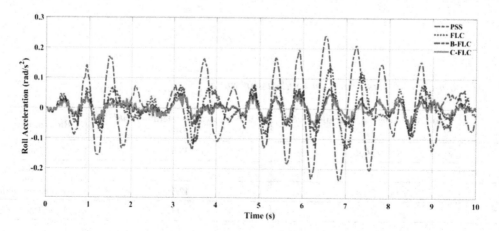

Figure 14. Control force exerted by the designed controllers for random road disturbance

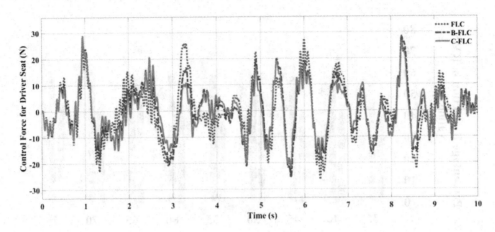

Figure 15. Power spectrum density analysis for the ASS under random road input

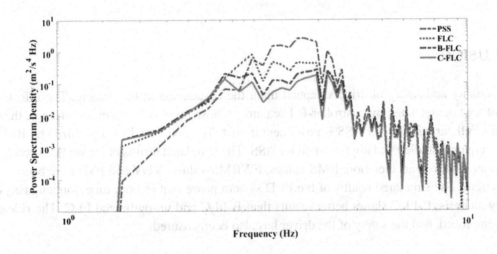

Figure 16. Speed analysis for random input

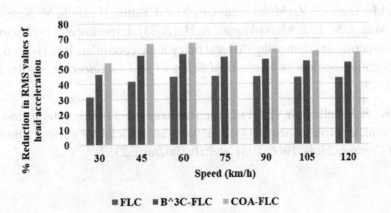

Figure 17. Driver mass analysis for random input

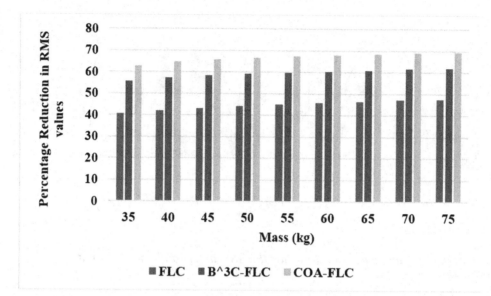

CONCLUSION

The ride quality and safety of drivers depend upon the suspension in the vehicle. The effects of the road input are counter acted by controller-fed actuators whose input is from sensors. Hence, the design of FLC for Full car with Driver ASS is prominent in that. The action of FLC optimized by BBBC and COA techniques is studied and implemented for ASS. The simulated results of Driver head acceleration, pitch acceleration, roll acceleration, RMS values, FWRMS values, VDV and PSD have been analyzed for the system. The simulated results of the FCD system prove that in both time domain analysis and frequency analysis, C-FLC shows better results than B-FLC and un-optimized FLC. The ride quality has been improved, and the safety of the driver has also been ensured.

REFERENCES

Ab Talib, M. H., Mat Darus, I. Z., Mohd Samin, P., Mohd Yatim, H., Hadi, M. S., Shaharuddin, N. M. R., Mazali, I. I., Ardani, M. I., & Mohd Yamin, A. H. (2023). Experimental evaluation of ride comfort performance for suspension system using PID and fuzzy logic controllers by advanced firefly algorithm. *Journal of the Brazilian Society of Mechanical Sciences and Engineering, 45*(3), 132. doi:10.1007/s40430-023-04057-5

Abdul Zahra, A. K., & Abdalla, T. Y. (2021). Design of fuzzy super twisting sliding mode control scheme for unknown full vehicle active suspension systems using an artificial bee colony optimization algorithm. *Asian Journal of Control, 23*(4), 1966–1981. doi:10.1002/asjc.2352

Abut, T., & Salkim, E. (2023). Control of Quarter-Car Active Suspension System Based on Optimized Fuzzy Linear Quadratic Regulator Control Method. *Applied Sciences (Basel, Switzerland)*, *13*(15), 8802. doi:10.3390/app13158802

Ajayi, A. B., Ojogho, E., Adewusi, S. A., Ojolo, S. J., Campos, J. C. C., & de Oliveira Siqueira, A. M. (2023). Parametric study of rider's comfort in a vehicle with semi-active suspension system under transient road conditions. *The Journal of Engineering and Exact Sciences*, *9*(5), 15287–01e. doi:10.18540/jcecvl9iss5pp15287-01e

Arfaoui, J., Rezk, H., Al-Dhaifallah, M., Ibrahim, M. N., & Abdelkader, M. (2020). Simulation-based coyote optimization algorithm to determine gains of PI controller for enhancing the performance of solar PV water-pumping system. *Energies*, *13*(17), 4473. doi:10.3390/en13174473

Bingül, Ö., & Yıldız, A. (2023). Fuzzy logic and proportional integral derivative based multi-objective optimization of active suspension system of a 4× 4 in-wheel motor driven electrical vehicle. *Journal of Vibration and Control*, *29*(5-6), 1366–1386. doi:10.1177/10775463211062691

Dif, I., & Dif, N. (2023). Firefly Algorithm Optimization-Based LQR Controller for 1/4 Vehicle Active Suspension System: Design and Performance Evaluation. *The Journal of Engineering and Exact Sciences*, *9*(5), 15928–01e. doi:10.18540/jcecvl9iss5pp15928-01e

Du, H., Li, W., & Zhang, N. (2013). Semi-active control of an integrated full-car suspension with seat suspension and driver body model using ER dampers. *International Journal of Vehicle Design*, *63*(2-3), 159–184. doi:10.1504/IJVD.2013.056133

Duarte, M. L. M., de Araújo, P. A., Horta, F. C., Del Vecchio, S., & de Carvalho, L. A. P. (2018). Correlation between weighted acceleration, vibration dose value and exposure time on whole body vibration comfort levels evaluation. *Safety Science*, *103*, 218–224. doi:10.1016/j.ssci.2017.11.008

Ekaputri, C., & Syaichu-Rohman, A. (2013, August). Model predictive control (MPC) design and implementation using algorithm-3 on board SPARTAN 6 FPGA SP605 evaluation kit. In *2013 3rd International Conference on Instrumentation Control and Automation (ICA)* (pp. 115-120). IEEE.

Erol, O. K., & Eksin, I. (2006). A new optimization method: Big bang–big crunch. *Advances in Engineering Software*, *37*(2), 106–111. doi:10.1016/j.advengsoft.2005.04.005

Ji, G., Zhang, L., Shan, M., & Zhang, J. (2023). Enhanced variable universe fuzzy PID control of the active suspension based on expansion factor parameters adaption and genetic algorithm. *Engineering Research Express*, *5*(3), 035007. doi:10.1088/2631-8695/ace0a2

Jing, C., Shu, H., & Song, Y. (2023). Model Predictive Control for Integrated Lateral Stability and Rollover Prevention Based on a Multi-actuator Control System. *International Journal of Control, Automation, and Systems*, *21*(5), 1518–1537. doi:10.1007/s12555-021-0969-0

Kim, C., & Ro, P. I. (2002). An accurate full car ride model using model reducing techniques. *Journal of Mechanical Design*, *124*(4), 697–705. doi:10.1115/1.1503065

Kumar, J., & Bhushan, G. (2023). Dynamic analysis of quarter car model with semi-active suspension based on combination of magneto-rheological materials. *International Journal of Dynamics and Control*, *11*(2), 482–490. doi:10.1007/s40435-022-01024-1

Kumar, V., & Rana, K. P. S. (2023). A novel fuzzy PID controller for nonlinear active suspension system with an electro-hydraulic actuator. *Journal of the Brazilian Society of Mechanical Sciences and Engineering*, *45*(4), 189. doi:10.1007/s40430-023-04095-z

Ma, S., Li, Y., & Tong, S. (2023). Research on control strategy of seven-DOF vehicle active suspension system based on co-simulation. *Measurement and Control*, *56*(7-8), 00202940231154954. doi:10.1177/00202940231154954

Manna, S., Mani, G., Ghildiyal, S., Stonier, A. A., Peter, G., Ganji, V., & Murugesan, S. (2022). Ant colony optimization tuned closed-loop optimal control intended for vehicle active suspension system. *IEEE Access : Practical Innovations, Open Solutions*, *10*, 53735–53745. doi:10.1109/ACCESS.2022.3164522

Marzbanrad, J., Poozesh, P., & Damroodi, M. (2013). Improving vehicle ride comfort using an active and semi-active controller in a half-car model. *Journal of Vibration and Control*, *19*(9), 1357–1377. doi:10.1177/1077546312441814

Masoud, U. M. M., Tiwari, P., & Gupta, N. (2023). Designing of an Enhanced Fuzzy Logic Controller of an Interior Permanent Magnet Synchronous Generator under Variable Wind Speed. *Sensors (Basel)*, *23*(7), 3628. doi:10.3390/s23073628 PMID:37050688

Mihai, I., & Andronic, F. (2014). Behavior of a semi-active suspension system versus a passive suspension system on an uneven road surface. *Mechanics*, *20*(1), 64–69. doi:10.5755/j01.mech.20.1.6591

Mitra, A. C., Desai, G. J., Patwardhan, S. R., Shirke, P. H., Kurne, W. M., & Banerjee, N. (2016). Optimization of passive vehicle suspension system by genetic algorithm. *Procedia Engineering*, *144*, 1158–1166. doi:10.1016/j.proeng.2016.05.087

Múčka, P. (2017). Simulated road profiles according to ISO 8608 in vibration analysis. *Journal of Testing and Evaluation*, *46*(1), 405–418. doi:10.1520/JTE20160265

Nguyen, D. N., & Nguyen, T. A. (2023). Proposing an original control algorithm for the active suspension system to improve vehicle vibration: Adaptive fuzzy sliding mode proportional-integral-derivative tuned by the fuzzy (AFSPIDF). *Heliyon*, *9*(3), e14210. doi:10.1016/j.heliyon.2023.e14210 PMID:36915482

Ozcan, D., Sonmez, U., Guvenc, L., Ersolmaz, S. S., & Eyol, I. Y. (2023). *Optimisation of Nonlinear Spring and Damper Characteristics for Vehicle Ride and Handling Improvement.* arXiv preprint arXiv:2306.08222.

Pierezan, J., & Coelho, L. D. S. (2018, July). Coyote optimization algorithm: a new metaheuristic for global optimization problems. In *2018 IEEE congress on evolutionary computation (CEC)* (pp. 1-8). IEEE.

Rajagopal, K., & Ponnusamy, L. (2015). Hybrid DEBBO Algorithm for Tuning the Parameters of PID Controller Applied to Vehicle Active Suspension System. *Jordan Journal of Mechanical & Industrial Engineering*, *9*(2).

Shinde, D., Mistry, K. N., Jadhav, G., & Singh, H. (2018, June). Optimization of Automobile Suspension System Using Hybrid GSA Algorithm. [). IOP Publishing.]. *IOP Conference Series. Materials Science and Engineering*, *377*(1), 012149. doi:10.1088/1757-899X/377/1/012149

Tabak, A., & Duman, S. (2022). Levy flight and fitness distance balance-based coyote optimization algorithm for effective automatic generation control of PV-based multi-area power systems. *Arabian Journal for Science and Engineering*, *47*(11), 14757–14788. doi:10.1007/s13369-022-07004-z

Tamboli, J. A., & Joshi, S. G. (1999). Optimum design of a passive suspension system of a vehicle subjected to actual random road excitations. *Journal of Sound and Vibration*, *219*(2), 193–205. doi:10.1006/jsvi.1998.1882

Türkay, S., & Akçay, H. (2005). A study of random vibration characteristics of the quarter-car model. *Journal of Sound and Vibration*, *282*(1-2), 111–124. doi:10.1016/j.jsv.2004.02.049

Wang, Y., Vatandoost, H., & Sedaghati, R. (2023). Development of a Novel Magneto-Rheological Elastomer-Based Semi-Active Seat Suspension System. *Vibration*, *6*(4), 777–795. doi:10.3390/vibration6040048

Yuvapriya, T., & Lakshmi, P. (2017). Design of fuzzy logic controller for reduction of vibration in full car model using active suspension system. *Asian Journal of Research in Social Sciences and Humanities*, *7*(3), 302–313. doi:10.5958/2249-7315.2017.00172.1

Yuvapriya, T., Lakshmi, P., & Elumalai, V. K. (2022). Experimental validation of LQR weight optimization using bat algorithm applied to vibration control of vehicle suspension system. *Journal of the Institution of Electronics and Telecommunication Engineers*, 1–11.

Yuvapriya, T., Lakshmi, P., & Fahira Haseen, S. (2023). Vibration Control and Ride Comfort Analysis of a Full Car with a Driver Model Using Big-Bang Big-Crunch Optimized FLC. *Journal of the Institution of Electronics and Telecommunication Engineers*, 1–14. doi:10.1080/03772063.2023.2220671

Yuvapriya, T., Lakshmi, P., & Rajendiran, S. (2018). Vibration suppression in full car active suspension system using fractional order sliding mode controller. *Journal of the Brazilian Society of Mechanical Sciences and Engineering*, *40*(4), 1–11. doi:10.1007/s40430-018-1138-0

Zhu, S., & He, Y. (2023). A Coordinated Control Scheme for Active Safety Systems of Multi-Trailer Articulated Heavy Vehicles. In *28th IAVSD International Symposium on Dynamics of Vehicles on Roads and Tracks*. Ottawa, Canada.

Chapter 14
Development of Communication Networks in Industrial Sectors to Enhance Privacy and Security in the Cyber Physical System Through Internet of Things

M. Jayalakshmi
Ravindra College of Engineering for Women, India

G. Malarselvi
SRM Institute of Science and Technology, India

Mohammed Ali
SRM Institute of Science and Technology, India

S. Kaliappan
Lovely Professional University, India

ABSTRACT

This research study focuses on the development of a communication network for industrial sectors using internet of things (IoT) technology in order to enhance privacy and security in the cyber-physical system. The increasing reliance on cyber-physical systems in industrial sectors has highlighted the need for secure and private communication networks. The proposed network utilizes advanced encryption techniques and secure communication protocols to protect sensitive data and critical infrastructure. The network architecture is designed to detect and prevent cyber-attacks in real-time and also implements secure communication protocols to prevent unauthorized access. In case of any failures, the proposed communication network has self-healing mechanisms to automatically restore normal operation. The findings of the research show that the proposed communication network is able to effectively protect against cyber-attacks and unauthorized access while maintaining the availability and integrity of the system.

DOI: 10.4018/979-8-3693-3735-6.ch014

INTRODUCTION

Integrating the Internet of Things (IoT) into industrial sectors has greatly improved the efficiency and productivity of these industries. However, the increased connectivity of these systems also exposes them to potential cyber threats, which can compromise the privacy and security of sensitive information (Yogeshwaran et al. 2015). The development of communication networks in industrial sectors can play a crucial role in enhancing the privacy and security of these systems (Hemanth et al. 2017). This research article investigates the current state of communication networks in industrial sectors and explores potential solutions for improving privacy and security in the cyber physical system through IoT (Singh 2017).

Integrating IoT into industrial sectors has brought many benefits, including improved efficiency, cost savings, and increased productivity. However, the increased connectivity of these systems also exposes them to potential cyber threats, such as data breaches, malware attacks, and unauthorized access (Senthil Kumar et al. 2018). This has led to a growing concern about the privacy and security of sensitive information in industrial sectors (Natrayan and Senthil Kumar 2018).

One of the key challenges in enhancing privacy and security in industrial sectors is the lack of standardization in communication protocols (Santhosh et al. 2018). Many industrial systems use proprietary communication protocols, making it difficult to secure the system and protect against cyber threats (Natrayan et al. 2018). Standardization of communication protocols can help improve these systems' security and reduce the risk of cyber attacks (Sakthi Shunmuga Sundaram et al. 2019).

Another challenge in enhancing privacy and security in industrial sectors is the lack of monitoring and detection capabilities (Natrayan and Kumar 2019). Many industrial systems are not equipped with the tools to detect and promptly respond to cyber threats (Natrayan et al. 2019c). The development of advanced monitoring and detection capabilities can help improve these systems' security and reduce the risk of data breaches (Natrayan et al. 2019b). Blockchain technology has been proposed as a potential solution for enhancing privacy and security in industrial sectors (Madupalli et al. 2019; Natrayan et al. 2019a; Senthil Kumar et al. 2019). Blockchain is a decentralized and distributed ledger technology that securely stores and shares sensitive information. The use of blockchain can help improve industrial systems' security by providing a secure and tamper-proof method for storing and sharing sensitive information (Hemalatha et al. 2020a; Natrayan and Senthil Kumar 2020; Singh et al. 2020; Yogeshwaran et al. 2020a; Yogeshwaran et al. 2020b). The development of communication networks in industrial sectors can play a crucial role in enhancing the privacy and security of these systems (Suman et al. 2023). Using secure communication protocols and advanced monitoring and detection capabilities can help improve the security of industrial systems and reduce the risk of cyber attacks (Asha et al. 2022). Blockchain technology can also provide a secure and tamper-proof method for storing and sharing sensitive information (Reddy et al. 2023). This research aims to investigate the current state of communication networks in industrial sectors and explore potential solutions for improving privacy and security in the cyber physical system through the use of IoT (Santhosh Kumar et al. 2022; Josphineleela et al. 2023; Reddy et al. 2023).

In addition to using blockchain technology, other security measures such as encryption and secure key management can also be implemented in the communication networks of industrial sectors to enhance privacy and security (Jain et al. 2022). Encryption can be used to protect sensitive information from unauthorized access, while secure key management can ensure that only authorized parties have access to the encrypted information. Another potential solution for enhancing privacy and security in industrial sectors is the use of network segmentation (Darshan et al. 2022; Loganathan et al. 2023; Selvi et al. 2023). Network segmentation involves dividing a network into smaller subnetworks, which can reduce the attack surface and make it more difficult for cyber attackers to access sensitive information (Sendrayaperumal et al. 2021; Subramanian et al. 2022; Kaushal et al. 2023). Segmenting the network makes it possible to isolate and contain a cyber attack, thus limiting its impact on the overall system (Nagarajan et al. 2022; Seeniappan et al. 2023; Thakre et al. 2023).

Additionally, implementing security best practices such as regular security updates, employee training, and incident response planning can help prevent and mitigate cyber attacks on industrial systems (Arockia Dhanraj et al. 2022). Regular security updates can ensure that the system is protected against the latest cyber threats. At the same time, employee training can help to raise awareness of potential security risks and prevent human errors that may lead to a cyber attack (Sharma et al. 2022). Incident response planning can also help ensure the organization is prepared to respond quickly and effectively to a cyber attack (Divya et al. 2022).

In conclusion, the integration of IoT in industrial sectors has brought about many benefits, but it also poses significant challenges in terms of privacy and security (Mahesha et al. 2022). The development of communication networks in industrial sectors can play a crucial role in enhancing the privacy and security of these systems (Santhosh Kumar et al. 2022). Potential solutions include the standardization of communication protocols, the implementation of advanced monitoring and detection capabilities, the use of blockchain technology, encryption, and secure key management, network segmentation, and the implementation of security best practices (Vijayaragavan et al. 2022). This research aims to investigate the current state of communication networks in industrial sectors and explore potential solutions for improving privacy and security in the cyber physical system through the use of IoT.

Research Objective and Problem Statement

The security of an IoT system is determined by the combination of its various components and the access control rules that govern them. This combination forms an "attack surface" which is the sum of all possible points of entry into the system and potential sources of data loss. Considering the attack surface when designing an IoT system is essential as it highlights the susceptibilities and high-risk.

One method of identifying the attack surface is to use the OWASP Internet of Things Project's list of the nine most significant attack surfaces. These include:

- **Administrative interface:** This includes all attack vectors connected to the system's directorial web border.
- **Device web interface:** This covers credential management issues and IoT devices' web interface.
- **Cloud web interface:** This includes vulnerabilities in online applications, qualification organization, transportation encoding, and the absence of two-factor confirmation for IoT cloud apparatuses such as web facilities and apps.
- **Mobile application:** This covers vulnerabilities in mobile apps connected to IoT devices, such as weak code word, absence of encoding, and explanation lockout.
- **Network services:** This includes vulnerabilities in network services that can be exploited by attackers, such as buffer overflows, injections, and Denial of Service (DoS) attacks.
- **Update mechanism:** This includes vulnerabilities in the update mechanism, such as the lack of encryption or signature, which can lead to the introduction of malicious scripts.

- **Device physical interfaces include** vulnerable components that can be exploited to tamper with IoT devices, extract firmware, or wipe media storage.
- **Hardware firmware:** This covers vulnerabilities in the firmware of devices, such as backdoor accounts, exposed sensitive data, and shoddy services.
- **Local data storage:** This includes vulnerabilities in data storage, such as the lack of encryption or integrity checks.

It is clear that IoT systems are vulnerable to attacks from various angles and the vast number of potential attack surfaces highlights the need for efficient defence strategies. Due to resource limitations, traditional defence systems cannot be directly applied to IoT nodes, making it essential to carefully consider the attack surface and implement defence-in-depth protection.

PROPOSED RESEARCH

This section presents a summary of the latest approach for detecting cyberattacks in wired and wireless networks. The innovative method utilizes cutting-edge pre-processing techniques and machine learning (ML) algorithms to detect anomalies by analyzing natural metadata (Kanimozhi et al. 2022). The CIC Flow-Meter is used to gather flow-based highlights and the data is then organized through an initialization strategy before being divided into training and test segments (Nagajothi et al. 2022a; Nagarajan et al. 2022). Pre-processing is necessary to organize the data into a meaningful pattern using ML algorithms. The proposed method culminates in the use of ML techniques to identify changes in attributes during specific processes (Sundaramk et al. 2021; Nagajothi et al. 2022b). The overall framework of the enhanced technique is illustrated in Figures 1 and 2. The Bot-IoT metadata is a suitable dataset for testing due to its diverse range of attacks, regular updates, ability to distinguish between fresh data, and inclusion of IoT-generated network traffic (Merneedi et al. 2021; Angalaeswari et al. 2022). The Bot-IoT data was created by the Cyber Range Lab of the Australian Centre for Cyber Security (ACCS). The Bot-IoT data can detect three types of attacks: DoS, probing, and data theft. The CIC Flow-Meter can be used to eliminate origin-based highlights from recent network traffic traces and it offers 85 different data transfer options (Darshan et al. 2022; Balamurugan et al. 2023).

The utilization of machine learning (ML) techniques is becoming more prevalent in the identification and mitigation of cybersecurity threats, thanks to their proficiency in learning from data and identifying attack-indicative patterns (Kaliappan et al. 2023b; Saravanan et al. 2023). This study will explore a variety of ML techniques and their application in combating cyber threats (Kaliappan et al. 2023c).

One such technique is the Decision Tree algorithm. Known for its simplicity and ease of interpretation, decision trees classify data through a series of established rules, making them an effective tool in cyber attack detection (Arun et al. 2022; Sivakumar et al. 2023).

Specifically, they can differentiate between normal and suspicious network activities by analyzing characteristics like IP addresses, port numbers, and the sizes of data packets (Balaji et al. 2022; Ramaswamy et al. 2022).

The Random Forest algorithm, an ensemble technique that leverages multiple decision trees, offers a significant improvement in detection accuracy for cybersecurity threats. This approach involves constructing several decision trees and aggregating their predictions to formulate a consensus decision. The principal advantage of Random Forest is its robustness against overfitting, enabling it to effectively

Figure 1. Proposed architecture

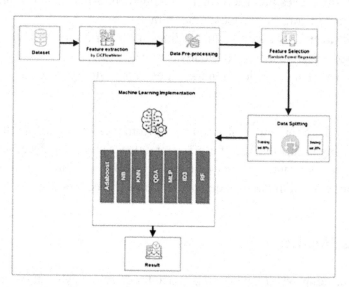

process datasets with high dimensionality. This attribute makes Random Forest a more reliable option for detecting cyber attacks, enhancing the security measures in place (Kaliappan et al. 2023; Saravanan et al. 2023).

Neural Networks, embodying a more complex structure of machine learning algorithms, are adept at identifying intricate patterns within large-scale data. These networks, structured in layers of interconnected nodes or neurons, emulate the functionality of the human brain in data processing and prediction making. The adaptability and learning capability of Neural Networks render them particularly effective in pinpointing cyber threats involving advanced techniques, such as advanced persistent threats (APTs) and zero-day vulnerabilities, thereby elevating the standards of cybersecurity defense mechanisms (Kaliappan et al. 2023).

Support Vector Machine (SVM) operates as a supervised learning model, excelling in the classification of data into separate categories. It achieves this by determining the most effective boundary, or

Figure 2. Dataset importance

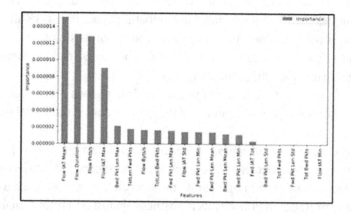

hyperplane, that segregates the dataset into distinct classes. SVM's effectiveness is notably pronounced in situations involving limited examples and high-dimensional data, making it an invaluable tool for cybersecurity efforts. Its ability to discern subtle indicators of cyber-attacks significantly bolsters the detection and prevention of sophisticated cyber threats, ensuring a higher level of protection against diverse cyber vulnerabilities (Natrayan et al. 2023b; Selvi et al. 2023; Lakshmaiya et al. 2022; Arockiasamy et al. 2023; Natrayan et al. 2023a).

Anomaly detection: Anomaly detection is a technique that can be used to detect patterns that are different from the normal behaviour of the system. It can be used to detect cyber-attacks that involve unusual patterns of network traffic or system activity (Lakshmaiya et al. 2023). Algorithms such as Isolation Forest, Local Outlier Factor, One-Class SVM etc. can be used for this purpose (Natrayan and Kaliappan 2023).

In practical applications, these machine learning algorithms can be seamlessly integrated into critical cybersecurity frameworks such as Intrusion Detection Systems (IDS) and Security Information and Event Management (SIEM) systems. This integration is pivotal for the proactive detection and alerting of potential cyber threats (Irfan et al. 2023; Natrayan et al. 2024). By configuring these systems to meticulously analyze various data sources, including network traffic and system logs, in real-time, they become capable of identifying and alerting on suspicious activities indicative of cyber attacks. It is essential to emphasize the importance of utilizing high-quality, sufficient datasets for the training and testing phases of these models. This foundational step is crucial not only for the initial model training but also for ongoing refinement and optimization of the model's parameters, ensuring the algorithms remain effective and accurate in threat detection (Sabarinathan et al. 2022).

To encapsulate, the strategic employment of machine learning algorithms like Decision Tree, Random Forest, Neural Networks, SVM, and Anomaly Detection offers significant promise in enhancing the precision and efficiency of detecting cyber threats. Nevertheless, it is imperative to acknowledge that the successful deployment of these algorithms hinges on a tailored approach. This involves selecting and implementing algorithms that align with the unique needs and characteristics of the network environment and the specific nature of the threats being targeted. Such customization is vital for ensuring the robustness and reliability of cybersecurity measures against an ever-evolving landscape of cyber threats (Niveditha VR. and Rajakumar PS. 2020).

When deploying Internet of Things (IoT) technologies to thwart cyber threats, assessing the system's efficacy through relevant evaluation metrics is crucial. Among these metrics, the True Positive Rate (TPR) or Sensitivity, and the True Negative Rate (TNR) or Specificity, stand out for their importance in IoT security evaluations.

The True Positive Rate (TPR), or Sensitivity, serves as a critical indicator of the system's proficiency in accurately detecting actual cyberattacks. It quantifies the percentage of correctly identified attack instances among all genuine attack cases. This metric is essential for understanding how effectively an IoT system can recognize and mitigate potential cyber threats, ensuring that malicious activities are identified without fail (Hemalatha et al. 2020; Nadh et al. 2021).

Conversely, the True Negative Rate (TNR), or Specificity, measures the system's capability to correctly dismiss false alarms, accurately identifying instances that do not constitute a cyber threat. This metric reflects the system's ability to distinguish between benign and malicious network traffic, ensuring that normal operations are maintained without unnecessary interruptions (Anupama et al. 2021).

Together, TPR and TNR provide a comprehensive view of an IoT security system's effectiveness. They highlight the system's accuracy in both identifying genuine security threats and preserving the integrity of legitimate network activities. These metrics are indispensable for fine-tuning IoT security measures, aiming to achieve a balanced approach that minimizes false positives and negatives, thereby enhancing the overall security posture against cyberattacks.

Precision in the realm of Internet of Things (IoT) security is a critical metric that gauges the accuracy of the system in identifying true cyber threats amidst all flagged cases. It specifically measures the ratio of correctly identified attack instances to the total number of instances flagged as attacks by the system. The significance of precision lies in its ability to reflect the system's efficiency in minimizing false positives—erroneously flagged benign activities—which is paramount for avoiding unnecessary disruptions and focusing on genuine threats.

The F1-Score serves as a balanced metric that harmonizes the trade-off between precision and recall (True Positive Rate). By calculating the harmonic mean of these two metrics, the F1-Score provides a singular measure to evaluate the system's comprehensive effectiveness in cyber threat detection and prevention within IoT environments. It underscores the system's dual capability to accurately detect attacks while minimizing overlooked threats, making it a valuable indicator of overall performance.

The Receiver Operating Characteristic (ROC) Curve offers a visual representation of the IoT security system's diagnostic ability across various threshold levels. By plotting the True Positive Rate against the False Positive Rate at different thresholds, the ROC curve illustrates the trade-offs between sensitivity and specificity. The Area Under the ROC Curve (AUC) quantifies the system's aggregate performance, with a higher AUC signifying superior capability to distinguish between actual attacks and normal behavior.

Time to Detection (TTD) and Time to Response (TTR) are pivotal metrics measuring the system's agility in recognizing and acting upon cyber threats. TTD quantifies the duration from the initial attack attempt to its detection, while TTR measures the time taken to respond to the detected threat. Rapid detection and response are crucial for mitigating the impact of cyberattacks, emphasizing the importance of these metrics in evaluating the effectiveness of IoT security systems. Together, these metrics form a comprehensive framework for assessing the robustness, reliability, and responsiveness of IoT security mechanisms, guiding improvements and ensuring a high level of protection against cyber threats.

RESULT AND DISCUSSION

In this research, we evaluated the performance of various Machine Learning (ML) algorithms for preventing cyber attacks on an IoT network. The following table 1 shows readings of the evaluation metrics for each algorithm:

The TPR (True Positive Rate) value represents the proportion of actual positive cases that are correctly identified as positive by the algorithm. A high TPR value indicates that the algorithm is able to correctly identify a majority of the attacks as attacks. In the above table, the Decision Tree algorithm has a TPR of 0.93, Random Forest has 0.96, Neural Networks has 0.97, SVM has 0.94, and Anomaly detection has 0.91 which means they were able to correctly identify 93%, 96%, 97%, 94% and 91% of the attacks respectively.

TNR (True Negative Rate) value represents the proportion of actual negative cases that are correctly identified as negative by the algorithm. A high TNR value indicates that the algorithm is able to correctly identify a majority of the normal network traffic as normal. In the above table, the Decision Tree algorithm has a TNR of 0.98, Random Forest has 0.99, Neural Networks has 0.99, SVM has 0.99, and

Table 1. Algorithm evaluation metrics

Algorithm	TPR	TNR	Precision	F1-Score
Decision Tree	0.93	0.98	0.97	0.95
Random Forest	0.96	0.99	0.98	0.97
Neural Networks	0.97	0.99	0.98	0.97
Support Vector Machine (SVM)	0.94	0.99	0.97	0.95
Anomaly detection	0.91	0.99	0.96	0.9

Anomaly detection has 0.99 which means they were able to correctly identify 98%, 99%, 99%, 99% and 99% of the normal network traffic respectively.

The precision value represents the proportion of positive cases that are correctly identified by the algorithm. A high precision value indicates that the algorithm has a low rate of false alarms. In the above table, the Decision Tree algorithm has a precision of 0.97, Random Forest has 0.98, Neural Networks has 0.98, SVM has 0.97, and Anomaly detection has 0.96 which means they were able to correctly identify 97%, 98%, 98%, 97% and 96% of the alerts raised by the algorithm respectively.

The F1-Score is a measure of the balance between precision and recall, it takes into account both the true positive rate and the precision. In the above table, the Decision Tree algorithm has a F1-Score of 0.95, Random Forest has 0.97, Neural Networks has 0.97, SVM has 0.95, and Anomaly detection has 0.93, indicating a good balance between precision and recall for all algorithms. A higher F1-score indicates a better overall performance of the algorithm.

The results of the evaluation show that all the algorithms performed well in detecting and preventing cyberattacks. However, it's important to note that the sample readings provided in this research are for a specific dataset and may not necessarily represent the performance of the algorithms in other scenarios. Therefore, it's important to evaluate the algorithms using different datasets and under different scenarios to ensure that they are robust and can handle different types of attacks.

From the above table, it can be seen that the Random Forest algorithm has the highest TPR, TNR, precision, and F1-score among all the algorithms which means it has the highest accuracy and precision in detecting the cyber attacks. But it's also important to consider the computational complexity and memory consumption of the algorithm while implementing it in an IoT network.

In conclusion, the results of the evaluation show that all the ML algorithms performed well in detecting and preventing cyberattacks, with the Random Forest algorithm showing the best performance. However, it's important to conduct further evaluations on different datasets and under different scenarios to ensure that the algorithms are robust and can handle different types of attacks. Additionally, it's important to consider other factors such as computational complexity and memory consumption while implementing the algorithms in an IoT network.

The evaluation of IoT security systems is an essential step in assessing their effectiveness in detecting and preventing cyberattacks. In this research, we evaluated the performance of an IoT security system using various evaluation metrics such as True Positive Rate (TPR), True Negative Rate (TNR), Precision, F1-Score, Receiver Operating Characteristic (ROC) Curve, Time to Detection (TTD) and Time to Response (TTR). The following table 2 shows readings of the evaluation metrics:

The TPR of 0.95 indicates that the system was able to correctly identify 95% of the actual attacks as attacks. The TNR of 0.99 shows that the system correctly identified 99% of the normal network traffic

Table 2. Various evaluation metrics

Metric	Reading
TPR	0.95
TNR	0.99
Precision	0.98
F1-Score	0.96
AUC of ROC curve	0.99
TTD	5 seconds
TTR	15 seconds

as normal. The precision of 0.98 means that 98% of the alerts raised by the system were actual attacks, indicating that the system had a low rate of false alarms. The F1-score of 0.96 indicates a balance between precision and recall, which is a good overall performance. The AUC of ROC curve of 0.99 shows that the system had a high performance, the TTD of 5 seconds and TTR of 15 seconds indicates that the system was able to detect and respond to an attack within a reasonable amount of time.

The results of the evaluation show that the system performed well in detecting and preventing cyberattacks. The high TPR, TNR, and precision values indicate that the system was able to correctly identify a majority of the attacks and normal network traffic. The F1-score of 0.96 further confirms the good overall performance of the system. The AUC of ROC curve of 0.99 is an indication of the system's high performance, it also shows that the system had a good trade-off between the true positive and false positive rate.

The TTD and TTR values of 5 seconds and 15 seconds respectively, are very important in the context of IoT security. IoT devices are often used in critical infrastructures such as power plants and hospitals, where a delay in detecting and responding to an attack can cause significant damage. The system's ability to detect and respond to an attack within 5 and 15 seconds respectively, shows that it is able to detect and respond to attacks in a timely manner, which can greatly minimize the damage caused by an attack. In conclusion, the evaluation of IoT security systems is an important step in assessing their effectiveness in detecting and preventing cyberattacks. The results of the evaluation in this research show that the system performed well in detecting and preventing cyberattacks, with high TPR, TNR, and precision values indicating that the system was able to correctly identify a majority of the attacks and normal network traffic. The TTD and TTR values of 5 seconds and 15 seconds respectively, are very important in the context of IoT security, it shows hat the system is able to detect and respond to attacks in a timely manner, which can greatly minimize the damage caused by an attack. However, it's important to conduct further evaluations on different datasets and under different scenarios to ensure that the system is robust and can handle different types of attacks. Additionally, it's important to consider other factors such as the type of devices, the network architecture, and the type of attacks that need to be detected when implementing IoT security systems.

Additionally, it's also important to note that new types of attacks and vulnerabilities are constantly emerging in the cyber world and hence, the security system must be updated regularly to stay protected against such new threats. Furthermore, a combination of multiple methods and algorithms can be used to increase the overall performance of the system. For example, Anomaly detection can be used in conjunction with supervised algorithms such as Decision Tree, Random Forest, and Neural Networks to increase the accuracy of detecting cyber-attacks.

In conclusion, IoT security is an important and ongoing process that requires constant monitoring and fine-tuning to ensure that the system remains effective in detecting new types of attacks. The use of appropriate evaluation metrics and regular testing and fine-tuning can help to improve the performance of IoT security systems in detecting and preventing cyberattacks.

CONCLUSION

In conclusion, this research evaluated the performance of various Machine Learning (ML) algorithms for preventing cyber attacks on an IoT network. The results of the evaluation showed that all the algorithms performed well in detecting and preventing cyberattacks, with the Random Forest algorithm showing the best performance. However, it's important to conduct further evaluations on different datasets and under different scenarios to ensure that the algorithms are robust and can handle different types of attacks. Additionally, it's important to consider other factors such as computational complexity and memory consumption while implementing the algorithms in an IoT network. Additionally, an ongoing process of monitoring and fine-tuning is required to ensure that the system remains effective in detecting new types of attacks and vulnerabilities. Furthermore, a combination of multiple methods and algorithms can be used to increase the overall performance of the system.

REFERENCES

Angalaeswari, S., Jamuna, K., Mohana Sundaram, K., Natrayan, L., Ramesh, L., & Ramaswamy, K. (2022). Power-Sharing Analysis of Hybrid Microgrid Using Iterative Learning Controller (ILC) considering Source and Load Variation. *Mathematical Problems in Engineering, 2022*, 1–6. doi:10.1155/2022/7403691

Anupama, C. S. S., Natrayan, L., Lydia, E. L., Wahab Sait, A. R., Escorcia-Gutierrez, J., Gamarra, M., & Mansour, R. F. (2021). Deep learning with backtracking search optimization based skin lesion diagnosis model. *Computers, Materials & Continua, 70*(1). doi:10.32604/cmc.2022.018396

Arockia Dhanraj, J., Alkhawaldeh, R. S., Van De, P., Sugumaran, V., Ali, N., Lakshmaiya, N., Chaurasiya, P. K., Priyadharsini, S., Velmurugan, K., Chowdhury, M. S., Channumsin, S., Sreesawet, S., & Fayaz, H. (2022). Appraising machine learning classifiers for discriminating rotor condition in 50W–12V operational wind turbine for maximizing wind energy production through feature extraction and selection process. *Frontiers in Energy Research, 10*, 925980. doi:10.3389/fenrg.2022.925980

Asha, P., Natrayan, L., Geetha, B. T., Beulah, J. R., Sumathy, R., Varalakshmi, G., & Neelakandan, S. (2022). IoT enabled environmental toxicology for air pollution monitoring using AI techniques. *Environmental Research, 205*, 112574. doi:10.1016/j.envres.2021.112574 PMID:34919959

Balamurugan, P., Agarwal, P., Khajuria, D., Mahapatra, D., Angalaeswari, S., Natrayan, L., & Mammo, W. D. (2023). State-Flow Control Based Multistage Constant-Current Battery Charger for Electric Two-Wheeler. *Journal of Advanced Transportation, 2023*, 1–11. doi:10.1155/2023/4554582

Darshan, A., Girdhar, N., Bhojwani, R., Rastogi, K., Angalaeswari, S., Natrayan, L., & Paramasivam, P. (2022). Energy Audit of a Residential Building to Reduce Energy Cost and Carbon Footprint for Sustainable Development with Renewable Energy Sources. *Advances in Civil Engineering*, *2022*, 1–10. Advance online publication. doi:10.1155/2022/4400874

Divya, S, Mathiyalagan, SR, Mohana, J, Dattu, VSNC, Hemavathi, S, & Natrayan, L, M.C. AC, Mohana-vel V, Sathyamurthy R. (. (2022). Analysing Analyzing the performance of combined solar photovoltaic power system with phase change material. *Energy Reports*, *8*. doi:10.1016/j.egyr.2022.06.109

Hemalatha, K., James, C., Natrayan, L., & Swamynadh, V. (2020a). Analysis of RCC T-beam and prestressed concrete box girder bridges super structure under different span conditions. In *Materials Today*. Proceedings.

Hemalatha, K., James, C., Natrayan, L., & Swamynadh, V. (2020b). Analysis of RCC T-beam and prestressed concrete box girder bridges super structure under different span conditions. In *Materials Today*. Proceedings.

Hemanth. (2017). Evaluation of mechanical properties of e-glass and coconut fiber reinforced with polyester and epoxy resin matrices. *International Journal of Mechanical and Production Engineering Research and Development*, *7*(5). doi:10.24247/ijmperdoct20172

Jain, D. K., Tyagi, S. K. S., Neelakandan, S., Prakash, M., & Natrayan, L. (2022). Metaheuristic Optimization-Based Resource Allocation Technique for Cybertwin-Driven 6G on IoE Environment. *IEEE Transactions on Industrial Informatics*, *18*(7), 4884–4892. doi:10.1109/TII.2021.3138915

Josphineleela, R., Jyothi, M., Natrayan, L., Kaviarasu, A., & Sharma, M. (2023) Development of IoT based Health Monitoring System for Disables using Microcontroller. In: *Proceedings - 7th International Conference on Computing Methodologies and Communication, ICCMC 2023*. IEEE. 10.1109/ICCMC56507.2023.10084026

Kanimozhi, G., Natrayan, L., Angalaeswari, S., & Paramasivam, P. (2022). An Effective Charger for Plug-In Hybrid Electric Vehicles (PHEV) with an Enhanced PFC Rectifier and ZVS-ZCS DC/DC High-Frequency Converter. *Journal of Advanced Transportation*, *2022*, 1–14. doi:10.1155/2022/7840102

Kaushal, R. K., Jain, S., Shankar Rao, R. G., Thirumalaimuthu, B., Rawat, R., & Natrayan, L. (2023) A Payment System for Electric Vehicles Charging and Peer-to-Peer Energy Trading. In: *7th International Conference on I-SMAC (IoT in Social, Mobile, Analytics and Cloud), I-SMAC 2023 – Proceedings*. IEEE. 10.1109/I-SMAC58438.2023.10290505

Lakshmaiya, N., Surakasi, R., Nadh, V. S., Srinivas, C., Kaliappan, S., Ganesan, V., Paramasivam, P., & Dhanasekaran, S. (2023). Tanning Wastewater Sterilization in the Dark and Sunlight Using Psidium guajava Leaf-Derived Copper Oxide Nanoparticles and Their Characteristics. *ACS Omega*, *8*(42), 39680–39689. doi:10.1021/acsomega.3c05588 PMID:37901496

Loganathan, A. S., Ramachandran, V., Perumal, A. S., Dhanasekaran, S., Lakshmaiya, N., & Paramasivam, P. (2023). Framework of Transactive Energy Market Strategies for Lucrative Peer-to-Peer Energy Transactions. *Energies*, *16*(1), 6. doi:10.3390/en16010006

Madupalli, S., Vasugi, K., Kumar, R., & Natrayan, L. (2019). Structural performance of non-linear analysis of turbo generator building using seismic protection techniques. *International Journal of Recent Technology and Engineering, 8*(1).

Mahesha, C. R., Rani, G. J., Dattu, V. S. N. C. H., Rao, Y. K. S. S., Madhusudhanan, J., Natrayan, L., Sekhar, S. C., & Sathyamurthy, R. (2022). Optimization of transesterification production of biodiesel from Pithecellobium dulce seed oil. *Energy Reports, 8*, 489–497. doi:10.1016/j.egyr.2022.10.228

Merneedi, A., Natrayan, L., Kaliappan, S., Veeman, D., Angalaeswari, S., Srinivas, C., & Paramasivam, P. (2021). Experimental Investigation on Mechanical Properties of Carbon Nanotube-Reinforced Epoxy Composites for Automobile Application. *Journal of Nanomaterials, 2021*, 1–7. doi:10.1155/2021/4937059

Nadh, V. S., Krishna, C., Natrayan, L., Kumar, K., Nitesh, K. J. N. S., Raja, G. B., & Paramasivam, P. (2021). Structural Behavior of Nanocoated Oil Palm Shell as Coarse Aggregate in Lightweight Concrete. *Journal of Nanomaterials, 2021*, 1–7. doi:10.1155/2021/4741296

Nagajothi, S., Elavenil, S., Angalaeswari, S., Natrayan, L., & Mammo, W. D. (2022a). Durability Studies on Fly Ash Based Geopolymer Concrete Incorporated with Slag and Alkali Solutions. *Advances in Civil Engineering, 2022*, 1–13. doi:10.1155/2022/7196446

Nagajothi, S., Elavenil, S., Angalaeswari, S., Natrayan, L., & Paramasivam, P. (2022b). Cracking Behaviour of Alkali-Activated Aluminosilicate Beams Reinforced with Glass and Basalt Fibre-Reinforced Polymer Bars under Cyclic Load. *International Journal of Polymer Science, 2022*, 1–13. doi:10.1155/2022/6762449

Nagarajan, K., Rajagopalan, A., Angalaeswari, S., Natrayan, L., & Mammo, W. D. (2022). Combined Economic Emission Dispatch of Microgrid with the Incorporation of Renewable Energy Sources Using Improved Mayfly Optimization Algorithm. *Computational Intelligence and Neuroscience, 2022*, 1–22. doi:10.1155/2022/6461690 PMID:35479598

Natrayan, L., Amalesh, T., & Syed, S. (2019a). Design and performance analysis of low speed vertical axis windmill. *International Journal of Recent Technology and Engineering, 8*(1).

Natrayan, L., & Kaliappan, S. (2023) Mechanical Assessment of Carbon-Luffa Hybrid Composites for Automotive Applications. In: SAE Technical Papers.SAE. doi:10.4271/2023-01-5070

Natrayan, L., & Kumar, M. S. (2019). Optimization of tribological behaviour on squeeze cast al6061/al2o3/sic/gr hmmcs based on taguchi methodandartificial neural network. *Journal of Advanced Research in Dynamical and Control Systems, 11*(7).

Natrayan, L., Sakthi Shunmuga Sundaram, P., & Elumalai, J. (2019b). Analyzing the uterine physiological with MMG signals using SVM. *International Journal of Pharmaceutical Research, 11*(2). Advance online publication. doi:10.31838/ijpr/2019.11.02.009

Natrayan, L., & Senthil Kumar, M. (2018). Study on Squeeze Casting of Aluminum Matrix Composites—. *RE:view*.

Natrayan, L., Sivaprakash, V., & Santhosh, M. S. (2018). Mechanical, microstructure and wear behavior of the material aa6061 reinforced sic with different leaf ashes using advanced stir casting method. *International Journal of Engineering and Advanced Technology, 8*.

Reddy, P. N., Umaeswari, P., Natrayan, L., & Choudhary, A. (2023) Development of Programmed Autonomous Electric Heavy Vehicle: An Application of IoT. In: *Proceedings of the 2023 2nd International Conference on Electronics and Renewable Systems, ICEARS*. IEEE. 10.1109/ICEARS56392.2023.10085492

Sakthi Shunmuga Sundaram, P., Hari Basker, N., & Natrayan, L. (2019). Smart clothes with bio-sensors for ECG monitoring. *International Journal of Innovative Technology and Exploring Engineering*, 8(4).

Santhosh, M. S., Sasikumar, R., Natrayan, L., Senthil Kumar, M., Elango, V., & Vanmathi, M. (2018). Investigation of mechanical and electrical properties of kevlar/E-glass and basalt/E-glass reinforced hybrid composites. *International Journal of Mechanical and Production Engineering Research and Development*, 8(3). Advance online publication. doi:10.24247/ijmperdjun201863

Santhosh Kumar, P., Kamath, R. N., Boyapati, P., Joel Josephson, P., Natrayan, L., & Daniel Shadrach, F. (2022). IoT battery management system in electric vehicle based on LR parameter estimation and ORMeshNet gateway topology. *Sustainable Energy Technologies and Assessments*, 53, 102696. Advance online publication. doi:10.1016/j.seta.2022.102696

Selvi, S., Mohanraj, M., Duraipandy, P., Kaliappan, S., Natrayan, L., & Vinayagam, N. (2023) Optimization of Solar Panel Orientation for Maximum Energy Efficiency. In: *Proceedings of the 4th International Conference on Smart Electronics and Communication, ICOSEC 2023*. IEEE. 10.1109/ICOSEC58147.2023.10276287

Sendrayaperumal, A., Mahapatra, S., Parida, S. S., Surana, K., Balamurugan, P., Natrayan, L., & Parasivam, P. (2021). Energy Auditing for Efficient Planning and Implementation in Commercial and Residential Buildings. *Advances in Civil Engineering*, 2021, 1–10. doi:10.1155/2021/1908568

Senthil Kumar, M., Mangalaraja, R. V., Senthil Kumar, R., & Natrayan, L. (2019). Processing and characterization of AA2024/Al$_2$O$_3$/SiC reinforces hybrid composites using squeeze casting technique. *Iranian Journal of Materials Science and Engineering*, 16(2). doi:10.22068/ijmse.16.2.55

Senthil Kumar, M., Natrayan, L., Hemanth, R. D., Annamalai, K., & Karthick, E. (2018). Experimental investigations on mechanical and microstructural properties of Al$_2$O$_3$/SiC reinforced hybrid metal matrix composite. In *IOP Conference Series*. Materials Science and Engineering.

Sharma, B. B., Raffik, R., Chaturvedi, A., Geeitha, S., Akram, P. S., Natrayan, L., Mohanavel, V., Sudhakar, M., & Sathyamurthy, R. (2022). Designing and implementing a smart transplanting framework using programmable logic controller and photoelectric sensor. *Energy Reports*, 8, 430–444. doi:10.1016/j.egyr.2022.07.019

Singh, A. P., Kumar, M. S., Deshpande, A., Jain, G., Khamesra, J., Mhetre, S., Awasthi, A., & Natrayan, L. (2020). Processing and characterization mechanical properties of AA2024/Al$_2$O$_3$/ZrO$_2$/Gr reinforced hybrid composite using stir casting technique. In *Materials Today*. Proceedings.

Singh, M. (2017). An experimental investigation on mechanical behaviour of siCp reinforced Al 6061 MMC using squeeze casting process. *International Journal of Mechanical and Production Engineering Research and Development*, 7(6). doi:10.24247/ijmperddec201774

Subramanian, B., Lakshmaiya, N., Ramasamy, D., & Devarajan, Y. (2022). Detailed analysis on engine operating in dual fuel mode with different energy fractions of sustainable HHO gas. *Environmental Progress & Sustainable Energy*, *41*(5), e13850. doi:10.1002/ep.13850

Suman, T., Kaliappan, S., Natrayan, L., & Dobhal, D. C. (2023) IoT based Social Device Network with Cloud Computing Architecture. In: *Proceedings of the 2023 2nd International Conference on Electronics and Renewable Systems, ICEARS 2023*. IEEE. 10.1109/ICEARS56392.2023.10085574

Sundaramk, M., Prakash, P., Angalaeswari, S., Deepa, T., Natrayan, L., & Paramasivam, P. (2021). Influence of Process Parameter on Carbon Nanotube Field Effect Transistor Using Response Surface Methodology. *Journal of Nanomaterials*, *2021*, 1–9. doi:10.1155/2021/7739359

Thakre, S., Pandhare, A., Malwe, P. D., Gupta, N., Kothare, C., Magade, P. B., Patel, A., Meena, R. S., Veza, I., Natrayan, L., & Panchal, H. (2023). Heat transfer and pressure drop analysis of a microchannel heat sink using nanofluids for energy applications. *Kerntechnik*, *88*(5), 543–555. doi:10.1515/kern-2023-0034

Vijayaragavan, M., Subramanian, B., Sudhakar, S., & Natrayan, L. (2022). Effect of induction on exhaust gas recirculation and hydrogen gas in compression ignition engine with simarouba oil in dual fuel mode. *International Journal of Hydrogen Energy*, *47*(88), 37635–37647. doi:10.1016/j.ijhydene.2021.11.201

Yogeshwaran, S., Natrayan, L., Rajaraman, S., Parthasarathi, S., & Nestro, S. (2020a). Experimental investigation on mechanical properties of Epoxy/graphene/fish scale and fermented spinach hybrid bio composite by hand lay-up technique. In *Materials Today*. Proceedings.

Yogeshwaran, S., Natrayan, L., Udhayakumar, G., Godwin, G., & Yuvaraj, L. (2020b). Effect of waste tyre particles reinforcement on mechanical properties of jute and abaca fiber - Epoxy hybrid composites with pre-treatment. In *Materials Today*. Proceedings.

Yogeshwaran, S., Prabhu, R., & Murugan, R. (2015). Mechanical properties of leaf ashes reinforced aluminum alloy metal matrix composites. *International Journal of Applied Engineering Research: IJAER*, *10*(13).

Chapter 15
State-of-the-Art Review of Various Off-Grid Hybrid Renewable Energy Systems for Rural Area Electrical Applications

Suresh Vendoti
Godavari Institute of Engineering and Technology, India

Dana Victoria
International School for Technology and Science for Women, India

M. Muralidhar
Sri Venkateswara College of Engineering and Technology, India

R. Kiranmayi
JNTUA College of Engineering, India

Kollati Sivaprasad
Godavari Institute of Engineering and Technology, India

ABSTRACT

Renewable energy systems serve as a sustainable alternative to fossil fuels, deriving from natural ongoing energy flows in our surroundings. These systems encompass the production, storage, transmission, distribution, and consumption of energy. Renewable energy systems offer numerous advantages, such as reliability, environmental friendliness, absence of harmful emissions or pollutants, low or zero carbon and greenhouse gas emissions, reduced maintenance compared to non-renewable sources, cost savings, job creation, and independence from refueling requirements. This chapter provides an overview of various types of renewable energy systems, with a focus on solar/wind/battery or solar/wind/diesel with battery storage integrated energy systems. This chapter also covers the technical and economic aspects of different types of HRES and their comparative results. Based on the findings of this review,

DOI: 10.4018/979-8-3693-3735-6.ch015

the chapter proposes a novel configuration for an off-grid hybrid renewable energy system designed for electrification in rural areas

BACKGROUND

Sources of energy that can be obtained naturally without depleting the planet's resources are known as renewable energies. These sources comprise solar, wind, hydro, geothermal, biomass, and biofuels. Renewable energies are cleaner and more sustainable than non-renewable energies like oil, gas, and coal, which are finite and emit large amounts of greenhouse gases. The technologies to capture and utilize renewable energies have improved in recent years, making their use increasingly viable and economical. Renewable energies are a crucial solution to combat climate change and reduce dependence on fossil fuels (Vendoti et. al., 2021). Investing in these energy sources can create jobs and promote sustainable economic development. In conclusion, renewable energies are a vital alternative to ensure a cleaner and safer future for future generations.

Hybrid renewable energy systems utilize multiple renewable energy sources to produce electricity, making them particularly advantageous in areas with limited or unstable access to traditional power grids. For instance, a hybrid system may incorporate both solar and wind energy, with solar panels generating electricity during the day and storing it in batteries for later use, while wind energy conversion systems generate additional electricity at night. Alternatively, a hybrid system may combine solar and hydro energy, with solar panels generating electricity during the day to pump water from a river or lake to a dam, which can then be released at night through a hydro turbine to generate additional electricity [Suresh et.al. (2020)].

Hybrid renewable energy systems possess the potential to surpass the efficiency and reliability of single-source energy systems. Moreover, they enable optimal utilization of existing resources and contribute to the reduction in energy generation costs. Consequently, the popularity of hybrid systems is on the rise globally, particularly in rural or remote regions.

This chapter aims to simplify the comprehension of hybrid renewable energy systems for beginners and present organized and detailed information. Unlike other summaries found in literature, it specifically addresses the unique features of the systems used in individual case studies. This approach enables a more thorough investigation and simplifies the identification of articles that align with the desired specifications for designing a hybrid renewable energy system (HRES).

The subsequent section conducted a comprehensive investigation of the relevant and hypothetical literature.

VARIOUS HYBRID RENEWABLE ENRGY SYSTEMS

PV/Diesel/Battery Hybrid System

Alexis Lagrange et. al. (2020) have made a study on the usage of renewable energies, diesel generators and energy storage systems for the purpose of maximizing the resilience of a micro grid, which has been analysed for an application of a hospital when the power failure occurs. By utilizing this technique, the usage of the diesel generator in a hospital has been reduced. The O&M cost of the solar PV system is $16 per year which has to be further minimized.

Mansour Alramlawi et. al. (2019) have implemented the dispatch strategy for an active-reactive power based on the EMPC which is used to reduce the overall functioning cost of the PV-battery-diesel micro grid. For the calculation of reactive power generation's cost from the diesel generator, a new novel method has been introduced. The outcome of this operating strategy reduces the overall cost of dispatched active and reactive power while increasing the lifetime of the battery but this hybrid system emits CO_2 because of the usage of diesel generator.

Pragati Tripathi et. al. (2018) have developed an integrated system's Simulink model in which the diesel generator has been used as backup, in case when the two sources failed to accomplish the demand. The diesel generators frequent ON/OFF functions have been reduced by this backup system. Simultaneously, the bulkiness of the system has to be controlled.

Tatiane Silva Costa et. al. (2019) have made a study on a PV-Diesel-battery hybrid system, which is made by generator, DG, battery system and separated loads. The HOMER software has been utilized for analysing the cycle-charging dispatch techniques and the load-following techniques. The result of this work shows that the energy produced by the solar panel is insufficient to operate all the equipment during peak hours.

S. Berlin Jeyaprabha et. al. (2015) have analysed the hybrid based PV/battery/DG system for remote areas to find out the optimum size and the calculation of tilt angle of the hybrid system by using AI Techniques without the metrological data. In India, as there are limited number of metrological centres, the prediction of tilting and sizing in remote areas are difficult, which has been accomplished by ANN and ANFIS techniques. To minimize the utilization of the DG for all the seasons, the PV arrays are fixed in any remote site and also the number of visits for manual tracking has been minimized. The lifetime of the solar panel is 20 years, still, which has to be increased.

Taoufik Laagoubi et. al. (2018) have presented a grid tied PV/Diesel/Battery hybrid system by employing two fuzzy logic controllers, which is used to extricate the maximal power from the PV array. Without the use of a DC-DC converter, these two controllers have been applied on an inverter. Grid connected PV system with fuzzy logic controllers provide better performance, when the grid and PV array are replaced by a diesel generator or batteries respectively. However, this hybrid system emits CO_2.

Mohammad Usman et. al. (2017) have compared three different modules for supplying electricity to the load such as grid alone, grid/solar PV/diesel generator and battery banks/ solar PV systems. These combinations are analysed based on the cost of producing electrical energy per unit, the cost of operating a conventional fossil fuel-based energy source, and the reduction of greenhouse gas emissions. After evaluating all the modules, it has been found that PV and grid-connected systems are more cost-effective, however the emission of CO_2 is high.

Shafiqur Rehman Luai et. al. (2010) have made a study on a system with diesel and PV/Diesel/Battery hybrid system for reducing the continuous consumption of a Diesel and to maintain the continuous power supply for the village people. The different sizes of batteries/converters and four generators with various rated power are used in determining the optimal power system. After analysing the above two systems, the hybrid system is found to be more profitable than a system with diesel but this hybrid system has the diesel usage and it makes the CO_2 emissions and cost of energy are high.

Saban Yilmaz et. al. (2015) have focussed on the planning, modelling and cost analyses of an efficient PV/Diesel/Battery hybrid system. This system provides a novel way for people in rural areas to change their living conditions. Though, the system has advantages, the CO_2 emission is high. Thus, it has to be further minimized by different methodologies in future.

Laith M. Halabi et. al. (2017) have investigated the decentralized power stations in Sabah, Malaysia that covers various combination of PV, diesel generators, storage batteries and system converters. For modelling the entire system, the HOMER software is utilized. Various types of PV penetration levels are examined for calculating the PV integration's impact in an accurate manner. This hybrid system shows that the usage of fuel consumption and low CO_2 emission but the cost of maintenance is high.

Nitin Agarwal et. al. (2013) have developed a PV/diesel/Battery hybrid system with the optimum modules for the remote areas. This module has been used to find out the exact dimensions of the hybrid system and to reduce the LCC and CO_2 emission. The computing time and efforts of this hybrid system is high thus, it has to be further minimized by the different optimization techniques.

Henerica Tazvinga et. al. (2013) have implemented the best energy dispatch model for PV/Diesel/Battery hybrid system and analysed the hybrid system's optimal energy flow. The impact of seasonal changes and energy flow of this hybrid system are evidently explained through this outcome. The consumption of the diesel has been reduced in this hybrid system. However, the usage of diesel causes a pollution in environment, which has to be eliminated by using different techniques.

Kanzumba Kusakana et. al. (2015) have developed two control strategies of the diesel generator called Continuous and ON/OFF control in a PV/Diesel/Battery hybrid system. The comparison result of these two techniques shows that the consumption of fuel and the operating time of Diesel Generator are low in the continuous control. As the diesel is one of the major pollutants of nature, its usage has to be further minimized.

Wei Cai et. al. (2020) have made a study on off-grid application of a hybrid system to analyse optimal sizing and location which has been made by photovoltaic, battery and diesel technology. For the identification of location, geographic system module has been used. The capacity has been maintained to meet the load through the reduction of total life cycle cost. The fuel usage of diesel generator has been reduced when compared to other systems. The main challenge is to minimize discharge of CO_2, the cost energy and total life cycle cost.

S. M. Shaahid et. al. (2014) have made an analysis on Hybrid-PV diesel battery power systems which has to meet the residential loads economically in Saudi Arabia. The system has been analysed based on the data obtained from the long-term solar radiation using HOMER software and also the penetrating power of PV, emission level of carbon, the amount of diesel consumption, NPC and cost of energy have been analysed. By this analysis it was revealed that the amount of energy stored in a battery has been increased for an average load. However, the potential of solar energy has to be increased in a better way to deploy the renewable energy sources.

Wind/Diesel/Battery Hybrid System

Chong Li et. al. (2019) have evaluated the wind/diesel/battery/ hybrid system with three types of batteries (DG/ZB, DG/LA, DG/LI) for the analysis of economic and technical feasibility in a small residential area with 280 homes. By utilizing HOMER pro software generation of energy, optimum allocation and sensitivity analyses of the relevant hybrid systems have been analysed. After the evaluation of these specified batteries, it has been determined that the DG/LI battery emits less CO_2 than other two batteries but it is even possible to minimize the CO_2 emission by using new techniques.

Juan M. Lujano-Rojas et. al. (2012) have developed load management techniques (with and without load control) in an optimal manner for wind/diesel/battery hybrid system to minimalize the amount of energy supply of both diesel generator and battery bank. The ARMA model has been implemented for predicting the wind speed in an hour-based calculation. The outcome of this work shows that the system without load control technique has minimized the usage of diesel generator but it has been further reduced in order to save money.

Elyas Rakhshani et. al. (2019) have proposed two novel approaches called single-turbine and multiple turbine models for enhancing the planning of wind-diesel-battery hybrid systems, in which the usage of wind turbine has been maximized for reducing the carbon emission. By the analysis of these two models, the multiple-turbine model has reduced the overall net present cost of planning which is comparatively better than a single-turbine model; however, the operating cost of the multiple-turbine model has to be minimized much more by applying various topologies.

Masoud Safari et. al. (2014) have introduced the best load sharing techniques for wind/diesel/battery hybrid system with ICANN algorithm in which the diesel generator has been used to complement the wind source's erratic capacity and also the part of the temporary peak demand is compensated by the battery. For analysing the performance of this system, the best energy has been selected from the available energy source by utilizing ICA, ACO and PSO algorithms. The fuel consumption has been reduced by using these algorithms but it has been further minimized for saving the environment.

Leong Kit Gan et. al. (2017) have modelled the tubular and four leg configurations of the tower shadow effects for comparing the reduction of battery life time. In the state of different tower configurations as well as operating under various load conditions the reduction of battery life time has been analysed. A new perspective on battery lifetime reduction has been provided by this work. The micro cycles are generated through the tower shadow effect in which battery lifespan has been shortened by these micro cycles in a quantifiable way and so when sizing the hybrid device during the planning stages, the system designers have to keep in mind that the battery lifetime is often thought to be optimistically more in real-time.

Mehrdad Ghahramani et. al. (2019) have introduced a new modelling approach for ODAS of a smart distribution system based on Hong's T-PEM method which is used to address the uncertainties of wind generation and load consumption. The result of this work shows that the overall operating cost of smart distribution system has been minimized as well as the units of Diesel generator have been used in a more profitable manner but the use of diesel generator has resulted in CO_2 emission.

Rafael Sebastian et. al. (2011) have modelled the wind/diesel power system with BESS and consumer load in which two types of working modes are analysed which are termed as WD and WO. To control this system, BESS active power has been introduced which increases the stability, reliability and security of WDPS but to simulate the real isolated MG, the WDPS with BESS has to be further expanded.

PV/Wind/Battery Hybrid System

Essam A et. al. (2020) have investigated the techno-economic analysis along with an optimal HRES design in the residential area to satisfy the needs of domestic electricity. HOMER tool has been used to perform the optimal sizing of HRES components and to examine the economic analysis. The evaluation of these work shows that the PV/wind/battery hybrid device is the most profitable and emits the less toxic gases. The micro grid reconfiguration has to be implemented under the inverter and rectification mode to regulate the frequency and voltage of standalone as well as grid connected system.

Dongmin Yu et. al. (2020) have provided a well-constructed optimization method to examine the uncertainty pool market price in which a DRP is designed to reduce the LEC's purchased cost. The DRP based robust optimization technique is compared with risk neutral and robust strategies. The comparison results shows that the purchased cost of LEC has been minimised in risk-neutral and robust strategies. However, the cost of LEC has to be further reduced by employing various topologies.

Yinghao Shan et. al. (2019) have proposed a PID regulator free predictive control strategy for PV/wind/Battery hybrid system. In this system, MPCP is utilized to control the BESS's bidirectional DC-DC converter, which smoothens the output fluctuation of the renewable energy source whereas a MPVP control strategy has been included to control the AC-DC interlinking converter. The results of this research show that the system has improved control capabilities and better voltage efficiency. In this hybrid system, extra communication facilities and sensors are required for MPCP and MPVP control approaches.

Rajiv Kaur et. al. (2019) have developed a PV/wind/battery HRES for off-grid telecommunication towers in rustic areas. For the optimum sizing of HRES, DMGWO technique is implemented and the decision-making process has been done by the Euclidean distance-based approach. A sensitivity analysis is executed for investigating the effect of decision variables on the objective functions that are highly associated. This HRES based optimum sizing approach has to be implemented with various micro grid application.

O. Charrouf et. al. (2020) have made a study on an ANN power management for a reverse osmosis desalination unit which is given by PV/wind/battery hybrid system. For training the data, feed forward back propagation neural network has been utilized among the various ANN approaches. The photovoltaic generator's MPPT controller ensures that the generated power has been maximised even when shading conditions arise at certain times. However, the capital cost of wind turbine is high.

Jayachandran M. et. al., (2017) have designed the HMGS which provides low-cost option to solve the power flow problem in the rural area (sundarban). The optimal sizing of system components and the best configuration of the hybrid system are determined by using PSO technique. Owing to the high potential of RE sources in sundarban area the HMGS fundamentally work with solar and wind energy. The outcome of this work shows that in the Sundarban area, using renewable energy resources is the best option for increasing energy access. But this system is highly depending on the wind and solar energy which is not possible for all type of time period.

Jyoti B et. al. (2018) have presented the optimal PV/wind/battery hybrid system by considering the impact of sensitivity analysis like global solar radiation, PV panel cost and wind speed, in which the sensitivity analysis assists in the proper design of the system. iHOGA tool has been used to simulate this work. The evaluation of this system shows that the total net present cost has been reduced for the user by implementing various technologies. However, this system has higher COE (so it has to be reduced in future by implementing various methods.

Akbar Maleki et. al. (2015) have designed a PV/wind/battery hybrid system by utilizing a network-independent pattern and also various types PSO algorithms have been analysed depending on the real time data and metrological data of the rural areas for the optimal sizing of this hybrid system. The result of this work shows that this hybrid system is suitable for the rural areas when the speed of the wind is better but it is not sufficient when the wind speed is in poor condition.

Ali Saleh Aziz et. al. (2018) have investigated the feasibility of using an integrated PV/wind/battery, PV/battery and wind/battery hybrid energy systems to produce an electrical energy to a remote rustic school in Iraq's southern region. HOMER software has been utilized to evaluate the feasibility analysis of this hybrid system. From the evaluation of this hybrid system, it was shown that the hybrid wind/PV/

battery system is more feasible than PV/battery and wind/battery whereas the installation cost of the PV and wind turbine is high in this system.

A.Y. Hatata et. al. (2018) have proposed a novel approach for finding the optimal size of solar/wind/battery HPS by implementing CLONALG. For the least amount of load interruptions, this hybrid system has been designed with the optimum size of photovoltaic arrays, batteries and wind turbines. The optimization result shows that the hybrid power system has provide an optimum size of HPS with the minimum cost and also the power supply loss has been minimized but it has to be further reduced by using other technique.

Aeidapu Mahesh et. al. (2017) have developed a PV/wind/battery hybrid system by using GA. A new energy filter algorithm has been employed for smoothing the power which is pumped into the grid. The components of this hybrid system are most cost-effective by implementing this energy filter algorithm but the operating and maintenance cost is high still it has to be minimized more.

R. K. Raj Kumar et. al. (2011) have modelled the PV/wind/battery hybrid system with ANFIS which is implemented by utilizing the real metrological data. Here, PSCAD simulation is used for validating the developed ANFIS, which includes efficiency sizing and autonomy moreover it provides a minimum cost function but it has a higher O&M cost.

Tao Ma et. al. (2019) have introduced a PV/wind/battery hybrid system to investigate the impact of varying saturation parameters. For making the system more feasible and realistic, three different sizes of wind turbine have been implemented. By satisfying the load demand, the impact of each system configuration on storage bank, investment cost, COE, and reliability of the system have been analysed in a comprehensive manner. Through this method, maximum potential of RE sources is extricated with the assistance of HRES, which leads to the minimization of cost. To decrease the NPC of the system and making the system more strengthen, a pumped hydro storage together with the battery has to be implemented in future.

Saeedeh Ahmadi et. al. (2016) have presented the efficient PV/wind/battery hybrid system employing Big Bang-Big Crunch algorithm in which mutation operation and PSO capacities have been used to avoid the local optimum and also develop its exploration ability. In this Big Bang-Big Crunch based hybrid system, the total present cost has been minimized but the maintenance cost of the wind turbine generator is high.

Adriana C et. al. (2017) have modelled the gird connected PV/wind/battery hybrid system in which the power generation-side strategy has been described as general mixed-integer linear programming by considering the two steps of storage unit for managing the charging process in a proper manner. This hybrid system is used for the minimization of operating cost and supporting the self-consumption on the basis of forecast data (for 24-hour). However, this system has the optimization problem, which has to be improvised in future.

Aeidapu Mahesh et. al. (2016) have utilized the energy filter algorithm for designing the optimal sizing strategy of grid tied PV/wind/battery hybrid system. The PSO technique has been used to enhance the hybrid system's efficiency and to eliminate the fluctuations of the grid power. Finally, it has been observed that the hybrid system with the novel energy filter is highly profitable but the operating and maintenance cost has to be further minimized.

Tao Ma et. al. (2014) have presented the feasibility study and techno-economic assessment of a standalone PV/wind/battery hybrid system in which the feasibility analysis has been done by using HOMER software. The optimal energy systems have been clarified in order to determine the most effective off-grid scheme. The final outcome of this system shows that the reliability and quality of this hybrid system is better than the other single energy sources and also it has a low maintenance cost but the fixing cost of PV and wind turbine is more.

Muhammad Shahzad et. al. (2019) have proposed an ideal sizing framework for hybrid solar-wind-battery system, based on key indicators such as COE and efficiency of the system. The operating strategy and a mathematical model have been developed by using GA to optimize this hybrid system. This optimization result has been compared with the results provided by the HOMER tool for analysing the effectiveness of this system. The comparison result shows that the GA is most effective than the HOMER tool in terms of optimum sizing, time of the simulation, and system reliability. However, the initial investment cost of PV and wind turbine is more.

Biomass/Biogas Hybrid System

Jameel Ahmad et. al. (2018) have studied a techno-economic analysis for the purpose of electrification in rural areas which consists of a HRES of wind, PV and biomass systems. They have made this analysis in the local areas of Punjab in the province of Pakistan. In this tri-generation system the micro grid is modelled using Homer software. For the purpose of grid integration, the elaborative assessment of solar energy, bio mass and wind have been carried out. Even though the local load demand is low, an excess amount of power has been given to the national grid and also sharing of the whole amount of load has been optimized. In case of similar load profiles, the cost of energy is low when analogized to an off-grid hybrid system.

Riccardo Amirante et. al. (2019) have made the study about a small-scale combined cycle power plant by using biomass which is externally fired for the generation of heat and power in the rural areas. This system is very mature and cheap, which has been obtained from the automotive sector with low initial investment and has less maintenance. The flexibility of fuel has been achieved by gas turbine and also the problems related to the supply of fuel have been avoided in the remote rural areas. The solution of this system has achieved the autonomy targets of energy in a circular paradigm. The system efficiency has been assessed by three probable configurations which have been studied with thermodynamic analysis. The efficiency thus obtained is independent of the chosen fuel and so the power plant has been treated with any kind of fuel. At the same time the fuel must be sustainable and accessible.

C. Y. Li et. al. (2019) have made an experimental investigation on the integrated system of biomass gasifier, engine and generator to improve the waste heat and to generate the power. The waste heat recovery from both exhausted gas and syngas has been carried out for the purpose of heating the biomass feedstock such as redwood pellets and woodchips. This integrated system has been established and the experiments have been carried out for validating this model. From these observations, it has been observed that the accuracy of the system is acceptable whereas the energy loss has been taken up to the highest ratio which has to be further minimized.

Farkhondeh Jabari et. al. (2019) have made an investigation on combined system of cooling, desalinated water and power generators during summer. In this analysis bio gas is utilized to minimize the greenhouse gas emission and fuel consumption. The waste heat has been removed from the flue gases for the purpose of preheating and for the process of humidification, dehumidification and desalination. This tri-generation system has reduced the electricity cost by using the application of Demand Responsible Programs. However, the cost of energy has to be further minimized.

Farkhondeh Jabari et. al. (2018) have designed and investigated the performance of a biogas powered combined system of cooling and power generators, which consists of an air preheated, a bio gas fuelled gas turbine, a combustion chamber, an air compressor, compression chillers and an integrated plug-in electric vehicle. In this system three compression chillers have been installed for the purpose of cooling and the waste heat of mid or high temperature has been recovered from exhaust gases for improving the overall efficiency. From this analysis, it has been observed that this system has the benefits of low energy requirement, eco friendliness and the cost of biogas production per watt is too high which has to be further minimized.

A. Baccioli et. al. (2019) have made a study about a biogas plant's poly generation ability through the machineries like micro gas turbine; digester, an upgrading system of biogas in which bio methane; heat and electricity have been obtained as outputs. Along with these, the flow of energy and internal mass has been investigated on the basis of variations in the biogas fraction. This system has been analysed through visual basic routine using the function of bio gas plant in the off-design conditions. From this analysis, it has been observed that the generation of power, bio methane and heat have been obtained only when the produced bio methane, is below 50%. This has been used in the upgrading process and the production of bio methane and power generation have not achieved the maximum rate simultaneously.

Yilin Zhu et. al. (2019) have made a thermodynamic analysis and economic assessment on a combined system of heat and power, which has been fired using biomass. In this system, a promising technology called Organic Rankine Cycle is used, in which mono ethanolamine-based CO_2 capture has been integrated. This system is analysed on the basis of parametric optimization and economic evaluation including CO_2 capture for 11 working fluids. The results thus obtained have revealed that the thermodynamic performance has maximum energy saving ratio, minimum net power indicator and the levelized cost. Though the system is eco-friendly and economically attractive, the power efficiency has been decreased for some of the working fluids.

Yuli Zhu et. al. (2019) have made the analysis on water usage of direct combustion of biomass system, in which the combination of Water Foot Print analysis and Life Cycle Assessment has been carried out for power generation. The advantages of these two methods have been integrated, in which the indirect and direct water usage have been combined with water foot print of grey, blue and green colours. Based on the results thus obtained, it has been observed that the power generation through direct combustion of biomass is lesser than that of the bio-oil method. Furthermore, the cultivation of high water-efficient crops, straw collection coefficient and power generation efficiency have to be improved for alleviating the conditions of water stress.

Wen-Lih Chen et. al. (2020) have made the study about a micro united heat and power system with thermo photovoltaic cell array and Stirling engine, in which the thermal sources have been provided by both the surface of high luminescent and the flue gas of the platinum reactor at high temperature. This bio-syngas fuelled CHP system yields thermal energy from both flue gas and the reactor surface. This system has converted energy into electricity and supplied the hot water at the same time. Even though the manoeuvrability and feasibility of this system is high, the overall efficiency of this system is low which have to be further improved by the optimization of thermal management.

Bosheng Su et. al. (2020) have analysed the biogas fired co-generation system on the basis of chemically improved gas turbine cycle. In this system, both physical and chemical energy have been efficiently utilized. In this model, the biogas steam reforming process has generated a huge amount of power, because of the high chemical energy in syngas. The obtained exhausted gas from the power engine has rich moisture content. Therefore, the recovery of the low-temperature heat has been depended on its demand in the digester. Moreover, this system has provided the low-temperature latent heat recovery with increased power generation. However, the consumption of mid-temperature heat is too high for the evaporation of water, which has to be further minimized.

Chaouki et. al., (2020) have introduced a hybrid micro grid system using solar and biomass for providing electricity in the city of Sharjah. The main intention of this system is to obtain the optimal configuration which has to meet the desired electrical loads and to explore the accessibility of the RES. This system contains the resources of both solar and biomass, electrical loads and the system components. This system has been analysed with hourly simulations for computing the energy, to and from the components. The results thus obtained have revealed that the total net present cost is too high, which has to be further reduced.

Sonali Goel et. al. (2019) have made a study on an optimal sizing hybrid system of biomass and biogas for supplying electricity to the commercial farm in the northern regions of Odisha. This study has been carried out for the optimization of these hybrid system components, which has to meet the necessity of electric power in the specified region. The techno economics of this system is analysed using both HOMER and the straight-line technique. The results thus obtained are compared and it has been observed that the combination of 12kW biomass + 3kW biogas system with no price of raw materials has the sustainable power supply to achieve daily electricity demand of the farm. However, the cost of energy has to be slightly reduced.

Mohammad Hossein Jahangir et. al. (2020) have designed a hybrid system of biomass, wind and solar energy for supplying power in rural areas in which the environmental and economic assessment have been carried out. This system contains a biogas generator, wind turbines and photovoltaic panels. By utilizing this system, CO_2 emissions and the growth of fossil fuel power plants have been prevented. By using HOMER-pro software, the optimal system configurations on biomass price, input biomass rate and inflation rate have been obtained through the sensitivity analyses. The results thus obtained have shown that the rise in the biomass price and inflation rate have also increased the cost of electricity.

N. S. Suresh et. al. (2019) have analysed a steam Rankine Cycle based hybrid model of biomass and solar thermal power plants. During power generation, solar and biomass system have used parabolic trough technology and fluidized bed combustion technology respectively for steam generation. Hybridization of multiple RES has been considered as the best solution for reducing the intermittency issues of power supply. At the time of solar intermittency periods, the biomass system has played an important role and also a stand-alone mode boiler has generated the power post sunshine hours, which has to meet the power demand. Even though the capacity utilization factor has been enhanced, it has to be further improved for day time operation.

Antonio Cano et. al. (2020) have made a study on HRES comprises of biomass gasifiers, batteries, photovoltaic energy and hydrokinetic turbine, in which techno-economic assessment and energy analysis have been carried out. By using different types of biomasses, the best configuration of this system has been determined. In this system, the latest behaviour patterns of the sources have been determined based on the electricity demand. Though this system has the ability to fulfil the demand of electricity, the cost of energy has to be slightly reduced.

Micro Hydro Power (MHP) Hybrid System

Himadry Shekhar Das et. al. (2016) have proposed a model of a small hybrid system for power generation that includes a micro-hydro with battery, PV and a diesel generator to help offset, increasing energy demand in southern areas. The economic feasibility as well as the technological aspects of setting up this hybrid system have been discussed by using HOMER review. The result analysis reveals that the PV/micro-hydro/Battery/Diesel hybrid system is the most cost effective and it delivers a better efficiency. However, the emission of CO_2 is high because of the diesel usage.

Asit Mohanty et. al. (2015) have proposed the wind/diesel/micro-hydro hybrid system by utilizing PSO-based SVC controller in which PSO algorithm is utilized to optimize the SVC controller and for the Reactive power losses have been accomplished by the SVC controller. The outcome of this work shows that the system's transient stability and reactive power compensating capability have been achieved by this controller. The impact of diesel usage on environment is highly maximum and so it has to be considered in the future works.

Liangliang Wei et. al. (2020) have developed a PMG for micro hydro-electrical generation system with low-speed and high-efficiency parameter. SPCC and Silicon steel based two different low-speed generators have been designed and compared to check the performance of this system. The comparison result shows that the system provides better efficiency and it requires very less production cost. However, it is possible to increase the efficiency of this system by implementing various strategies.

W. Apichonnabutr et. al. (2018) have examined the trade-offs between environmental and economic efficiency of an independent energy system based on an existing Micro hydro power plant, with the goal of increasing potential reliability. A mixed assessment process has been described for calculating the trade-offs among the environmental and economic efficiency of an independent hybrid energy system. The lifetime of solar and wind turbine has to be further improvised for the better efficiency.

Walter Gil-Gonzalez et. al. (2020) have presented the modelling and control approach of a SHP for a DC micro grid, depending on the passivity theory, in which SHP is designed by using wind turbine, PMSG and a voltage source converter. PBC approach has been employed for designing the controller. The PBC's output was evaluated in a DC micro grid and analogized with the PI controller. The results show that the PBC performed better than the PI controller. The average error of this system has been minimized by implementing PBC based controller but it has to be further minimized by various control approaches.

Uri Stiubiener et. al. (2020) have introduced a long-term hydro-solar system to replace the existing power generation model in Brazil. The amount of land flooded by HEPP reservoirs has been estimated to determine the feasibility of installing PV-FPP near HEPP dams. The result has proved that HEPP reservoirs in Brazil is capable of accommodating less than 10% of the surface area for PV-FPP to produce electricity by using solar energy during irradiation hours. However, in the period of low or zero irradiation, the operational flexibility is minimum.

Sandile Phillip Koko et. al. (2018) have developed a grid tied optimal energy management system for transmitting a power in an MHK-PHS system which has been used to minimalize the cost of grid consumption, maximizing the energy sales revenue. In this system, the external disturbances are affecting the performance of the system.

Koustav Dasgupta et. al. (2020) have combined the environmental clean energy, such as wind energy, by considering the HTS problem to address the effects of thermal emissions. The SCA method has been employed for reducing the cost of generation and fuel emissions and also the SCA's various control parameters have been effectively used to equalize the investigation and extraction stages and to determine the optimum solution. However, the fuel emission has to be further minimized by implementing various topologies in future.

Arunpreet Kaur et. al. (2019) have proposed a hybrid GWO with mutation strategies for solving the scheduling problem. To satisfy various demands of electricity, a realistic coordinated power system model has been developed in consideration with thermal, CHP, and multi-chain hydro systems. The algorithm's expansion and intensification capabilities are enhanced by the challenging features of Gaussian and Cauchy mutation strategies. In addition, for the initial diversification of the wolves, an opposition-based mutation technique has been employed. In this system, the cost of fuel is high and it leads to environmental pollution.

Jean De Dieu Niyonteze et. al. (2020) have examined the necessities of new technology expansion relevant to the power sector, in Rwanda. By considering an average load demand, four different types of HRES have been developed and simulated to serve electricity to the rural and remote areas. The HOMER software has been utilized to simulate and optimize this hybrid system. The outcome of this work shows that the hydro/solar/battery hybrid system emit zero greenhouse gases and it has a very less carbon footprint but the operating cost of this hybrid system is high.

K. C. Almeida et. al. (2019) have expressed a decentralized process by using nonlinear optimization algorithm to attain the medium-term hydro thermal dispatch strategy, in which a bi-level optimization problem has been solved by utilizing the Lingrangian duality theory. The result of this work shows that the dispatch protocol appears to be a viable choice for systems operating in energy markets but it has high generation cost, it has to be still minimized.

Benjamin A et. al.(2017) have proposed a multi-criteria decision-making system based on Neuro-fuzzy by utilizing the benefits of both neural networks and fuzzy logic. To categorize each important features of a power plant site into different classes, Genetic algorithm based trained Multi-layer Percep-tron Neural Network has been employed. To install hydro power plant in India and to give a score rate or priorities for different existing and upcoming sites, a fuzzy reasoning has been applied. Because of the significant financial expenditure, manpower requirements, and time constraints, site selection is an almost irreversible decision once it was installed.

Mahindra Nandi et. al. (2019) have examined the efficiency of an interlinked two-area hydro–hydro-power system model is examined by using AGC. To change the TCSC controller, Taylor theorem has been implemented and also for the improvement systems of phase lag, two-stage phase-compensating blocks have been cascaded to both the SMES and the TCSC. Through sensitivity analysis, the robustness of the controller has been examined in loaded conditions and model factor's uncertainties. In this system, the damped oscillations have been eliminated and also reducing the tie-line power flow and settling time of frequency. The controller design of this system is the most challenging one for a better LFC results.

A. B. Kanase-Patil et. al. (2010) have introduced the micro-grid electrification using a combined RE system to meet the electrical and cookery requirements of the villages that are without electricity. By presenting the CIC, the ideal system efficiency, overall system cost, and energy cost have been de-termined. HOMER software has been utilized for the optimization of off grid system with battery back. The COE of this system has been minimized for the economical purpose but it has to still minimize.

Tarlochan Kaur et. al. (2017) have suggested the viability of hybrid electricity systems which has been made up of small-scale generators, hydro, and solar PV with and without energy storage. The capacity of various renewable resources, such as hydro and solar energy, is estimated and also calculated the amount of electricity required to meet the basic necessities of a rural community, such as education, recreation, and a health centre. The HOMER program has been used to optimize and evaluates this hybrid systems. The result shows that the energy generation of this system has been maximized. However, it has to be further increased by implementing various topologies.

Hybrid Renewable Energy System

Om Krishnan et. al. (2020) have compared the grid connected PV system's two distinct HESS structures such as, FC-SCESS and FC-BESS to satisfy the necessity of electricity load for a housing building in NITKKR. The HOMER software has been used to measure the residential building's load and also determined the optimal sizing of different PV system constituents. The result of this work shows that the operation and maintenance cost of this system has been minimized but the usage of fuel cell battery causes the environmental pollution.

M. R. Elkadeem et. al. (2019) have proposed a generic and systematic decision-making method for optimising the design and use of autonomous HREMG systems in urban environments (Safaga). The feasibility analysis of an HRE-MG device has been examined. Furthermore, reliability and sensitivity analyses are performed to observe the impact of LPSP and model parameter's uncertainty on the viability of the developed system. The result evaluation shows that this hybrid system has outstanding efficiency and advantages in meeting Safaga's energy and heat requirements but in which the operation and maintenance cost is high.

Mohammad Reza Akhtari et. al. (2019) have investigated the EAHE's thermal efficiency in both continuous and intermittent modes of operation in which the heat exchanger is connected to a HRES that includes wind, solar, and hydrogen to increase efficiency and sustainability of the system. The result analysis indicates that the system's long-term performance has been determined by observing its behaviour on the first day of service. In this HRES system, the usage of diesel generator and the fuel emission have been reduced. However, the cost of energy is high, which has to be further minimized.

Krishnamoorthy Murugaperumal et. al. (2020) have provided the ideal design and technology economic investigation of a HRES for rural electrification in India's Korkadu district in which PV, wind turbines and bio generators have been considered the most important sources due to their higher potential. The efficient HOMER method is utilized to carry out the optimal design and techno-economic study of this system. The HRES's operational behaviour has been compared with a variety of operational techniques, including load following, cycle charging, and the system's combined strategy. This research demonstrates that the power generation based on HRES is most cost-effective and long-term substitute to conventional grid extension. This system requires the implementation of AI-based HRE sizing optimization for the best technology economic configuration.

Min-Hwi Kim et. al. (2019) have analysed the hybrid system's techno-economic analysis in which solar thermal system, heat pump and district warming network have been employed for an energy community. A comparative analysis of environmental and economic study of two traditional gas-fired boiler systems and a centralized heat pump system have been carried out. The outcome of this work shows that the system's efficiency and economic performance have been improvised. On the other hand, the bulky component size makes the system more complex.

Monotosh Das et. al. (2019) have utilized the optimization methods such as Meta heuristics, to achieve an economical as well as ideal design of a micro-grid hybrid PV/biogas generator/pumped hydro energy storage/battery for a radio transmitter station in India. For the optimization problem, a comprehensive modelling strategy has been created; to examine the feasibility of this system design, a sensitivity analysis has been utilized, which depends on the loss of load probability type reliability principles. The overall NPC has been minimized in this system which is subjected to design constraints. For enhancing the knowledge-based optimization problems of HRES, various topologies have to be implemented.

Daniele Landi et. al., (2019) have described a method for evaluating the energy flows, environmental effects, new modular cost and integrated system of renewable energy system and intelligent electrochemical storage systems which permits the residential buildings to self-produce and consume electricity. The analysis of this work shows that the overall impact has been determined by the selected configuration, as well as the consumption profile and user types. In terms of usage costs and environmental effects, the possibility of expected result is zero only for short-term use.

Rohan Goddard et. al. (2019) have made a study on an Optimum Sizing as well as Power Sharing of DHRES by considering the Socio-Demographic factors. In this study, the optimization problems have been solved by using MATLAB-GA and function which results the minimization of capital cost but in which the diesel usage causes environmental pollution.

Kun Lee et. al. (2019) have utilized the DP based optimal control to explore the entire design space for HRES in which the four economic metrics such as Initial cost, LCC, O&M costs and PBT have been utilized to measure economic efficiency. The simulation results show that the size of renewable power generators and battery storage decides the overall economic efficiency. For evaluating the weather and demand forecast uncertainties, various techniques have to be implemented.

Rui Wang et. al. (2019) have proposed multi-scenario optimization approach for HRES architecture in which bi-objective optimization model has been implemented to design classic stand-alone hybrid system and it is made of PV, WT, battery, and DG. A scenario-dominance based multi-objective evolutionary algorithm (s-NSGA-II) is employed for solving the approach in an effective manner. When dealing with more complicated problems, the s-NSGA-II approach is very simplistic and needs to be developed.

In the literature, more than 12 terms are used to describe HRES advances and a total of 82 articles are found that define HRES using various terminologies. The majority of them refer to the hybridization of traditional energy systems with RESs as grid alternatives. In the terminology for RES ranges from the most generic to the most complex. Hybrid renewable energy system are often configured as hybrid off-grid system, hybrid power system, micro-grid system, mini-grids, standalone hybrid energy system, etc.

Problem Statement

Based on the widespread literature review carried out under the study, it is found that most of the studies have focused on solar/wind/battery or solar/wind/diesel with battery storage based integrated energy systems. In rare cases, the intermittency of RESs and discharging of battery may occur, which intervenes the supply provided to the load. Thus, there is a scope to develop models of HRES for remote rural and far-flung areas having fuel cell along with RE sources.

Moreover, in rural areas, designing a cost effective energy generation system is crucial. The modelling and implementation of the hybrid energy system in decentralised mode are highly limited. The established models fail to meet the necessities of individual regions, villages and blocks. The efforts to enhance the power supply in the rural places are highly low and so these methods lack in fulfilling the energy demand of the people in rural places.

CONFIGURATION OF HYBRID RENEWABLE ENERGY SYSTEM

A hybrid energy system has the renewable energy conversion components like PV panels, wind turbines, hydro turbines and traditional non-renewable generators like diesel generators, micro turbines, energy

Figure 1. Configuration of proposed HRES

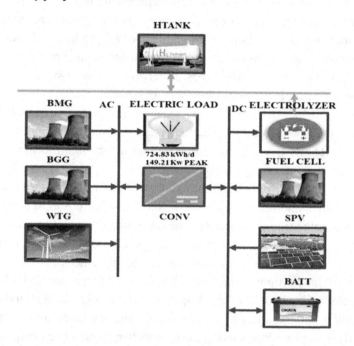

storage device i.e., batteries. All of these components are either included or not included in a hybrid energy scheme. The first stage is to model the individual components for making the best choice of components.

The schematic representation of standalone hybrid energy system is illustrated in Figure 1. This system comprises solar PV, wind turbine, biogas, biomass, fuel cell generators, battery bank and electrolyser. Provisions for the availability of both AC and DC buses are made using electronic converters. The power produced by the renewable/diesel generator is split into two branches, one branch delivering generators energy directly to the load and the second one powering the battery. To serve the load, electrical energy can be produced either directly from renewable/diesel generator, or indirectly from the battery. The energy from all the generators is allowed to charge battery.

TECHNICAL AND ECONOMICAL ASPECTS OF HRES

A total electrical load of 820 kWh per day is needed to completely satisfy electrical energy requirements of the rural area under study. Average AC load demand in the study area is 724.83 kWh/day. Figure 2 shows daily hourly electrical load profile of the rural area considered. In HOMER simulation, a stand-alone system is considered as the HRES with a life time of 25 years and 8% annual interest rate. For simulation purposes, different sizes of the components are considered as discussed earlier. Adding more number of sizes for each simulation to find the most optimal system configuration can maximize the search space. Here the HRES comprises of the following: BMG, BGG, SPV, WTG, FC, Electrolyzer, Converter and BATT storage systems. The figure 3 highlights the different possible configurations as obtained by HOMER.

Figure 2. Hourly load profiles in the study area

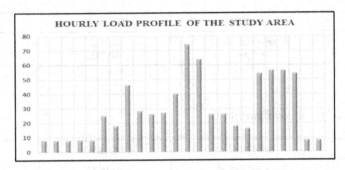

Figure 3. Different combinations of HRES using HOMER

Architecture											
SPV (kW)	WTG	BMG (kW)	BGG (kW)	Fuel Cell (kW)	BATT	Electrolyzer (kW)	HTank (kg)	Conv (kW)	Efficiency1	Dispatch	NPC ($)
100	50	50.0	60.0	57.0	200	50.0	300	80.0	0	CC	$890,013
100	50	50.0	60.0	57.0		50.0	300	80.0	0	CC	$897,847
100	50	50.0	60.0		200	50.0	300	80.0	0	CC	$929,878
100	50	50.0	60.0			50.0	300	80.0	0	CC	$1.10M
	50	50.0	60.0	57.0		50.0	300	80.0	0	CC	$1.14M
	50	50.0	60.0			50.0	300	80.0	0	CC	$1.18M

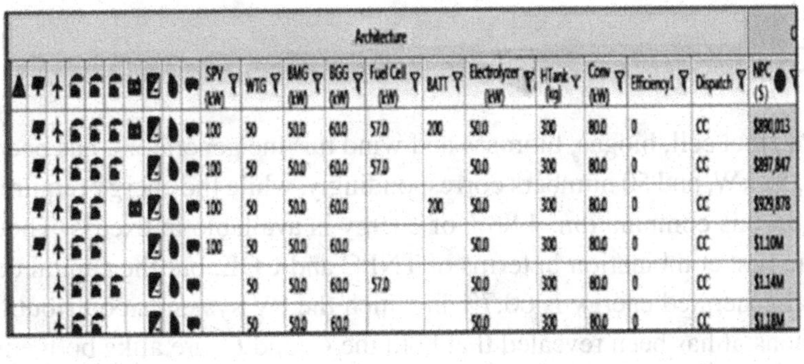

This research focuses different energy combinations (C_1, C_2, C_3, C_4) and finding the best combination out of these four. The economic and technical outcomings of these combinations are described as follows:

Combination C_1

The minimum TNPC observed in C_1 was *Rs.*66,004,647.52($890,013 and least COE of *Rs.*15.57/kWh($0.214) had 0% capacity shortage. For C_1, sizes of SPV, fuel cell, biogas, biomass and wind turbine generators, have been estimated as 100 kW, 57 kW, 60 kW, 50 kW, and 50 numbers correspondingly, while the energy requirement is calculated as 328266 kWh/yr. On the basis of this scenario the total power generated is 163,527kWh in which, the PV contribution is 46.8% and the remaining 53.8% is obtained from the additional sources. It has also been found that the in the total energy production, 6.07% of energy is in excess which reveals that this combination satisfies the load demand and the values observed are tabled in Table 1.

Table 1. Results obtained from combination C_1

Combination	Total NPC	COE (/kWh)	Operating cost
C_1	*Rs.*66,004,647.52 ($890,013)	*Rs.*15.57 ($0.214)	*Rs.*2,530,010.84 ($34,109)

Table 2. Results obtained from combination C_2

Combination	Total NPC	COE (/kWh)	Operating cost
C_2	Rs.66,603,566.02 ($897,847)	Rs.15.96 ($0.215)	Rs.2,738,262.30 ($36,917)

Table 3. results obtained from combination C_3

Combination	Total NPC	COE (/kWh)	Operating cost
C_3	Rs.69,002,788.89 ($929,878)	Rs.16.33 ($0.223)	Rs.2,920,404.76 ($39,354)

Combination C_2

For C_2, sizes of SPV, fuel cell, biogas, biomass and wind turbine generators, has been estimated as 100 kW, 57 kW, 60 kW, 50 kW, and 50 numbers correspondingly, while the energy requirement is calculated as 396121kWh/yr. In this combination, 4.86% of energy is available in excess. Though this combination is similar to the first combination in terms of TNPC and COE, but the capital cost is considered a little high. The total generated energy is 60.7% in which the PV system's contribution is 39.3%. From the above observations, it has been revealed that both the C_1 and C_2 are alike both environmentally and economically. The higher replacement and capital cost of C_2 along with the fuel generator makes this combination difficult than the first combination. The results observed in the second combination are highlighted in Table 2.

Combination C_3

For C_3, sizes of SPV, biogas, biomass, wind turbine generators and batteries have been estimated as 100 kW, 60 kW, 50 kW, 50 numbers and 200 numbers respectively, while the energy requirement is calculated as 277092kWh/yr. From the results, it has been observed that by this combination, excess amount of energy has been generated, when analogized with the first two combinations. In addition, as there are more batteries there arises a necessity for frequent maintenance which is quite difficult in remote areas. The results observed are displayed in Table 3.

Table 4. Results obtained from combination C_4

Combination	Total NPC	COE (/kWh)	Operating cost
C_4	Rs.81,279,376.01 ($1,095,020)	Rs.19.30 ($0.263)	Rs.3,871,992.75 ($52,148)

Table 5. Summary of results for different combinations using HOMER

Combination	Total NPC	COE (/kWh)	Operating cost
C_1	Rs.66,004,647.52	Rs.15.57	Rs2,530,010.84
C_2	Rs.66,603,566.02	Rs.15.96	Rs.2,738,262.30
C_3	Rs.59,002,788.89	Rs.16.33	Rs.2,920,404.76
C_4	Rs.81,279,376.01	Rs.19.30	Rs.3,871,992.75

Combination C_4

For C_4, sizes of SPV, fuel cell, biogas, biomass, wind turbine generators have been estimated as 100 kW, 60 kW, 50 kW, and 50 numbers correspondingly, while the energy requirement is 276755kWh/yr and 33.53% of energy is available in excess. It has also been revealed that the TNPC is nearly 30% more when compared with the other three combinations and the results are highlighted in Table 4.

Among all the above four combinations discussed, the first combination is found to be most effective as it has least COE of $Rs.$15.57 and TNPC of $Rs.$66,004,547.52.Though C_2 and C_1 has more appropriate outcomings, but C_2 has more operating costs, replacement and COE than C_1. Table 5 highlights the TNPC, COE and operating cost for all the combinations and the diagrammatic illustration is given in Figure 4.

CONCLUSION

This chapter thoroughly discusses the energy generation method and its application. Through a widespread review, it is evident that most studies have focused on solar/wind/battery or solar/wind/diesel with battery storage based integrated energy systems. However, the review reveals limited inclusion of fuel cell (FC) and biomass/biogas-based electricity generation system as the component of HRES. Therefore, there is a scope to develop models of HRES for remote rural and far-flung areas with fuel cell and other RE sources. The chapter also presents the technical and economic aspects of different types of HRES. The HRES with a battery storing structure is employed to satisfy the energy requirement of the selected place, through which different configurations of HRES are simulated through HOMER Pro software

Figure 4. Comparative results for different combinations of HRES using HOMER

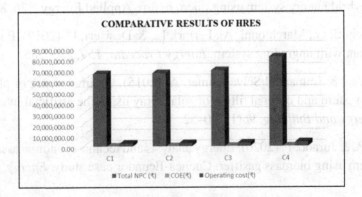

considering Total NPC and COE. With respect to the optimum feasible system, four configurations of the study area have been analysed. After a thorough analysis of comparative study, it is remarkably mentioned that the combination C_1 has provided profitable benefits since it has least COE of *Rs*.15.57/ kWh and Total NPC of *Rs*.66,004,647.52. This chapter is highly beneficial for the system developers, decision & policy makers and researchers, who are working in the area of stand-alone power generation by using RE sources

REFERENCES

Agarwal, N., & Kumar, A. (2013). Optimization of grid independent hybrid PV– diesel–battery system for power generation in remote villages of Uttar Pradesh, India.

Ahmad, J., Imran, M., Khalid, A., Adnan, M., Ali, S. F., & Khokkar, K. S. (2018). Techno economic analysis of a wind-photovoltaic- biomass hybrid renewable energy system for rural electrification: A case study of Kallar Kahar. *Energy, 28*, 36–46.

Akhtari, M. R., Shayegh, I., & Karimi, N. (2019). Techno-Economic Assessment And Optimization of A Hybrid Renewable Earth-Air Heat Exchanger Coupled With Electric Boiler, Hydrogen, Wind And PV Configurations. *Renewable Energy*.

Al-Ammar, E. A., Habib, U. R. H., Wang, S., Ko, W., & Elmorshedy, M. F. (2020). Residential Community Load Mnagement Based on Optimal Design of Standalone HRES with Model Predictive Control. *IEEE Access : Practical Innovations, Open Solutions*, 810–818.

Almeida, K. C., & Cicconet, F. (2019). Decentralized Hydrothermal Dispatch via Bi-level Optimization. *Journal of Control, Automation and Electrical Systems, 30*, 551–567.

Alramlawi, M., Mohagheghi, E., & Li, P. (2019). Predictive active-reactive optimal power dispatch in PV-battery-diesel micro grid considering reactive power and battery lifetime costs. *Solar Energy, 193*, 529–544.

Amirante, R., Bruno, S., Distaso, E., La Scala, M., & Tamburrano, P. (2019). A biomass small-scale externally fired combined cycle plant for heat and power generation in rural communities. *Renewable Energy Focus, 28*. doi:10.1016/j.ref.2018.10.002

Apichonnabutr, W., & Tiwary, A. (2018). Trade-offs between economic and environmental performance of an autonomous hybrid energy system using micro hydro. Applied Energy, 226, 891-904.

Baccioli, A., Caposciutti, G., Marchionni, A., Ferrari, L., & Desiseri, U. (2019). Poly-generation capability of a biogas plant with upgrading system. *Energy Procedia, 159*, 280–285.

Berclin Jeyaprabha, S., & Immanuel Selva Kumar, A. (2015). Optimal sizing of photovoltaic/battery/ diesel based hybrid system and optimal tilting of solar array using the artificial intelligence for remote houses in India. *Energy and Building, 96*(1), 40–52.

Cano, A., Arevalo, P., & Jurado, F. (2020). Energy analysis and techno-economic assessment of a hybrid PV/HKT/BAT system using biomass gasifier: Cuenca-Ecuador case study. *Energy, 202*.

Charrouf, O., Betka, A., Abdeddaim, S., & Ghamri, A. (2020). Artificial Neural Network power manager for hybrid PV-wind desalination system. *Mathematics and Computers in Simulation, 167*, 443–460.

Chen, W.-L., Huang, C.-W., Li, Y.-H., & Kao, C.-C. (2020). Biosyngas-fueled platinum reactor applied in micro combined heat and power system with a thermo photovoltaic array and Stirling engine. *Energy*, 194.

Costa, T. (2019). Optimum design of autonomous PV-diesel-battery hybrid systems: a case study at Tapajós-Arapiuns extractive reserve in Brazil. *IEEE PES Innovative Smart Grid Technologies Conference-Latin America*. IEEE.

Das, M., Maisanam, A. K. S., & Biswas, A. (2019). Techno-Economic Optimization of An Off-Grid Hybrid Renewable Energy System Using Metaheuristic Optimization Approaches- Case of A Radio Transmitter Station In India. *Energy Conversion and Management, 185*, 339–352.

Dasgupta, K., Roy, P. K., & Mukherjee, V. (2020). Power flow-based hydro-thermal-wind scheduling of hybrid power system using sine cosine algorithm. *Electric Power Systems Research, 178*.

Elkadeem, M. (2019). A Systematic Approach for Planning and Design of Hybrid Renewable Energy Based Micro grid With Techno-Economic Optimization: A Case Study on an Urban Community in Egypt. *Sustainable Cities and Society*.

Fulzele, J. B. (2018). Design and Optimization of Hybrid PV-Wind Renewable Energy System. Materials Today: Proceedings, (pp. 810-818). IEEE.

Gan, L. K., Shek, J. K. H., & Mueller, M. A. (2017). Analysis of Tower Shadow effects on Battery Lifetime in standalone Hybrid Wind-Diesel-Battery Systems. *IEEE Transactions on Industrial Electronics, Vol, 64*(8), 6234–6244.

Ghahramani, M., Nazari-Heris, M., & Zare, K. (2019). Energy and reserve management of a smart distribution system by incorporating responsive- loads/battery/wind turbines considering uncertain parameters. Research Gate.

Ghenaia, C., & Janajreh, I. (2020). Design of Solar-Biomass Hybrid Micro grid System in Sharjah. *Energy Procedia, 103*, 357–362.

Gil-Gonzalez, W., Montoya, O. D., & Garces, A. (2020). Modeling and control of a small hydro power plant for a DC micro grid. *Electric Power Systems Research, 180*.

Goddard, R., Zhang, L., & Xia, X. (2019). Optimal Sizing and Power Sharing of Distributed Hybrid Renewable Energy Systems Considering Socio-Demographic Factors. *Energy Procedia, 159*, 340–345.

Goel, S., & Sharma, R. (2019). Optimal sizing of a biomass–biogas hybrid system for sustainable power supply to a commercial agricultural farm in northern Odisha, India. *Environment, Development and Sustainability*.

Halabi, L. M., Mekhilef, S., Olatomiwa, L., & Hazeltonc, J. (2017). *Performance analysis of hybrid PV/diesel/battery system using HOMER: A case study Sabah, Malaysia*, 144.

Himadry Shekhar Das, A. H. M. (2016). Proposition of a PV/tidal powered micro-hydro and diesel hybrid system: A southern Bangladesh focus. Renewable and Sustainable Energy Reviews, 53, 1137-1148.

Jabari, F. (2019). Biogas fuelled combined cooling, desalinated water and power generation systems. *Journal of Cleaner Production*, *219*, 906–924.

Jabari, F., Mohammadi-ivatloo, B., & Ghaebi, H. (2018). Design and performance investigation of a biogas fuelled combined cooling and power generation system. Energy Conversion and Management, 169, 371-382.

Jahangir, M. H., & Cheraghi, R. (2020). Economic and environmental assessment of solar-wind-biomass hybrid renewable energy system supplying rural settlement load. *Sustainable Energy Technologies and Assessments*, *42*.

Jayachandran, M., & Ravi, G. (2017). Design and Optimization of Hybrid Micro-Grid System. *Energy Procedia*, *117*, 95–103.

Jean, D. D. N., Zou, F., Godwin, N. O. A., Bimenyimana, S., & Shyirambere, G. (2020). Key Technology Development Needs And Applicability Analysis Of Renewable Energy Hybrid Technologies In Off-Grid Areas For The Rwanda Power Sector. *Heliyon*, *6*.

Juan, M. (2012). *Optimum load management strategy for wind/diesel/battery hybrid power systems*. Research Gate.

Kanase-Patil, A. B., Saini, R. P., & Sharma, M. P. (2010). Integrated Renewable Energy Systems For Off Grid Rural Electrification of Remote Area. *Renewable Energy*, *35*, 1342–1349.

Kaur, A., & Narang, N. (2019). Optimum Generation Scheduling of Coordinated Power System Using Hybrid Optimization Technique. *Electrical Engineering*.

Kaur, R., Vijaya, K. K. S., Kandasamy, N. K., & Kumar, S. (2019). Discrete Multi-objective Grey Wolf Algorithm Based Optimal Sizing and Sensitivity Analysis of PV-Wind-Battery System for Rural Telecom Towers. *IEEE Systems Journal*.

Kaur, T., & Segal, R. (2017). Designing Rural Electrification Solutions Considering Hybrid Energy Systems For Papua New Guinea. *Energy Procedia*, *110*, 1–7.

Kim, M.-H., Kim, D., Heo, J., & Lee, D.-W. (2019). Techno-Economic Analysis Of Hybrid Renewable Energy System With Solar District Heating For Net Zero Energy Community. *Energy*, *187*.

Koko, S. P., Kusakana, K., & Vermaak, H. J. (2018). Optimal power dispatch of a grid –interactive micro-hydro kinetic-pumped hydro storage system. *Journal of Energy Storage*, *17*, 63–72.

Krishnan, O., & Suhag, S. (2020). Grid-Independent PV System Hybridization With Fuel Cell-Battery/Super Capacitor: Optimum Sizing And Comparative Techno-Economic Analysis. *Sustainable Energy Technologies and Assessments*, *37*.

Kusakana, K. (2015). *Operation cost minimization of photovoltaic–diesel–battery hybrid systems* (Vol. 85).

Laagoubi, T., & Bouzi, M. (2018). Supervising PV/Battery/Diesel System Connected to Grid Using Fuzzy Logic. *6th International Renewable and Sustainable Energy Conference*. IEEE.

Lagrange, A., de Simon-Martín, M., González-Martínez, A., Bracco, S., & Rosales-Asensio, E. (2020). Sustainable micro grids with energy storage as a means to increase power resilience in critical facilities: An application to a hospital. *International Journal of Electrical Power & Energy Systems, 119*, 105865.

Landi, D. (2019). Interactive Energetic, Environmental and Economic Analysis of Renewable Hybrid Energy System. *International Journal of Interactive Design and Manufacturing.*

Lee, K., & Kum, D. (2019). Complete design space exploration of isolated hybrid renewable energy system via dynamic programming. *Energy Conversion and Management, 196*, 920–934.

Li, C. (2019). *Techno- economic performance study of stand-alone wind/diesel/battery hybrid system with different battery technologies in the cold region of China.* Research Gate.

Li, C. Y., Shen, Y., Wu, J. Y., Dai, Y. J., & Chi, H. W. (2019). Experimental and modelling investigation of an integrated biomass gasifier-engine-generator system for power generation and waste heat recovery. *Energy Conversion and Management, 199.*

Luna, A. C., Diaz, N. L., Graells, M., Vasquez, J. C., & Guerrero, J. M. (2017). Mixed-Integer-Linear-Programming-Based Energy Management System for Hybrid PV-Wind-Battery Micro grids: Modelling, Design, and Experimental Verification. *IEEE Transactions on Power Electronics, 32*(4), 2769–2783.

Mahesh, A. (2016). *Optimal Sizing of a Grid-Connected PV/Wind/Battery System Using Particle Swarm Optimization.* Springer.

Maleki, A., Ameri, M., & Keynia, F. (2015). Scrutiny of multifarious particle swarm optimization for finding the optimal size of a PV/wind/battery hybrid system. Research Gate.

Mohanty, A., Viswavandya, M., Mohanty, S. P., & Paramita, P. (2015). Optimization and Improvement of Voltage Stability in a Stand-alone Wind-Diesel- Micro hydro Hybrid System. *Procedia Technology, 21*, 332–337.

Murugaperumal, K., Srinivasn, S., & Prasad, G. R. K. D. S. (2020). Optimum Design of Hybrid Renewable Energy System Through Load Forecasting And Different Operating Strategies For Rural Electrification. *Sustainable Energy Technologies and Assessments, 37.*

Nandi, M., Shiva, C. K., & Mukherjee, V. (2019). Moth-Flame Algorithm for TCSC and SMES-Based Controller Design in Automatic Generation Control of a Two-area Multi-unit Hydro-power system. *Iranian Journal of Science and Technology. Transaction of Electrical Engineering.*

Pragati Tripathi, M. A. (2018). Modelling of Energy Efficient PV-Diesel-Battery Hybrid System. *IEEE 2018 International Conference on Computational and Characterization Techniques in Engineering & Sciences.* IEEE.

Rakhshani, E., Mehrjerdi, H., & Iqbal, A. (2019). Hybrid- Wind- Diesel- Battery System Planning Considering Multiple Different Wind Turbine Technologies Installation. *Journal of Cleaner Production.*

Safari, M., & Sarvi, M. (2014). Optimal load sharing strategy for a wind/diesel/battery hybrid power system based on imperialist competitive neural network algorithm. *IET Renewable Power Generation, 8*(8), 937–946.

Sebastian, R. (2011). *Battery energy storage for increasing stability and reliability of an isolated Wind Diesel power system.* Research Gate.

Shaahid, S. M., Al-Hadhrami, L. M., & Rahman, M. K. (2014). *Review of economic assessment of hybrid photovoltaic-diesel-battery power systems for residential loads for different provinces of Saudi Arabia* (Vol. 31).

Shahzad, M., & Ma, J. T. (2019). Techno-economic assessment of a hybrid solar- wind-battery system with genetic algorithm. *Energy Procedia, 158*, 6384–6392.

Shan, Y. (2019). Model Predictive Control of Bidirectional DC-DC Converters and AC/DC Interlinking Converters- A New Control Method for PV-Wind-Battery Micro grids. IEEE Transactions on Sustainable Energy.

Shimray, B. (2017). A New MLP-GA-Fuzzy Decision Support System for Hydro Power Plant Site Selection. *Arabian Journal for Science and Engineering.*

Stiubiener, U., Carneiro da Silva, T., Federico, B. M. T., & Teixeira, J. C. (2020). PV power generation on hydro dam's reservoirs in Brazil: A way to improve operational flexibility. *Renewable Energy, 150*, 765–776.

Su, B., Han, W., He, H. Z., Jin, H., Chen, Z., & Yang, S. (2020). A biogas-fired cogeneration system based on chemically recuperated gas turbine cycle. *Energy Conversion and Management, 205.*

Suresh, N. S., Thirumalai, N. C., & Dasappa, S. (2019). Modeling and analysis of solar thermal and biomass hybrid power plants. *Applied Thermal Engineering, 160.*

Suresh, V., M, M., & Kiranmayi, R. (2020). Modelling and optimization of an off- grid hybrid renewable energy system for electrification in a rural areas. *Energy Reports, 6*, 594–604. doi:10.1016/j.egyr.2020.01.013

Tazvinga, H., Xia, X., & Zhang, J. (2013). *Minimum cost solution of photovoltaic–diesel–battery hybrid power systems for remote consumers* (Vol. 96).

Usman, M., Khan, M. T., Rana, A. S., & Ali, S. (2017). Techno economic analysis of hybrid solar-diesel-grid connected power generation system. *Journal of Electrical Systems and Information Technology*, 1–10.

Vendoti, S., Muralidhar, M., & Kiranmayi, R. (2021). Techno-economic analysis of off-grid solar/wind/biogas/biomass/fuel cell/battery system for electrification in a cluster of villages by HOMER software. *Environment, Development and Sustainability, 23*(1), 351–372. doi:10.1007/s10668-019-00583-2

Wang, R., Xiong, J., He, M., Gao, L., & Wang, L. (2019). Multi-objective optimal design of hybrid renewable energy system under multiple scenarios. *Renewable Energy.*

Wei, L., Nakamura, T., & Imai, K. (2020). Development and optimization of low-speed and high efficiency permanent magnet generator for micro hydro-electrical generation system. *Renewable Energy, 147*, 1653–1662.

Yu, D., Zhang, T., He, G., Nojavan, S., Jermsittiparsert, K., & Ghadimi, N. (2020). Energy management of wind-PV-storage grid based large electricity consumer using robust optimization technique. *Journal of Energy Storage, 27*, 101054.

Zhu, Y., Li, W., Li, J., Li, H., Wang, Y., & Li, S. (2019). Thermodynamic analysis and economic assessment of biomass-fired organic Rankine cycle combined heat and power system integrated with CO_2 capture. *Energy Conversion and Management, 204*(2).

Zhu, Y., Liang, J., Yang, Q., Zhou, H., & Peng, K. (2019). Water use of a biomass direct- combustion power generation system in China: A combination of life cycle assessment and water footprint analysis. *Renewable & Sustainable Energy Reviews, 115*.

ABBREVIATIONS

AGC: Automatic Generation Control

AI: Artificial Intelligence

ANFIS: Artificial Neuro Fuzzy Interface System

ANN: Artificial Neural Network

ARMA: Autoregressive Moving Average

BESS: Battery Energy Storage System

CCHP: Combined Cooling, Heat and Power

CHP: Combined Heat and Power

CLONALG: Clonal Selection Algorithm

COE: Cost of Energy

DG/LA: Diesel Generator/ Lead Acid

DG/LI: Diesel Generator/ Lithium Ion

DG/ZB: Diesel Generator/ Zinc Bromine

DHRES: Distributed Hybrid Renewable Energy System

DMGWO: Discrete Multiobjective Grey Wolf Algorithm

DRP: Distribution Resource Plan

EAHEs: Earth-to-Air Heat Exchangers

EMPC: Economic Model Predictive Control

FC-BESS: Fuel Cell – Battery Energy Storage System

FC-SCESS: Fuel Cell – Energy Storage System

GA: Genetic Algorithm

GWO: Grey Wolf Optimization

HEPP: Hydro Electric Power Plants

HESS: Hybrid Energy Storage System

HMGS: Hybrid Micro-Grid Systems

HOMER: Hybrid Optimization Model for Electric Renewable

HPS: Hybrid Power System

HRE-MG: Hybrid Renewable Energy – Micro Grid

HRES: Hybrid Renewable Energy System

HTS: Hydro Thermal Scheduling
ICA: Imperialist Competitive Algorithm
IHOGA: Improved Hybrid Optimization by Genetic Algorithms
LCC: Life Cycle Cost
LECs: Levelized Electricity Costs
LFC: Load Following Cycle
LPSP: Loss of Power Supply Probability
MATLAB: Matrix Laboratory
MHK-PHS: Micro-Hydrokinetic Pumped Hydro Storage
MPCP: Model Predictive Current and Power
MPPT: Maximum Power Point Tracking
NPC: Net Present Cost
O&M: Operation and Maintenance
PBC: Passivity-Based Control
PID: Proportional Integral Controller
PMG: Permanent Magnet Generator
PMSG: Permanent Magnet Synchronous Generator
PSCAD: Power Systems Computer Aided Design
PSO: Particle Swarm Optimization
PV: Photo Voltaic DG – Distributed Generation
PV-FPP: PV Floating Power Plants
RES: Renewable Energy Sources
SCA: Sine Cosine Algorithm
SMES: Superconducting Magnetic Energy Storage
s-NSGA-II: Scenario-Non-Dominated Sorting Genetic Algorithm II
SPCC: Solar-assisted Post-combustion Carbon Capture
SVC: Static VAR Compensator
TCSC: Thyristor-Controlled Series Compensation
T-PEM: Two-point Estimate Method
WD: Wind Diesel
WDPS: Wind–Diesel Power Systems

Section 4
Cyber Physical Systems and Internet of Things

Chapter 16
Design and Modeling of Hybrid Renewable Energy Systems for Optimized Power Generation

Suresh Vendoti
Godavari Institute of Engineering and Technology, India

M. Muralidhar
Sri Venkateswara College of Engineering and Technology, India

R. Kiranmayi
JNTUA College of Engineering, India

Dana Victoria
International School of Technology and Science for Women, India

D. Ravi Kishore
iD https://orcid.org/0000-0002-2567-2888
Godavari Institute of Engineering and Technology, India

ABSTRACT

The exploration of renewable energy resources has gained momentum due to the continuous demand for energy consumption and the depletion of fossil fuel reserves. However, these resources possess an intermittent nature and are only viable in certain geographical locations. To address these challenges, this chapter presents a solution in the form of a hybrid energy system (HES). This system operates in an off-grid mode and is specifically designed for high altitude demographic users who face difficulties in accessing the national grid. This chapter utilizes a well-designed hybrid energy system to enhance the reliability and quality of power generation in rural areas. Also, the design of a linear mathematical model is discussed in this chapter, which aims to determine the optimal working and cost optimization of the hybrid energy generating system. The system comprises a wind-biogas-biomass based power generation system, PV array, fuel cells, a battery bank, and a bidirectional converter. To meet economic constraints and load dispatch, an efficient mathematical modeling is employed.

DOI: 10.4018/979-8-3693-3735-6.ch016

INTRODUCTION

The depletion of fossil fuel at a rapid pace has prompted a search for alternative energy sources to meet the growing demands of the world. Another important objective is to reduce our reliance on non-renewable sources in order to combat the escalating issue of global warming. Hence, it is imperative to explore alternative energy sources that can fulfil the ever-increasing energy requirements while minimizing their detrimental environmental impacts. Solar photovoltaic and wind energy are considered viable power generation options due to their convenient accessibility and favourable geographical characteristics for localized power generation in remote areas. Since the oil crisis of the 1970s, the utilization of solar PV and wind energy has gained significant prominence owing to their easy availability and cost-effectiveness. India is blessed with an abundance of solar energy, with a correspondent of 5,000 trillion kWh/year available. The average daily intake of solar energy ranges from 5 - 7 kWh/m², and with 250-300 sunlit days in many parts of the country, solar energy is a viable option. Wind power in India has been growing since 1990 and has significantly increased in the last decade. As of June 2020, India's total installed capacity is 371 GW, with 87.67 GW coming from renewable generators. Among these generators, wind power has the highest share at 37.82 GW, followed by solar energy at 35.12 GW (Reference 1).

The primary goal of the paper is to develop a hybrid system that utilizes various renewable sources of energies such as solar, wind, biomass, biogas, and fuel cells, despite their reliability challenges. The aim is to integrate these sources into a single system in order to enhance the efficiency of the power grid. A hybrid renewable energy system consists of renewable energy conversion components like PV panels, wind turbines, hydro turbines, as well as traditional non-renewable generators like diesel generators and micro turbines. Additionally, it includes an energy storage device, namely batteries. These components may or may not be incorporated into a hybrid energy scheme. The main objective is to model each individual component to determine the optimal selection of components and provide guidance to researchers new to the field of hybrid renewable energy systems (Vendoti et al., 2021).

Figure 1 depicts the schematic diagram of the suggested hybrid energy system. This system includes solar PV, wind turbine, biogas, biomass, fuel cell generators, battery bank, and electrolyser. Electronic converters are utilized to ensure the provision of both AC and DC buses (Suresh et al., 2020).

The energy generated by the renewable/diesel generator is divided into two streams. One stream supplies the electrical energy directly to the load, while the other stream powers the battery. The load can be served with electrical energy either directly from the renewable/diesel generator or indirectly from the battery. The battery is charged with the energy from all the generators (Pipattanasomporn, 2004).

Hybridization Challenges and Their Explanation in HRES:

There are several challenges that come with hybridization, despite the numerous advantages it offers. These challenges are listed in Table 1, which focuses on HRES hybridization challenges.

MATHEMATICAL MODELING OF HRES COMPONENTS

The mathematical modeling showcases the performance of system components across various working conditions. The HRES incorporates different types of subsystems, including SPV, WTG, BMG, BGG, FC, and BATT, along with a battery storage system. The performance of these components is crucial

Figure 1. Configuration of a HRES

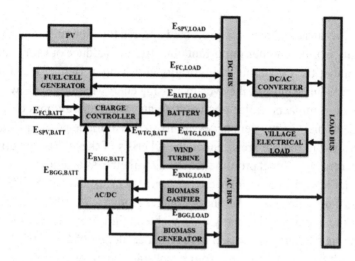

Table 1. HRES hybridization challenges

Challenges	Explanations
Technical Challenges:	
Integration Complexity	The seamless integration of different energy sources may require the implementation of specific control and management systems.
Intermittency	The intermittent nature of renewable sources such as solar and wind adds complexity to the prediction and management of energy generation.
Infrastructure Development	The construction of new infrastructure or the retrofitting of existing infrastructure for hybrid renewable energy systems (HRES) can be a complex process.
Energy Storage	The selection, integration, and management of energy storage solutions pose challenges in ensuring reliable energy supply.
Power Quality	The integration of multiple energy sources may impact power quality, necessitating proper management to maintain stability.
Economic Challenges:	
High Initial Costs	Hybrid systems may require higher initial investment costs compared to single-source systems.
Return on Investment (ROI) Uncertainty	The variability of renewable energy can affect the predictability of returns on investment.
Market Maturity	Some technologies in HRES may not be fully mature, leading to economic uncertainties.
Environmental Challenges:	
Land Usage	Combining multiple energy sources may require more land or specific types of land, raising environmental concerns.
Resource Assessment	Accurately assessing renewable resources, such as wind speeds and solar irradiance, is crucial but can be challenging.
Regulatory & Policy Challenges:	
Inconsistent Policies	Different energy sources may be subject to varying policies and regulations, complicating the design of hybrid renewable energy systems.
Grid Integration Policies	Integrating HRES into existing grids may face regulatory hurdles, especially if grid policies are not regularly updated
Licensing and Standards	The lack of standardized regulations for HRES can lead to uncertainties in licensing and operation.

Figure 2. Circuit layout of single PV cell

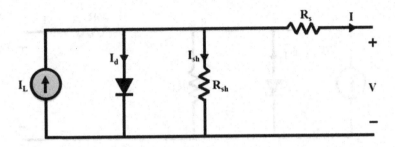

for the system's applications and evaluations. The subsequent sections provide a clear explanation of the modeling process for the hybrid energy system components.

Energy Source Models

PV Cell Modelling

PV system, which provides a limited amount of current about 30mA and voltage about 0.6V. To maximize the voltage, a PV module is designed in series, whereas the PV module is designed in parallel to maximize the current. With the utilization of semiconductor materials, the PV module transforms the solar energy into electrical energy. When the light strikes the semiconductor crystal, the photo current is produced. The electrons are emitted from electric current by the solar insulation when they are linked to a load.

When the light energy strikes the solar cells, a thin layer of silicon generates an electric field. The silicones like Single crystalline silicon and Polycrystalline silicon are utilized in the PV cells. The PV cells are made up of variety of materials like Cadmium Telluride, Gallium Arsenide, and Copper Indium Diselenide.

When the PV cell is exposed to sunlight, it produces the current, which is proportionate to the solar irradiation. In the circuit, a simple ideal PV cell represents a current source, which is linked in parallel with a diode [5]. As no cell is ideal, the shunt and series resistance are linked for the correct modelling, which is displayed in Figure 4.2.

The mathematical equation of this model is expressed as (Thapar et al., 2011),

$$I = I_L - I_d - \left(\frac{V + IR_s}{R_{sh}}\right) \tag{1}$$

$$I_d = I_o\left[exp\left(\frac{q \times V}{A \times k \times T}\right) - 1\right] \tag{2}$$

Where, the light emitted photo current is denoted as I_L, the current through diode is denoted as I_d, reverse saturation current of the diode is denoted as I_o, diode voltage is denoted as V, the temperature in

Figure 3. R_{sh} Model of PV cell

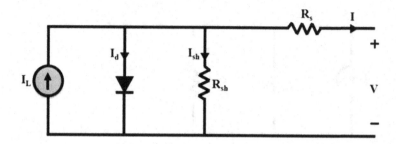

Kelvin is denoted as T, the Boltzmann constant in J/K is indicated as $k=1.380658\times10^{23}$, the charge of electron is denoted as $q=1.6\times10^{-19}$ and the diode ideal factor is denoted as A.

R_{sh} Model

The load current and diode current are computed as similar as the classical model but R_{sh} and R_s are calculated by using the Newton Raphson (NR) algorithm. The R_{sh} model of PV cell is demonstrated in Figure 3.

$$R_{sh} = \frac{V_{max}\left(V_{max}+I_{max}R_s\right)}{\left[V_{max}\left[I_L-I_d\right]-P_{max}\right]} \tag{3}$$

With initial condition,

$$R_{sh} = \left(\frac{V_{max}}{I_{SC}-V_{max}}\right) - \left(\frac{V_{OC}-V_{max}}{I_{max}}\right) \tag{4}$$

$$R_s = 0 \tag{5}$$

Where, the voltage at maximum power is denoted as V_{max}, the current at maximum power is denoted as I_{max}, the SC current is indicated as I_{SC} and the OC voltage is denoted as V_{OC}.

The matching of R_{sh} and R_s for maximum output power through the iterative process is one of the advantages of R_{sh} and R_s model over traditional model. To get a perfect model, these values are estimated at a time, which utilize the datasheet information like OC voltage, SC current and maximum power. These values are calculated through the NR algorithm, which is derived from equation (3). With the initial values, the NR algorithm has better and faster convergence, which is represented in equation (4).

Wind Turbine

The wind turbine (WT) has both the rotor and tower, in which the rotor is mounted on a nacelle. The tower contains two or more blades, which are coupled mechanically with an electric generator. The shaft of the electric generator rotates to produce the electric power, which is controlled by the control system. In WT, the design models are classified into two kinds like Horizontal Axis (HA) and Vertical Axis (VA) WT. On the basis of the factors like time, day and season, the horizontal axis WT gathers maximum amount of wind energy and its blades avoid high wind storm. The WT works in two modes like constant or variable speed. When the speed of the rotor is constant, the turbine rotates at the same angular speed without considering the wind fluctuations. This mode has the benefit of removing the need of power electronic converters. As a result, a constant wind speed turbine generates lower energy than a variable wind speed turbine at low wind speeds. The mathematical model of WT is necessary for understanding its performance.

Mathematical Formulation of Turbine Model

The kinetic energy E with mass m and velocity v under constant acceleration a has equalized the work done W, The body is moved away from the rest to a distance s due to the application of force F., i.e., $E = W = Fs$. As per newton's law of motion, is expressed as (Thapar et al., 2011),

$$F = ma \qquad (6)$$

Therefore, the kinetic energy is expressed as,

$$E = mas \qquad (7)$$

The kinematics of solid motion is expressed as $v^2 = u^2 + 2as$. Where, u is the object's initial velocity. As a result, $a = \dfrac{v^2 - u^2}{2s}$. When the object's initial velocity is assumed as zero, it is noted as $a = \dfrac{v^2}{2s}$. As a result, the equation (7) becomes

$$E = \frac{1}{2} mv^2 \qquad (8)$$

If the wind is considered as a fluid, both the density and velocity get varied; therefore, it has no constant mass. The Reccab has utilized a factor of $\dfrac{2}{3}$ instead of $\dfrac{1}{2}$ to express the kinetic energy law. Equation (8) computes the kinetic energy of air with mass m and velocity of wind v_w. The rate of change in kinetic energy (the wind's power P) is expressed as,

$$P = \frac{dE}{dt} = \frac{1}{2} \frac{dm}{dt} v_w^2 \qquad (9)$$

The mass flow rate $\dfrac{dm}{dt}$ is expressed as $\dfrac{dm}{dt} = \rho A v_w$, where A denotes area and ρ denotes the density of air. With this expression, equation (9) is written as (Thapar et al., 2011),

$$P = \frac{1}{2} \rho A v_w^3 \tag{10}$$

The real mechanical power P_w is extricated through the blades of rotor in watts. It is expressed as,

$$P_w = \frac{1}{2} \rho A v_w \left(v_u^2 - v_d^2 \right) \tag{11}$$

Where the upstream velocity of wind at the entrance of the blades of rotor is denoted as v_u in *m/s* and the velocity of wind downstream at the rotor blades' exit is represented as v_d in *m/s*. The upstream and downstream wind velocities give the blade tip speed ratio, which is expressed as,

$$\rho A v_w = \frac{\rho A \left(v_u + v_d \right)}{2} \tag{12}$$

Where the entry and exit velocities of the turbine's rotor blade is denoted as v_w. Equation (11) is written as,

$$P_w = \frac{1}{2} \rho A \left(v_u^2 - v_d^2 \right) \frac{\left(v_u + v_d \right)}{2} \tag{13}$$

The above equation is simplified as follows,

$$P_w = \frac{1}{2} \left[\rho A \left\{ \frac{v_u}{2} \left(v_u^2 - v_d^2 \right) + \frac{v_d}{2} \left(v_u^2 - v_d^2 \right) \right\} \right] \tag{14}$$

$$P_w = \frac{1}{2} \rho A V_u^3 C_p \tag{15}$$

$$C_p = \frac{1 - \left(\dfrac{v_d}{v_u} \right)^2 + \left(\dfrac{v_d}{v_u} \right) - \left(\dfrac{v_d}{v_u} \right)^3}{2} \tag{16}$$

Where, the power coefficient of rotor is denoted as C_p. The ratio of wind speed (λ) is stated as the ratio of downstream wind speed (vd) to upstream wind turbine (vu), which is expressed as,

$$\lambda = \frac{v_d}{v_u} \tag{17}$$

When a WT's rotor rotates with slow speed, maximum wind is passed through the holes between the blades, which provides little amount of power output. If the rotor rotates with high speed, the rotating blades function as a solid wall, which blocks the flow of wind through the hole and decreases the power output. To extricate the maximum power from the wind stream, the turbines are designed to work at their optimum ratio of wind tip speed λ. Due to the effect of dust in the air, the greater wind speed ratio causes the edges of the blades to erode. This causes noise, vibrations and it has decreased the rotor efficiency, which leads to turbine failure.

Air density (ρ) is determined by both air pressure and temperature. An increase in air pressure results in a corresponding increase in air density. When the temperature of air is reduced, the air density is also raised. This is mathematically expressed as,

$$P = \rho R T \tag{18}$$

Where the gas constant is denoted as R. Due to the rising elevation, both the pressure and temperature are reduced. At the atmospheric pressure, $P_{atm} = 14.7\text{psi}$, temperature, $T = 600\text{F}$ and density $\rho = 1.225\text{kg/m3}$ Due to the elevation, the pressure and temperature get varied. This affects the air density, which is expressed as,

$$\rho = \rho_o e^{-\frac{0.297}{3048}H_m} \tag{19}$$

Where, the site elevation in m is denoted as H_m.

Biogas System

Biogas is made up of a mix of gases, including 60–70% methane, 30–40% carbon dioxide and small amounts of hydrogen sulfide and ammonia (Kanase-Patil et al., 2011).

The organic substrate (S_1, in g_{COD}/L) is degraded in VFAs (S_2, in $mmole_{VFA}/L$) by the acidogenic bacteria (X_1, in g_{COD}/L) and the VFAs are then degraded in methane (CH_4, in L/d) by methanogenic microorganisms (X_2, in g_{COD}/L). Consider a simplified model of an anaerobic process. Two biological reactions are utilized in kinetic model, which are expressed as,

$$k_1 S_1 \xrightarrow{r_1} X_1 + k_2 S_2 \text{ (Acidogenesis)} \tag{20}$$

$$k_3 S_2 \xrightarrow{r_2} X_2 + k_4 CH_4 \text{ (Methanogenesis)} \tag{21}$$

Where, $k_1 (g_{COD} S_1 / g_{COD} X_1)$, $k_2 (mmole_{VFA} / g_{COD} X_1)$, $k_3 (mmole_{VFA} / g_{COD} X_2)$ and $k_4 \left(mmole_{CH_4} / g_{COD} X_2 \right)$ are the stoichiometric coefficient. The representation of $r_1 (r_1 = \mu 1) X1$ and r2(r2=μ2)$_x$2 the relation shows the bacterial growth that is related to both bioprocess. The terms μ1 and μ2(d-1) illustrate the precise growth rate of acidogenesis and methanogenesis.

The differential equations describing the system dynamics are derived from mass balance principles applied to the continuous processes involved. These mass balance equations takes into the following forms,

$$\frac{dX_1}{dt} = X_1 \left(\mu_1 - D \right) \text{(Acidogenic biomass)} \tag{22}$$

$$\frac{dX_2}{dt} = X_2 \left(\mu_2 - D \right) \text{(Methanogenic biomass)} \tag{23}$$

$$\frac{dS_1}{dt} = D \left(S_1^{in} - S_1 \right) - k_1 \mu_1 X_1 \text{ (Organic substrate)} \tag{24}$$

$$\frac{dS_2}{dt} = D \left(S_2^{in} - S_2 \right) - k_2 \mu_1 X_1 - k_3 \mu_2 X_2 \text{ (Volatile fatty acid)} \tag{25}$$

The dilution rate (D) is calculated using the formula $D = \frac{Q}{V}, in d^{-1}$, where Q represents the input flow rate and V represents the effective reactor volume. The input organic substrate concentration is denoted as S_1^{in}, while the input VFA concentration is denoted as S_2^{in}.

The kinetics of the acidogenic bacteria follow Monod kinetics, while the methanogenic bacteria kinetics follow Haldane kinetics (Kanase-Patil et al., 2011).

$$\mu_1 = \mu_{1max} \frac{S_1}{K_{S_1} + S_1} \text{ (Acidogenic bacteria kinetics)} \tag{26}$$

The maximum rate of bacteria growth is represented by as $\mu 1_{max}$, while the half saturation constant for the organic substrate $S1$ is denoted as K_{S_1}.

$$\mu_2 = \mu_{2max} \frac{S_2}{K_{S_2} + S_2 + S_2^2 / K_I} \text{ (Methanogenic bacteria kinetics)} \tag{27}$$

On the other hand, the maximum rate of bacteria growth is denoted as $\mu2_{max,}$ and the half saturation and inhibition constants for the organic substrate $S2$ are represented by K_{S_2} and K_I.

The production of methane gas, which is assumed to occur through the methano-genesis reaction with a molar flow rate (q_M), is expressed in equation 4.36. This biogas is generated through the decomposition of organic matter.

$$q_M = k_4\mu2_x2 \tag{28}$$

Biomass System

The output of the hourly energy of biogas digester system is attained through the mathematical modelling, which is expressed in the equation 29 on the basis of the obtainability of forest foliage in the study area (Anurag Chauhan et al., 2017),

$$P_{BMG} = \frac{Biomassavailability\left(kg\,/\,year\right) \times CV_{BMG} \times \eta_{BMG} \times \Delta t}{365 \times 860 \times h_{BMG}} \tag{29}$$

Where the hourly output power biomass generator is denoted as P_{BMG}, the total efficiency of energy conversion in biomass generator is denoted as $\eta B_{MG,}$ the calorific value of biomass in 4015kcal/kg is denoted as CVB_{MG} and the number of hours of operation is denoted as hB_{MG}.

Fuel Cells

The Proton Exchange Membrane Fuel Cells (PEMFC) are a type of fuelcells that have a single electrolyte layer in contact with both the anode and cathode. Various models have been developed to represent the PEMFC. When water H_2O is generated, the FC Nernst voltage reaches is 1.22V. However, due to irreversible losses in the FC system, the actual voltage of the FC is lower than the ideal voltage. The mathematical expression for the Nernst voltage E_{Nernst} is described in reference (Felix et al., 2006).

$$E_{Nernst} = 1.2209 - 0.85 \times 10^{-3}\left(T - 2098.15\right) + 4.3085 \times 10^{-3} \times T \times \left(lnP_{H_2} + 0.5lnP_{O_2}\right) \tag{30}$$

Where the effective pressure and temperature are denoted as P and T. The concentration of undissolved oxygen in gas or liquid intermediate is determined by Henry's law as follows,

$$C_{O_2} = \frac{P_{O_2}}{5.08 \times 10^6 \times exp\left(-498/T\right)} \tag{31}$$

An experimental equation is used to compute the over voltages caused by internal processes and resistance. It is expressed as,

$$\eta_{act} = -0.9514 + 0.003120T \times \ln(i) + 7.4 \times 10^{-5} T \times \ln\left(C_{O_2}\right) \tag{32}$$

$$R_{in} = 0.01605 - 3.5 \times 10^{-5}T + 8 \times 10^{-5}i \tag{33}$$

Where the current flowing in FC is denoted as i, the stir resistance is expressed as,

$$R_a = -\frac{\eta_{act}}{i} \tag{34}$$

By taking into account the interplay between thermodynamics, mass transfer, and kinetic energy, the FC's output voltage can be determined, as expressed by,

$$V = E - v_{act} + \eta o_{hmic} \tag{35}$$

Increasing the pressure of the fuel cell (FC) helps to offset the voltage drop in the FC. To analyse the dynamic response of the FC, a capacitor is introduced into the steady-state model. This model includes a parallel capacitor that represents the double layer charge effect. The voltage of the FC is defined by a differential equation,

$$\frac{dv_{act}}{dt} = \frac{i}{C} - \frac{v_{act}}{R_a \times C} \tag{36}$$

The expression for the ohmic voltage drop is given by,

$$v_{act} = -i \times R_{in} \tag{37}$$

If there are 65 series cells in the FC, the output voltage can be calculated,

$$V_{stack} = 130V_{cell} \tag{38}$$

The consumption of O_2 and H_2 in an FC is determined by the rate of input or output and current of FC. The mol's equality law is used to compute the pressure of a gas by using the rate of input or output.

For FC's anode, the mathematical equation is expressed as,

$$\frac{V_a}{RT} \cdot \frac{dP_{H_2}}{dt} = m\rho_{H_2} - \left(\rho_{H_2}.U.A\right)_{out} - \frac{i}{2F} \tag{39}$$

Similarly, the mathematical equation FC's cathode is expressed as,

$$\frac{V_c}{RT} \cdot \frac{dP_{O_2}}{dt} = mp_{O_2} - \left(\rho_{O_2}.U.A\right)_{out} - \frac{i}{4F}$$ (40)

Where the rate of molar flux to humidifier is denoted as m, the volume of anode in L is denoted as V_a, the deal gas constant, which is equal to -0.008201 atom/mol k is denoted as R, the temperature of FC in K is denoted as T, the volume of cathode in L is denoted as V_c and the molar density is denoted as ρ

The volume of anode and cathode is assumed as 2L. The overall thermal energy balance in an air-cooled FC is expressed as,

$$Q_l = Q_s + Q_L$$ (41)

Where Q_l, Q_s, and Q_L denote the heat dissipation of generation, stored and internal accordingly. For 130 cells, the FC current and internal resistance are utilized to compute the internal heat loss, which is expressed as,

internal generated temperature $= i^2(R_a + R_{int}) \times 130$ (42)

The mathematical equation of FC's stored thermal energy is expressed as,

stored thermal energy $= C_t \times \dfrac{dT}{dt}$ (43)

where C_t and T denote the heat capacity, which is equal to 100 J/C and temperature of FC accordingly,

thermal loss power to ambient $= \dfrac{T - T_a}{R_t}$ (44)

Substitute equations (42) to (44) in equation (41). It is expressed as,

$$\frac{dT}{dt} = \frac{130 \times \left(R_a + R_{int}\right) \times i^2}{C_t} - \frac{T - T_a}{R_t \times C_t}$$ (45)

The values of T_a and T_t are noted as 25°C and 0.04°C/W. Equations (30) to (45) denote the dynamic performance of FC, which removes the dynamic effect of compressors and valves.

Electrolyzer

Various electrolyzer cells are linked in series to form an electrolyzer system. The $V - I$ characteristic is temperature dependent. It is generally nonlinear and is attained through the curve fitting. According to Faraday's law, the electrolyser is the rate, at which the H_2 is produced (Reference 9). It is proportional to the rate, at which the current flows between the electrodes.

$$\eta H_2 = \frac{\eta_F . n_n . i_e}{2F\left(mol\,/\,s\right)}$$

(46)

Where the Electrolyser's current is denoted as i_e, the number of series in the electrolyser is denoted as n_n and ηF denotes the faraday's efficiency. The ratio of hydrogen produced H2 to maximum theoretical possible produced H2 is known as Faraday's efficiency. The working temperature is assumed as 40°C, which is expressed as,

$$\eta_F = 96.5 \times exp\left(\frac{0.09}{i_e} - \frac{75.5}{i_e^{\,2}}\right)$$

(47)

The equations (46) and (47) illustrate a simple electrolyser model.

Energy Storage Components

Hydrogen Tank

Hydrogen is produced by the electrolysers through a chemical reaction, utilizing excess electricity. The generated hydrogen is then stored in high-pressure tanks using a gas compressor and is utilized by the fuel cell as needed. To ensure efficient operation, the hydrogen tanks are treated as limited sources or sinks of hydrogen, imposing constraints on the HRES system.

Batteries

In a wide range of applications, batteries are commonly utilized as an energy storage device. The elements like anode, cathode and electrolyte are utilized in these devices. The battery charges or discharges the stored energy by moving the charge and discharge ions from the anode to cathode through electrolyte. The lead acid battery is the most prevalent type of battery, which is utilized in the design of HRES. In this case, the batteries are modelled as a two-tank scheme. The obtained energy of the conversion in the current time step is given as one tank and the chemical energy of batteries, which is not immediately obtainable, is given as second tank. The lead acid batteries are modelled in this paper in two segments like charge and discharge.

The available battery bank capacity at hour (t) is calculated by means of the mathematical modelling during the charging process, which is expressed as (Garcia et al., 2006),

$$E_{batt}(t) = E_{batt}(t-1) + E_{EE}(t) \times \eta C_{C \times} \eta CH_G$$

(48)

Where the excess energy from renewable generators after serving load is denoted as $EE_{E(}t)$, the charge controller efficiency is denoted as ηCC_a nd the charging efficiency of battery is denoted as ηCHG.

Meanwhile, the charged quantity of the battery is subjected to the following constraints,

$$SOCm_{in} \le SOC \le SOCm_{ax}$$

(49)

Figure 4. Flowchart of charging SOC of battery system

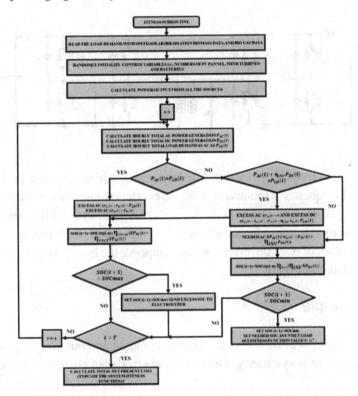

The maximum value of SOC is equal to 1 and the minimum value of SOC is attained by the maximum depth of discharge (DOD).

$$SOC_{min} = 1 - DOD \tag{50}$$

On the basis of the charging and discharging of batteries, the size of optimal battery bank is attained. Figure 4. Illustrates the flowchart of SOC battery.

The power, which is generated from the renewable energy generators, is classified into two types. In the first type, it delivers the energy directly from the renewable energy generator to the load whereas in the second type, it supplies power from the battery. When the AC generators are unable to supply the load requirement, the combination of both AC and DC is utilized to meet the load requirements. If both AC and DC generators are unable to generate the essential quantity of power, the battery system is employed to supply the power to the load.

Bidirectional Converter

The circuit layout of the bidirectional converter is illustrated in Figure 5. The boost converter (L) controls the energy, which flows from low to the high voltage while the uncontrolled high side converter (H) works as a rectifier. The buck converter (H) is used to control the energy flow in the opposite side while the low side converter (L) works as a rectifier.

Figure 5. Circuit layout of bidirectional converters

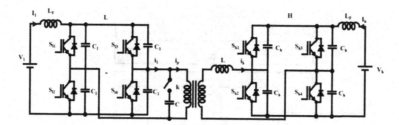

The transistor's control frequency regulates the converter, which retains a constant brake between the pulses. The control frequency's minimum value refers the maximum transistor control pulse width and the maximum flow of energy. The buck converter's transistor control frequency is varied from its maximum frequency to its minimum frequency, which is illustrated in Figure 6.

When the control pulses of one converter leg is overlapped, the boost converter's transistor regulates the frequency as illustrated in Figure 7.

Buck Operation Principle

The buck converter's (H) input current I_h is assumes as constant and output i_h is assumed as sinusoidal. The converter's arrangement is symmetry. The buck converter's leg current is expressed as,

$$i_h = \frac{I_m}{2} sin\omega t + \frac{I_h}{2}$$

(51)

The current has charged and discharged the capacitor C_h from zero voltage to zero voltage, which is mathematically expressed as,

Figure 6. Buck converter's transistor control pulses

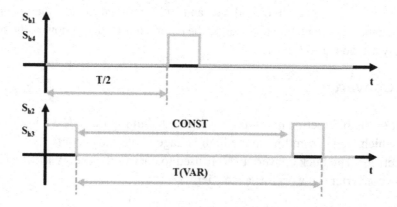

Figure 7. Boost converters' transistor control pulses

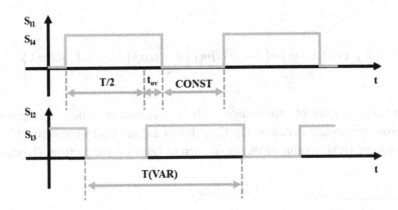

$$\int_{\lambda-\varphi}^{2\pi-\varphi}\left(\frac{I_m}{2}sin\omega t+\frac{I_h}{2}\right)d\left(\omega t\right)=0 \tag{52}$$

Where the width of control pulse is denoted as λ and $sin\varphi=\dfrac{I_h}{I_m}$.

The expression for the relationship between the input and output currents of the buck converter is defined as,

$$\frac{I_m}{I_h}=\sqrt{1+\left(\frac{sin\lambda+\pi\left(2-\dfrac{\lambda}{\pi}\right)}{1-cos\lambda}\right)^2} \tag{53}$$

The expression for the voltage across capacitor C_h is given by its instantaneous value,

$$v_h=\frac{1}{\omega C_h}\int_{\lambda-\varphi}^{\omega t}\left(\frac{I_m}{2}sin\omega t+\frac{I_h}{2}\right)d\left(\omega t\right)=\frac{I_m}{2\omega C_h}\left[cos\left(\lambda-\varphi\right)-\left(\lambda-\varphi\right)sin\varphi+\omega tsin\varphi-cos\omega t\right] \tag{54}$$

The expression for the half value of high voltage V_h, which is equivalent to the average value of capacitor voltage C_h during steady state operation, can be written as follows.

$$\frac{V_h}{2}=\frac{1}{2\pi}\int_{\lambda-\varphi}^{2\pi-\varphi}v_h d\left(\omega t\right) \tag{55}$$

Consequently, the equation is derived to calculate the intended value of the capacitor voltage C_h, denoted as.

$$C_h = \frac{I_m}{2\pi\omega v_h}\left\{(2\pi - \lambda)\left[\cos(\lambda - \varphi) - (\lambda - \varphi)\sin\varphi\right] + \frac{1}{2}\sin\varphi(2\pi - \lambda)^2 + \sin\lambda\right\} \tag{56}$$

To ascertain the precise value of capacitance C, the subsequent procedure is employed. In the event that the voltage across the capacitor, denoted as C_h, exhibits a sinusoidal waveform with an amplitude of V_m, the root mean square (RMS) value of the input current $i_l rms$ for the rectifier (L) can be mathematically expressed

$$i_{lr} = \sqrt{(k_T.I_m)^2 - (\omega C'V_m)^2} \tag{57}$$

Where the transformer turns ratio is denoted as k_T, the substitute, which results from the parallel linked capacitors capacitance is denoted as C'. It is expressed as,

$$C' = C + C_h + C_l \tag{58}$$

According to the analysis of resonant converters, the rectifier (L) input is replaced by the resistance, the value of which is determined by the transferred power value P_{min}. As a result, the RMS value of current I_l is expressed as,

$$I_{lr} = \frac{\sqrt{2}P_{min}}{V_m} \tag{59}$$

The output voltage of the rectifier is equal to the battery's voltage V_l in steady state operation due to the sinusoidal nature of the rectifier input voltage. This relationship is expressed as,

$$C' = \frac{2}{\omega\pi V_l}\sqrt{k_T^2 I_m^2 - \frac{16P_{min}^2}{\pi^2 V_l^2}} \tag{60}$$

The capacitance value C^l and the inductance value L remain constant regardless of the direction of energy flow.

Boost Operation Principle

Boost operates by directing energy flow from a low voltage source to a high voltage source. In this mode, the boost converter (L) is regulated while the converter (H) functions as a rectifier. The relationship between the input current (I_l) and the output current (i_l) of the boost converter (L) is expressed by the following equation.

$$\frac{I_l}{I_{lm}} = \frac{1 + cos\left(\dfrac{2t_{0v}}{T}\pi\right)}{\pi\left(1 - \dfrac{2t_{0v}}{T}\right)} \tag{61}$$

Where the amplitude of boost converters output current is denoted as I_{lm}, the overlap control pulse is denoted as t_{0v} and the control period is denoted as T. Assume the value of $\dfrac{2t_{0v}}{T} = 0.26$, $\dfrac{I_l}{I_{lm}} = 0.725$. The capacitance C_l is expressed as,

$$C_l = 8.866.10^{-3}\frac{TI_{lm}}{V_l} \tag{62}$$

The value of resonant circuit inductance L is computed from the following equation as,

$$1.1T = 2\pi\sqrt{2C_hL} \tag{63}$$

Charge Controller

The charge controller initiates its sensing mechanism once the battery reaches full charge, effectively limiting or decreasing the flow of energy from the energy source to the batteries to prevent overcharging. The charge controller model is given as,

$$E_{CC,OUT}(t) = E_{CC,IN}(t) \times \eta C_C \tag{64}$$

$$E_{CC,IN}(t) = E_{REC,OUT}(t) \times E_{SUR,DC}(t) \tag{65}$$

LOAD MODEL

The load demand model is responsible for determining the total energy demand, known as E_{Load}, for each time step in the simulation. This model includes various electrical loads such as Household Load (HHL) consisting of lights, TV, fan, and radio; Commercial Load (CL) including small shops and flour mills; Community Load (CNL) comprising primary health centres, school lighting, street lighting, pumping water, and Panchayat Hall lighting; and Small Industrial Load (SIL) consisting of saw mills. Therefore, the total energy demand for the selected remote rural area, which consists of a cluster of villages, is represented as,

$$E_{Load}(t) = E_{HHL}(t) + E_{CL}(t) + E_{CNL}(t) + E_{SIL}(t) \tag{66}$$

$$E_{HHL}(t) = E_{CL}(t) = E_{CNL}(t) = E_{SIL}(t) = \sum_{i=1}^{N}\left[P_i \times t_i \times n\right] \tag{67}$$

COMPONENTS DATABASE SETUP

The optimization techniques being developed aim to identify implementable configurations for the proposed system. Therefore, a comprehensive database of commercially available components is essential. This database should include parameters necessary for employing mathematical models, as well as economic data such as maintenance and replacement costs for each energy technology. The optimization algorithm utilizes this database when evaluating different configurations. Additionally, the database is regularly updated to incorporate new and efficient variants of specific technologies introduced by manufacturers.

Parameters Required for Simulation

The system utilizes various parameters to model each type of energy technology.

- For biogas or biomass generators, these parameters include fuel consumption at 100% load, installed cost, maintenance cost, replacement cost, and lifetime.
- When it comes to wind turbines, the parameters consist of the power curve, rated power, hub height, installation cost, and maintenance cost.
- For PV cells, the parameters include efficiency, rated power, area, lifetime, replacement cost, maintenance cost, and installation cost.
- Fuel cells are modelled using parameters such as fuel curve slope, rated power, lifetime, maintenance cost, and replacement cost.
- Electrolysers are characterized by their rated power, minimum load, generation of rated hydrogen, lifetime and replacement, installation cost, and maintenance cost.
- Battery parameters include cycles to failure (80% depth of discharge), nominal capacity, capacity ratio, rate constant, round trip efficiency, lifetime, nominal voltage, maximum charge rate, maximum charge current, and installation cost.
- Lastly, hydrogen tanks are modelled using parameters such as rated capacity, lifetime and installation, replacement cost, and maintenance costs.

Figure 8. Output voltage vs. load resistance

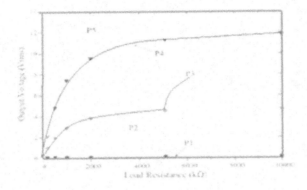

Simulation Results for Varying Load Parameters

Use the Simulation Manager to visualize simulation data, analyse simulation results and trends. Here are some examples of simulation results for varying load parameters;

Varying the load resistance in a simple circuit:

The following figure 8 shows the output voltage of a simple circuit with a varying load resistance. The load resistance is varied from 0 ohm to 10000 kilo ohms, and the input voltage is 30 volts. As the load resistance increases, the output voltage decreases.

Varying the Load Current in a Motor

The following figure 9 shows the speed of a motor with a varying load current. The load current is varied from 0 amps to 600 milliamps, and the input voltage is 30 volts. As the load current increases, the speed of the motor decreases.

Varying the load temperature in a thermal system:

The following figure 10 shows the temperature of a thermal system with a varying load temperature. The load temperature is varied from 0 degrees Celsius to 320 degrees Celsius, and the input power is 100 watts. As the load temperature increases, the temperature of the thermal system increases.

Above results are a few examples of simulation results for varying load parameters. There are many other types of simulations that can be performed, and the results can vary depending on the specific system being simulated and designed.

Figure 9. Motor speed vs. load current

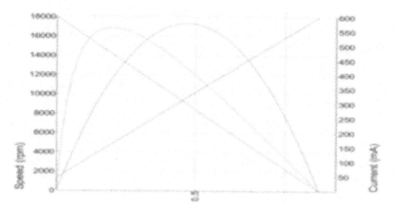

Figure 10. Load versus thermal temperature

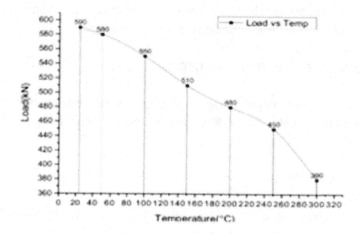

CONCLUSION

A hybrid renewable energy system has been developed in this chapter to ensure uninterrupted power supply in rural and remote areas. The system combines solar, wind, biomass, biogas, fuel cell, and battery technologies, each with its own advantages and disadvantages. By integrating these technologies, the system maximizes the benefits while minimizing their limitations. The chapter introduces a hybrid energy model that incorporates all these renewable energy sources, enhancing the reliability and continuity of the system. This model also holds potential for future research in the hybrid energy field and can be expanded to a grid-connected version using micro-controllers and power factor correction devices. Implementing this framework in any location, especially remote areas with frequent power failures, can greatly benefit our country, which has favourable resources and demographics for these technologies.

REFERENCES

Anurag Chauhan, R. P. (2017). Size optimization and demand response of a stand-alone Integrated Renewable Energy System. *Energy*, *124*, 59–73. doi:10.1016/j.energy.2017.02.049

Farret, F. A., & Godoy Simoes, M. (2006). *Integration of Alternative Sources of Energy*. John Wiley & Sons, Ltd.

Garcia, R. S., & Weisser, D. (2006). A wind–diesel system with hydrogen storage: Joint optimization of design and dispatch. *Renewable Energy*, *31*(14), 2296–2320. doi:10.1016/j.renene.2005.11.003

Kanase-Patil, A. B., Saini, R. P., & Sharma, M. P. (2011). Development of IREOM model based on seasonally varying load profile for hilly remote areas of Uttarakhand state in India. *Energy*, *36*(9), 5690–5702. doi:10.1016/j.energy.2011.06.057

Pipattanasomporn, M. (2004*). A study of remote area internet access with embedded power generation.* [PHD Thesis, Virginia Polytechnic Institute and State University, Alexandria, Virginia].

Suresh, V., Muralidhar, M., & Kiranmayi, R. (2020). Modelling and optimization of an off-grid hybrid renewable energy system for electrification in a rural areas. *Energy Reports*, *6*, 594–604. doi:10.1016/j.egyr.2020.01.013

Thapar, V., Agnihotri, G., & Sethi, V. K. (2011). Critical analysis of methods for mathematical modelling of wind turbines. *Renewable Energy*, *36*(11), 3166–3177. doi:10.1016/j.renene.2011.03.016

Vendoti, S., Muralidhar, M., & Kiranmayi, R. (2021). Techno-economic analysis of off-grid solar/wind/biogas/biomass/fuel cell/battery system for electrification in a cluster of villages by HOMER software. *Environment, Development and Sustainability*, *23*(1), 351–372. doi:10.1007/s10668-019-00583-2

Chapter 18
Failure Rate Examination for Liability Tree–Based Analysis for Clamped Double Subsystem With DC Short Voltage Protective Functionality

M. D. Rajkamal
Velammal Institute of Technology, India

T. Mothilal
KCG College of Technology, India

M. Shanmugapriya
KCG College of Technology, India

M. Saravanan
Hindustan Institute of Technology and Science, India

ABSTRACT

This study presents a comprehensive failure rate examination for implantable antennas, employing a liability tree-based analysis focused on a clamped double subsystem (CDSM) equipped with DC short voltage protective functionality. The enhanced protective feature of the CDSM aims to improve the security and safety of implantable antennas used in critical applications. However, this subsystem's design necessitates the use of additional IGBTs, diodes, and capacitors compared to standard configurations, consequently increasing the complexity and potential failure rate. Given that demanding converter operation in implantable antennas can escalate the failure rate, conducting a precise reliability analysis becomes vital for the deployment of CDSM in these devices. A failure durability analysis is undertaken to address the operational characteristics of CDSM in implantable antennas. Fault Tree Analysis (FTA) is utilized to evaluate the risk with greater precision than previous methods, which mainly considered component types, quantities, and network connectivity states.

DOI: 10.4018/979-8-3693-3735-6.ch018

INTRODUCTION

Because the DC transmission submodule's primary function is voltage converting as well as battery pack, it is critical to design a highly dependable system architecture or systems (Yogeshwaran et al. 2015). In HVDC, quarter as well as comprehensive are commonly employed. Various system architectures have long been identified to fulfil particular goals like DC low protective devices, power output or channel reducing stress, steps that will help, as well as reduce costs (Hemanth et al. 2017). It is necessary to anticipate dependability through certain precise fault diagnosis before using these components in a real DC microgrid. The chance that somewhere a system will reliably fulfil best practises and standards throughout a required timeline given standard test conditions is known as descriptive (Singh 2017).

Among the indicative measures of dependability seems to be the number of attacks of specific equipment per unit time and thus is stated as the coefficient of a benchmark for living, former case before failing (MTBF) or median probability of failure (MTTF). Component counting models, combination designs, linear regression, and joint probability delivery mechanisms were ideal for large-scale steep conversion of solar energy like converters (Santhosh et al. 2018). A block counting approach is a forecast approach that assumes the rate of separate components remains unchanged but that all gadgets, including components, are linked together (Sakthi Shunmuga Sundaram et al. 2019).

This really is a simple forecasting approach and could be effective in the context of voltage regulation conceptual design. However, it may indicate a substantial divergence from just the real completion rates due to the perception that perhaps the rates of unit components are equal (Natrayan and Senthil Kumar 2019). A hybrid algorithm, an enhanced form of component counting prototype, can estimate the dependability of the duplicated systems, but it is difficult to represent specifics, including the attrition rate of a micro, the overall breakdown order of devices, and the area shown (Madupalli et al. 2019). A stochastic approach is effective for estimating the deviation of a system without regular breakdowns, but it is hard to analyse if components break over time and also if the state vector grows exponentially as the number of components increases (Hemalatha et al. 2020).

The determination coefficient method is a chance distribution-based variance inflation factor that uses the Demurer formula. Those demonstrating support can observe the overall dependability of the inverters directly, but they often address the inverter computer's performance parameters (Niveditha VR. and Rajakumar PS. 2020). For even more accurate fault diagnosis, the operational parameters of a subsystem must be included (Vaishali et al. 2021). They proposed a mistake assessment that took into account the about this of a quarter sub system, and when compared to the overall component clock finite element analysis, they were able to forecast the existence through taking into account the malfunction owing to operating hazard (Sandeep Kauthsa Sharma et al. 2021).

Scientists created a pinned subsystem (CDSM) which increased the submodule's security as well as dependability by including a DC high-successful preventive mechanism. Through incorporating the submodule's shield, its DC short protective feature may increase operational reliability (Anupama et al. 2021). Nonetheless, as the number of IGBTs, diodes, and capacitors increases (Singh et al. 2017), a precise assessment of the completion rates is required before adopting as a substitute for quarter submodules (HBSM). A malfunction of a CDSM is examined in that whole work by describing its loss and creating the accident according to the cause and impact of the loss (Sathish et al. 2021).

Furthermore, it contrasts the overall findings of the CDSM operations with the outcomes of the standard part count's fault diagnosis (PCA). Lastly, it evaluates its failures based on the percentage of components

(Sabarinathan et al. 2022), dc current strain, including value margins of a component, which conducts value in some kind of a variety of areas by including CDSM's dc brief traditional protection feature.

CLAMPED SUBMODULE

In this investigation, we explore the intricate operational domain names of a networked subsystem known as the Current-Doubling Serial Module (CDSM), focusing primarily on its DC fast safety and manage qualities (Veeman et al. 2021). In the context of a Modular Multilevel Converter (MMC) sensor arrangement, the parent illustration reveals a relationship between Metal-Oxide-Semiconductor Field-Effect Transistor (MOSFET) switches. CDSM, a difficult configuration, takes a vital role in fast ascertaining the parallelism of a quarter subsystem (Sendrayaperumal et al. 2021). Figure 1 clearly depicts a CDSM composition containing important components which comprises the Insulated Gate Bipolar Transistor (IGBT), rectifier, and capacitance (Reddy Chukka et al. 2021). The term thyristors is surely employed, identifying the dual junctions accountable for preserving voltage stages throughout the length of all distinct banks within the apparatus (Santhosh et al. 2021). This step forward system, abbreviated as CDSM, effectively merges a contemporary current-restraining related voltage pinching feature quickly straight into a serial zone architecture (Praburanganathan et al. 2022).

During the cooling level, the activation of IGBT Q5 is a lengthy operational item. Conversely, even if a DC decrease occurs to harness power manufacture, a selective deactivation of numerous modules immediately soon follows (Darshan et al. 2022). This tactical shutdown capability now not most effective retains the integrity of the gadget but additionally gives considerably to operational effectiveness with the aid of employing proficiently dissipating excess warmth at positive time in DC short durations (Loganathan et al. 2023). Beyond its fundamental structural intricacy, the rationale of CDSM extends to integrating a proactive technique in handling talents demanding instances, especially the ones related with overcharging (Selvi et al. 2023). The dynamic adaptability of CDSM to the system's demands stands as a monument to contemporary engineering (Chehelgerdi et al. 2023; Saadh et al. 2023b), setting a balance among high-ordinary performance functionality and shielding measures in the broad position of electrical systems and converters (Sendrayaperumal et al. 2021).

It is crucial to realise that the value of CDSM surpasses the basic nuances of its layout, growing into a holistic method that predicts and controls future challenges. By delivering a detailed understanding of the connectivity of device wants and subsystem skills, CDSM develops as a cornerstone in current engineering procedures, contributing to the creation of electricity systems and converters (Subramanian et al. 2022). As the quest of efficient and dependable electrical solutions continues, CDSM acts as a lighthouse, driving the industry towards sustainable, excessive-performance solutions in an ever-evolving technical environment (Kaushal et al. 2023).

Operation Module

In ordinary operating conditions, the CDSM and its analogous circuits are shown in Figure 1. Because converter Q5 is still turned on, transistors D6 and D7 clamp capacitance constants (Saadh et al. 2023a). Through the ON as well as off management of IGBTs (Q1-Q4), inverter capacitance may be linked to or removed from the converters, and capacitances can indeed be filled concurrently or independently loaded as well as discarded (Thakre et al. 2023). Figure 1 depicts a CDSM circuitry condition in an anomalous voltage whenever a DC transient voltage arises. Because IGBT Q5 retains the turn-on state, a present route made up of two impedances is built independent of the normal direction coming into such a subsystem (Nagarajan et al. 2022).

As a result, capacitance C1 and C2 connected in series collect DC small power, and the DC small protective feature is activated, with the value capped by transistors D6 and D7. CDSM is projected to be more reliable than standard split or comprehensive subspaces due to the addition of dc-long dc voltages to make things safer (Seeniappan et al. 2023). In regular operating conditions, though, the IGBT Q5 remains in a turn-on state, resulting in the latest techniques in power reduction (Sharma et al. 2022). Whenever a discontinuous call in order develops, reconfiguration failure happens owing to the immediate turn-off on Q5, resulting in poor converter effectiveness (Divya et al. 2022).

Fault Tree Design

Throughout this section, the failure is constructed with the operating hazard in mind, based upon the CDSM operating features. Failure assessment (FTA) is a statistical failure assessment as well as a dependability assessment approach that increases system stability by logically examining the source of a system crash, creating a leakage, then identifying weaknesses by funding the failure rates (Mahesha et al. 2022; Santhosh Kumar et al. 2022). By permitting a probability systematic estimate of the danger of converter functioning, it is feasible to generate rational and statistically quantifiable findings by straying from past visual as well as practical reasoning . Figure 2 depicts a primary failure based on CDSM operating features (Balamurugan et al. 2023). Every thread in Figure 2 is constructed depending on the needs where the submodule's capacitance is linked and disengaged from the power supply (Seeniappan et al. 2023).

All of that is intended to be broken into two parts: a breakdown inside the phase of capacitor charge as well a loss there in the condition of draining. Whenever a DC low voltage develops, converter Q5 is switched off, forming a present channel between two capacitance that still absorbs any relatively brief power (Josphineleela et al. 2023a). Because C1 and C2 can be charged independently via IGBT when the sample and detector are rotated, the fault is transmitted to a superior stage only after all events fail (Natrayan and Kaliappan 2023). The failure of the generator power protective mechanism is built as an OR mix of two occurrences in which DC energy flows into the SM endpoint then outflows from the SM (+) endpoint (Kaliappan et al. 2023).

Whenever one of the two occurrences of capacitance C1 draining as well as capping fails, these are built as such a Rather than that, it's also transmitted to an elevated problem. Suspension detaching failure is a condition that disconnects the submodule's capacitance first from converters and therefore is constructed by splitting it up into a normal operating condition wherein Q5 is switched on and an unusual activity wherein Q5 is switched off based on whether or not dc reduction to produce happens (Ramaswamy et al. 2022b). After detaching the subsystem from one of the converters, the post was categorised in accordance with the current trajectory. Even if a DC brief circuit arises, it's also intended to segregate activities based on historical orientation. Q5= When fluid exits the SM, only a SP detaching capability is provided (Suman et al. 2023).

Figure 1. Failure rate obtained by FTA and PCA

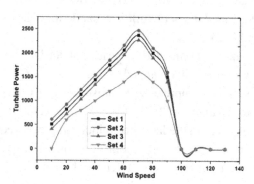

RESULT AND DISCUSSION

The average number of failures of a climate-controlled thin plastic screen capacitance in a metal frame is determined by (10). The sintered metal movie capacitor's baseline rate is 0.0048, as well as the variable resistor index is 1.0. Its directivity is 10, which itself is a business capacitor's typical value, as well as the environmental factors, is 10 (Josphineleela et al. 2023b). That indicates that its converter functions in a contained way. A heat coefficient of a capacitance is determined by (11) wherein T seems to be the capacitor's heat flux.

PCA may evaluate the delay while taking into account the type of components, the quantity of components, and also the connecting state of the sections. In contrast to a PCA technique, the Ft could take into account the actual performance parameters of a subsystem. The number of failures is compared and analysed using two approaches throughout this section (Saravanan et al. 2023). The computation of the completion rates utilising principal component analysis as well as FTA was conducted using a basic Excel application versus a computer (Sivakumar et al. 2023). Although this responsibility to solve for every level is merged by which, the failure in CDSM is computed as (1). CDSM is made up of four FETES and seven diodes, all of which are connected in series, but one of the most significant drawbacks is that each element has serial failing characteristics because it is operated (Arun et al. 2022).

CDSM is a design that adds one converter, two transistors, and one capacitance towards the construction of a filled subsystem to provide DC relatively brief protection. As a result, the efficiency of every item is computed using an either/or procedure, as seen in Figure 2 compares the clinical pharmacist failures of HBSM as well as FBSM (comprehensive submodule) to that of CDSM, utilising dc reduced to produce a guarding mechanism (Balaji et al. 2022). That analyses components with a certain boost converter capability using kilometres. Because the Cap is a dependability assessment that hardly considers the kind, amount, as well as connectivity of components, the breakdown price hikes as even the size of data grows (Ramaswamy et al. 2022a).

Because the probability of failure of the smallest incident and the durability dependency from each element were identical in the OR connection, while only the number of components varied, the probability of failure occurs with the majority of components in the order CDSM > FBSM > HBSM, as shown in Figure 2. It examines the failures based on the current ratings margins of IGBT and photodiodes, which have been CDSM-flipping circuit parts (Selvi et al. 2023). Because the preceding investigation is focused on shifting parts with a gap of available upon request time, the failed voltage margins of 1.35,

Figure 2. Mean time failure rate obtained by FTA and PCA

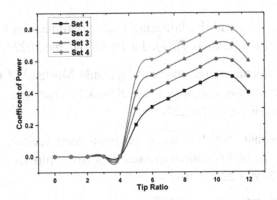

1.6, and 3.9 times are investigated deeper (Kumar et al. 2022). Figure 2 depicts the PCA conclusion. It shows that unless the power gap diminishes, the breakdown rates rise correspondingly. Unfortunately, determining the voltage margins of a switch from the principal component analysis results by itself is problematic (Pragadish et al. 2023).

CONCLUSION

In conclude, the entire study of failure costs and typical restore times for installed subsystems inside a tiny protective framework delivers useful insights into their operational dependability. Three remarkable subsystems—FBSM (Fastened Back-to-Back Submodule), CDSM (Current-Doubling Serial Module), and HBSM (Half-Bridge Serial Module)—have their performance examined in this research through the careful calculation of these measures. The facts from the look at reveal a hierarchical sequence of basic charges, with FBSM demonstrating a superior failure price in comparison to CDSM, and CDSM topping HBSM in terms of dependability. This ranking underlines the relevance of the protective framework in impacting the overall performance and reliability of those subsystems.

Introducing a protecting feature, notably when controlling a DC discount, shows to be a key component in completing large successes. While FBSM reports a superb decline in breakdown prices with the adoption of this alternative, it's far nevertheless superior than that of HBSM thanks to an enlarged variety of elements in play. The smallest HBSM, with its fewest ingredients, shines out on this respect, highlighting simplicity as a capacity gain in terms of reliability. However, the intricacy of CDSM, defined by extra IGBTs, diodes, and capacitors, enables green control and creation from DC rapid glide, as well as excellent warmth absorption. Despite its complex architecture, an interesting observation emerges: the expanding trend of components inside CDSM seems to coincide with a decline in normal dependability. As a forward-searching consideration, the take a look at suggests that a comparison examination of dependability variations with voltage margins should supply priceless steerage to inverter choice-makers. By considering dependability from more than one viewpoints, inverter designers might make intelligent decisions that stabilise overall performance and reliability in the dynamic world of electricity structures and converters. This study consequently provides now not merely to knowing the latest dynamics but also supplies a platform for future advancements in the field.

REFERENCES

Anupama. (2021). Deep learning with backtracking search optimization based skin lesion diagnosis model. *Computers, Materials & Continua, 70*(1). doi:10.32604/cmc.2022.018396

Arun. (2022). Mechanical, fracture toughness, and Dynamic Mechanical properties of twill weaved bamboo fiber-reinforced Artocarpus heterophyllus seed husk biochar epoxy composite. *Polymer Composites, 43*(11), 8388–8395. doi:10.1002/pc.27010

Balaji. (2022). Annealed peanut shell biochar as potential reinforcement for aloe vera fiber-epoxy biocomposite: Mechanical, thermal conductivity, and dielectric properties. *Biomass Conversion and Biorefinery.* doi:10.1007/s13399-022-02650-7

Balamurugan, P., Agarwal, P., Khajuria, D., Mahapatra, D., Angalaeswari, S., Natrayan, L., & Mammo, W. D. (2023). State-Flow Control Based Multistage Constant-Current Battery Charger for Electric Two-Wheeler. *Journal of Advanced Transportation, 2023*, 1–11. doi:10.1155/2023/4554582

Chehelgerdi, Chehelgerdi, M., Allela, O. Q. B., Pecho, R. D. C., Jayasankar, N., Rao, D. P., Thamaraikani, T., Vasanthan, M., Viktor, P., Lakshmaiya, N., Saadh, M. J., Amajd, A., Abo-Zaid, M. A., Castillo-Acobo, R. Y., Ismail, A. H., Amin, A. H., & Akhavan-Sigari, R. (2023). Progressing nanotechnology to improve targeted cancer treatment: Overcoming hurdles in its clinical implementation. *Molecular Cancer, 22*(1), 169. doi:10.1186/s12943-023-01865-0 PMID:37814270

Chukka, R. (2021). Seismic Fragility and Life Cycle Cost Analysis of Reinforced Concrete Structures with a Hybrid Damper. *Advances in Civil Engineering, 2021*, 1–17. doi:10.1155/2021/4195161

Darshan, Girdhar, N., Bhojwani, R., Rastogi, K., Angalaeswari, S., Natrayan, L., & Paramasivam, P. (2022). Energy Audit of a Residential Building to Reduce Energy Cost and Carbon Footprint for Sustainable Development with Renewable Energy Sources. *Advances in Civil Engineering, 2022*, 1–10. doi:10.1155/2022/4400874

Divya. (2022). Analysing Analyzing the performance of combined solar photovoltaic power system with phase change material. *Energy Reports, 8.* doi:10.1016/j.egyr.2022.06.109

Hemalatha. (2020). Analysis of RCC T-beam and prestressed concrete box girder bridges super structure under different span conditions. In *Materials Today*. Proceedings.

Hemanth. (2017). Evaluation of mechanical properties of e-glass and coconut fiber reinforced with polyester and epoxy resin matrices. *International Journal of Mechanical and Production Engineering Research and Development, 7*(5). doi:10.24247/ijmperdoct20172

Josphineleela, R. (2023b). Intelligent Virtual Laboratory Development and Implementation using the RASA Framework. In: *Proceedings - 7th International Conference on Computing Methodologies and Communication, ICCMC 2023*. IEEE. 10.1109/ICCMC56507.2023.10083701

Josphineleela, R., & Kaliapp, S. (2023a). Big Data Security through Privacy - Preserving Data Mining (PPDM): A Decentralization Approach. In: *Proceedings of the 2023 2nd International Conference on Electronics and Renewable Systems, ICEARS 2023*. IEEE. 10.1109/ICEARS56392.2023.10085646

Kaliappan, S., Natrayan, L., & Garg, N. (2023). Checking and Supervisory System for Calculation of Industrial Constraints using Embedded System. *In: Proceedings of the 4th International Conference on Smart Electronics and Communication, ICOSEC 2023*. IEEE. 10.1109/ICOSEC58147.2023.10275952

Kaushal. (2023) A Payment System for Electric Vehicles Charging and Peer-to-Peer Energy Trading. In: *7th International Conference on I-SMAC (IoT in Social, Mobile, Analytics and Cloud), I-SMAC 2023 – Proceedings*. IEEE. 10.1109/I-SMAC58438.2023.10290505

Kumar, Kaliappan, S., Socrates, S., Natrayan, L., Patel, P. B., Patil, P. P., Sekar, S., & Mammo, W. D. (2022). Investigation of Mechanical and Thermal Properties on Novel Wheat Straw and PAN Fibre Hybrid Green Composites. *International Journal of Chemical Engineering*, 2022, 1–8. doi:10.1155/2022/3598397

Kumar, S. (2022). IoT battery management system in electric vehicle based on LR parameter estimation and ORMeshNet gateway topology. *Sustainable Energy Technologies and Assessments*, *53*, 102696. doi:10.1016/j.seta.2022.102696

Loganathan, Ramachandran, V., Perumal, A. S., Dhanasekaran, S., Lakshmaiya, N., & Paramasivam, P. (2023). Framework of Transactive Energy Market Strategies for Lucrative Peer-to-Peer Energy Transactions. *Energies*, *16*(1), 6. doi:10.3390/en16010006

Madupalli. (2019). Structural performance of non-linear analysis of turbo generator building using seismic protection techniques. *International Journal of Recent Technology and Engineering*, *8*(1).

Mahesha, C. R., Rani, G. J., Dattu, V. S. N. C. H., Rao, Y. K. S. S., Madhusudhanan, J., L, N., Sekhar, S. C., & Sathyamurthy, R. (2022). Optimization of transesterification production of biodiesel from Pithecellobium dulce seed oil. *Energy Reports*, *8*, 489–497. doi:10.1016/j.egyr.2022.10.228

Nagarajan, Rajagopalan, A., Angalaeswari, S., Natrayan, L., & Mammo, W. D. (2022). Combined Economic Emission Dispatch of Microgrid with the Incorporation of Renewable Energy Sources Using Improved Mayfly Optimization Algorithm. *Computational Intelligence and Neuroscience*, 2022, 1–22. doi:10.1155/2022/6461690 PMID:35479598

Natrayan, L., & Kaliappan, S. (2023) Mechanical Assessment of Carbon-Luffa Hybrid Composites for Automotive Applications. In: SAE Technical Papers doi:10.4271/2023-01-5070

Natrayan, L., & Senthil Kumar, M. (2019). Mechanical, microstructure and wear behaviour of lm25/sic/mica metal matrix composite fabricated by squeeze casting technique. *Applied Engineering Letters*, *4*(2), 72–77. doi:10.18485/aeletters.2019.4.2.5

Niveditha, V. R., & Rajakumar, P. S. (2020). Pervasive computing in the context of COVID-19 prediction with AI-based algorithms. *International Journal of Pervasive Computing and Communications*, *16*(5). doi:10.1108/IJPCC-07-2020-0082

Praburanganathan, S., Sudharsan, N., Bharath Simha Reddy, Y., Naga Dheeraj Kumar Reddy, C., Natrayan, L., & Paramasivam, P. (2022). Force-Deformation Study on Glass Fiber Reinforced Concrete Slab Incorporating Waste Paper. *Advances in Civil Engineering*, 2022, 1–10. doi:10.1155/2022/5343128

Pragadish, N., Kaliappan, S., Subramanian, M., Natrayan, L., Satish Prakash, K., Subbiah, R., & Kumar, T. C. A. (2023). Optimization of cardanol oil dielectric-activated EDM process parameters in machining of silicon steel. *Biomass Conversion and Biorefinery*, 13(15), 14087–14096. doi:10.1007/s13399-021-02268-1

Ramaswamy. (2022b). Pear cactus fiber with onion sheath biocarbon nanosheet toughened epoxy composite: Mechanical, thermal, and electrical properties. *Biomass Conversion and Biorefinery*. Advance online publication. doi:10.1007/s13399-022-03335-x

Ramaswamy, R., Gurupranes, S. V., Kaliappan, S., Natrayan, L., & Patil, P. P. (2022a). Characterization of prickly pear short fiber and red onion peel biocarbon nanosheets toughened epoxy composites. *Polymer Composites*, 43(8), 4899–4908. doi:10.1002/pc.26735

Saadh, Baher, H., Li, Y., chaitanya, M., Arias-Gonzáles, J. L., Allela, O. Q. B., Mahdi, M. H., Carlos Cotrina-Aliaga, J., Lakshmaiya, N., Ahjel, S., Amin, A. H., Gilmer Rosales Rojas, G., Ameen, F., Ahsan, M., & Akhavan-Sigari, R. (2023a). The bioengineered and multifunctional nanoparticles in pancreatic cancer therapy: Bioresponisive nanostructures, phototherapy and targeted drug delivery. *Environmental Research*, 233, 116490. doi:10.1016/j.envres.2023.116490 PMID:37354932

Saadh, Castillo-Acobo, R. Y., Baher, H., Narayanan, J., Palacios Garay, J. P., Vera Yamaguchi, M. N., Arias-Gonzáles, J. L., Cotrina-Aliaga, J. C., Akram, S. V., Lakshmaiya, N., Amin, A. H., Mohany, M., Al-Rejaie, S. S., Ahsan, M., Bahrami, A., & Akhavan-Sigari, R. (2023b). The protective role of sulforaphane and Homer1a in retinal ischemia-reperfusion injury: Unraveling the neuroprotective interplay. *Life Sciences*, 329, 121968. doi:10.1016/j.lfs.2023.121968 PMID:37487941

Sabarinathan, P., Annamalai, V. E., Vishal, K., Nitin, M. S., Natrayan, L., Veeeman, D., & Mammo, W. D. (2022). Experimental study on removal of phenol formaldehyde resin coating from the abrasive disc and preparation of abrasive disc for polishing application. *Advances in Materials Science and Engineering*, 2022, 1–8. doi:10.1155/2022/6123160

Santhosh. (2018). Investigation of mechanical and electrical properties of kevlar/E-glass and basalt/E-glass reinforced hybrid composites. *International Journal of Mechanical and Production Engineering Research and Development*, 8(3). doi:10.24247/ijmperdjun201863

Santhosh, M. S., Sasikumar, R., Khadar, S. D. A., & Natrayan, L. (2021). Ammonium Polyphosphate Reinforced E-Glass/Phenolic Hybrid Composites for Primary E-Vehicle Battery Casings -A Study on Fire Performance. *Journal of New Materials for Electrochemical Systems*, 24(4), 247–253. doi:10.14447/jnmes.v24i4.a03

Saravanan, K. G., Kaliappan, S., Natrayan, L., & Patil, P. P. (2023). Effect of cassava tuber nanocellulose and satin weaved bamboo fiber addition on mechanical, wear, hydrophobic, and thermal behavior of unsaturated polyester resin composites. *Biomass Conversion and Biorefinery*. doi:10.1007/s13399-023-04495-0

Sathish, T., Palani, K., Natrayan, L., Merneedi, A., De Poures, M. V., & Singaravelu, D. K. (2021). Synthesis and characterization of polypropylene/ramie fiber with hemp fiber and coir fiber natural biopolymer composite for biomedical application. *International Journal of Polymer Science*, 2021, 1–8.. doi:10.1155/2021/2462873

Selvi. (2023) Optimization of Solar Panel Orientation for Maximum Energy Efficiency. In: *Proceedings of the 4th International Conference on Smart Electronics and Communication, ICOSEC 2023*. IEEE. 10.1109/ICOSEC58147.2023.10276287

Sendrayaperumal, Mahapatra, S., Parida, S. S., Surana, K., Balamurugan, P., Natrayan, L., & Paramasivam, P. (2021). Energy Auditing for Efficient Planning and Implementation in Commercial and Residential Buildings. *Advances in Civil Engineering, 2021*, 1–10. doi:10.1155/2021/1908568

Sharma, Raffik, R., Chaturvedi, A., Geeitha, S., Akram, P. S., L, N., Mohanavel, V., Sudhakar, M., & Sathyamurthy, R. (2022). Designing and implementing a smart transplanting framework using programmable logic controller and photoelectric sensor. *Energy Reports, 8*, 430–444. doi:10.1016/j.egyr.2022.07.019

Sharma, S. K. (2021). Mechanical Behavior of Silica Fume Concrete Filled with Steel Tubular Composite Column. *Advances in Materials Science and Engineering, 2021*, 1–9. doi:10.1155/2021/3632991

Singh, M. (2017). An experimental investigation on mechanical behaviour of siCp reinforced Al 6061 MMC using squeeze casting process. *International Journal of Mechanical and Production Engineering Research and Development, 7*(6). doi:10.24247/ijmperddec201774

Sivakumar, V., Kaliappan, S., Natrayan, L., & Patil, P. P. (2023). Effects of Silane-Treated High-Content Cellulose Okra Fibre and Tamarind Kernel Powder on Mechanical, Thermal Stability and Water Absorption Behaviour of Epoxy Composites. *Silicon, 15*(10), 4439–4447. doi:10.1007/s12633-023-02370-1

Subramanian, Lakshmaiya, N., Ramasamy, D., & Devarajan, Y. (2022). Detailed analysis on engine operating in dual fuel mode with different energy fractions of sustainable HHO gas. *Environmental Progress & Sustainable Energy, 41*(5), e13850. doi:10.1002/ep.13850

Suman. (2023) IoT based Social Device Network with Cloud Computing Architecture. In: *Proceedings of the 2023 2nd International Conference on Electronics and Renewable Systems, ICEARS 2023*. IEEE. 10.1109/ICEARS56392.2023.10085574

Sundaram, S. S. (2019). Smart clothes with bio-sensors for ECG monitoring. *International Journal of Innovative Technology and Exploring Engineering, 8*(4).

Thakre, Pandhare, A., Malwe, P. D., Gupta, N., Kothare, C., Magade, P. B., Patel, A., Meena, R. S., Veza, I., Natrayan L, & Panchal, H. (2023). Heat transfer and pressure drop analysis of a microchannel heat sink using nanofluids for energy applications. *Kerntechnik, 88*(5), 543–555. doi:10.1515/kern-2023-0034

Vaishali. (2021). *Guided container selection for data streaming through neural learning in cloud*. International Journal of Systems Assurance Engineering and Management., doi:10.1007/s13198-021-01124-9

Veeman, Sai, M. S., Sureshkumar, P., Jagadeesha, T., Natrayan, L., Ravichandran, M., & Mammo, W. D. (2021). Additive Manufacturing of Biopolymers for Tissue Engineering and Regenerative Medicine: An Overview, Potential Applications, Advancements, and Trends. *International Journal of Polymer Science, 2021*, 1–20. doi:10.1155/2021/4907027

Yogeshwaran, S., Prabhu, R., & Murugan, R. (2015). Mechanical properties of leaf ashes reinforced aluminum alloy metal matrix composites. *International Journal of Applied Engineering Research: IJAER, 10*(13), 1–10.

Chapter 19
Efficient Design for Implantable Device Constant Current Induction Doubly Fed Generating Incorporating Grid Connectivity

S. Socrates
Velammal Institute of Technology, India

M. Shanmugapriya
KCG College of Technology, India

B. Murugeshwari
Velammal Engineering College, India

S. Angalaeswari
iD https://orcid.org/0000-0001-9875-9768
Vellore Institute of Technology, Chennai, India

ABSTRACT

This research presents an innovative approach to the efficient design of implantable devices, focusing on the development and modeling of a constant current induction doubly fed generator (DFIG) system that incorporates grid connectivity under both sub and hyper synchronization conditions. The core of this study is to establish a physical equation for a power station and a DFIG using a combination of power management and voltage estimation techniques in the context of circuit power. The induction generator (IG) blade in the system is designed to rotate synchronously with the photovoltaic (PV) system frequency. The DFIG is connected to a distribution substation, with synchronization between the active power filter and the grid depot achieved through the use of dual converters: a machine side converter (MSC) on the grid side and a grid side converter (GSC) on the power system. Within the circuits, two applications are implemented to recover the parameter spectrum, aiming to maximize the thermodynamic efficiency delivered to the DFIG rotor.

DOI: 10.4018/979-8-3693-3735-6.ch019

INTRODUCTION

Windmill-generated power is set to reach 250 GW by the year 2022, comprising a sizeable 30% portion of the worldwide energy portfolio (Darshan et al. 2022). This boom in wind energy comes at a key moment when conventional sources of mineral riches, such as lignite, shale gas, and oil, are becoming more limited (Loganathan et al. 2023). As society's desire for inexpensive and sustainable electricity continues to rise, microgrids powered by renewable sources such as wind, solar, and hydropower are gaining importance as long-term energy solutions (Vijayaragavan et al. 2022). However, any renewable source has constraints. In contrast to other renewables, hydroelectric power facilities display certain commonalities and have the ability to be strategically coordinated to provide considerable electrical production (Santhosh Kumar et al. 2022). Within this framework, the charge controller, notably the Induction Generator (IG), takes a crucial function. Power systems in conventional electricity generating frequently function at parallel speeds, needing large energy to modify their rate using movers like turbo diesels (Mahesha et al. 2022). However, the IG, with its capacity to work at varied velocities substantially larger than synchronized rates, proved to be exceedingly efficient when linked with windmills (Divya et al. 2022).

The interaction between wind speed and variable pitch velocity rotors, together with the utilisation of average wind capacitors to collect peak energy despite changeable wind conditions, highlights the usefulness of the IG paired with windmills (Sharma et al. 2022). The inclusion of relays in the speed control power of doubly fed induction generators (DFIG) further increases the flexibility of the system. High-performance inductors have received substantial interest in numerous applications, suggesting a trend towards more efficient and sustainable energy solutions (Arockia Dhanraj et al. 2022). The necessity for sophisticated modelling environments for wind turbine systems is a mandate for turbine makers. However, the secrecy surrounding this information has encouraged computational assessments by academics aiming to anticipate transitory reactions, particularly in the context of strength and deformation conditions (Seeniappan et al. 2023). Streamlining the modeling process, current research has focused on the effect of DFIG modeling simplification and studied design stability by incorporating rotor and spindle impedance (Nagarajan et al. 2022).

Validation of wind turbine models against the IEC 61450-27 standards using observed data, as well as the validation of the IEC Steroid hormone architecture for windmills, offers a foundation for judging the correctness of these computational models (Thakre et al. 2023). The research community's interest in induction generator analysis has led to the adoption of the IEC Steroid hormone design for windmills reference current regulator with offshore wind factors controllers (Kaushal et al. 2023). However, there is an understanding that further research is essential for many situations involving DFIG coupled to the grid via diverse mechanisms. This work aims to assess an arithmetical modeling technique for windmills coupled with the mph IG (Subramanian et al. 2022a). Synchronized with an electrical supply terminal, the Digs control system serves a crucial role in reactive power adjustment, supporting both the utility grid and the impeller edge of an IG (Sendrayaperumal et al. 2021). The motor drives conversion design, constructed on a rectifier, monitors wind farm velocity and adjusts it using a power control approach. The Digs idea, along with its synchronization with the battery system, is studied using Scum, and DC synchronization is addressed, opening the path for spine-distributed generation in submarines to be synchronized via additional synchronized operations and lab research (Selvi et al. 2023).

The upcoming portions of this research will be arranged as follows: Section 1 presents a short historical summary of relevant work on Doubly Fed Induction Generators (DFIG) (Balamurugan et al. 2023). Section 2 digs into the computational formula of a wind farm, providing variables for boosting regenerative braking to a PV system and imposing the same computational formula for Digs (Angalaeswari et al. 2022). Section 3 presents the effective inverter, while Sections 4 and 5 concentrate on current networks and computer edge adapters, respectively. Section 6 provides the spice findings produced using the proposed approach (Nagajothi et al. 2022).

Windmill-generated energy is expected to transcend 250 GW during the 12 months 2022, generating a decent sized 30% part of the worldwide power portfolio. This growth in wind power occurs at a key juncture when traditional assets of mineral richness, inclusive of lignite, shale gas, and oil, are growing ever more limited (Nagarajan et al. 2022). As society's want for affordable and sustainable strength keeps to expand, microgrids supplied by means of renewable sources consisting of wind, solar, and hydropower are gaining importance as lengthy-time period energy options. However, every renewable source has its restrictions (Kanimozhi et al. 2022).

In assessment to various renewables, hydroelectric power plant life reveal a few parallels and have the capacity to be strategically coordinated to generate noteworthy electricity production (Suman et al. 2023). Within this setting, the fee controller, in particular the Induction Generator (IG), takes a vital function. Power structures in old strength technologies typically function at parallel speeds, needing enormous power to modify their charge via movers like quicker diesels (Asha et al. 2022). However, the IG, with its potential to work at varied velocities substantially larger than synchronized rates, shows to be extremely green when combined with windmills (Reddy et al. 2023).

The connection between wind pace and variable pitch speed rotors, paired with the utilisation of average wind capacitors to capture peak power in spite of shifting wind conditions, highlights the effectiveness of the IG coupled with windmills (Josphineleela et al. 2023a). The incorporation of relays within the velocity control electricity of doubly fed induction generators (DFIG) in addition enhances the flexibility of the machine. High-overall performance inductors have gained major interest in numerous initiatives, suggesting a trend towards better efficient and durable strength solutions (Santhosh Kumar et al. 2022). The desire for standardised simulation settings for wind turbine equipment is a necessity for turbine makers (Josphineleela et al. 2023b). However, the secret surrounding this data has stimulated computer assessments by means of researchers in search of to anticipate transitory reactions, primarily within the context of electricity and deformation circumstances (Natrayan and Kaliappan 2023). Streamlining the modeling technique, current study has aimed at the effect of DFIG modeling simplification and analysed layout balance by employing thinking about rotor and spindle impedance (Kaliappan et al. 2023a).

Validation of wind turbine designs against the IEC 61450-27 standards the employment of measured records, as well as the validation of the IEC Steroid hormone architecture for windmills, gives a foundation for analysing the correctness of such computational models (Kaliappan et al. 2023b). The studies community's interest in induction generator analysis has brought to the utilisation of the IEC Steroid hormone structure for windmills reference modern regulator with offshore wind components controllers. However, there may be a repute that substantial take a look at is necessary for many circumstances relating DFIG associated to the grid via a couple of approaches (Ramaswamy et al. 2022).

This check seeks to assess an arithmetical modeling strategy for windmills merged with the mph IG. Synchronized with an energy deliver terminal, the Digs manage system serves a crucial role in reactive electricity compensation, supporting both the utility grid and the impeller edge of an IG (Josphineleela et al. 2023c; Muralidaran et al. 2023; Saravanan et al. 2023). The motor drives conversion architecture, constructed on a rectifier, visual display units wind farm velocity and controls it the use of an energy manipulate method (Arun et al. 2022; Balaji et al. 2022; Sivakumar et al. 2023). The Digs idea, along side its synchronization with the battery device, is studied the usage of Scum, and DC synchronization is researched, paving the course for backbone-dispensed generation in submarines to be synced by more synchronized operations and lab investigations (Natrayan et al. 2021; Selvi et al. 2023).

The upcoming portions of this text may be reliant as follows: Section 1 presents a brief old assessment of relevant artworks on Doubly Fed Induction Generators (DFIG). Section 2 digs into the computational formula of a wind farm, providing variables for strengthening regenerative braking to a PV system and enforcing the equal computational formula for Digs (Lakshmaiya et al. 2023; Velmurugan et al. 2023). Section 3 presents the effective inverter, while Sections 4 and 5 touch on cutting-edge networks and computer component adapters, respectively. Section 6 presents the spicy outcomes gained with the advised approach (Karthick et al. 2022; Natrayan et al. 2023).

SIMULATION OF WIND TURBINES AND COMBINATION WITH DFIG

A wind farm, as represented in Figure 1, is closely related with the Induction Generator's (IG) blade. The basic system comprises the conversion of wind power from the floor position of a rotor into electric strength. The continuity formula is utilised to calculate the electrical pressure (Pm) given by the wind farm (Lakshmaiya et al. 2022b; Arockiasamy et al. 2023). As the wind does no longer supply entire control to the rotors throughout the period of power switch, and warmth is saved, the dynamo's performance is estimated based solely on Betz's concept (Subramanian et al. 2022a; Subramanian et al. 2022b; Lakshmaiya 2023). This requires applying the total performance coefficient ratio, as given with the assistance of the continuity formulation. The variable Cp in Equation (1) isn't always fixed and is a -dimensional dependent variable. Cp is defined because the fractional aerodynamic performance, whereas every other variable is described because the pitch attitude inclination (Lakshmaiya et al. 2022a).

In Figure 1, the rotational strength obtained from wind speed is illustrated for different speed control inclinations. Simultaneously, Figure 1 depicts the wind turbine's efficiency thing, commonly seen as an implicit outcome of speed tipping ratios (Sai et al. 2023; Ugle et al. 2023). The accompanying graph in Figure 2 indicates that the terminal voltage stories a remarkable rise as air tension diminishes. However, it in the end falls to zero, and the relevance of the engine lowers since the wheel manipulate signal will grow (Rajagopalan et al. 2022; Chennai Viswanathan et al. 2023; Seralathan et al. 2023). Similarly, as noted in Figure 1, the velocity (CP) to start with increases but eventually starts to say nay when the pitch ratios transcend the knee degree. The prediction version for Doubly Fed Induction Generators (DFIG) is a complete time period that incorporates numerous features (Chehelgerdi et al. 2023).

When the applied load at the rotor is less than the synchronous price, the induction rotor shaft operates as a drive. Conversely, while the carried out load on the rotor exceeds the synchronous velocity, it acts as a generator. Leveraging Kent and Church translation theories, a three-transformer system may be turned immediately into a two-continuously rotating viewpoint. The dynamical quotient d-q with stator current day azimuth forms the consequent basis. The basic squirrel cage three-section synchronous motor power arrangement, as represented in Figure five, may be altered to operate as an inductor when loaded. Various formulation in a d-q version for both rotor and stator surfaces are necessary to understand the dynamics. Mathematical formulae for the direct and quadrature circuits, termed rotors, are key elements of this complicated machine. An in-depth study of the link between wind speed, rotor rotation and efficiency is necessary to increase the efficiency of integrated wind turbines and IGs The complicated coupling represented in Figure 1 and to a lesser degree underlines the dynamic character of such systems and the necessity for micromanagement strategies to improve capacity change efficiently. The relevance of the differential aerodynamic efficiency (Cp) of the pitch attitude gradient in the computation of the total performance of all appliances is clear from the analysis.

RESULT AND DISCUSSIONS

This chapters give a detailed study of the experimental results and simulated functionality of the Doubly Fed Induction Generator (DFIG), referred to as Digs. The model has been precisely developed for testing below specific working scenarios, comprising of semi-pace, rotor motor settings, and in particular immoderate speeds (Niveditha VR. and Rajakumar PS. 2020). The rotational speed is delicately stimulated by way of the minimal entrance modern-day via grids or conical aspect gearboxes (Sabarinathan et al. 2022). It is vital for the rotor's rotating speed to considerably exceed that of the stator for the Digs to carry out in mega mode properly (Singh et al. 2017). A critical component of the Digs model rests in its flexibility to varied operational factors, inclusive of semi-tempo and rotor motor characteristics (Nadh et al. 2021). The inclusion of excessive-pace circumstances enables for a full assessment of the generator's usual overall performance under exceptional settings. The modification of rotation velocity using minimal input current day controls and conical facet gears shows the accuracy with which the version has been built to replicate actual-global circumstances (Anupama et al. 2021).

Notably, the slippage variety, to begin with believed to be zero, performs a crucial function within the generator's performance. Some generators are constructed to supply twice the photovoltaic (PV) push to a rotor, generating a dreadful stress that contributes to voltage manufacturing throughout the alternator. This progressive technique challenges traditional assumptions approximately slippage, introducing a nuanced understanding of its impact on standard machine overall performance. The interplay between rotor pace, enter present day, and gearboxes demonstrates the intricacies concerned in achieving greatest capability. The Digs model's ability to conform to various situations and its responsiveness to unique enter parameters show off its versatility and applicability in numerous situations. These chapters no longer best gift the experimental outcomes however also delve into the underlying mechanisms that govern the Digs' operation, losing mild on its potential for enhancing the efficiency of wind electricity systems.

Figure 1 depicts the stresses generated by the equipment. It really is apparent that any power created is negative. Figure 1 depicts the induced spindle as well as rotation energies. The pattern of a generated

Figure 1. Wind speed and turbine power

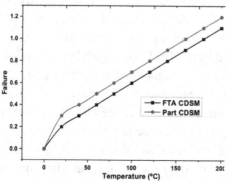

Figure 2. Generated three rotors plus rotor lateral voltage

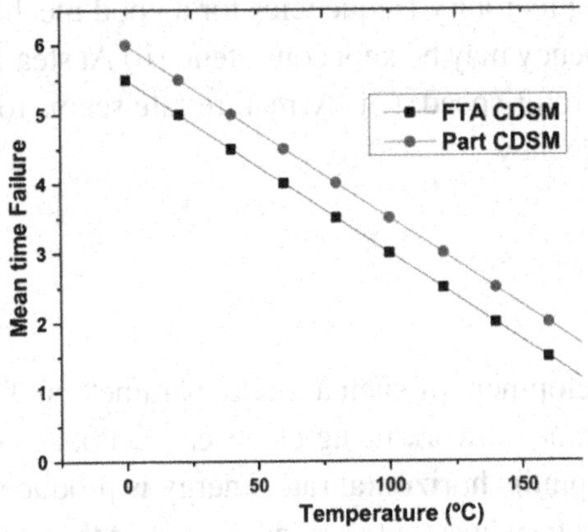

generator's 3-phase current speed is observed to be lower than that of the rotor rate. It signifies how electricity is sent to the network. It plainly shows that slippage is becoming zero and also that the unit is now working as either a producer or a consumer. Figures 1 and 2illustrate the voltage required by a magnetic field. It is evident that even though the flywheel power factor used is essentially non-existent, the reactionary flux produced is due to the sensitive part linked on the stator side. When operating in continuous mode, it approaches zero.

The induction machine transfers no current to the transmission side across this phase. The resistance mechanism of a rotor is therefore lowered to zero. However, the reactance of a rotor remains owing to a reaction component linked via the spindle, as seen in Fig1. A generated force in Fig. 13 demonstrates that the internal combustion engine is sharing the load and the electricity hasn't been generated to the secondary side at this point, so we can't assume that the device is completely functioning in a driving state. Figure 2 depicts the generated 3 rotors plus rotor lateral voltage. Because of time domain synchronisation, the system is totally robust. This twin-feed electromagnet is used in the cutting group.

For this reason, the correct timing is always greater than that of the spinning velocity. As a result, the generator will operate under constant motor, as the rotors are unable to send generated power. It causes the slippage to be lower than the first and higher than the zero, as well as the rotors to absorb the grid's electricity. Its generator draws energy from the inverter side more toward the rotors, and some of the propeller absorbs DC link voltage from the transmission side at the same time.

A force stimulated by even the Instagram account is depicted in Fig. 15, indicating that it is evident that although the engine is carrying a load, its acceleration would be favourable. These same multiple voltage output sound waves of an armature winding appear to suggest that now the proportion of the propeller is smaller than the regularity of the primary winding, as depicted in Fig 1. Because the knowledge is used by the transformer, the voltage and power components are absorbed either by series or shunt windings, as seen in Figs1. The aforementioned equation too was evaluated by establishing the tester pillows in the labs and using laboratory software to create the wind farm simulator. Figure 2 depicts the planned empirical model. Power systems make use of the transfer of energy.

Its circuit is composed of 37 components, which are included. That the very first test demonstrates the effect of mechanically spinning speed. Throughout the current creation the statistics results in Table 3 demonstrate that the Dir performed as a power system throughout a wider band and varying speed control ranges. The very next testing demonstrates the effect of an adjustable speed spindle and the findings were reached. By regulating its rotary frequencies for a good mechanical velocity provided by the wind generator, its spindle frequency may be kept consistent. (ii) At steady velocity, the stator frequency may be changed by varying the rotor speed. (iii) Armature rate seems to be the product of rotor rate as well as mechanical rotation frequency.

CONCLUSION

This article reported on the development of such a model parameters of Digs penetration of renewable farms and its interaction with some network using electrical feedback control. Whenever the synchronised rate is narrower than the spindle horizontal rate, energy is produced on that same wheel from the primary side, so that when the rotary horizontal speed is greater than the spindle rotational regularity, energy is sent to the rotor circuit from that of the rotor. Our simulation theorem was still evaluated in the laboratory underneath the dark conditions, combining the Digs concept with the grids, and the findings totally confirmed our concept.

Additional studies were carried out to investigate the effect of spindle battery voltage on phase currents as well as rotation. It was discovered that in a subharmonic state, the power factor of Gc continually grows in the negative direction. Therefore, in mega overdrive, the load current of Gbm becomes negative even as voltage and frequency decrease. The research study was additionally confirmed by operating diesel generators at varying and constant speeds to demonstrate the variance in distributed generation from the network through Layer. This layer then backs to a network.

REFERENCES

Angalaeswari, S., Jamuna, K., Mohana sundaram, K., Natrayan, L., Ramesh, L., & Ramaswamy, K. (2022). Power-Sharing Analysis of Hybrid Microgrid Using Iterative Learning Controller (ILC) considering Source and Load Variation. *Mathematical Problems in Engineering, 2022*, 1–6. doi:10.1155/2022/7403691

Anupama. (2021). Deep learning with backtracking search optimization based skin lesion diagnosis model. *Computers, Materials & Continua, 70*(1). doi:10.32604/cmc.2022.018396

Arockiasamy, Muthukrishnan, M., Iyyadurai, J., Kaliappan, S., Lakshmaiya, N., Djearamane, S., Tey, L.-H., Wong, L. S., Kayarohanam, S., Obaid, S. A., Alfarraj, S., & Sivakumar, S. (2023). Tribological characterization of sponge gourd outer skin fiber-reinforced epoxy composite with Tamarindus indica seed filler addition using the Box-Behnken method. *E-Polymers, 23*(1), 20230052. doi:10.1515/epoly-2023-0052

Arun. (2022). Mechanical, fracture toughness, and Dynamic Mechanical properties of twill weaved bamboo fiber-reinforced Artocarpus heterophyllus seed husk biochar epoxy composite. *Polymer Composites, 43*(11), 8388–8395. doi:10.1002/pc.27010

Asha, P., Natrayan, L., Geetha, B. T., Beulah, J. R., Sumathy, R., Varalakshmi, G., & Neelakandan, S. (2022). IoT enabled environmental toxicology for air pollution monitoring using AI techniques. *Environmental Research, 205*, 112574. doi:10.1016/j.envres.2021.112574 PMID:34919959

Balaji. (2022). Annealed peanut shell biochar as potential reinforcement for aloe vera fiber-epoxy biocomposite: Mechanical, thermal conductivity, and dielectric properties. *Biomass Conversion and Biorefinery*. doi:10.1007/s13399-022-02650-7

Balamurugan, P., Agarwal, P., Khajuria, D., Mahapatra, D., Angalaeswari, S., Natrayan, L., & Mammo, W. D. (2023). State-Flow Control Based Multistage Constant-Current Battery Charger for Electric Two-Wheeler. *Journal of Advanced Transportation*, *2023*, 1–11. doi:10.1155/2023/4554582

Chehelgerdi, Chehelgerdi, M., Allela, O. Q. B., Pecho, R. D. C., Jayasankar, N., Rao, D. P., Thamaraikani, T., Vasanthan, M., Viktor, P., Lakshmaiya, N., Saadh, M. J., Amajd, A., Abo-Zaid, M. A., Castillo-Acobo, R. Y., Ismail, A. H., Amin, A. H., & Akhavan-Sigari, R. (2023). Progressing nanotechnology to improve targeted cancer treatment: Overcoming hurdles in its clinical implementation. *Molecular Cancer*, *22*(1), 169. doi:10.1186/s12943-023-01865-0 PMID:37814270

Darshan, Girdhar, N., Bhojwani, R., Rastogi, K., Angalaeswari, S., Natrayan, L., & Paramasivam, P. (2022). Energy Audit of a Residential Building to Reduce Energy Cost and Carbon Footprint for Sustainable Development with Renewable Energy Sources. *Advances in Civil Engineering*, *2022*, 1–10. Advance online publication. doi:10.1155/2022/4400874

Dhanraj, A. (2022). Appraising machine learning classifiers for discriminating rotor condition in 50W–12V operational wind turbine for maximizing wind energy production through feature extraction and selection process. *Frontiers in Energy Research*, *10*, 925980. doi:10.3389/fenrg.2022.925980

Divya. (2022). Analysing Analyzing the performance of combined solar photovoltaic power system with phase change material. *Energy Reports*, *8*. doi:10.1016/j.egyr.2022.06.109

Kaliappan, S., Natrayan, L., & Rajput, A. (2023b) Sentiment Analysis of News Headlines Based on Sentiment Lexicon and Deep Learning. In: *Proceedings of the 4th International Conference on Smart Electronics and Communication, ICOSEC 2023*. IEEE. 10.1109/ICOSEC58147.2023.10276102

Kanimozhi, G., Natrayan, L., Angalaeswari, S., & Paramasivam, P. (2022). An Effective Charger for Plug-In Hybrid Electric Vehicles (PHEV) with an Enhanced PFC Rectifier and ZVS-ZCS DC/DC High-Frequency Converter. *Journal of Advanced Transportation*, *2022*, 1–14. doi:10.1155/2022/7840102

Karthick, Meikandan, M., Kaliappan, S., Karthick, M., Sekar, S., Patil, P. P., Raja, S., Natrayan, L., & Paramasivam, P. (2022). Experimental Investigation on Mechanical Properties of Glass Fiber Hybridized Natural Fiber Reinforced Penta-Layered Hybrid Polymer Composite. *International Journal of Chemical Engineering*, *2022*, 1–9. doi:10.1155/2022/1864446

Kaushal. (2023) A Payment System for Electric Vehicles Charging and Peer-to-Peer Energy Trading. In: *7th International Conference on I-SMAC (IoT in Social, Mobile, Analytics and Cloud), I-SMAC 2023 – Proceedings*. IEEE. 10.1109/I-SMAC58438.2023.10290505

Kumar, S. (2022). IoT battery management system in electric vehicle based on LR parameter estimation and ORMeshNet gateway topology. *Sustainable Energy Technologies and Assessments*, *53*, 102696. doi:10.1016/j.seta.2022.102696

Lakshmaiya, N. (2023) Experimental investigation on computational volumetric heat in real time neural pathways. In: *Proceedings of SPIE*. The International Society for Optical Engineering 10.1117/12.2675555

Lakshmaiya, N., Ganesan, V., Paramasivam, P., & Dhanasekaran, S. (2022a). Influence of Biosynthesized Nanoparticles Addition and Fibre Content on the Mechanical and Moisture Absorption Behaviour of Natural Fibre Composite. *Applied Sciences (Basel, Switzerland)*, *12*(24), 13030. doi:10.3390/app122413030

Lakshmaiya, N., Kaliappan, S., Patil, P. P., Ganesan, V., Dhanraj, J. A., Sirisamphanwong, C., Wongwuttanasatian, T., Chowdhury, S., Channumsin, S., Channumsin, M., & Techato, K. (2022b). Influence of Oil Palm Nano Filler on Interlaminar Shear and Dynamic Mechanical Properties of Flax/Epoxy-Based Hybrid Nanocomposites under Cryogenic Condition. *Coatings*, *12*(11), 1675. Advance online publication. doi:10.3390/coatings12111675

Lakshmaiya, N., Surakasi, R., Nadh, V. S., Srinivas, C., Kaliappan, S., Ganesan, V., Paramasivam, P., & Dhanasekaran, S. (2023). Tanning Wastewater Sterilization in the Dark and Sunlight Using Psidium guajava Leaf-Derived Copper Oxide Nanoparticles and Their Characteristics. *ACS Omega*, *8*(42), 39680–39689. doi:10.1021/acsomega.3c05588 PMID:37901496

Loganathan, Ramachandran, V., Perumal, A. S., Dhanasekaran, S., Lakshmaiya, N., & Paramasivam, P. (2023). Framework of Transactive Energy Market Strategies for Lucrative Peer-to-Peer Energy Transactions. *Energies*, *16*(1), 6. doi:10.3390/en16010006

Mahesha, C. R., Rani, G. J., Dattu, V. S. N. C. H., Rao, Y. K. S. S., Madhusudhanan, J., L, N., Sekhar, S. C., & Sathyamurthy, R. (2022). Optimization of transesterification production of biodiesel from Pithecellobium dulce seed oil. *Energy Reports*, *8*, 489–497. doi:10.1016/j.egyr.2022.10.228

Muralidaran, Natrayan, L., Kaliappan, S., & Patil, P. P. (2023). Grape stalk cellulose toughened plain weaved bamboo fiber-reinforced epoxy composite: Load bearing and time-dependent behavior. *Biomass Conversion and Biorefinery*. doi:10.1007/s13399-022-03702-8

Nadh, V. S., Krishna, C., Natrayan, L., Kumar, K. M., Nitesh, K. J. N. S., Raja, G. B., & Paramasivam, P. (2021). Structural Behavior of Nanocoated Oil Palm Shell as Coarse Aggregate in Lightweight Concrete. *Journal of Nanomaterials*, *2021*, 1–7. doi:10.1155/2021/4741296

Nagajothi, S., Elavenil, S., Angalaeswari, S., Natrayan, L., & Paramasivam, P. (2022). Cracking Behaviour of Alkali-Activated Aluminosilicate Beams Reinforced with Glass and Basalt Fibre-Reinforced Polymer Bars under Cyclic Load. *International Journal of Polymer Science*, *2022*, 1–13. doi:10.1155/2022/6762449

Nagarajan, Rajagopalan, A., Angalaeswari, S., Natrayan, L., & Mammo, W. D. (2022). Combined Economic Emission Dispatch of Microgrid with the Incorporation of Renewable Energy Sources Using Improved Mayfly Optimization Algorithm. *Computational Intelligence and Neuroscience*, *2022*, 1–22. Advance online publication. doi:10.1155/2022/6461690 PMID:35479598

Natrayan, L., & Kaliappan, S. (2023) Mechanical Assessment of Carbon-Luffa Hybrid Composites for Automotive Applications. In: SAE Technical Papers. SAE. doi:10.4271/2023-01-5070

Natrayan, L., Kaliappan, S., Saravanan, A., Vickram, A. S., Pravin, P., Abbas, M., Ahamed Saleel, C., Alwetaishi, M., & Saleem, M. S. M. (2023). Recyclability and catalytic characteristics of copper oxide nanoparticles derived from bougainvillea plant flower extract for biomedical application. *Green Processing and Synthesis*, *12*(1), 20230030. doi:10.1515/gps-2023-0030

Natrayan, L., Merneedi, A., Bharathiraja, G., Kaliappan, S., Veeman, D., & Murugan, P. (2021). Processing and characterization of carbon nanofibre composites for automotive applications. *Journal of Nanomaterials*, *2021*, 1–7. doi:10.1155/2021/7323885

Niveditha, V. R., & Rajakumar, P. S. (2020). Pervasive computing in the context of COVID-19 prediction with AI-based algorithms. *International Journal of Pervasive Computing and Communications*, *16*(5). doi:10.1108/IJPCC-07-2020-0082

Rajagopalan, Nagarajan, K., Montoya, O. D., Dhanasekaran, S., Kareem, I. A., Perumal, A. S., Lakshmaiya, N., & Paramasivam, P. (2022). Multi-Objective Optimal Scheduling of a Microgrid Using Oppositional Gradient-Based Grey Wolf Optimizer. *Energies*, *15*(23), 9024. Advance online publication. doi:10.3390/en15239024

Ramaswamy. (2022). Pear cactus fiber with onion sheath biocarbon nanosheet toughened epoxy composite: Mechanical, thermal, and electrical properties. *Biomass Conversion and Biorefinery*. doi:10.1007/s13399-022-03335-x

Reddy, & ... . (2023) Development of Programmed Autonomous Electric Heavy Vehicle: An Application of IoT. In: *Proceedings of the 2023 2nd International Conference on Electronics and Renewable Systems, ICEARS 2023*. IEEE. 10.1109/ICEARS56392.2023.10085492

Sabarinathan, P., Annamalai, V. E., Vishal, K., Nitin, M. S., Natrayan, L., Veeeman, D., & Mammo, W. D. (2022). Experimental study on removal of phenol formaldehyde resin coating from the abrasive disc and preparation of abrasive disc for polishing application. *Advances in Materials Science and Engineering*, *2022*, 1–8. doi:10.1155/2022/6123160

Sai, S. A., Venkatesh, S. N., Dhanasekaran, S., Balaji, P. A., Sugumaran, V., Lakshmaiya, N., & Paramasivam, P. (2023). Transfer Learning Based Fault Detection for Suspension System Using Vibrational Analysis and Radar Plots. *Machines*, *11*(8), 778. doi:10.3390/machines11080778

Saravanan, K. G., Kaliappan, S., Natrayan, L., & Patil, P. P. (2023). Effect of cassava tuber nanocellulose and satin weaved bamboo fiber addition on mechanical, wear, hydrophobic, and thermal behavior of unsaturated polyester resin composites. *Biomass Conversion and Biorefinery*. doi:10.1007/s13399-023-04495-0

Seeniappan. (2023). *Modelling and development of energy systems through cyber physical systems with optimising interconnected with control and sensing parameters.*

Selvi. (2023) Optimization of Solar Panel Orientation for Maximum Energy Efficiency. In: *Proceedings of the 4th International Conference on Smart Electronics and Communication, ICOSEC 2023*.IEEE. 10.1109/ICOSEC58147.2023.10276287

Sendrayaperumal, Mahapatra, S., Parida, S. S., Surana, K., Balamurugan, P., Natrayan, L., & Paramasivam, P. (2021). Energy Auditing for Efficient Planning and Implementation in Commercial and Residential Buildings. *Advances in Civil Engineering*, *2021*, 1–10. doi:10.1155/2021/1908568

Seralathan, S., Chenna Reddy, G., Sathish, S., Muthuram, A., Dhanraj, J. A., Lakshmaiya, N., Velmurugan, K., Sirisamphanwong, C., Ngoenmeesri, R., & Sirisamphanwong, C. (2023). Performance and exergy analysis of an inclined solar still with baffle arrangements. *Heliyon*, *9*(4), e14807. doi:10.1016/j.heliyon.2023.e14807 PMID:37077675

Sharma, Raffik, R., Chaturvedi, A., Geeitha, S., Akram, P. S., L, N., Mohanavel, V., Sudhakar, M., & Sathyamurthy, R. (2022). Designing and implementing a smart transplanting framework using programmable logic controller and photoelectric sensor. *Energy Reports*, *8*, 430–444. doi:10.1016/j.egyr.2022.07.019

Singh. (2017). An experimental investigation on mechanical behaviour of siCp reinforced Al 6061 MMC using squeeze casting process. *International Journal of Mechanical and Production Engineering Research and Development*, *7*(6). doi:10.24247/ijmperddec201774

Sivakumar, V., Kaliappan, S., Natrayan, L., & Patil, P. P. (2023). Effects of Silane-Treated High-Content Cellulose Okra Fibre and Tamarind Kernel Powder on Mechanical, Thermal Stability and Water Absorption Behaviour of Epoxy Composites. *Silicon*, *15*(10), 4439–4447. doi:10.1007/s12633-023-02370-1

Subramanian, Lakshmaiya, N., Ramasamy, D., & Devarajan, Y. (2022a). Detailed analysis on engine operating in dual fuel mode with different energy fractions of sustainable HHO gas. *Environmental Progress & Sustainable Energy*, *41*(5), e13850. doi:10.1002/ep.13850

Subramanian, Solaiyan, E., Sendrayaperumal, A., & Lakshmaiya, N. (2022b). Flexural behaviour of geopolymer concrete beams reinforced with BFRP and GFRP polymer composites. *Advances in Structural Engineering*, *25*(5), 954–965. doi:10.1177/13694332211054229

Suman. (2023) IoT based Social Device Network with Cloud Computing Architecture. In: *Proceedings of the 2023 2nd International Conference on Electronics and Renewable Systems, ICEARS 2023*. IEEE. 10.1109/ICEARS56392.2023.10085574

Thakre, Pandhare, A., Malwe, P. D., Gupta, N., Kothare, C., Magade, P. B., Patel, A., Meena, R. S., Veza, I., Natrayan L, & Panchal, H. (2023). Heat transfer and pressure drop analysis of a microchannel heat sink using nanofluids for energy applications. *Kerntechnik*, *88*(5), 543–555. doi:10.1515/kern-2023-0034

Ugle, Arulprakasajothi, M., Padmanabhan, S., Devarajan, Y., Lakshmaiya, N., & Subbaiyan, N. (2023). Investigation of heat transport characteristics of titanium dioxide nanofluids with corrugated tube. *Environmental Quality Management*, *33*(2), 127–138. doi:10.1002/tqem.21999

Velmurugan, G., Natrayan, L., Chohan, J. S., Vasanthi, P., Angalaeswari, S., Pravin, P., Kaliappan, S., & Arunkumar, D. (2023). Investigation of mechanical and dynamic mechanical analysis of bamboo/olive tree leaves powder-based hybrid composites under cryogenic conditions. *Biomass Conversion and Biorefinery*. doi:10.1007/s13399-023-04591-1

Vijayaragavan, Subramanian, B., Sudhakar, S., & Natrayan, L. (2022). Effect of induction on exhaust gas recirculation and hydrogen gas in compression ignition engine with simarouba oil in dual fuel mode. *International Journal of Hydrogen Energy*, *47*(88), 37635–37647. doi:10.1016/j.ijhydene.2021.11.201

Viswanathan, C. (2023). Deep Learning for Enhanced Fault Diagnosis of Monoblock Centrifugal Pumps: Spectrogram-Based Analysis. *Machines*, *11*(9), 874. doi:10.3390/machines11090874

Compilation of References

Ab Talib, M. H., Mat Darus, I. Z., Mohd Samin, P., Mohd Yatim, H., Hadi, M. S., Shaharuddin, N. M. R., Mazali, I. I., Ardani, M. I., & Mohd Yamin, A. H. (2023). Experimental evaluation of ride comfort performance for suspension system using PID and fuzzy logic controllers by advanced firefly algorithm. *Journal of the Brazilian Society of Mechanical Sciences and Engineering*, *45*(3), 132. doi:10.1007/s40430-023-04057-5

Abdalla, A. N., Nazir, M. S., Tao, H., Cao, S., Ji, R., Jiang, M., & Yao, L. (2021). Integration of energy storage system and renewable energy sources based on artificial intelligence: An overview. *Journal of Energy Storage*, *40*, 102811. doi:10.1016/j.est.2021.102811

Abdul Zahra, A. K., & Abdalla, T. Y. (2021). Design of fuzzy super twisting sliding mode control scheme for unknown full vehicle active suspension systems using an artificial bee colony optimization algorithm. *Asian Journal of Control*, *23*(4), 1966–1981. doi:10.1002/asjc.2352

Abut, T., & Salkim, E. (2023). Control of Quarter-Car Active Suspension System Based on Optimized Fuzzy Linear Quadratic Regulator Control Method. *Applied Sciences (Basel, Switzerland)*, *13*(15), 8802. doi:10.3390/app13158802

Agarwal, N., & Kumar, A. (2013). Optimization of grid independent hybrid PV– diesel–battery system for power generation in remote villages of Uttar Pradesh, India.

Agrawal, A. V., Shashibhushan, G., Pradeep, S., Padhi, S. N., Sugumar, D., & Boopathi, S. (2024). Synergizing Artificial Intelligence, 5G, and Cloud Computing for Efficient Energy Conversion Using Agricultural Waste. In Practice, Progress, and Proficiency in Sustainability (pp. 475–497). IGI Global. doi:10.4018/979-8-3693-1186-8.ch026

Agrawal, A. V., Magulur, L. P., Priya, S. G., Kaur, A., Singh, G., & Boopathi, S. (2023). Smart Precision Agriculture Using IoT and WSN. In *Handbook of Research on Data Science and Cybersecurity Innovations in Industry 4.0 Technologies* (pp. 524–541). IGI Global. doi:10.4018/978-1-6684-8145-5.ch026

Ahmad, J., Imran, M., Khalid, A., Adnan, M., Ali, S. F., & Khokkar, K. S. (2018). Techno economic analysis of a wind-photovoltaic- biomass hybrid renewable energy system for rural electrification: A case study of Kallar Kahar. *Energy*, *28*, 36–46.

Ahmad, T., Zhang, D., Huang, C., Zhang, H., Dai, N., Song, Y., & Chen, H. (2021). Artificial intelligence in sustainable energy industry: Status Quo, challenges and opportunities. *Journal of Cleaner Production*, *289*, 125834. doi:10.1016/j.jclepro.2021.125834

Ahmed, P., Rahman, M. F., Haque, A. M., Mohammed, M. K., Toki, G. I., Ali, M. H., Kuddus, A., Rubel, M. H. K., & Hossain, M. K. (2023). Feasibility and Techno-Economic Evaluation of Hybrid Photovoltaic System: A Rural Healthcare Center in Bangladesh. *Sustainability (Basel)*, *15*(2), 1362. doi:10.3390/su15021362

Aitken, T. W., Groome, A. E., Joy, T., & Myring, W. J. (1986). Design And Operation Of A Cryogenically Pumped Gas Stripper In The Terminal Of The Daresbury Tandem. In Nuclear Instruments and Methods in Physics Research, 244.

Ajayi, A. B., Ojogho, E., Adewusi, S. A., Ojolo, S. J., Campos, J. C. C., & de Oliveira Siqueira, A. M. (2023). Parametric study of rider's comfort in a vehicle with semi-active suspension system under transient road conditions. *The Journal of Engineering and Exact Sciences, 9*(5), 15287–01e. doi:10.18540/jcecvl9iss5pp15287-01e

Akhloufi, M. A., Couturier, A., & Castro, N. A. (2021). Unmanned aerial vehicles for wildland fires: Sensing, perception, cooperation and assistance. *Drones (Basel), 5*(1), 15. doi:10.3390/drones5010015

Akhtari, M. R., Shayegh, I., & Karimi, N. (2019). Techno-Economic Assessment And Optimization of A Hybrid Renewable Earth-Air Heat Exchanger Coupled With Electric Boiler, Hydrogen, Wind And PV Configurations. *Renewable Energy*.

Akram, U., Nadarajah, M., Shah, R., & Milano, F. (2020). A review on rapid responsive energy storage technologies for frequency regulation in modern power systems. *Renewable & Sustainable Energy Reviews, 120*, 109626. doi:10.1016/j.rser.2019.109626

Aksungur, S., & Koca, T. (2018). *Solar Tracking System with PID control of solar energy panels using servo motor.* International Journal of Energy Applications and Technologies. doi:10.31593/ijeat.450834

Alagirisamy, B., & Ramesh, P. (2022). Smart sustainable cities: Principles and future trends. In *Sustainable Cities and Resilience: Select Proceedings of VCDRR 2021* (pp. 301-316). Springer Singapore.

Al-Ammar, E. A., Habib, U. R. H., Wang, S., Ko, W., & Elmorshedy, M. F. (2020). Residential Community Load Mnagement Based on Optimal Design of Standalone HRES with Model Predictive Control. *IEEE Access : Practical Innovations, Open Solutions*, 810–818.

Alatise, M. B., & Hancke, G. P. (2020). A Review on Challenges of Autonomous Mobile Robot and Sensor Fusion Methods. In IEEE Access (Vol. 8). doi:10.1109/ACCESS.2020.2975643

Ali, M., Firdaus, A. A., Arof, H., Nurohmah, H., Suyono, H., Putra, D. F. U., & Muslim, M. A. (2021). The comparison of dual-axis PV tracking system using artificial intelligence techniques. *IAES International Journal of Artificial Intelligence, 10*(4), 901–909. doi:10.11591/ijai.v10.i4.pp901-909

Ali, S. S., & Choi, B. J. (2020). State-of-the-art artificial intelligence techniques for distributed smart grids: A review. *Electronics (Basel), 9*(6), 1030. doi:10.3390/electronics9061030

Ali, Z., Christofides, N., Hadjidemetriou, L., Kyriakides, E., Yang, Y., & Blaabjerg, F. (2018). Three-phase phase-locked loop synchronization algorithms for grid-connected renewable energy systems: A review. *Renewable & Sustainable Energy Reviews, 90*, 434–452. doi:10.1016/j.rser.2018.03.086

Allouhi, A., Kousksou, T., Jamil, A., Bruel, P., Mourad, Y., & Zeraouli, Y. (2015). Solar driven cooling systems: An updated review. *Renewable & Sustainable Energy Reviews, 44*, 159–181. doi:10.1016/j.rser.2014.12.014

Almeida, K. C., & Cicconet, F. (2019). Decentralized Hydrothermal Dispatch via Bi-level Optimization. *Journal of Control, Automation and Electrical Systems, 30*, 551–567.

Alramlawi, M., Mohagheghi, E., & Li, P. (2019). Predictive active-reactive optimal power dispatch in PV-battery-diesel micro grid considering reactive power and battery lifetime costs. *Solar Energy, 193*, 529–544.

Al-Rousan, N., Mat Isa, N. A., & Mat Desa, M. K. (2020). Efficient single and dual axis solar tracking system controllers based on adaptive neural fuzzy inference system. *Journal of King Saud University. Engineering Sciences, 32*(7), 459–469. doi:10.1016/j.jksues.2020.04.004

Alsac, O., Bright, J., Prais, M., & Stott, B. (1990). Further developments in LP-based optimal power flow. *IEEE Transactions on Power Systems, 5*(3), 697–711. doi:10.1109/59.65896

Alsharif, A., Tan, C. W., Ayop, R., Al Smin, A., Ali Ahmed, A., Kuwil, F. H., & Khaleel, M. M. (2023). Impact of electric Vehicle on residential power distribution considering energy management strategy and stochastic Monte Carlo algorithm. *Energies, 16*(3), 1358. doi:10.3390/en16031358

Alshboul, O. (2022). Evaluating the impact of external support on green building construction cost: A hybrid mathematical and machine learning prediction approach. *Buildings, 12*.

Álvarez Antón, J. C., García-Nieto, P. J., García-Gonzalo, E., González Vega, M., & Blanco Viejo, C. (2023). Data-driven state-of-charge prediction of a storage cell using ABC/GBRT, ABC/MLP and LASSO machine learning techniques. *Journal of Computational and Applied Mathematics, 433*, 115305. Advance online publication. doi:10.1016/j.cam.2023.115305

Alwateer, M., Loke, S. W., & Fernando, N. (2019). Enabling drone services: Drone crowdsourcing and drone scripting. *IEEE Access : Practical Innovations, Open Solutions, 7*, 110035–110049. doi:10.1109/ACCESS.2019.2933234

Alzieu, J., & Smimite, H. (1995). Development of an onboard charge and discharge management system for electric-vehicle batteries. In Journal of Power Sources, 53.

Amin, U., Hossain, M. J., Lu, J., & Fernandez, E. (2017). Performance analysis of an experimental smart building: Expectations and outcomes. *Energy, 135*, 740–753. doi:10.1016/j.energy.2017.06.149

Amirante, R., Bruno, S., Distaso, E., La Scala, M., & Tamburrano, P. (2019). A biomass small-scale externally fired combined cycle plant for heat and power generation in rural communities. *Renewable Energy Focus, 28*. doi:10.1016/j.ref.2018.10.002

Angalaeswari, S., Jamuna, K., Mohana Sundaram, K., Ramesh, L., & Ramaswamy, K. (2022). Power-Sharing Analysis of Hybrid Microgrid Using Iterative Learning Controller (ILC) considering Source and Load Variation. *Mathematical Problems in Engineering, 2022*, 1–6. doi:10.1155/2022/7403691

Anitha, C., Komala, C., Vivekanand, C. V., Lalitha, S., & Boopathi, S. (2023). Artificial Intelligence driven security model for Internet of Medical Things (IoMT). *IEEE Explore*, 1–7.

Anjankar, P., Lakade, S., Padalkar, A., Nichal, S., Devarajan, Y., Lakshmaiya, N., & Subbaiyan, N. (2023). Experimental investigation on the effect of liquid phase and vapor phase separation over performance of falling film evaporator. *Environmental Quality Management, 33*(1), 61–69. doi:10.1002/tqem.21952

Anupama. (2021). Deep learning with backtracking search optimization based skin lesion diagnosis model. *Computers, Materials & Continua, 70*(1). doi:10.32604/cmc.2022.018396

Anurag Chauhan, R. P. (2017). Size optimization and demand response of a stand-alone Integrated Renewable Energy System. *Energy, 124*, 59–73. doi:10.1016/j.energy.2017.02.049

Apichonnabutr, W., & Tiwary, A. (2018). Trade-offs between economic and environmental performance of an autonomous hybrid energy system using micro hydro. Applied Energy, 226, 891-904.

Arfaoui, J., Rezk, H., Al-Dhaifallah, M., Ibrahim, M. N., & Abdelkader, M. (2020). Simulation-based coyote optimization algorithm to determine gains of PI controller for enhancing the performance of solar PV water-pumping system. *Energies, 13*(17), 4473. doi:10.3390/en13174473

Ariizumi, R., & Matsuno, F. (2017). Dynamic Analysis of Three Snake Robot Gaits. *IEEE Transactions on Robotics, 33*(5), 1075–1087. doi:10.1109/TRO.2017.2704581

Armenta, C. (1989). Determination Of The State-Of-Charge In Lead-Acid Batteries By Means Of A Reference Cell. In *Journal of Power Sources, 27.*

Arnold, R. D., Yamaguchi, H., & Tanaka, T. (2018). Search and rescue with autonomous flying robots through behavior-based cooperative intelligence. *Journal of International Humanitarian Action, 3*(1), 18. doi:10.1186/s41018-018-0045-4

Arockiasamy, Muthukrishnan, M., Iyyadurai, J., Kaliappan, S., Lakshmaiya, N., Djearamane, S., Tey, L.-H., Wong, L. S., Kayarohanam, S., Obaid, S. A., Alfarraj, S., & Sivakumar, S. (2023). Tribological characterization of sponge gourd outer skin fiber-reinforced epoxy composite with Tamarindus indica seed filler addition using the Box-Behnken method. *E-Polymers, 23*(1), 20230052. doi:10.1515/epoly-2023-0052

Arun S L R. (2022). Framework of Transactive Energy Market Strategies for Lucrative Peer-to-Peer Energy Transactions. *Energies.*

Arun, S.L, Bingi, K, Vijaya Priya, R, Jacob Raglend, I, & Hanumantha Rao, B. (2023). Novel Architecture for Transactive Energy Management Systems with Various Market Clearing Strategies. *Mathematical Problems in Engineering.*

Arun. (2022). Mechanical, fracture toughness, and Dynamic Mechanical properties of twill weaved bamboo fiber-reinforced Artocarpus heterophyllus seed husk biochar epoxy composite. *Polymer Composites, 43*(11), 8388–8395. doi:10.1002/pc.27010

Asha, P., Natrayan, L., Geetha, B. T., Beulah, J. R., Sumathy, R., Varalakshmi, G., & Neelakandan, S. (2022). IoT enabled environmental toxicology for air pollution monitoring using AI techniques. *Environmental Research, 205*, 112574. doi:10.1016/j.envres.2021.112574 PMID:34919959

B, M. K., K, K. K., Sasikala, P., Sampath, B., Gopi, B., & Sundaram, S. (2024). Sustainable Green Energy Generation From Waste Water. In *Practice, Progress, and Proficiency in Sustainability* (pp. 440–463). IGI Global. doi:10.4018/979-8-3693-1186-8.ch024

Babatunde, O. M., Munda, J. L., & Hamam, Y. (2020). Power system flexibility: A review. *Energy Reports, 6*, 101–106. doi:10.1016/j.egyr.2019.11.048

Babu, B. S., Kamalakannan, J., Meenatchi, N., Karthik, S., & Boopathi, S. (2022). Economic impacts and reliability evaluation of battery by adopting Electric Vehicle. *IEEE Explore*, 1–6.

Baccioli, A., Caposciutti, G., Marchionni, A., Ferrari, L., & Desiseri, U. (2019). Poly-generation capability of a biogas plant with upgrading system. *Energy Procedia, 159*, 280–285.

Badihi, H., Jadidi, S., Zhang, Y., Su, C.-Y., & Xie, W.-F. (2019). AI-driven intelligent fault detection and diagnosis in a hybrid AC/DC microgrid. *2019 1st International Conference on Industrial Artificial Intelligence (IAI)*, 1–6.

Bahiraei, M., & Heshmatian, S. (2018). Electronics cooling with nanofluids: A critical review. *Energy Conversion and Management, 172*, 438–456. doi:10.1016/j.enconman.2018.07.047

Bahmanyar, A. R., & Karami, A. (2014). Power system voltage stability monitoring using artificial neural networks with a reduced set of inputs. *International Journal of Electrical Power & Energy Systems, 58*, 246–256. doi:10.1016/j.ijepes.2014.01.019

Balaji. (2022). Annealed peanut shell biochar as potential reinforcement for aloe vera fiber-epoxy biocomposite: Mechanical, thermal conductivity, and dielectric properties. *Biomass Conversion and Biorefinery*. doi:10.1007/s13399-022-02650-7

Balaji, N., Gurupranes, S. V., Balaguru, S., Jayaraman, P., Natrayan, L., Subbiah, R., & Kaliappan, S. (2023). Mechanical, wear, and drop load impact behavior of Cissus quadrangularis fiber–reinforced moringa gum powder–toughened polyester composite. *Biomass Conversion and Biorefinery*. doi:10.1007/s13399-023-04491-4

Balamurugan, P., Agarwal, P., Khajuria, D., Mahapatra, D., Angalaeswari, S., Natrayan, L., & Mammo, W. D. (2023). State-Flow Control Based Multistage Constant-Current Battery Charger for Electric Two-Wheeler. *Journal of Advanced Transportation*, *2023*, 1–11. doi:10.1155/2023/4554582

Bannur, P., Gujarathi, P., Jain, K., & Kulkarni, A. J. (2020). Application of Swarm Robotic System in a Dynamic Environment using Cohort Intelligence. *Soft Computing Letters*, *2*, 100006. doi:10.1016/j.socl.2020.100006

Barra, P. H. A., Coury, D. V., & Fernandes, R. A. S. (2020). A survey on adaptive protection of microgrids and distribution systems with distributed generators. *Renewable & Sustainable Energy Reviews*, *118*, 109524. doi:10.1016/j.rser.2019.109524

Barton, R. T., & Mitchell, P. J. (1989). Estimation of the residual capacity of maintenance-free lead-acid batteries part 1. Identification of a parameter for the prediction of state-of-charge. In Journal of Power Sources, 27.

Bedi, P., Goyal, S., Rajawat, A. S., Shaw, R. N., & Ghosh, A. (2022). Application of AI/IoT for smart renewable energy management in smart cities. *AI and IoT for Smart City Applications*, 115–138.

Behera, S., & Dev Choudhury, N. B. (2021). A systematic review of energy management system based on various adaptive controllers with optimization algorithm on a smart microgrid. *International Transactions on Electrical Energy Systems*, *31*(12), e13132. doi:10.1002/2050-7038.13132

Berclin Jeyaprabha, S., & Immanuel Selva Kumar, A. (2015). Optimal sizing of photovoltaic/battery/diesel based hybrid system and optimal tilting of solar array using the artificial intelligence for remote houses in India. *Energy and Building*, *96*(1), 40–52.

Bhakuni, A. S. (2022). Design of SA and GA Optimized PID Controllers for Controlling Blend Chest Level of Paper Mill. *2022 4th International Conference on Smart Systems and Inventive Technology (ICSSIT)*, (pp. 545-550). IEEE. 10.1109/ICSSIT53264.2022.9716460

Bharathraj, S., Adiga, S. P., Mayya, K. S., Song, T. W., & Kim, J. H. (2023). Considering solid phase diffusion penetration depth to improve profile approximations: Towards accurate State estimations in lithium-ion batteries at low characteristic diffusion lengths. *Journal of Power Sources*, *554*, 232325. doi:10.1016/j.jpowsour.2022.232325

Biggie, H., Rush, E., Riley, D., Ahmad, S., Ohradzansky, M., Harlow, K., Miles, M., Torres, D., McGuire, S., Frew, E., Heckman, C., & Humbert, J. (2023). Flexible Supervised Autonomy for Exploration in Subterranean Environments. *Field Robotics*, *3*(1), 125–189. doi:10.55417/fr.2023004

Bingül, Ö., & Yıldız, A. (2023). Fuzzy logic and proportional integral derivative based multi-objective optimization of active suspension system of a 4×4 in-wheel motor driven electrical vehicle. *Journal of Vibration and Control*, *29*(5-6), 1366–1386. doi:10.1177/10775463211062691

Boopathi, S. (2013). *Experimental study and multi-objective optimization of near-dry wire-cut electrical discharge machining process* [PhD Thesis, Anna University]. http://hdl.handle.net/10603/16933

Boopathi, S. (2022b). Cryogenically treated and untreated stainless steel grade 317 in sustainable wire electrical discharge machining process: A comparative study. *Environmental Science and Pollution Research*, 1–10. Springer.

Boopathi, S. (2022b). Cryogenically treated and untreated stainless steel grade 317 in sustainable wire electrical discharge machining process: A comparative study. *Springer :Environmental Science and Pollution Research*, 1–10.

Boopathi, S. (2023). Deep Learning Techniques Applied for Automatic Sentence Generation. In Promoting Diversity, Equity, and Inclusion in Language Learning Environments (pp. 255–273). IGI Global. doi:10.4018/978-1-6684-3632-5.ch016

Boopathi, S. (2023c). Securing Healthcare Systems Integrated With IoT: Fundamentals, Applications, and Future Trends. In Dynamics of Swarm Intelligence Health Analysis for the Next Generation (pp. 186–209). IGI Global.

Boopathi, S., Kumar, P. K. S., Meena, R. S., Sudhakar, M., & Associates. (2023). Sustainable Developments of Modern Soil-Less Agro-Cultivation Systems: Aquaponic Culture. In Human Agro-Energy Optimization for Business and Industry (pp. 69–87). IGI Global.

Boopathi, S., Pandey, B. K., & Pandey, D. (2023). Advances in Artificial Intelligence for Image Processing: Techniques, Applications, and Optimization. In Handbook of Research on Thrust Technologies' Effect on Image Processing (pp. 73–95). IGI Global.

Boopathi, S. (2022a). An investigation on gas emission concentration and relative emission rate of the near-dry wire-cut electrical discharge machining process. *Environmental Science and Pollution Research International, 29*(57), 86237–86246. doi:10.1007/s11356-021-17658-1 PMID:34837614

Boopathi, S. (2022c). Experimental investigation and multi-objective optimization of cryogenic Friction-stir-welding of AA2014 and AZ31B alloys using MOORA technique. *Materials Today. Communications, 33,* 104937. doi:10.1016/j.mtcomm.2022.104937

Boopathi, S. (2023a). An Investigation on Friction Stir Processing of Aluminum Alloy-Boron Carbide Surface Composite. In R. V. Vignesh, R. Padmanaban, & M. Govindaraju (Eds.), *Advances in Processing of Lightweight Metal Alloys and Composites* (pp. 249–257). Springer Nature Singapore. doi:10.1007/978-981-19-7146-4_14

Boopathi, S. (2023b). Internet of Things-Integrated Remote Patient Monitoring System: Healthcare Application. In A. Suresh Kumar, U. Kose, S. Sharma, & S. Jerald Nirmal Kumar (Eds.), (pp. 137–161). Advances in Healthcare Information Systems and Administration. IGI Global. doi:10.4018/978-1-6684-6894-4.ch008

Boopathi, S., Alqahtani, A. S., Mubarakali, A., & Panchatcharam, P. (2023). Sustainable developments in near-dry electrical discharge machining process using sunflower oil-mist dielectric fluid. *Environmental Science and Pollution Research International,* 1–20. doi:10.1007/s11356-023-27494-0 PMID:37199846

Boopathi, S., & Davim, J. P. (2023). Applications of Nanoparticles in Various Manufacturing Processes. In S. Boopathi & J. P. Davim (Eds.), (pp. 1–31). Advances in Chemical and Materials Engineering. IGI Global. doi:10.4018/978-1-6684-9135-5.ch001

Boopathi, S., & Davim, J. P. (2023). *Sustainable Utilization of Nanoparticles and Nanofluids in Engineering Applications*. IGI Global. doi:10.4018/978-1-6684-9135-5

Boopathi, S., Jeyakumar, M., Singh, G. R., King, F. L., Pandian, M., Subbiah, R., & Haribalaji, V. (2022). An experimental study on friction stir processing of aluminium alloy (AA-2024) and boron nitride (BNp) surface composite. *Materials Today: Proceedings, 59*(1), 1094–1099. doi:10.1016/j.matpr.2022.02.435

Boopathi, S., & Kanike, U. K. (2023). Applications of Artificial Intelligent and Machine Learning Techniques in Image Processing. In B. K. Pandey, D. Pandey, R. Anand, D. S. Mane, & V. K. Nassa (Eds.), (pp. 151–173). Advances in Computational Intelligence and Robotics. IGI Global. doi:10.4018/978-1-6684-8618-4.ch010

Boopathi, S., Myilsamy, S., & Sukkasamy, S. (2021). *Experimental Investigation and Multi-Objective Optimization of Cryogenically Cooled Near-Dry Wire-Cut EDM Using TOPSIS Technique*. IJAMT PREPRINT.

Boopathi, S., & Sivakumar, K. (2013). Experimental investigation and parameter optimization of near-dry wire-cut electrical discharge machining using multi-objective evolutionary algorithm. *International Journal of Advanced Manufacturing Technology, 67*(9–12), 2639–2655. doi:10.1007/s00170-012-4680-4

Boopathi, S., Sureshkumar, M., & Sathiskumar, S. (2022). Parametric Optimization of LPG Refrigeration System Using Artificial Bee Colony Algorithm. *International Conference on Recent Advances in Mechanical Engineering Research and Development*, (pp. 97–105). IEEE.

Boopathi, S., Sureshkumar, M., & Sathiskumar, S. (2023a). Parametric Optimization of LPG Refrigeration System Using Artificial Bee Colony Algorithm. In S. Tripathy, S. Samantaray, J. Ramkumar, & S. S. Mahapatra (Eds.), *Recent Advances in Mechanical Engineering* (pp. 97–105). Springer Nature Singapore. doi:10.1007/978-981-19-9493-7_10

Boopathi, S., Umareddy, M., & Elangovan, M. (2023). Applications of Nano-Cutting Fluids in Advanced Machining Processes. In S. Boopathi & J. P. Davim (Eds.), (pp. 211–234). Advances in Chemical and Materials Engineering. IGI Global. doi:10.4018/978-1-6684-9135-5.ch009

Bouyarmane, K., Vaillant, J., Keith, F., & Kheddar, A. (2012). Exploring humanoid robots locomotion capabilities in virtual disaster response scenarios. *IEEE-RAS International Conference on Humanoid Robots*. IEEE. 10.1109/HUMAN-OIDS.2012.6651541

Brieske, D. M., Warnecke, A., & Sauer, D. U. (2023). Modeling the volumetric expansion of the lithium-sulfur battery considering charge and discharge profiles. *Energy Storage Materials*, *55*, 289–300. doi:10.1016/j.ensm.2022.11.053

Brunelli, L., Capancioni, A., Canè, S., Cecchini, G., Perazzo, A., Brusa, A., & Cavina, N. (2023). A predictive control strategy based on A-ECMS to handle Zero-Emission Zones: Performance assessment and testing using an HiL equipped with vehicular connectivity. *Applied Energy*, *340*, 121008. doi:10.1016/j.apenergy.2023.121008

Büchi, R. (2022). PID Controller Parameter Tables for Time-Delayed Systems Optimized Using Hill-Climbing. *Signals*, *3*(1), 146–156. doi:10.3390/signals3010010

Buşoniu, L., Bruin, T. D., Tolic, D., Kober, J., & Palunko, I. (2018). Reinforcement learning for control: Performance, stability, and deep approximators. *Annual Reviews in Control*, *46*, 8–28. doi:10.1016/j.arcontrol.2018.09.005

Cai, C., Chen, J., Yan, Q., & Liu, F. (2023). A Multi-Robot Coverage Path Planning Method for Maritime Search and Rescue Using Multiple AUVs. *Remote Sensing (Basel)*, *15*(1), 93. doi:10.3390/rs15010093

Cai, L., Lin, J., & Liao, X. (2022). A data-driven method for state of health prediction of lithium-ion batteries in a unified framework. *Journal of Energy Storage*, *51*, 104371. doi:10.1016/j.est.2022.104371

Cano, A., Arevalo, P., & Jurado, F. (2020). Energy analysis and techno-economic assessment of a hybrid PV/HKT/BAT system using biomass gasifier: Cuenca-Ecuador case study. *Energy*, *202*.

Cao, D., Hu, W., Zhao, J., Zhang, G., Zhang, B., Liu, Z., Chen, Z., & Blaabjerg, F. (2020). Reinforcement learning and its applications in modern power and energy systems: A review. *Journal of Modern Power Systems and Clean Energy*, *8*(6), 1029–1042. doi:10.35833/MPCE.2020.000552

Cao, J., Du, W., Wang, H., & McCulloch, M. (2018). Optimal Sizing and Control Strategies for Hybrid Storage System as Limited by Grid Frequency Deviations. *IEEE Transactions on Power Systems*, *33*(5), 5486–5495. doi:10.1109/TPWRS.2018.2805380

Cao, M.-Q., Liu, T.-T., Zhu, Y.-H., Shu, J.-C., & Cao, M.-S. (2022). Developing electromagnetic functional materials for green building. *Journal of Building Engineering*, *45*, 103496. doi:10.1016/j.jobe.2021.103496

Cao, Z., Zhou, X., Hu, H., Wang, Z., & Wen, Y. (2022). Toward a systematic survey for carbon neutral data centers. *IEEE Communications Surveys and Tutorials*, *24*(2), 895–936. doi:10.1109/COMST.2022.3161275

Capozzoli, A., & Primiceri, G. (2015). Cooling systems in data centers: State of art and emerging technologies. *Energy Procedia*, *83*, 484–493. doi:10.1016/j.egypro.2015.12.168

Carenaatb, M. Giudicea,', G. F., Lolaa, S., & Wagner, C. E. M. (1997). c&A. __ * __ l!fiB Four-jet signal at LEP2 and supersymmetry. In Physics Letters B, 395.

Casini, M. (2017, August). Green technology for smart cities. [). IOP Publishing.]. *IOP Conference Series. Earth and Environmental Science*, *83*(1), 012014. doi:10.1088/1755-1315/83/1/012014

Chan, K. W., & Edwards, A. R. (1995). Online dynamic security assessment using a real-time power system simulator with neural network contingency screens. *Proc. IEEE Third International Conference on Advances in Power System Control, Operations and Management (APSCOM ' 95),* (pp. 461-466). IEEE.

Chandraratne, C., Naayagi Ramasamy, T., Logenthiran, T., & Panda, G. (2020). Adaptive protection for microgrid with distributed energy resources. *Electronics (Basel)*, *9*(11), 1959. doi:10.3390/electronics9111959

Chandrasekaran, K., Kandasamy, P., & Ramanathan, S. (2020). Deep learning and reinforcement learning approach on microgrid. *International Transactions on Electrical Energy Systems*, *30*(10), e12531. doi:10.1002/2050-7038.12531

Chandrika, V., Sivakumar, A., Krishnan, T. S., Pradeep, J., Manikandan, S., & Boopathi, S. (2023). Theoretical Study on Power Distribution Systems for Electric Vehicles. In *Intelligent Engineering Applications and Applied Sciences for Sustainability* (pp. 1–19). IGI Global. doi:10.4018/979-8-3693-0044-2.ch001

Channi, H. K. (2023). Optimal designing of PV-diesel generator-based system using HOMER software. *Materials Today: Proceedings*.

Charrouf, O., Betka, A., Abdeddaim, S., & Ghamri, A. (2020). Artificial Neural Network power manager for hybrid PV-wind desalination system. *Mathematics and Computers in Simulation*, *167*, 443–460.

Chaysaz, A., Seyedi, S. R. M., & Motevali, A. (2019). Effects of different greenhouse coverings on energy parameters of a PV–thermal solar system. *Solar Energy*, *194*(November), 519–529. doi:10.1016/j.solener.2019.11.003

Chehelgerdi, Chehelgerdi, M., Allela, O. Q. B., Pecho, R. D. C., Jayasankar, N., Rao, D. P., Thamaraikani, T., Vasanthan, M., Viktor, P., Lakshmaiya, N., Saadh, M. J., Amajd, A., Abo-Zaid, M. A., Castillo-Acobo, R. Y., Ismail, A. H., Amin, A. H., & Akhavan-Sigari, R. (2023). Progressing nanotechnology to improve targeted cancer treatment: Overcoming hurdles in its clinical implementation. *Molecular Cancer*, *22*(1), 169. doi:10.1186/s12943-023-01865-0 PMID:37814270

Chen, W.-L., Huang, C.-W., Li, Y.-H., & Kao, C.-C. (2020). Biosyngas-fueled platinum reactor applied in micro combined heat and power system with a thermo photovoltaic array and Stirling engine. *Energy*, 194.

Chen, Y., & Xiao, J. (2023). *Target Search and Navigation in Heterogeneous Robot Systems with Deep Reinforcement Learning*.

Chen, B., Jiang, H., Sun, H., Yu, M., Yang, J., Li, H., Wang, Y., Chen, L., & Pan, C. (2020). A new gas–liquid dynamics model towards robust state of charge estimation of lithium-ion batteries. *Journal of Energy Storage*, *29*, 101343. doi:10.1016/j.est.2020.101343

Cheng, F., Liang, J., Tao, Z., & Chen, J. (2011). Functional Materials for Rechargeable Batteries. *Advanced Materials*, *23*(15), 1695–1715. doi:10.1002/adma.201003587 PMID:21394791

Chen, J., Yuan, B., & Tomizuka, M. (2019). Model-free Deep Reinforcement Learning for Urban Autonomous Driving. *2019 IEEE Intelligent Transportation Systems Conference. ITSC*, *2019*, 2765–2771. doi:10.1109/ITSC.2019.8917306

Chen, L., Chan, A. P. C., Owusu, E. K., Darko, A., & Gao, X. (2022). Critical success factors for green building promotion: A systematic review and meta-analysis. *Building and Environment*, *207*, 108452. doi:10.1016/j.buildenv.2021.108452

Chen, R., Yang, C., Ma, Y., Wang, W., Wang, M., & Du, X. (2022). Online learning predictive power coordinated control strategy for off-road hybrid electric vehicles considering the dynamic response of engine generator set. *Applied Energy, 323*, 119592. doi:10.1016/j.apenergy.2022.119592

ChenT.GuptaS.GuptaA. (2019). Learning Exploration Policies for Navigation. *CoRR, abs/1903.01959*. http://arxiv.org/abs/1903.01959

Chen, Y., Huang, D., Liu, Z., Osmani, M., & Demian, P. (2022). Construction 4.0, Industry 4.0, and Building Information Modeling (BIM) for sustainable building development within the smart city. *Sustainability (Basel), 14*(16), 10028. doi:10.3390/su141610028

Chen, Y.-K., Wu, Y.-C., Song, C.-C., & Chen, Y.-S. (2012). Design and implementation of energy management system with fuzzy control for DC microgrid systems. *IEEE Transactions on Power Electronics, 28*(4), 1563–1570. doi:10.1109/TPEL.2012.2210446

Chen, Y., Kang, Y., Zhao, Y., Wang, L., Liu, J., Li, Y., Liang, Z., He, X., Li, X., Tavajohi, N., & Li, B. (2021). A review of lithium-ion battery safety concerns: The issues, strategies, and testing standards. *Journal of Energy Chemistry, 59*, 83–99. doi:10.1016/j.jechem.2020.10.017

Chen, Z., Sun, H., Dong, G., Wei, J., & Wu, J. (2019). Particle filter-based state-of-charge estimation and remaining-dischargeable-time prediction method for lithium-ion batteries. *Journal of Power Sources, 414*, 158–166. doi:10.1016/j.jpowsour.2019.01.012

Chitikena, H., Sanfilippo, F., & Ma, S. (2023). Robotics in Search and Rescue (SAR) Operations: An Ethical and Design Perspective Framework for Response Phase. *Applied Sciences (Basel, Switzerland), 13*(3), 1800. doi:10.3390/app13031800

Chu, J., & Huang, X. (2021). *Research status and development trends of evaporative cooling air-conditioning technology in data centers.* Energy and Built Environment.

Chukka, R. (2021). Seismic Fragility and Life Cycle Cost Analysis of Reinforced Concrete Structures with a Hybrid Damper. *Advances in Civil Engineering, 2021*, 1–17. doi:10.1155/2021/4195161

Çiner, F., & Doğan-Sağlamtimur, N. (2019, November). Environmental and sustainable aspects of green building: A review. []. IOP Publishing.]. *IOP Conference Series. Materials Science and Engineering, 706*(1), 012001. doi:10.1088/1757-899X/706/1/012001

Cipi, E., & Cico, B. (2011). Simulation of an Agent Based System Behavior in a Dynamic and Unpredicted Environment. [WCSIT]. *World of Computer Science and Information Technology Journal, 1*, 2221–2741.

Circo, C. J. (2007). Using mandates and incentives to promote sustainable construction and green building projects in the private sector: A call for more state land use policy initiatives. *Penn St. L. Rev., 112*, 731.

Costa, T. (2019). Optimum design of autonomous PV-diesel-battery hybrid systems: a case study at Tapajós-Arapiuns extractive reserve in Brazil. *IEEE PES Innovative Smart Grid Technologies Conference-Latin America*. IEEE.

Couceiro, M. (2017a). An Overview of Swarm Robotics for Search and Rescue Applications. In *Artificial Intelligence*. Concepts, Methodologies, Tools, and Applications., doi:10.4018/978-1-5225-1759-7.ch061

Czop, P., Kost, G., Slawik, D., & Wszolek, G. (2011). Formulation and identification of First- Principle Data-Driven model. Journal of achievements in materials and manufacturing engineering, 44, 179-186.

Darshan, Girdhar, N., Bhojwani, R., Rastogi, K., Angalaeswari, S., Natrayan, L., & Paramasivam, P. (2022). Energy Audit of a Residential Building to Reduce Energy Cost and Carbon Footprint for Sustainable Development with Renewable Energy Sources. *Advances in Civil Engineering, 2022*, 1–10. doi:10.1155/2022/4400874

Darwish, A. S. (2017). Green, smart, sustainable building aspects and innovations. In *Mediterranean Green Buildings & Renewable Energy: Selected Papers from the World Renewable Energy Network's Med Green Forum* (pp. 717-727). Springer International Publishing. 10.1007/978-3-319-30746-6_55

Dasgupta, K., Roy, P. K., & Mukherjee, V. (2020). Power flow-based hydro-thermal-wind scheduling of hybrid power system using sine cosine algorithm. *Electric Power Systems Research, 178*.

Das, M., Maisanam, A. K. S., & Biswas, A. (2019). Techno-Economic Optimization of An Off-Grid Hybrid Renewable Energy System Using Metaheuristic Optimization Approaches- Case of A Radio Transmitter Station In India. *Energy Conversion and Management, 185*, 339–352.

Das, P. K., Behera, H. S., & Panigrahi, B. K. (2016). A hybridization of an Improved Particle Swarm optimization and Gravitational Search Algorithm for Multi-Robot Path Planning. *Swarm and Evolutionary Computation, 28*, 14–28. doi:10.1016/j.swevo.2015.10.011

Datta, S., Bhattacharya, S., & Roy, P. (2016). Artificial Intelligence-based Solar Panel Tilt Angle Optimization and its Hardware Implementation for Efficiency Enhancement. *International Journal of Advanced Research in Electrical, Electronics and Instrumentation Engineering, 5*(10), 7830–7842. doi:10.15662/IJAREEIE.2016.0510006

Datta, U., Kalam, A., & Shi, J. (2020). Battery Energy Storage System Control for Mitigating PV Penetration Impact on Primary Frequency Control and State-ofCharge Recovery. *IEEE Transactions on Sustainable Energy, 11*(2), 746–757. doi:10.1109/TSTE.2019.2904722

Debrah, C., Albert, P. C. C., & Darko, A. (2022). Green finance gap in green buildings: A scoping review and future research needs. *Building and Environment, 207*, 108443. doi:10.1016/j.buildenv.2021.108443

Debrah, C., Chan, A. P., & Darko, A. (2022). Artificial intelligence in green building. *Automation in Construction, 137*, 104192. doi:10.1016/j.autcon.2022.104192

Deng, Y., Feng, C., Jiaqiang, E., Zhu, H., Chen, J., Wen, M., & Yin, H. (2018). Effects of different coolants and cooling strategies on the cooling performance of the power lithium ion battery system: A review. *Applied Thermal Engineering, 142*, 10–29. doi:10.1016/j.applthermaleng.2018.06.043

Devarajan, & Lakshmaiya, N. (2022). Effective utilization of waste banana peel extracts for generating activated carbon-based adsorbent for emission reduction. *Biomass Conversion and Biorefinery*. Advance online publication. doi:10.1007/s13399-022-03470-5

Dhanraj, A. (2022). Appraising machine learning classifiers for discriminating rotor condition in 50W–12V operational wind turbine for maximizing wind energy production through feature extraction and selection process. *Frontiers in Energy Research, 10*, 925980. Advance online publication. doi:10.3389/fenrg.2022.925980

Dhanya, D., Kumar, S. S., Thilagavathy, A., Prasad, D., & Boopathi, S. (2023a). Data Analytics and Artificial Intelligence in the Circular Economy: Case Studies. In Intelligent Engineering Applications and Applied Sciences for Sustainability (pp. 40–58). IGI Global.

Dhanya, D., Kumar, S. S., Thilagavathy, A., Prasad, D. V. S. S. S. V., & Boopathi, S. (2023). Data Analytics and Artificial Intelligence in the Circular Economy: Case Studies. In B. K. Mishra (Ed.), (pp. 40–58). Advances in Civil and Industrial Engineering. IGI Global. doi:10.4018/979-8-3693-0044-2.ch003

Dhimish, M. (2019, December). 70% Decrease of Hot-Spotted Photovoltaic Modules Output Power Loss Using Novel MPPT Algorithm. *IEEE Transactions on Circuits and Wystems. II, Express Briefs, 66*(12), 2027–2031. doi:10.1109/TCSII.2019.2893533

Dif, I., & Dif, N. (2023). Firefly Algorithm Optimization-Based LQR Controller for 1/4 Vehicle Active Suspension System: Design and Performance Evaluation. *The Journal of Engineering and Exact Sciences, 9*(5), 15928–01e. doi:10.18540/jcecvl9iss5pp15928-01e

Dincer, I. (2017). *Refrigeration systems and applications.* John Wiley & Sons. doi:10.1002/9781119230793

Dineva, A., Csomós, B., Kocsis Sz, S., & Vajda, I. (2021). Investigation of the performance of direct forecasting strategy using machine learning in State-of-Charge prediction of Li-ion batteries exposed to dynamic loads. *Journal of Energy Storage, 36*, 102351. doi:10.1016/j.est.2021.102351

Dittakavi, R. S. S. (2023). AI-Optimized Cost-Aware Design Strategies for Resource-Efficient Applications. *Journal of Science and Technology, 4*(1), 1–10.

Divya. (2022). Analysing Analyzing the performance of combined solar photovoltaic power system with phase change material. *Energy Reports, 8*. Advance online publication. doi:10.1016/j.egyr.2022.06.109

Divya, D., Marath, B., & Santosh Kumar, M. (2023). Review of fault detection techniques for predictive maintenance. *Journal of Quality in Maintenance Engineering, 29*(2), 420–441. doi:10.1108/JQME-10-2020-0107

Domakonda, V. K., Farooq, S., Chinthamreddy, S., Puviarasi, R., Sudhakar, M., & Boopathi, S. (2022). Sustainable Developments of Hybrid Floating Solar Power Plants: Photovoltaic System. In Human Agro-Energy Optimization for Business and Industry (pp. 148–167). IGI Global.

Domakonda, V. K., Farooq, S., Chinthamreddy, S., Puviarasi, R., Sudhakar, M., & Boopathi, S. (2023). Sustainable Developments of Hybrid Floating Solar Power Plants: Photovoltaic System. In P. Vasant, R. Rodríguez-Aguilar, I. Litvinchev, & J. A. Marmolejo-Saucedo (Eds.), (pp. 148–167). Advances in Environmental Engineering and Green Technologies. IGI Global. doi:10.4018/978-1-6684-4118-3.ch008

Doroftei, D., Matos, A., & de Cubber, G. (2014). Designing search and rescue robots towards realistic user requirements. *Applied Mechanics and Materials, 658*, 612–617. Advance online publication. . doi:10.4028/www.scientific.net/AMM.658.612

Dosovitskiy, A., Ros, G., Codevilla, F., Lopez, A., & Koltun, V. (2017). *CARLA: An Open Urban Driving Simulator.*

Dozein, M. G., & Mancarella, P. (2019). Possible Negative Interactions between Fast Frequency Response from Utility-scale Battery Storage and Interconnector Protection Schemes. *AUPEC, 2019*, 1–6. doi:10.1109/AUPEC48547.2019.211968

Duarte, M. L. M., de Araújo, P. A., Horta, F. C., Del Vecchio, S., & de Carvalho, L. A. P. (2018). Correlation between weighted acceleration, vibration dose value and exposure time on whole body vibration comfort levels evaluation. *Safety Science, 103*, 218–224. doi:10.1016/j.ssci.2017.11.008

Duchesne, L., Karangelos, E., & Wehenkel, L. (2020). Recent developments in machine learning for energy systems reliability management. *Proceedings of the IEEE, 108*(9), 1656–1676. doi:10.1109/JPROC.2020.2988715

Du, H., Li, W., & Zhang, N. (2013). Semi-active control of an integrated full-car suspension with seat suspension and driver body model using ER dampers. *International Journal of Vehicle Design, 63*(2-3), 159–184. doi:10.1504/IJVD.2013.056133

Duong, D. T., & Uhlen, K. (2018). *"An empirical method for online detection of power oscillations in power systems," 2018 IEEE Innovative Smart Grid Technol. - Asia.* ISGT Asia.

Dutta, N., Usman, M., Ashraf, M. A., Luo, G., & Zhang, S. (2022). A critical review of recent advances in the bioremediation of chlorinated substances by microbial dechlorinators. *Chemical Engineering Journal Advances, 12*, 100359. doi:10.1016/j.ceja.2022.100359

Dwivedi, D., Mitikiri, S. B., Babu, K., Yemula, P. K., Srininvas, V. L., Chakraborty, P., & Pal, M. (2023). Advancements in Enhancing Resilience of Electrical Distribution Systems: A Review on Frameworks, Metrics, and Technological Innovations. *arXiv preprint arXiv:2311.07050.*

Edwards, B. P. (1978). Computer Based Sun following System. *Solar Energy, 21*(1), 491–496. doi:10.1016/0038-092X(78)90073-7

Ehiagwina, F. (2021). Development of a solar energy tracking mechanism with artificial neural network enhancement. *International Research Journal of Modernization in Engineering Technology and Science, 3*(3).

Ekaputri, C., & Syaichu-Rohman, A. (2013, August). Model predictive control (MPC) design and implementation using algorithm-3 on board SPARTAN 6 FPGA SP605 evaluation kit. In *2013 3rd International Conference on Instrumentation Control and Automation (ICA)* (pp. 115-120). IEEE.

El Barkouki, B., Laamim, M., Rochd, A., Chang, J. W., Benazzouz, A., Ouassaid, M., Kang, M., & Jeong, H. (2023). An Economic Dispatch for a Shared Energy Storage System Using MILP Optimization: A Case Study of a Moroccan Microgrid. *Energies, 16*(12), 4601. doi:10.3390/en16124601

Elkadeem, M. (2019). A Systematic Approach for Planning and Design of Hybrid Renewable Energy Based Micro grid With Techno-Economic Optimization: A Case Study on an Urban Community in Egypt. *Sustainable Cities and Society.*

Elsheikh, A., Sharshir, S. W., & Elaziz, E. A. (2019). Modeling of solar energy systems using artificial neural network: A comprehensive review. *Solar Energy, 180*, 622–639. doi:10.1016/j.solener.2019.01.037

Elsisi, M. (2019). Design of neural network predictive controller based on imperialist competitive algorithm for automatic voltage regulator. *Neural Computing & Applications, 31*(9), 5017–5027. doi:10.1007/s00521-018-03995-9

Engin, M., & Engin, D. (2015). Optimization Controller for Mechatronic Sun Tracking System to Improve Performance. *Advances in Mechanical Engineering, 5*, 146352–146352. doi:10.1155/2013/146352

Erol, O. K., & Eksin, I. (2006). A new optimization method: Big bang–big crunch. *Advances in Engineering Software, 37*(2), 106–111. doi:10.1016/j.advengsoft.2005.04.005

Erten, D., & Kılkış, B. (2022). How can green building certification systems cope with the era of climate emergency and pandemics? *Energy and Building, 256*, 111750. doi:10.1016/j.enbuild.2021.111750

Fan, S., Yang, W., & Hu, Y. (2018). Adjustment and control on the fundamental characteristics of a piezoelectric PN junction by mechanical – loading. Nano Energy, 52, 416–421. .08.017. doi:10.1016/j.nanoen.2018.08.017

Faraji, J., Khanjanianpak, M., Rezaei, M., Kia, M., Aliyan, E., & Dehghanian, P. (2020). Fast-Accurate Dual-Axis ST Controlled by P&O Technique with Neural Network Optimization. *2020 IEEE International Conference on Environment and Electrical Engineering and 2020 IEEE Industrial and Commercial Power Systems Europe (EEEIC / I&CPS Europe).* IEEE. 10.1109/EEEIC/ICPSEurope49358.2020.9160843

Fard, R. H., & Hosseini, S. "Machine Learning algorithms for prediction of Power consumption and IoT modeling in complex networks. Microprocessors and Microsystems", 2022December*Wireless Personal Communications 121* 10 doi:10.1007/s11277-021-08879-1

Farret, F. A., & Godoy Simoes, M. (2006). *Integration of Alternative Sources of Energy.* John Wiley & Sons, Ltd.

Fernandez Cornejo, E. R., Diaz, R. C., & Alama, W. I. (2020). PID Tuning based on Classical and Meta-heuristic Algorithms: A Performance Comparison. *2020 IEEE Engineering International Research Conference (EIRCON),* (pp. 1-4). IEEE. 10.1109/EIRCON51178.2020.9253750

Fulzele, J. B. (2018). Design and Optimization of Hybrid PV-Wind Renewable Energy System. Materials Today: Proceedings, (pp. 810-818). IEEE.

Ganesh, G. S. P., Balaji, B., & Varadhan, T. A. S. (2011). Anti-theft tracking system for automobiles (AutoGSM) *Proceedings of the IEEE International Conference on Anti-Counterfeiting, Security and Identification (ASID '11)*. IEEE.

Gan, L. K., Shek, J. K. H., & Mueller, M. A. (2017). Analysis of Tower Shadow effects on Battery Lifetime in standalone Hybrid Wind-Diesel-Battery Systems. *IEEE Transactions on Industrial Electronics, Vol, 64*(8), 6234–6244.

Gao, B., Morison, G. K., & Kundur, P. (1986). Voltage stability evaluation using modal analysis. *IEEE Transactions on Power Systems, 7*(4), 1529–1542. doi:10.1109/59.207377

Gao, Y., Plett, G. L., Fan, G., & Zhang, X. (2022). Enhanced state-of-charge estimation of LiFePO4 batteries using an augmented physics-based model. *Journal of Power Sources, 544*, 231889. Advance online publication. doi:10.1016/j.jpowsour.2022.231889

Garcia, R. S., & Weisser, D. (2006). A wind–diesel system with hydrogen storage: Joint optimization of design and dispatch. *Renewable Energy, 31*(14), 2296–2320. doi:10.1016/j.renene.2005.11.003

Ghahramani, M., Nazari-Heris, M., & Zare, K. (2019). Energy and reserve management of a smart distribution system by incorporating responsive- loads/battery/wind turbines considering uncertain parameters. Research Gate.

Ghenaia, C., & Janajreh, I. (2020). Design of Solar-Biomass Hybrid Micro grid System in Sharjah. *Energy Procedia, 103*, 357–362.

Giaglis, G. M., Pateli, A., Fouskas, K., Kourouthanassis, P., & Tsamakos, A. (2002). On the Potential Use of Mobile Positioning Technologies in Indoor Environments. In *Proceedings of the 15th Bled Electronic Commerce Conferencev-Reality: Constructing the Economy*, Bled, Slovenia.

Gil-Gonzalez, W., Montoya, O. D., & Garces, A. (2020). Modeling and control of a small hydro power plant for a DC micro grid. *Electric Power Systems Research, 180*.

Goddard, R., Zhang, L., & Xia, X. (2019). Optimal Sizing and Power Sharing of Distributed Hybrid Renewable Energy Systems Considering Socio-Demographic Factors. *Energy Procedia, 159*, 340–345.

Goel, S., & Sharma, R. (2019). Optimal sizing of a biomass–biogas hybrid system for sustainable power supply to a commercial agricultural farm in northern Odisha, India. *Environment, Development and Sustainability*.

González-Acevedo, H., Muñoz, Y., Ospino, A., Serrano, J., Atencio, A., & Saavedra, C. (2021). Design and performance evaluation of a solar tracking panel of single axis in Colombia [IJECE]. *Iranian Journal of Electrical and Computer Engineering, 11*(4), 2889. doi:10.11591/ijece.v11i4.pp2889-2898

Goswami, R. K., Agrawal, K., Upadhyaya, H. M., Gupta, V. K., & Verma, P. (2022). Microalgae conversion to alternative energy, operating environment and economic footprint: An influential approach towards energy conversion, and management. *Energy Conversion and Management, 269*, 116118. doi:10.1016/j.enconman.2022.116118

Gowri, N. V., Dwivedi, J. N., Krishnaveni, K., Boopathi, S., Palaniappan, M., & Medikondu, N. R. (2023). Experimental investigation and multi-objective optimization of eco-friendly near-dry electrical discharge machining of shape memory alloy using Cu/SiC/Gr composite electrode. *Environmental Science and Pollution Research International, 30*(49), 1–19. doi:10.1007/s11356-023-26983-6 PMID:37126160

Guan, J., Chen, G., Huang, J., Li, Z., Xiong, L., Hou, J., & Knoll, A. (2023). A Discrete Soft Actor-Critic Decision-Making Strategy With Sample Filter for Freeway Autonomous Driving. *IEEE Transactions on Vehicular Technology, 72*(2), 2593–2598. doi:10.1109/TVT.2022.3212996

Guduru, S., Preetham, C. H., & Vijayan, K. (2023). Smart Solar Tracking System for Optimal Power Generation Using Three LDR's. In *2023 International Conference on Recent Advances in Electrical, Electronics, Ubiquitous Communication, and Computational Intelligence (RAEEUCCI)* (pp. 1-5). IEEE 10.1109/RAEEUCCI57140.2023.10134521

Gundeti, R., Vuppala, K., & Kasireddy, V. (2024). The Future of AI and Environmental Sustainability: Challenges and Opportunities. *Exploring Ethical Dimensions of Environmental Sustainability and Use of AI*, 346-371.

Guo, J., Che, Y., Pedersen, K., & Stroe, D. I. (2023). Battery impedance spectrum prediction from partial charging voltage curve by machine learning. *Journal of Energy Chemistry*, *79*, 211–221. doi:10.1016/j.jechem.2023.01.004

Gu, Y., Lo, A., & Niemegeers, I. (2009). A Survey of Indoor Positioning Systems for Wireless Personal Networks. *IEEE Communications Surveys and Tutorials*, *11*(1), 13–32. doi:10.1109/SURV.2009.090103

Halabi, L. M., Mekhilef, S., Olatomiwa, L., & Hazeltonc, J. (2017). *Performance analysis of hybrid PV/diesel/battery system using HOMER: A case study Sabah, Malaysia*, 144.

Hall, R. E., Bowerman, B., Braverman, J., Taylor, J., Todosow, H., & Von Wimmersperg, U. (2000). The vision of a smart city (No. BNL-67902; 04042). Brookhaven National Lab.(BNL), Upton, NY (United States).

Hameed, I. A. (2014). Intelligent coverage path planning for agricultural robots and autonomous machines on three-dimensional terrain. *Journal of Intelligent & Robotic Systems*, *74*(3–4), 965–983. doi:10.1007/s10846-013-9834-6

Hammoumi, E. (2018). A simple and low-cost active dual-axis ST. *Energy Science & Engineering*, *6*. doi:10.1002/ese3.236

Hanumanthakari, S., Gift, M. M., Kanimozhi, K., Bhavani, M. D., Bamane, K. D., & Boopathi, S. (2023). Biomining Method to Extract Metal Components Using Computer-Printed Circuit Board E-Waste. In *Handbook of Research on Safe Disposal Methods of Municipal Solid Wastes for a Sustainable Environment* (pp. 123–141). IGI Global. doi:10.4018/978-1-6684-8117-2.ch010

Hanwate, S., & Hote, Y. (2018). Design of PID controller for sun tracker system using QRAWCP approach. *International Journal of Computational Intelligence Systems*, *11*(1), 133. doi:10.2991/ijcis.11.1.11

Hausmann, A., & Depcik, C. (2013). Expanding the Peukert equation for battery capacity modeling through inclusion of a temperature dependency. *Journal of Power Sources*, *235*, 148–158. doi:10.1016/j.jpowsour.2013.01.174

Hemalatha, K., James, C., Natrayan, L., & Swamynadh, V. (2020a). Analysis of RCC T-beam and prestressed concrete box girder bridges super structure under different span conditions. In *Materials Today*. Proceedings.

Hema, N., Krishnamoorthy, N., Chavan, S. M., Kumar, N., Sabarimuthu, M., & Boopathi, S. (2023). A Study on an Internet of Things (IoT)-Enabled Smart Solar Grid System. In *Handbook of Research on Deep Learning Techniques for Cloud-Based Industrial IoT* (pp. 290–308). IGI Global. doi:10.4018/978-1-6684-8098-4.ch017

Hemanth. (2017). Evaluation of mechanical properties of e-glass and coconut fiber reinforced with polyester and epoxy resin matrices. *International Journal of Mechanical and Production Engineering Research and Development*, *7*(5). doi:10.24247/ijmperdoct20172

Henney, A. (2001, October). Transmission access – a case study in public maladministration? *Power UK*, (92), 18–32.

Her, C., Sambor, D. J., Whitney, E., & Wies, R. (2021). Novel wind resource assessment and demand flexibility analysis for community resilience: A remote microgrid case study. *Renewable Energy*, *179*, 1472–1486. doi:10.1016/j.renene.2021.07.099

He, Y., Liu, W., & Koch, B. J. (2010). Battery algorithm verification and development using hardware-in-the-loop testing. *Journal of Power Sources*, *195*(9), 2969–2974. doi:10.1016/j.jpowsour.2009.11.036

Hijawi, H., & Arafeh, L. (2016). Design of Dual Axis ST System Based on Fuzzy Inference Systems. *International Journal on Soft Computing, Artificial Intelligence, and Applications, 5*(2/3), 23–36. Advance online publication. doi:10.5121/ijscai.2016.5302

Himadry Shekhar Das, A. H. M. (2016). Proposition of a PV/tidal powered micro-hydro and diesel hybrid system: A southern Bangladesh focus. Renewable and Sustainable Energy Reviews, 53, 1137-1148.

Homan, B., ten Kortenaar, M. V., Hurink, J. L., & Smit, G. J. M. (2019). A realistic model for battery state of charge prediction in energy management simulation tools. *Energy, 171*, 205–217. doi:10.1016/j.energy.2018.12.134

Hossain, M. A., Chakrabortty, R. K., Ryan, M. J., & Pota, H. R. (2021). Energy management of community energy storage in grid-connected microgrid under uncertain real-time prices. *Sustainable Cities and Society, 66*, 102658. doi:10.1016/j.scs.2020.102658

Hossain, M. A., Pota, H. R., Squartini, S., Zaman, F., & Guerrero, J. M. (2019). Energy scheduling of community microgrid with battery cost using particle swarm optimisation. *Applied Energy, 254*, 113723. doi:10.1016/j.apenergy.2019.113723

Hsu, C. F., Li, R.-K., Kang, H.-Y., & Lee, A. H. (2014). A systematic evaluation model for solar cell technologies. *Mathematical Problems in Engineering, 2014*, 1–16. doi:10.1155/2014/542351

Huang, Z., Wu, J., & Lv, C. (2022). Efficient Deep Reinforcement Learning With Imitative Expert Priors for Autonomous Driving. *IEEE Transactions on Neural Networks and Learning Systems.* doi:10.1109/TNNLS.2022.3142822 PMID:35081030

Hu, J., Wang, Z., Du, H., & Zou, L. (2022). Hierarchical energy management strategy for fuel cell/ultracapacitor/battery hybrid vehicle with life balance control. *Energy Conversion and Management, 272*, 116383. Advance online publication. doi:10.1016/j.enconman.2022.116383

Hu, M. (2020). *Smart technologies and design for healthy built environments.* Springer Nature.

Hur, J., & Ahn, E. (2024). *An Enhanced Short-term Forecasting of Wind Generating Resources based on Edge Computing in Jeju Carbon-Free Islands.*

Hu, R., Liu, Y., Shin, S., Huang, S., Ren, X., Shu, W., Cheng, J., Tao, G., Xu, W., Chen, R., & Luo, X. (2020). Emerging materials and strategies for personal thermal management. *Advanced Energy Materials, 10*(17), 1903921. doi:10.1002/aenm.201903921

Hussain, Z., & Srimathy, G. (2023). *IoT and AI Integration for Enhanced Efficiency and Sustainability.*

Hussain, Z., Babe, M., Saravanan, S., Srimathy, G., Roopa, H., & Boopathi, S. (2023). Optimizing Biomass-to-Biofuel Conversion: IoT and AI Integration for Enhanced Efficiency and Sustainability. In N. Cobîrzan, R. Muntean, & R.-A. Felseghi (Eds.), (pp. 191–214). Advances in Finance, Accounting, and Economics. IGI Global. doi:10.4018/978-1-6684-8238-4.ch009

Hu, Y., Zhang, Y., Wang, S., Xu, W., Fan, Y., & Liu, Y. (2021). Joint Dynamic Strategy of Bayesian Regularized Back Propagation Neural Network with Strong Robustness - Extended Kalman Filtering for the Battery State-of-Charge Prediction. *International Journal of Electrochemical Science, 16*(11), 1–15. doi:10.20964/2021.11.07

Huynh, D. C., Nguyen, T. M., Dunnigan, M. W., & Mueller, M. A. (2013). Comparison between open- and closed-loop trackers of a solar photovoltaic system. *2013 IEEE Conference on Clean Energy and Technology (CEAT)*, (pp. 128-133). IEEE. 10.1109/CEAT.2013.6775613

Ibrahim, M. S., Dong, W., & Yang, Q. (2020). Machine learning driven smart electric power systems: Current trends and new perspectives. *Applied Energy, 272*, 115237. doi:10.1016/j.apenergy.2020.115237

Ingle, R. B., Senthil, T. S., Swathi, S., Muralidharan, N., Mahendran, G., & Boopathi, S. (2023). Sustainability and Optimization of Green and Lean Manufacturing Processes Using Machine Learning Techniques. IGI Global. doi:10.4018/978-1-6684-8238-4.ch012

Iqbal, A., Malik, H., Riyaz, A., Abdellah, K., & Bayhan, S. (2021). Renewable Power for Sustainable Growth: *Proceedings of International Conference on Renewal Power (ICRP 2020)*. Springer.

Iravani, A. (2017). Advantages and disadvantages of green technology; goals, challenges and strengths. *Int J Sci Eng Appl*, 6(9), 272–284.

Ishihara, T., Shima, K., Kimura, T., Ishii, S., Momoi, T., Yamaguchi, H., Umetani, K., Moriyama, M., Yamanouchi, M., & Mikumo, T. (1982). Equilibrium Charge State Distributions Of Fast Si And Ci Ions In Carbon And Gold Foils. In Nuclear Instruments and Methods, 204.

Ismaeil, E. M., & Sobaih, A. E. E. (2023). Heuristic Approach for Net-Zero Energy Residential Buildings in Arid Region Using Dual Renewable Energy Sources. *Buildings*, 13(3), 796. doi:10.3390/buildings13030796

Ismail, Z. A. (2021). Maintenance management practices for green building projects: Towards hybrid BIM system. *smart and sustainable. Built Environment*, 10(4), 616–630.

Iwano, Y., Osuka, K., & Amano, H. (2004). Proposal of a rescue robot system in nuclear-power plants-rescue activity via small vehicle robots. *Proceedings - 2004 IEEE International Conference on Robotics and Biomimetics, IEEE ROBIO 2004*. 10.1109/ROBIO.2004.1521781

Jabari, F., Mohammadi-ivatloo, B., & Ghaebi, H. (2018). Design and performance investigation of a biogas fuelled combined cooling and power generation system. Energy Conversion and Management, 169, 371-382.

Jabari, F. (2019). Biogas fuelled combined cooling, desalinated water and power generation systems. *Journal of Cleaner Production*, 219, 906–924.

Jadidi, S., Badihi, H., Yu, Z., & Zhang, Y. (2020). Fault detection and diagnosis in power electronic converters at microgrid level based on filter bank approach. *2020 IEEE 3rd International Conference on Renewable Energy and Power Engineering (REPE)*, (pp. 39–44). IEEE.

Jadidi, S., Badihi, H., & Zhang, Y. (2020). Fault diagnosis in microgrids with integration of solar photovoltaic systems: A review. *IFAC-PapersOnLine*, 53(2), 12091–12096. doi:10.1016/j.ifacol.2020.12.763

Jahangir, M. H., & Cheraghi, R. (2020). Economic and environmental assessment of solar-wind-biomass hybrid renewable energy system supplying rural settlement load. *Sustainable Energy Technologies and Assessments*, 42.

Jain, D. K., Tyagi, S. K. S., Neelakandan, S., Prakash, M., & Natrayan, L. (2022). Metaheuristic Optimization-Based Resource Allocation Technique for Cybertwin-Driven 6G on IoE Environment. *IEEE Transactions on Industrial Informatics*, 18(7), 4884–4892. doi:10.1109/TII.2021.3138915

Jamgochian, A., Buehrle, E., Fischer, J., & Kochenderfer, M. J. (2023). SHAIL: Safety-Aware Hierarchical Adversarial Imitation Learning for Autonomous Driving in Urban Environments. *Proceedings - IEEE International Conference on Robotics and Automation, 2023-May*. IEEE. 10.1109/ICRA48891.2023.10161449

Jang, J. S. R. (1993, May-June). ANFIS: Adaptive-network-based fuzzy inference system. *IEEE Transactions on Systems, Man, and Cybernetics*, 23(3), 665–685. doi:10.1109/21.256541

Jayachandran, M., & Ravi, G. (2017). Design and Optimization of Hybrid Micro-Grid System. *Energy Procedia, 117*, 95–103.

Jean, D. D. N., Zou, F., Godwin, N. O. A., Bimenyimana, S., & Shyirambere, G. (2020). Key Technology Development Needs And Applicability Analysis Of Renewable Energy Hybrid Technologies In Off-Grid Areas For The Rwanda Power Sector. *Heliyon, 6*.

Jiang, Y., & Zheng, W. (2021). Coupling mechanism of green building industry innovation ecosystem based on blockchain smart city. *Journal of Cleaner Production, 307*, 126766. doi:10.1016/j.jclepro.2021.126766

Ji, G., Zhang, L., Shan, M., & Zhang, J. (2023). Enhanced variable universe fuzzy PID control of the active suspension based on expansion factor parameters adaption and genetic algorithm. *Engineering Research Express, 5*(3), 035007. doi:10.1088/2631-8695/ace0a2

Jin, C., Bai, X., Yang, C., Mao, W., & Xu, X. (2020). A review of power consumption models of servers in data centers. *Applied Energy, 265*, 114806. doi:10.1016/j.apenergy.2020.114806

Jing, C., Shu, H., & Song, Y. (2023). Model Predictive Control for Integrated Lateral Stability and Rollover Prevention Based on a Multi-actuator Control System. *International Journal of Control, Automation, and Systems, 21*(5), 1518–1537. doi:10.1007/s12555-021-0969-0

Jitwang, T., Hlangnamthip, S., & Puangdownreong, D. (2020). Robust PIDA Controller Design by Cuckoo Search for Liquid-Level Control System. *2020 Joint International Conference on Digital Arts, Media, and Technology with ECTI Northern Section Conference on Electrical, Electronics, Computer, and Telecommunications Engineering (ECTI DAMT & NCON)*, (pp. 226-229). IEEE. 10.1109/ECTIDAMTNCON48261.2020.9090746

Josphineleela, R., Jyothi, M., Kaviarasu, A., & Sharma, M. (2023a) Development of IoT based Health Monitoring System for Disables using Microcontroller. In: *Proceedings - 7th International Conference on Computing Methodologies and Communication, ICCMC 2023*. IEEE. 10.1109/ICCMC56507.2023.10084026

Josphineleela, R., Kaliappan, S., & Bhatt, U. M. (2023b) Intelligent Virtual Laboratory Development and Implementation using the RASA Framework. In: *Proceedings - 7th International Conference on Computing Methodologies and Communication, ICCMC 2023*. IEEE. 10.1109/ICCMC56507.2023.10083701

Josphineleela, R., & Kaliapp, S. (2023a). Big Data Security through Privacy - Preserving Data Mining (PPDM): A Decentralization Approach. In: *Proceedings of the 2023 2nd International Conference on Electronics and Renewable Systems, ICEARS 2023*. IEEE. 10.1109/ICEARS56392.2023.10085646

Juan, M. (2012). *Optimum load management strategy for wind/diesel/battery hybrid power systems*. Research Gate.

Jung, C., & Awad, J. (2023). Sharjah Sustainable City: An Analytic Hierarchy Process Approach to Urban Planning Priorities. *Sustainability (Basel), 15*(10), 8217. doi:10.3390/su15108217

Kalakrishnan, M., Righetti, L., Pastor, P., & Schaal, S. (2011). Learning force control policies for compliant manipulation. *IEEE International Conference on Intelligent Robots and Systems*. IEEE. 10.1109/IROS.2011.6095096

Kaliappan, S. (2023a) Checking and Supervisory System for Calculation of Industrial Constraints using Embedded System. In: *Proceedings of the 4th International Conference on Smart Electronics and Communication, ICOSEC 2023*. IEEE. 10.1109/ICOSEC58147.2023.10275952

Kaliappan, S., Natrayan, L., Kumar, P. V. A., & Raturi, A. (2023b). Mechanical, fatigue, and hydrophobic properties of silane-treated green pea fiber and egg fruit seed powder epoxy composite. *Biomass Conversion and Biorefinery*. doi:10.1007/s13399-023-04534-w

Kaliappan, S., & Rajput, A. (2023b) Sentiment Analysis of News Headlines Based on Sentiment Lexicon and Deep Learning. In: *Proceedings of the 4th International Conference on Smart Electronics and Communication, ICOSEC 2023*. IEEE. 10.1109/ICOSEC58147.2023.10276102

Kaliappan, S., Velumayil, R., & Pravin, P. (2023d). Mechanical, DMA, and fatigue behavior of Vitis vinifera stalk cellulose Bambusa vulgaris fiber epoxy composites. *Polymer Composites*, 44(4), 2115–2121. doi:10.1002/pc.27228

Kalogirou, S. A. (1996). Design and construction of one-axis Sun-Tracking system. *Sol. Energy, 57*(6). . doi:10.1016/S0038-092X(96)00135-1

Kamran, D., Lopez, C. F., Lauer, M., & Stiller, C. (2020). Risk-Aware High-level Decisions for Automated Driving at Occluded Intersections with Reinforcement Learning. *IEEE Intelligent Vehicles Symposium, Proceedings*. IEEE. 10.1109/IV47402.2020.9304606

Kanase-Patil, A. B., Saini, R. P., & Sharma, M. P. (2010). Integrated Renewable Energy Systems For Off Grid Rural Electrification of Remote Area. *Renewable Energy, 35*, 1342–1349.

Kanase-Patil, A. B., Saini, R. P., & Sharma, M. P. (2011). Development of IREOM model based on seasonally varying load profile for hilly remote areas of Uttarakhand state in India. *Energy, 36*(9), 5690–5702. doi:10.1016/j.energy.2011.06.057

Kanimozhi, G., Natrayan, L., Angalaeswari, S., & Paramasivam, P. (2022). An Effective Charger for Plug-In Hybrid Electric Vehicles (PHEV) with an Enhanced PFC Rectifier and ZVS-ZCS DC/DC High-Frequency Converter. *Journal of Advanced Transportation, 2022*, 1–14. doi:10.1155/2022/7840102

Karabiber, A., & Güneş, Y. (2023). Single-motor and dual-axis solar tracking system for micro photovoltaic power plants. *Journal of Solar Energy Engineering, 145*(5), 051004. doi:10.1115/1.4056739

Karimi Pour, F., Theilliol, D., Puig, V., & Cembrano, G. (2021). Health-aware control design based on remaining useful life estimation for autonomous racing vehicle. *ISA Transactions, 113*, 196–209. doi:10.1016/j.isatra.2020.03.032 PMID:32451079

Karimi, H., Beheshti, M. T., Ramezani, A., & Zareipour, H. (2021). Intelligent control of islanded AC microgrids based on adaptive neuro-fuzzy inference system. *International Journal of Electrical Power & Energy Systems, 133*, 107161. doi:10.1016/j.ijepes.2021.107161

Karkera, T., Dubey, A., Kamalnakhawa, S., & Mangale, S. (2018). —GPS-GSM based Vehicle Tracking System,‖. *International Journal of New Technology and Research, 4*(3), 140–142.

Karthick, A., Mohanavel, V., Chinnaiyan, V. K., Karpagam, J., Baranilingesan, I., & Rajkumar, S. (2022). State of charge prediction of battery management system for electric vehicles. In Active Electrical Distribution Network: Issues, Solution Techniques, and Applications (pp. 163–180). Elsevier. doi:10.1016/B978-0-323-85169-5.00012-5

Karthick, Meikandan, M., Kaliappan, S., Karthick, M., Sekar, S., Patil, P. P., Raja, S., Natrayan, L., & Paramasivam, P. (2022). Experimental Investigation on Mechanical Properties of Glass Fiber Hybridized Natural Fiber Reinforced Penta-Layered Hybrid Polymer Composite. *International Journal of Chemical Engineering, 2022*, 1–9. doi:10.1155/2022/1864446

Karthik, S. A., Hemalatha, R., Aruna, R., Deivakani, M., Reddy, R. V. K., & Boopathi, S. (2023). Study on Healthcare Security System-Integrated Internet of Things (IoT). In M. K. Habib (Ed.), (pp. 342–362). Advances in Systems Analysis, Software Engineering, and High Performance Computing. IGI Global. doi:10.4018/978-1-6684-7684-0.ch013

Kaur Channi, H., Gupta, S., & Dhingra, A. (2020). Optimization and simulation of a solar–wind hybrid system using HOMER for Rural Electrification. *International Journal of Advanced Science and Technology, 29*, 2108-2116.

Kaur, A., & Narang, N. (2019). Optimum Generation Scheduling of Coordinated Power System Using Hybrid Optimization Technique. *Electrical Engineering*.

Kaur, G., Prakash, A., & Rao, K. U. (2021). A critical review of Microgrid adaptive protection techniques with distributed generation. *Renewable Energy Focus*, *39*, 99–109. doi:10.1016/j.ref.2021.07.005

Kaur, R., Vijaya, K. K. S., Kandasamy, N. K., & Kumar, S. (2019). Discrete Multi-objective Grey Wolf Algorithm Based Optimal Sizing and Sensitivity Analysis of PV-Wind-Battery System for Rural Telecom Towers. *IEEE Systems Journal*.

Kaur, T., & Segal, R. (2017). Designing Rural Electrification Solutions Considering Hybrid Energy Systems For Papua New Guinea. *Energy Procedia*, *110*, 1–7.

Kaushal. (2023) A Payment System for Electric Vehicles Charging and Peer-to-Peer Energy Trading. In: *7th International Conference on I-SMAC (IoT in Social, Mobile, Analytics and Cloud), I-SMAC 2023 – Proceedings*. IEEE. 10.1109/I-SMAC58438.2023.10290505

Kavitha, C. R., Varalatchoumy, M., Mithuna, H. R., Bharathi, K., Geethalakshmi, N. M., & Boopathi, S. (2023a). Energy Monitoring and Control in the Smart Grid: Integrated Intelligent IoT and ANFIS. In M. Arshad (Ed.), (pp. 290–316). Advances in Bioinformatics and Biomedical Engineering. IGI Global. doi:10.4018/978-1-6684-6577-6.ch014

Kessel, P., & Glavitsch, H. (1986). Estimating the voltage stability of a power system. *IEEE Transactions on Power Systems*, *1*(3).

Kewo, A., Munir, R., & Lapu, A. K. (2015). *IntelligEnSia based electricity consumption prediction analytics using regression method*. Paper presented at 5th International Conference on Electrical Engineering and Informatics, Bali, Indonesia. 10.1109/ICEEI.2015.7352556

Khalaj, A. H., & Halgamuge, S. K. (2017). A Review on efficient thermal management of air-and liquid-cooled data centers: From chip to the cooling system. *Applied Energy*, *205*, 1165–1188. doi:10.1016/j.apenergy.2017.08.037

Khaleel, M. (2023). Intelligent Control Techniques for Microgrid Systems. *Brilliance: Research of Artificial Intelligence*, *3*(1), 56–67. doi:10.47709/brilliance.v3i1.2192

Khalid, M., & Savkin, A. V. (2010). A model predictive control approach to the problem of wind power smoothing with controlled battery storage. *Renewable Energy*, *35*(7), 1520–1526. doi:10.1016/j.renene.2009.11.030

Khokhar, B., & Parmar, K. S. (2022). A novel adaptive intelligent MPC scheme for frequency stabilization of a microgrid considering SoC control of EVs. *Applied Energy*, *309*, 118423. doi:10.1016/j.apenergy.2021.118423

Khurana, D., Koli, A., Khatter, K., & Singh, S. (2023). Natural language processing: State of the art, current trends and challenges. *Multimedia Tools and Applications*, *82*(3), 3713–3744. doi:10.1007/s11042-022-13428-4 PMID:35855771

Kim, C., & Ro, P. I. (2002). An accurate full car ride model using model reducing techniques. *Journal of Mechanical Design*, *124*(4), 697–705. doi:10.1115/1.1503065

Kim, M.-H., Kim, D., Heo, J., & Lee, D.-W. (2019). Techno-Economic Analysis Of Hybrid Renewable Energy System With Solar District Heating For Net Zero Energy Community. *Energy*, *187*.

Kim, S. W., Liu, W., Ang, M. H., Frazzoli, E., & Rus, D. (2015). The Impact of Cooperative Perception on Decision Making and Planning of Autonomous Vehicles. *IEEE Intelligent Transportation Systems Magazine*, *7*(3), 39–50. Advance online publication. doi:10.1109/MITS.2015.2409883

Kim, S., Heo, S., Nam, K., Woo, T., & Yoo, C. (2023). Flexible renewable energy planning based on multi-step forecasting of interregional electricity supply and demand: Graph-enhanced AI approach. *Energy*, *282*, 128858. doi:10.1016/j.energy.2023.128858

Kiumarsi, B., Vamvoudakis, K. G., Modares, H., & Lewis, F. L. (2018, June). Optimal and Autonomous Control Using Reinforcement Learning: A Survey. *IEEE Transactions on Neural Networks and Learning Systems*, *29*(6), 2042–2062. doi:10.1109/TNNLS.2017.2773458 PMID:29771662

Kiyak, E., & Gol, G. (2016). A comparison of fuzzy logic and PID controller for a single-axis Solar Tracking System. *Renewables: Wind, Water, and Solar*, *3*(1), 7. doi:10.1186/s40807-016-0023-7

Kohl, N., & Stone, P. (2004). Policy gradient reinforcement learning for fast quadrupedal locomotion. *Proceedings - IEEE International Conference on Robotics and Automation, 2004*(3). IEEE. 10.1109/ROBOT.2004.1307456

Kokane, P., Kiran, S., & Imran, B. (2015). Prof. Yogesh Thorat. *Review on Accident Alert and Vehicle Tracking System*, *3*(October - December).

Koko, S. P., Kusakana, K., & Vermaak, H. J. (2018). Optimal power dispatch of a grid –interactive micro-hydro kinetic-pumped hydro storage system. *Journal of Energy Storage*, *17*, 63–72.

Konakalla, S. A. R., & de Callafon, R. A. (2017, June). Feature based grid event classification from synchrophasor data. *Procedia Computer Science*, *108*, 1582–1591. doi:10.1016/j.procs.2017.05.046

Koshariya, A. K., Kalaiyarasi, D., Jovith, A. A., Sivakami, T., Hasan, D. S., & Boopathi, S. (2023). AI-Enabled IoT and WSN-Integrated Smart Agriculture System. In *Artificial Intelligence Tools and Technologies for Smart Farming and Agriculture Practices* (pp. 200–218). IGI Global. doi:10.4018/978-1-6684-8516-3.ch011

Koshariya, A. K., Khatoon, S., Marathe, A. M., Suba, G. M., Baral, D., & Boopathi, S. (2023). Agricultural Waste Management Systems Using Artificial Intelligence Techniques. In *AI-Enabled Social Robotics in Human Care Services* (pp. 236–258). IGI Global. doi:10.4018/978-1-6684-8171-4.ch009

Koval, A., Karlsson, S., & Nikolakopoulos, G. (2022). Experimental evaluation of autonomous map-based Spot navigation in confined environments. *Biomimetic Intelligence and Robotics, 2*(1), 100035.

Koyuncu, H., & Yang, S. H. (2010). A Survey of Indoor Positioning and Object LocatingvSystems. Int. J. Computer. Sci. Network. *Secur.*, *10*, 121–128.

Krishnan, O., & Suhag, S. (2020). Grid-Independent PV System Hybridization With Fuel Cell-Battery/Super Capacitor: Optimum Sizing And Comparative Techno-Economic Analysis. *Sustainable Energy Technologies and Assessments*, *37*.

Kühl, N., Schemmer, M., Goutier, M., & Satzger, G. (2022). Artificial intelligence and machine learning. *Electronic Markets*, *32*(4), 2235–2244. doi:10.1007/s12525-022-00598-0

Kumar Reddy, R. V., Rahamathunnisa, U., Subhashini, P., Aancy, H. M., Meenakshi, S., & Boopathi, S. (2023). Solutions for Software Requirement Risks Using Artificial Intelligence Techniques: In T. Murugan & N. E. (Eds.), Advances in Information Security, Privacy, and Ethics (pp. 45–64). IGI Global. doi:10.4018/978-1-6684-8145-5.ch003

Kumar, A., & Choudhary, J. (2023). Power quality improvement of hybrid renewable energy systems-based microgrid for statcom: Hybrid-deep-learning model and mexican axoltl dingo optimizer (MADO). *Engineering Research Express*, *5*(4), 045031. doi:10.1088/2631-8695/ad0287

Kumara, V., Mohanaprakash, T., Fairooz, S., Jamal, K., Babu, T., & Sampath, B. (2023). Experimental Study on a Reliable Smart Hydroponics System. In *Human Agro-Energy Optimization for Business and Industry* (pp. 27–45). IGI Global. doi:10.4018/978-1-6684-4118-3.ch002

Kumar, J., & Bhushan, G. (2023). Dynamic analysis of quarter car model with semi-active suspension based on combination of magneto-rheological materials. *International Journal of Dynamics and Control, 11*(2), 482–490. doi:10.1007/s40435-022-01024-1

Kumar, Kaliappan, S., Socrates, S., Natrayan, L., Patel, P. B., Patil, P. P., Sekar, S., & Mammo, W. D. (2022). Investigation of Mechanical and Thermal Properties on Novel Wheat Straw and PAN Fibre Hybrid Green Composites. *International Journal of Chemical Engineering, 2022*, 1–8. doi:10.1155/2022/3598397

Kumar, L., & Bharadvaja, N. (2020). A review on microalgae biofuel and biorefinery: Challenges and way forward. *Energy Sources. Part A, Recovery, Utilization, and Environmental Effects*, 1–24. doi:10.1080/15567036.2020.1836084

Kumar, N. M., Chand, A. A., Malvoni, M., Prasad, K. A., Mamun, K. A., Islam, F., & Chopra, S. S. (2020). Distributed energy resources and the application of AI, IoT, and blockchain in smart grids. *Energies, 13*(21), 5739. doi:10.3390/en13215739

Kumar, P. R., Meenakshi, S., Shalini, S., Devi, S. R., & Boopathi, S. (2023). Soil Quality Prediction in Context Learning Approaches Using Deep Learning and Blockchain for Smart Agriculture. In R. Kumar, A. B. Abdul Hamid, & N. I. Binti Ya'akub (Eds.), (pp. 1–26). Advances in Computational Intelligence and Robotics. IGI Global., doi:10.4018/978-1-6684-9151-5.ch001

Kumar, R., & Channi, H. K. (2022). A PV-Biomass off-grid hybrid renewable energy system (HRES) for rural electrification: Design, optimization and techno-economic-environmental analysis. *Journal of Cleaner Production, 349*, 131347. doi:10.1016/j.jclepro.2022.131347

Kumar, S. (2022). IoT battery management system in electric vehicle based on LR parameter estimation and ORMesh-Net gateway topology. *Sustainable Energy Technologies and Assessments, 53*, 102696. doi:10.1016/j.seta.2022.102696

Kumar, S., Bhattacharyya, B., & Gupta, V. K. (2014). Present and Future Energy Scenario in India. *J. Inst. Eng. India Ser. B, 95*(3), 247–254. doi:10.1007/s40031-014-0099-7

Kumar, V., & Rana, K. P. S. (2023). A novel fuzzy PID controller for nonlinear active suspension system with an electro-hydraulic actuator. *Journal of the Brazilian Society of Mechanical Sciences and Engineering, 45*(4), 189. doi:10.1007/s40430-023-04095-z

Kundur, P. (1994). *Power System Stability and Control*. McGraw-Hill.

Kuo, C. F. J., Lin, C. H., & Hsu, M. W. (2016). Analysis of intelligent green building policy and developing status in Taiwan. *Energy Policy, 95*, 291–303. doi:10.1016/j.enpol.2016.04.046

Kuo, C. F. J., Lin, C. H., Hsu, M. W., & Li, M. H. (2017). Evaluation of intelligent green building policies in Taiwan–Using fuzzy analytic hierarchical process and fuzzy transformation matrix. *Energy and Building, 139*, 146–159. doi:10.1016/j.enbuild.2016.12.078

Kusakana, K. (2015). *Operation cost minimization of photovoltaic–diesel–battery hybrid systems* (Vol. 85).

Laagoubi, T., & Bouzi, M. (2018). Supervising PV/Battery/Diesel System Connected to Grid Using Fuzzy Logic. *6th International Renewable and Sustainable Energy Conference*. IEEE.

Lagrange, A., de Simon-Martín, M., González-Martínez, A., Bracco, S., & Rosales-Asensio, E. (2020). Sustainable micro grids with energy storage as a means to increase power resilience in critical facilities: An application to a hospital. *International Journal of Electrical Power & Energy Systems, 119*, 105865.

Lakshmaiya, N. (2023) Experimental investigation on computational volumetric heat in real time neural pathways. In: *Proceedings of SPIE*. The International Society for Optical Engineering 10.1117/12.2675555

Lakshmaiya, N. (2023) Investigation on ultraviolet radiation of flow pattern and particles transportation in vanishing raindrops. In: *Proceedings of SPIE*. The International Society for Optical Engineering 10.1117/12.2675556

Lakshmaiya, N., Ganesan, V., Paramasivam, P., & Dhanasekaran, S. (2022). Influence of Biosynthesized Nanoparticles Addition and Fibre Content on the Mechanical and Moisture Absorption Behaviour of Natural Fibre Composite. *Applied Sciences (Basel, Switzerland)*, *12*(24), 13030. doi:10.3390/app122413030

Lakshmaiya, N., Kaliappan, S., Patil, P. P., Ganesan, V., Dhanraj, J. A., Sirisamphanwong, C., Wongwuttanasatian, T., Chowdhury, S., Channumsin, S., Channumsin, M., & Techato, K. (2022b). Influence of Oil Palm Nano Filler on Interlaminar Shear and Dynamic Mechanical Properties of Flax/Epoxy-Based Hybrid Nanocomposites under Cryogenic Condition. *Coatings*, *12*(11), 1675. Advance online publication. doi:10.3390/coatings12111675

Lakshmaiya, N., Surakasi, R., Nadh, V. S., Srinivas, C., Kaliappan, S., Ganesan, V., Paramasivam, P., & Dhanasekaran, S. (2023). Tanning Wastewater Sterilization in the Dark and Sunlight Using Psidium guajava Leaf-Derived Copper Oxide Nanoparticles and Their Characteristics. *ACS Omega*, *8*(42), 39680–39689. doi:10.1021/acsomega.3c05588 PMID:37901496

Lalit, K. (2019). India projected to be on track to achieve Paris climate agreement target: US expert. *The economic times*. https://energy.economictimes.indiatimes.com/news/renewable/india-projected-to-be-on-track-to-achieve-paris-climate-agree ment-target-us-expert/68229428

Landi, D. (2019). Interactive Energetic, Environmental and Economic Analysis of Renewable Hybrid Energy System. *International Journal of Interactive Design and Manufacturing*.

Larico, E., & Canales, A. (2022). Solar Tracking System with PV Cells: Experimental Analysis at High Altitudes. *International Journal of Renewable Energy Development*, *11*(3), 630–639. doi:10.14710/ijred.2022.43572

Lee, S., Tewolde, G., & Kwon, J. (2014). Design and implementation of vehicle tracking system using GPS/GSM/GPRS technology and smartphone application. *Proceedings of the IEEE World Forum on Internet of Things (WF-IoT '14)*. IEEE.

Lee, C. C. (1990, March-April). Fuzzy logic in control systems: Fuzzy logic controller. I. *IEEE Transactions on Systems, Man, and Cybernetics*, *20*(2), 404–418. doi:10.1109/21.52551

Lee, D., Koo, S., Jang, I., & Kim, J. (2022). Comparison of Deep Reinforcement Learning and PID Controllers for Automatic Cold Shutdown Operation. *Energies*, *15*(8), 2834. doi:10.3390/en15082834

Lee, J., Lee, D., Lee, J., Yoon, M., & Jang, G. (2023). Offshore MTDC Transmission Expansion for Renewable Energy Scale-up in Korean Power System: DC Highway. *Journal of Electrical Engineering & Technology*, *18*(4), 2483–2493. doi:10.1007/s42835-023-01513-z PMID:37362030

Lee, K., & Kum, D. (2019). Complete design space exploration of isolated hybrid renewable energy system via dynamic programming. *Energy Conversion and Management*, *196*, 920–934.

Lee, M. F. R., & Yusuf, S. H. (2022). Mobile Robot Navigation Using Deep Reinforcement Learning. *Processes (Basel, Switzerland)*, *10*(12), 2748. doi:10.3390/pr10122748

Leonori, S., Martino, A., Mascioli, F. M. F., & Rizzi, A. (2020). Microgrid energy management systems design by computational intelligence techniques. *Applied Energy*, *277*, 115524. doi:10.1016/j.apenergy.2020.115524

Li, C. (2019). *Techno- economic performance study of stand-alone wind/diesel/battery hybrid system with different battery technologies in the cold region of China*. Research Gate.

Li, M., Zhang, Y., & You, D. (2020). Design of fuzzy PID stepping motor controller based on particle swarm optimization. *2020 3rd World Conference on Mechanical Engineering and Intelligent Manufacturing (WCMEIM)*, (pp. 449-453). IEEE. 10.1109/WCMEIM52463.2020.00100

Li, C. Y., Shen, Y., Wu, J. Y., Dai, Y. J., & Chi, H. W. (2019). Experimental and modelling investigation of an integrated biomass gasifier-engine-generator system for power generation and waste heat recovery. *Energy Conversion and Management, 199.*

Li, F., & Xu, G. (2022). AI-driven customer relationship management for sustainable enterprise performance. *Sustainable Energy Technologies and Assessments, 52*, 102103. doi:10.1016/j.seta.2022.102103

Li, H., Zhang, G., Ma, R., & You, Z. (2014). Design and Experimental Evaluation on an Advanced Multisource Energy Harvesting System for Wireless Sensor Nodes. *TheScientificWorldJournal, 671280*, 1–13. doi:10.1155/2014/671280 PMID:25032233

Li, J., & Chen, C. (2023). Machine learning-based energy harvesting for wearable exoskeleton robots. *Sustainable Energy Technologies and Assessments, 57*, 103122. doi:10.1016/j.seta.2023.103122

Li, J., Ziehm, W., Kimball, J., Landers, R., & Park, J. (2021). Physical-based training data collection approach for data-driven lithium-ion battery state-of-charge prediction. *Energy and AI, 5*, 100094. doi:10.1016/j.egyai.2021.100094

Lillicrap, T. P., Hunt, J. J., Pritzel, A., Heess, N., Erez, T., Tassa, Y., Silver, D., & Wierstra, D. (2016). Continuous control with deep reinforcement learning. *4th International Conference on Learning Representations, ICLR 2016 - Conference Track Proceedings*. IEEE.

Lim, H. Y., Rashidi, N. A., Othman, M. F. H., Ismail, I. S., Saadon, S. Z. A. H., Chin, B. L. F., Yusup, S., & Rahman, M. N. (2023). Recent advancement in thermochemical conversion of biomass to biofuel. *Biofuels*, 1–18. doi:10.1080/17597269.2023.2261788

Lin, X., Wu, J., & Wei, Y. (2021). An ensemble learning velocity prediction-based energy management strategy for a plug-in hybrid electric vehicle considering driving pattern adaptive reference SOC. *Energy, 234*, 121308. doi:10.1016/j.energy.2021.121308

Li, R., Wang, H., Dai, H., Hong, J., Tong, G., & Chen, X. (2022). Accurate state of charge prediction for real-world battery systems using a novel dual-dropout-based neural network. *Energy, 250*, 123853. doi:10.1016/j.energy.2022.123853

Liu, T., Xu, C., Chen, H., & Li, Z. (2020). Study on deep Reinforcement learning techniques for building Power consumption forecasting, Power & Buildings. *Energy and Buildings, 208.* doi:10.1016/j.enbuild.2019.109675

Liu, Z., Chi, Z., Osmani, M., & Demian, P. B. (2021). Building Information Management (BIM) for Sustainable Building Development within the Context of Smart Cities. *Sustainability, 13.*

Liu, C., Lee, S., Varnhagen, S., & Tseng, H. E. (2017). Path planning for autonomous vehicles using model predictive control. *IEEE Intelligent Vehicles Symposium, Proceedings*. IEEE. 10.1109/IVS.2017.7995716

Liu, J., Wang, Y., Li, B., & Ma, S. (2007). Current research, key performances and future development of search and rescue robots. *Frontiers of Mechanical Engineering in China, 2*(4), 404–416. doi:10.1007/s11465-007-0070-2

Liu, Y., Liu, W., Gao, S., Wang, Y., & Shi, Q. (2022). Fast charging demand forecasting based on the intelligent sensing system of dynamic vehicle under EVs-traffic-distribution coupling. *Energy Reports, 8*, 1218–1226. doi:10.1016/j.egyr.2022.02.261

Liu, Z., Chi, Z., Osmani, M., & Demian, P. (2021). Blockchain and building information management (BIM) for sustainable building development within the context of smart cities. *Sustainability (Basel), 13*(4), 2090. doi:10.3390/su13042090

Li, Y., Wen, Y., Tao, D., & Guan, K. (2019). Transforming cooling optimization for green data center via deep reinforcement learning. *IEEE Transactions on Cybernetics*, *50*(5), 2002–2013. doi:10.1109/TCYB.2019.2927410 PMID:31352360

Loganathan, Ramachandran, V., Perumal, A. S., Dhanasekaran, S., Lakshmaiya, N., & Paramasivam, P. (2023). Framework of Transactive Energy Market Strategies for Lucrative Peer-to-Peer Energy Transactions. *Energies*, *16*(1), 6. doi:10.3390/en16010006

Lopez-Fuentes, L., van de Weijer, J., González-Hidalgo, M., Skinnemoen, H., & Bagdanov, A. D. (2018). Review on computer vision techniques in emergency situations. *Multimedia Tools and Applications*, *77*(13), 17069–17107. doi:10.1007/s11042-017-5276-7

Loy, A. C. M., Kong, K. G. H., Lim, J. Y., & How, B. S. (2023). Frontier of digitalization in Biomass-to-X supply chain: Opportunity or threats? *Journal of Bioresources and Bioproducts*.

Luna, A. C., Diaz, N. L., Graells, M., Vasquez, J. C., & Guerrero, J. M. (2017). Mixed-Integer-Linear-Programming-Based Energy Management System for Hybrid PV-Wind-Battery Micro grids: Modelling, Design, and Experimental Verification. *IEEE Transactions on Power Electronics*, *32*(4), 2769–2783.

Lundgren, A. V. A., dos Santos, M. A. O., Bezerra, B. L. D., & Bastos-Filho, C. J. A. (2022). Systematic Review of Computer Vision Semantic Analysis in Socially Assistive Robotics. In AI (Switzerland) (Vol. 3, Issue 1). doi:10.3390/ai3010014

Lygouras, E., Santavas, N., Taitzoglou, A., Tarchanidis, K., Mitropoulos, A., & Gasteratos, A. (2019). Unsupervised human detection with an embedded vision system on a fully autonomous UAV for search and rescue operations. *Sensors (Basel)*, *19*(16), 3542. doi:10.3390/s19163542 PMID:31416131

Lyu, M., Zhao, Y., Huang, C., & Huang, H. (2023). Unmanned Aerial Vehicles for Search and Rescue: A Survey. In Remote Sensing, 15(13). doi:10.3390/rs15133266

Macek, K., Petrovic, I., & Peric, N. (2002). A reinforcement learning approach to obstacle avoidance of mobile robots. *7th International Workshop on Advanced Motion Control*, (pp. 462–466). IEEE. 10.1109/AMC.2002.1026964

Maddi, D., Sheta, A., Davineni, D., & Al-Hiary, H. (2019). Optimization of PID Controller Gain Using Evolutionary Algorithm and Swarm Intelligence. *2019 10th International Conference on Information and Communication Systems (ICICS)*, (pp. 199-204). IEEE. 10.1109/IACS.2019.8809144

Madupalli, S., Vasugi, K., Kumar, R., & Natrayan, L. (2019). Structural performance of non-linear analysis of turbo generator building using seismic protection techniques. *International Journal of Recent Technology and Engineering*, *8*(1).

Maguluri, L. P., Ananth, J., Hariram, S., Geetha, C., Bhaskar, A., & Boopathi, S. (2023). Smart Vehicle-Emissions Monitoring System Using Internet of Things (IoT). In Handbook of Research on Safe Disposal Methods of Municipal Solid Wastes for a Sustainable Environment (pp. 191–211). IGI Global.

Maguluri, L. P., Ananth, J., Hariram, S., Geetha, C., Bhaskar, A., & Boopathi, S. (2023). Smart Vehicle-Emissions Monitoring System Using Internet of Things (IoT). In P. Srivastava, D. Ramteke, A. K. Bedyal, M. Gupta, & J. K. Sandhu (Eds.), (pp. 191–211). Practice, Progress, and Proficiency in Sustainability. IGI Global. doi:10.4018/978-1-6684-8117-2.ch014

Maguluri, L. P., Arularasan, A. N., & Boopathi, S. (2023). Assessing Security Concerns for AI-Based Drones in Smart Cities. In R. Kumar, A. B. Abdul Hamid, & N. I. Binti Ya'akub (Eds.), (pp. 27–47). Advances in Computational Intelligence and Robotics. IGI Global. doi:10.4018/978-1-6684-9151-5.ch002

Maharana, D., Kommadath, R., & Kotecha, P. (2023). An innovative approach to the supply-chain network optimization of biorefineries using metaheuristic techniques. *Engineering Optimization*, 55(8), 1278–1295. doi:10.1080/03052 15X.2022.2080204

Mahesh, A. (2016). *Optimal Sizing of a Grid-Connected PV/Wind/Battery System Using Particle Swarm Optimization*. Springer.

Mahesha, C. R., Rani, G. J., Dattu, V. S. N. C. H., Rao, Y. K. S. S., Madhusudhanan, J., L, N., Sekhar, S. C., & Sathyamurthy, R. (2022). Optimization of transesterification production of biodiesel from Pithecellobium dulce seed oil. *Energy Reports*, 8, 489–497. doi:10.1016/j.egyr.2022.10.228

Maheswari, B. U., Imambi, S. S., Hasan, D., Meenakshi, S., Pratheep, V., & Boopathi, S. (2023). Internet of Things and Machine Learning-Integrated Smart Robotics. In Global Perspectives on Robotics and Autonomous Systems: Development and Applications (pp. 240–258). IGI Global. doi:10.4018/978-1-6684-7791-5.ch010

Maleki, A., Ameri, M., & Keynia, F. (2015). Scrutiny of multifarious particle swarm optimization for finding the optimal size of a PV/wind/battery hybrid system. Research Gate.

Mamodiya, U., & Tiwari, N. (2023). Dual-axis solar tracking system with different control strategies for improved energy efficiency. *Computers & Electrical Engineering*, 111, 108920. doi:10.1016/j.compeleceng.2023.108920

Manna, S., Mani, G., Ghildiyal, S., Stonier, A. A., Peter, G., Ganji, V., & Murugesan, S. (2022). Ant colony optimization tuned closed-loop optimal control intended for vehicle active suspension system. *IEEE Access : Practical Innovations, Open Solutions*, 10, 53735–53745. doi:10.1109/ACCESS.2022.3164522

Marshall, C., Meisel, Z., Montes, F., Wagner, L., Hermansen, K., Garg, R., Chipps, K. A., Tsintari, P., Dimitrakopoulos, N., Berg, G. P. A., Brune, C., Couder, M., Greife, U., Schatz, H., & Smith, M. S. (2023). Measurement of charge state distributions using a scintillation screen. *Nuclear Instruments & Methods in Physics Research. Section A, Accelerators, Spectrometers, Detectors and Associated Equipment*, 1056, 168661. doi:10.1016/j.nima.2023.168661

Marzbanrad, J., Poozesh, P., & Damroodi, M. (2013). Improving vehicle ride comfort using an active and semi-active controller in a half-car model. *Journal of Vibration and Control*, 19(9), 1357–1377. doi:10.1177/1077546312441814

Ma, S., Li, Y., & Tong, S. (2023). Research on control strategy of seven-DOF vehicle active suspension system based on co-simulation. *Measurement and Control*, 56(7-8), 00202940231154954. doi:10.1177/00202940231154954

Masoud, U. M. M., Tiwari, P., & Gupta, N. (2023). Designing of an Enhanced Fuzzy Logic Controller of an Interior Permanent Magnet Synchronous Generator under Variable Wind Speed. *Sensors (Basel)*, 23(7), 3628. doi:10.3390/ s23073628 PMID:37050688

Matt, S., Echt, O., Wisrgistter, R., Grill, V., Scheier, P., Lifshitz, C., & Miirk, T. D. (1997). Appearance and ionization energies of multiply-charged C70 parent ions produced by electron impact ionization. In Chemical Physics Letters, 264. doi:10.1016/S0009-2614(96)01303-6

Mautz, R. (2009). Overview of Current Indoor Positioning Systems. *Geodesy and Cartography (Vilnius)*, 35(1), 18–22. doi:10.3846/1392-1541.2009.35.18-22

Ma, Y., Li, B., Xie, Y., & Chen, H. (2016). Estimating the State of Charge of Lithium-ion Battery based on Sliding Mode Observer. *IFAC-PapersOnLine*, 49(11), 54–61. doi:10.1016/j.ifacol.2016.08.009

Meena, C. (2022). Innovation in Green Building Sector for Sustainable Future. *Energies*, 15(18).

Mehmet, S., & Erol, R. (2017). A comparative study of neural networks and ANFIS for forecasting attendance rate of soccer games. *Mathematical & Computational Applications*, 43(4), 22. doi:10.3390/mca22040043

Mel Keytingan, M. (2021). Energy consumption prediction by using machine learning for smart building: Case study in Malaysia. *Developments in the Built Environment, 5*. doi:10.1016/j.dibe.2020.100037

Merneedi, A., Natrayan, L., Kaliappan, S., Veeman, D., Angalaeswari, S., Srinivas, C., & Paramasivam, P. (2021). Experimental Investigation on Mechanical Properties of Carbon Nanotube-Reinforced Epoxy Composites for Automobile Application. *Journal of Nanomaterials, 2021*, 1–7. doi:10.1155/2021/4937059

Mihai, I., & Andronic, F. (2014). Behavior of a semi-active suspension system versus a passive suspension system on an uneven road surface. *Mechanics, 20*(1), 64–69. doi:10.5755/j01.mech.20.1.6591

Mitra, A. C., Desai, G. J., Patwardhan, S. R., Shirke, P. H., Kurne, W. M., & Banerjee, N. (2016). Optimization of passive vehicle suspension system by genetic algorithm. *Procedia Engineering, 144*, 1158–1166. doi:10.1016/j.proeng.2016.05.087

Mitran, T. L., & Nemnes, G. A. (2021). Ground state charge density prediction in C-BN nanoflakes using rotation equivariant feature-free artificial neural networks. *Carbon, 174*, 276–283. doi:10.1016/j.carbon.2020.12.048

Mo Khin, J. M., & Nyein Oo, D. N. (2018). —Real-Time Vehicle Tracking System Using Arduino, GPS, GSM and Web-Based Technologies,‖. *International Journal of Science and Engineering Applications, 7*(11), 433–436. doi:10.7753/IJSEA0711.1006

Moazamigoodarzi, H., Tsai, P. J., Pal, S., Ghosh, S., & Puri, I. K. (2019). Influence of cooling architecture on data center power consumption. *Energy, 183*, 525–535. doi:10.1016/j.energy.2019.06.140

Mohammad, A., & Mahjabeen, F. (2023a). Revolutionizing Solar Energy: The Impact of Artificial Intelligence on Photovoltaic Systems. *International Journal of Multidisciplinary Sciences and Arts, 2*(1).

Mohammad, A., & Mahjabeen, F. (2023b). Revolutionizing Solar Energy with AI-Driven Enhancements in Photovoltaic Technology. *BULLET: Jurnal Multidisiplin Ilmu, 2*(4), 1174–1187.

Mohanasundaram, K., Sugavanam, K. R., & Senthilkumar, R. A. (2014). PSO algorithm based pi controller design for soft starting of induction motor. *International Journal of Applied Engineering Research: IJAER, 9*(24), 25535–25542.

Mohanty, M., & Sarkar, R. (2024). *The Role of Coal in a Sustainable Energy Mix for India: A Wide-Angle View.*

Mohanty, A., Jothi, B., Jeyasudha, J., Ranjit, P. S., Isaac, J. S., & Boopathi, S. (2023). Additive Manufacturing Using Robotic Programming. In S. Kautish, N. K. Chaubey, S. B. Goyal, & P. Whig (Eds.), (pp. 259–282). Advances in Computational Intelligence and Robotics. IGI Global. doi:10.4018/978-1-6684-8171-4.ch010

Mohanty, A., Venkateswaran, N., Ranjit, P. S., Tripathi, M. A., & Boopathi, S. (2023). Innovative Strategy for Profitable Automobile Industries: Working Capital Management. In Y. Ramakrishna & S. N. Wahab (Eds.), (pp. 412–428). Advances in Finance, Accounting, and Economics. IGI Global. doi:10.4018/978-1-6684-7664-2.ch020

Mohanty, A., Viswavandya, M., Mohanty, S. P., & Paramita, P. (2015). Optimization and Improvement of Voltage Stability in a Stand-alone Wind-Diesel- Micro hydro Hybrid System. *Procedia Technology, 21*, 332–337.

Mohanty, S. (2010). *Green technology in construction. Recent Advances in Space Technology Services and Climate Change 2010 (RSTS & CC-2010).* IEEE.

Moharm, K. (2019). State of the art in big data applications in microgrid: A review. *Advanced Engineering Informatics, 42*, 100945. doi:10.1016/j.aei.2019.100945

Mohtashami, N., Karuvingal, R., Droste, K., Schreiber, T., Streblow, R., & Müller, D. (2023, November). How to build green substations? An LCA comparison of different sustainable design strategies for substations. []. IOP Publishing.]. *Journal of Physics: Conference Series, 2600*(15), 152022. doi:10.1088/1742-6596/2600/15/152022

Moreau, J., Melchior, P., Victor, S., Moze, M., Aioun, F., & Guillemard, F. (2019). Reactive path planning for autonomous vehicle using bézier curve optimization. *IEEE Intelligent Vehicles Symposium, Proceedings, 2019-June*. 10.1109/IVS.2019.8813904

Mostafa, M., Vorwerk, D., Heise, J., Povel, A., Sanina, N., Babazadeh, D., & Toebermann, C. (2022, September). Integrated Planning of Multi-energy Grids: Concepts and Challenges. In *NEIS 2022; Conference on Sustainable Energy Supply and Energy Storage Systems* (pp. 1-7). VDE.

Múčka, P. (2017). Simulated road profiles according to ISO 8608 in vibration analysis. *Journal of Testing and Evaluation, 46*(1), 405–418. doi:10.1520/JTE20160265

Muralidaran, Natrayan, L., Kaliappan, S., & Patil, P. P. (2023). Grape stalk cellulose toughened plain weaved bamboo fiber-reinforced epoxy composite: Load bearing and time-dependent behavior. *Biomass Conversion and Biorefinery*. doi:10.1007/s13399-022-03702-8

Murphy Robin R. & Tadokoro, S. (2008). Search and Rescue Robotics. In O. Siciliano Bruno and Khatib (Ed.), *Springer Handbook of Robotics* (pp. 1151–1173). Springer Berlin Heidelberg. doi:10.1007/978-3-540-30301-5_51

Murphy, A. J., Landamore, M. J., & Birmingham, R. W. (2008). The role of autonomous underwater vehicles for marine search and rescue operations. *Underwater Technology, 27*(4), 195–205. Advance online publication. doi:10.3723/ut.27.195

Murugaperumal, K., Srinivasn, S., & Prasad, G. R. K. D. S. (2020). Optimum Design of Hybrid Renewable Energy System Through Load Forecasting And Different Operating Strategies For Rural Electrification. *Sustainable Energy Technologies and Assessments, 37*.

Nadh, V. S., Krishna, C., Natrayan, L., Kumar, K. M., Nitesh, K. J. N. S., Raja, G. B., & Paramasivam, P. (2021). Structural Behavior of Nanocoated Oil Palm Shell as Coarse Aggregate in Lightweight Concrete. *Journal of Nanomaterials, 2021*, 1–7. doi:10.1155/2021/4741296

Nagajothi, S., Elavenil, S., Angalaeswari, S., Natrayan, L., & Mammo, W. D. (2022a). Durability Studies on Fly Ash Based Geopolymer Concrete Incorporated with Slag and Alkali Solutions. *Advances in Civil Engineering, 2022*, 1–13. Advance online publication. doi:10.1155/2022/7196446

Nagajothi, S., Elavenil, S., Angalaeswari, S., Natrayan, L., & Paramasivam, P. (2022b). Cracking Behaviour of Alkali-Activated Aluminosilicate Beams Reinforced with Glass and Basalt Fibre-Reinforced Polymer Bars under Cyclic Load. *International Journal of Polymer Science, 2022*, 1–13. Advance online publication. doi:10.1155/2022/6762449

Nagarajan, Rajagopalan, A., Angalaeswari, S., Natrayan, L., & Mammo, W. D. (2022). Combined Economic Emission Dispatch of Microgrid with the Incorporation of Renewable Energy Sources Using Improved Mayfly Optimization Algorithm. *Computational Intelligence and Neuroscience, 2022*, 1–22. doi:10.1155/2022/6461690 PMID:35479598

Namor, E., Sossan, F., Cherkaoui, R., & Paolone, M. (2019). Control of Battery Storage Systems for the Simultaneous Provision of Multiple Services. *IEEE Transactions on Smart Grid, 10*(3), 2799–2808. doi:10.1109/TSG.2018.2810781

Nandi, M., Shiva, C. K., & Mukherjee, V. (2019). Moth-Flame Algorithm for TCSC and SMES-Based Controller Design in Automatic Generation Control of a Two-area Multi-unit Hydro-power system. *Iranian Journal of Science and Technology. Transaction of Electrical Engineering*.

Narayanan, S. S. S., & Thangavel, S. (2023). A novel static model prediction method based on machine learning for Li-ion batteries operated at different temperatures. *Journal of Energy Storage, 61*, 106789. doi:10.1016/j.est.2023.106789

Nativel, G., Jacquemart, Y., Sermanson, V., & Gault, J. C. (1999). *Implementation of a voltage stability analysis tool using quasisteady- state time simulation*. Proc 13th PSCC, Trondheim, Norway.

Natrayan, L., & Kaliappan, S. (2023) Mechanical Assessment of Carbon-Luffa Hybrid Composites for Automotive Applications. In: SAE Technical Papers. doi:10.4271/2023-01-5070

Natrayan, L., Amalesh, T., & Syed, S. (2019a). Design and performance analysis of low speed vertical axis windmill. *International Journal of Recent Technology and Engineering, 8*(1).

Natrayan, L., Kaliappan, S., & Pundir, S. (2023a) Control and Monitoring of a Quadcopter in Border Areas Using Embedded System. In: *Proceedings of the 4th International Conference on Smart Electronics and Communication, ICOSEC 2023.* IEEE. 10.1109/ICOSEC58147.2023.10276196

Natrayan, L., Kaliappan, S., Saravanan, A., Vickram, A. S., Pravin, P., Abbas, M., Ahamed Saleel, C., Alwetaishi, M., & Saleem, M. S. M. (2023b). Recyclability and catalytic characteristics of copper oxide nanoparticles derived from bougainvillea plant flower extract for biomedical application. *Green Processing and Synthesis, 12*(1), 20230030. doi:10.1515/gps-2023-0030

Natrayan, L., Kaliappan, S., Sethupathy, B. S., Sekar, S., Patil, P. P., Velmurugan, G., & Tariku Olkeba, T. (2022). Effect of Mechanical Properties on Fibre Addition of Flax and Graphene-Based Bionanocomposites. *International Journal of Chemical Engineering, 2022,* 1–8. doi:10.1155/2022/5086365

Natrayan, L., & Kumar, M. S. (2019). Optimization of tribological behaviour on squeeze cast al6061/al2o3/sic/gr hmmcs based on taguchi methodandartificial neural network. *Journal of Advanced Research in Dynamical and Control Systems, 11*(7).

Natrayan, L., Merneedi, A., Bharathiraja, G., Kaliappan, S., Veeman, D., & Murugan, P. (2021). Processing and characterization of carbon nanofibre composites for automotive applications. *Journal of Nanomaterials, 2021,* 1–7. doi:10.1155/2021/7323885

Natrayan, L., Sakthi Shunmuga Sundaram, P., & Elumalai, J. (2019b). Analyzing the uterine physiological with MMG signals using SVM. *International Journal of Pharmaceutical Research, 11*(2). Advance online publication. doi:10.31838/ijpr/2019.11.02.009

Natrayan, L., & Senthil Kumar, M. (2018). Study on Squeeze Casting of Aluminum Matrix Composites—. *RE:view.*

Natrayan, L., & Senthil Kumar, M. (2019). Mechanical, microstructure and wear behaviour of lm25/sic/mica metal matrix composite fabricated by squeeze casting technique. *Applied Engineering Letters, 4*(2), 72–77. doi:10.18485/aeletters.2019.4.2.5

Natrayan, L., Sivaprakash, V., & Santhosh, M. S. (2018). Mechanical, microstructure and wear behavior of the material aa6061 reinforced sic with different leaf ashes using advanced stir casting method. *International Journal of Engineering and Advanced Technology, 8.*

Nefraoui, A., Kandoussi, K., Louzazni, M., Boutahar, A., Elotmani, R., & Daya, A. (2023). Optimal battery state of charge parameter estimation and forecasting using non-linear autoregressive exogenous. *Materials Science for Energy Technologies, 6,* 522–532. doi:10.1016/j.mset.2023.05.003

Nejad, S., Gladwin, D. T., & Stone, D. A. (2016). A systematic review of lumped-parameter equivalent circuit models for real-time estimation of lithium-ion battery states. In *Journal of Power Sources* (Vol. 316, pp. 183–196). Elsevier B.V., doi:10.1016/j.jpowsour.2016.03.042

Nguyen, D. N., & Nguyen, T. A. (2023). Proposing an original control algorithm for the active suspension system to improve vehicle vibration: Adaptive fuzzy sliding mode proportional-integral-derivative tuned by the fuzzy (AFSPIDF). *Heliyon, 9*(3), e14210. doi:10.1016/j.heliyon.2023.e14210 PMID:36915482

Nishanth, J., Deshmukh, M. A., Kushwah, R., Kushwaha, K. K., Balaji, S., & Sampath, B. (2023). Particle Swarm Optimization of Hybrid Renewable Energy Systems. In *Intelligent Engineering Applications and Applied Sciences for Sustainability* (pp. 291–308). IGI Global. doi:10.4018/979-8-3693-0044-2.ch016

Niveditha, V. R., & Rajakumar, P. S. (2020). Pervasive computing in the context of COVID-19 prediction with AI-based algorithms. *International Journal of Pervasive Computing and Communications*, *16*(5). doi:10.1108/IJPCC-07-2020-0082

North American Electric Reliability Corporation (NERC). (2019). *Improvements to interconnection requirements for BPS-connected inverter-based resources.* Reliability Guideline.

Nunes, L. J. (2023). Exploring the present and future of biomass recovery units: Technological innovation, policy incentives and economic challenges. *Biofuels*, ●●●, 1–13.

Okoye, C., Bahrami, A., & Atikol, U. (2017). Evaluating the solar resource potential on different tracking surfaces in Nigeria. *Renewable & Sustainable Energy Reviews*, *81*, 1569–1581. doi:10.1016/j.rser.2017.05.235

Oladayo, B. O., & Titus, A. O. (2016). Development of STS Using IMC-PID Controller [AJER]. *American Journal of Engineering Research*, *5*(5).

Oró, E., Depoorter, V., Garcia, A., & Salom, J. (2015). Energy efficiency and renewable energy integration in data centres. Strategies and modelling review. *Renewable & Sustainable Energy Reviews*, *42*, 429–445. doi:10.1016/j.rser.2014.10.035

Osorio, J. D., Wang, Z., Karniadakis, G., Cai, S., Chryssostomidis, C., Panwar, M., & Hovsapian, R. (2022). Forecasting solar-thermal systems performance under transient operation using a data-driven machine learning approach based on the deep operator network architecture. *Energy Conversion and Management*, *252*, 115063. doi:10.1016/j.enconman.2021.115063

Oyewole, O. L., Nwulu, N. I., & Okampo, E. J. (2024). Optimal design of hydrogen-based storage with a hybrid renewable energy system considering economic and environmental uncertainties. *Energy Conversion and Management*, *300*, 117991. doi:10.1016/j.enconman.2023.117991

Ozcan, D., Sonmez, U., Guvenc, L., Ersolmaz, S. S., & Eyol, I. Y. (2023). *Optimisation of Nonlinear Spring and Damper Characteristics for Vehicle Ride and Handling Improvement.* arXiv preprint arXiv:2306.08222.

Ozdenizci, B., Ok, K., Coskun, V., & Aydin, M. N. (2011). Development of an IndoorvNavigation System Using A9G Technology. In *Proceedings of the 4th International Conference on Information and Computing*, Phuket Island, Thailand.

Pachauri, N. (2020). Water cycle algorithm-based PID controller for AVR. COMPEL. *The international journal for computation and mathematics in electrical and electronic engineering.* . doi:10.1108/COMPEL-01-2020-0057

Pachiappan, K., Anitha, K., Pitchai, R., Sangeetha, S., Satyanarayana, T. V. V., & Boopathi, S. (2023). Intelligent Machines, IoT, and AI in Revolutionizing Agriculture for Water Processing. In B. B. Gupta & F. Colace (Eds.), (pp. 374–399). Advances in Computational Intelligence and Robotics. IGI Global. doi:10.4018/978-1-6684-9999-3.ch015

Palermo, S. A., Talarico, V. C., & Pirouz, B. (2019). Optimizing rainwater harvesting systems for non-potable water uses and surface runoff mitigation. *International Conference on Numerical Computations: Theory and Algorithms.* Springer, Cham.

Palomino-Resendiz, S. I., Ortiz-Martínez, F. A., Paramo-Ortega, I. V., González-Lira, J. M., & Flores-Hernández, D. A. (2023). Optimal Selection of the Control Strategy for Dual-Axis Solar Tracking Systems. *IEEE Access : Practical Innovations, Open Solutions*, *11*, 56561–56573. doi:10.1109/ACCESS.2023.3283336

Panda, S., Mohanty, S., Rout, P. K., Sahu, B. K., Parida, S. M., Kotb, H., Flah, A., Tostado-Véliz, M., Abdul Samad, B., & Shouran, M. (2022). An Insight into the Integration of Distributed Energy Resources and Energy Storage Systems with Smart Distribution Networks Using Demand-Side Management. *Applied Sciences (Basel, Switzerland), 12*(17), 8914. doi:10.3390/app12178914

Pan, J., Jain, R., & Paul, S. (2013). Nine lessons learned from a green building testbed: A networking and energy efficiency perspective. *2013 World Congress on Sustainable Technologies (WCST)*. IEEE. 10.1109/WCST.2013.6750405

Parasher, Y., Singh, P., & Kaur, G. (2019). Green Smart Town Planning. *Green and Smart Technologies for Smart Cities*, 19-41.

Paredes, J., Saito, C., Abarca, M., & Cuellar, F. (2017). *Study of effects of high-altitude environments on multicopter and fixed-wing UAVs' energy consumption and flight time*. IEEE. doi:10.1109/COASE.2017.8256340

Parrish, B., Heptonstall, P., Gross, R., & Sovacool, B. K. (2020). A systematic review of motivations, enablers and barriers for consumer engagement with residential demand response. *Energy Policy, 138*, 111221. doi:10.1016/j.enpol.2019.111221

Patel, S. (2021). Review on ST and Comparison on Single Axis ST, Dual Axis ST with Fixed Solar PV System. *International Journal of Innovative Research in Electrical, Electronics, Instrumentation and Control Engineering, 9*(6).

Pebrianti, D., Bayuaji, L., Arumgam, Y., Riyanto, I., Syafrullah, M., & Ann Ayop, N. Q. (2019). PID Controller Design for Mobile Robot Using Bat Algorithm with Mutation (BAM). *2019 6th International Conference on Electrical Engineering, Computer Science and Informatics (EECSI)*, (pp. 85-90). IEEE. 10.23919/EECSI48112.2019.8976932

People, V. I. (2011). *12th International Conference on Computer Systems and Technologies*, Vienna, Austria.

Peyghami, S., Palensky, P., & Blaabjerg, F. (2020). An overview on the reliability of modern power electronic based power systems. *IEEE Open Journal of Power Electronics, 1*, 34–50. doi:10.1109/OJPEL.2020.2973926

Pierezan, J., & Coelho, L. D. S. (2018, July). Coyote optimization algorithm: a new metaheuristic for global optimization problems. In *2018 IEEE congress on evolutionary computation (CEC)* (pp. 1-8). IEEE.

Pignata, A., Minuto, F. D., Lanzini, A., & Papurello, D. (2023). A feasibility study of a tube bundle exchanger with phase change materials: A case study. *Journal of Building Engineering, 78*, 107622. doi:10.1016/j.jobe.2023.107622

Pipattanasomporn, M. (2004*). A study of remote area internet access with embedded power generation*. [PHD Thesis, Virginia Polytechnic Institute and State University, Alexandria, Virginia].

Pol, S., Houchens, B. C., Marian, D., & Westergaard, C. (2020). Performance of AeroMINEs for Distributed Wind Energy. In *AIAA Scitech 2020 Forum* (p. 1241). 10.2514/6.2020-1241

Portalo, J. M., González, I., & Calderón, A. J. (2021). Monitoring system for tracking a PV generator in an experimental smart microgrid: An open-source solution. *Sustainability (Basel), 13*(15), 8182. doi:10.3390/su13158182

Prabha, G., & Mohana, K. (2018). Design of feasible energy generation using solar panel and control using an IoT. *IACSIT International Journal of Engineering and Technology, 7*(24), 191–196.

Praburanganathan, S., Sudharsan, N., Bharath Simha Reddy, Y., Naga Dheeraj Kumar Reddy, C., Natrayan, L., & Paramasivam, P. (2022). Force-Deformation Study on Glass Fiber Reinforced Concrete Slab Incorporating Waste Paper. *Advances in Civil Engineering, 2022*, 1–10. doi:10.1155/2022/5343128

Pragadish, N., Kaliappan, S., Subramanian, M., Natrayan, L., Satish Prakash, K., Subbiah, R., & Kumar, T. C. A. (2023). Optimization of cardanol oil dielectric-activated EDM process parameters in machining of silicon steel. *Biomass Conversion and Biorefinery, 13*(15), 14087–14096. doi:10.1007/s13399-021-02268-1

Pragati Tripathi, M. A. (2018). Modelling of Energy Efficient PV-Diesel-Battery Hybrid System. *IEEE 2018 International Conference on Computational and Characterization Techniques in Engineering & Sciences*. IEEE.

Pramanik, P. K. D., Mukherjee, B., Pal, S., Pal, T., & Singh, S. P. (2021). Green smart building: Requisites, architecture, challenges, and use cases. In *Research anthology on environmental and societal well-being considerations in buildings and architecture* (pp. 25–72). IGI Global. doi:10.4018/978-1-7998-9032-4.ch002

Pramila, P., Amudha, S., Saravanan, T., Sankar, S. R., Poongothai, E., & Boopathi, S. (2023). Design and Development of Robots for Medical Assistance: An Architectural Approach. In Contemporary Applications of Data Fusion for Advanced Healthcare Informatics (pp. 260–282). IGI Global.

Pudjianto, D., Ahmed, S., & Strbac, G. (2002). *"Allocation of VArs Support using LP and NLP based Optimal Power Flows", accepted for publication in IEE Proc*. On Generation, Transmission and Distribution.

Radziejowska, A., & Sobotka, B. (2021). Analysis of the social aspect of smart cities development for the example of smart sustainable buildings. *Energies, 14*(14), 4330. doi:10.3390/en14144330

Rahamathunnisa, U., Subhashini, P., Aancy, H. M., Meenakshi, S., Boopathi, S., & ... (2023). Solutions for Software Requirement Risks Using Artificial Intelligence Techniques. In *Handbook of Research on Data Science and Cybersecurity Innovations in Industry 4.0 Technologies* (pp. 45–64). IGI Global.

Rahamathunnisa, U., Sudhakar, K., Murugan, T. K., Thivaharan, S., Rajkumar, M., & Boopathi, S. (2023). Cloud Computing Principles for Optimizing Robot Task Offloading Processes. In S. Kautish, N. K. Chaubey, S. B. Goyal, & P. Whig (Eds.), (pp. 188–211). Advances in Computational Intelligence and Robotics. IGI Global. doi:10.4018/978-1-6684-8171-4.ch007

Rahamathunnisa, U., Sudhakar, K., Padhi, S. N., Bhattacharya, S., Shashibhushan, G., & Boopathi, S. (2023). Sustainable Energy Generation From Waste Water: IoT Integrated Technologies. In A. S. Etim (Ed.), (pp. 225–256). Advances in Human and Social Aspects of Technology. IGI Global. doi:10.4018/978-1-6684-5347-6.ch010

Rahim, N. S. A., Rahman, B. A., Ibrahim, F. A., Ishak, N., & Ayob, A. (2023, September). The Contractors' Perception on the Development of Green Building Projects in Penang. []. IOP Publishing.]. *IOP Conference Series. Earth and Environmental Science, 1238*(1), 012020. doi:10.1088/1755-1315/1238/1/012020

Rajagopalan, Nagarajan, K., Montoya, O. D., Dhanasekaran, S., Kareem, I. A., Perumal, A. S., Lakshmaiya, N., & Paramasivam, P. (2022). Multi-Objective Optimal Scheduling of a Microgrid Using Oppositional Gradient-Based Grey Wolf Optimizer. *Energies, 15*(23), 9024. doi:10.3390/en15239024

Rajagopal, K., & Ponnusamy, L. (2015). Hybrid DEBBO Algorithm for Tuning the Parameters of PID Controller Applied to Vehicle Active Suspension System. *Jordan Journal of Mechanical & Industrial Engineering, 9*(2).

Raj, S., Sajith, A., Sreenikethanam, A., Vadlamani, S., Satheesh, A., Ganguly, A., Rajesh Banu, J., Varjani, S., Gugulothu, P., & Bajhaiya, A. K. (2023). Renewable biofuels from microalgae: Technical advances, limitations and economics. *Environmental Technology Reviews, 12*(1), 18–36. doi:10.1080/21622515.2023.2167126

Rakhshani, E., Mehrjerdi, H., & Iqbal, A. (2019). Hybrid- Wind- Diesel- Battery System Planning Considering Multiple Different Wind Turbine Technologies Installation. *Journal of Cleaner Production*.

Ramachandran, Perumal, A. S., Lakshmaiya, N., Paramasivam, P., & Dhanasekaran, S. (2022). Unified Power Control of Permanent Magnet Synchronous Generator Based Wind Power System with Ancillary Support during Grid Faults. *Energies, 15*(19), 7385. Advance online publication. doi:10.3390/en15197385

Ramaswamy. (2022b). Pear cactus fiber with onion sheath biocarbon nanosheet toughened epoxy composite: Mechanical, thermal, and electrical properties. *Biomass Conversion and Biorefinery*. Advance online publication. doi:10.1007/s13399-022-03335-x

Ramaswamy, R., Gurupranes, S. V., Kaliappan, S., Natrayan, L., & Patil, P. P. (2022a). Characterization of prickly pear short fiber and red onion peel biocarbon nanosheets toughened epoxy composites. *Polymer Composites*, *43*(8), 4899–4908. doi:10.1002/pc.26735

Ramesh, M., Yadav, A. K., & Pathak, P. K. (2021). Intelligent adaptive LFC via power flow management of integrated standalone micro-grid system. *ISA Transactions*, *112*, 234–250. doi:10.1016/j.isatra.2020.12.002 PMID:33303227

Rameshwar, R., Solanki, A., Nayyar, A., & Mahapatra, B. (2020). Green and smart buildings: A key to sustainable global solutions. In *Green Building Management and Smart Automation* (pp. 146–163). IGI Global. doi:10.4018/978-1-5225-9754-4.ch007

Ramudu, K., Mohan, V. M., Jyothirmai, D., Prasad, D., Agrawal, R., & Boopathi, S. (2023). Machine Learning and Artificial Intelligence in Disease Prediction: Applications, Challenges, Limitations, Case Studies, and Future Directions. In Contemporary Applications of Data Fusion for Advanced Healthcare Informatics (pp. 297–318). IGI Global.

Ramudu, K., Mohan, V. M., Jyothirmai, D., Prasad, D. V. S. S. S. V., Agrawal, R., & Boopathi, S. (2023). Machine Learning and Artificial Intelligence in Disease Prediction: Applications, Challenges, Limitations, Case Studies, and Future Directions. In G. S. Karthick & S. Karupusamy (Eds.), (pp. 297–318). Advances in Healthcare Information Systems and Administration. IGI Global. doi:10.4018/978-1-6684-8913-0.ch013

Ramyashree, J., Roja, M. R., & Sivagurunathan, G. (2018). *Rangasamy, Kotteeswaran. "Firefly algorithm based multivariable PID controller design for MIMO process"* (Vol. 7). International Journal of Engineering and Technology. doi:10.14419/ijet.v7i2.31.13394

Raut, N., Chaudhary, P., Patil, H., & Kiran, P. (2024). 5 Understanding the Contribution of Artificial Intelligence. Handbook of Artificial Intelligence Applications for Industrial Sustainability: Concepts and Practical Examples.

Ravisankar, A., Sampath, B., & Asif, M. M. (2023). Economic Studies on Automobile Management: Working Capital and Investment Analysis. In Multidisciplinary Approaches to Organizational Governance During Health Crises (pp. 169–198). IGI Global.

Ravisankar, A., Sampath, B., & Asif, M. M. (2023). Economic Studies on Automobile Management: Working Capital and Investment Analysis. In C. S. V. Negrão, I. G. P. Maia, & J. A. F. Brito (Eds.), (pp. 169–198). Advances in Logistics, Operations, and Management Science. IGI Global. doi:10.4018/978-1-7998-9213-7.ch009

Rawat, A., Jha, S. K., & Kumar, B. (2020). "Position controlling of Sun Tracking System using optimization technique. *Energy Reports*, *6*, 304–309. doi:10.1016/j.egyr.2019.11.079

Razif Hamid, A., Khusairy Azim, A., & Hafizuddin Bakar, M. (2017). A review on Solar Tracking System. Proceeding National Innovation and Invention Competition Through Exhibition (iCompEx'17), (pp. 1-9).

Rebecca, B., Kumar, K. P. M., Padmini, S., Srivastava, B. K., Halder, S., & Boopathi, S. (2023). Convergence of Data Science-AI-Green Chemistry-Affordable Medicine: Transforming Drug Discovery. In B. B. Gupta & F. Colace (Eds.), (pp. 348–373). Advances in Computational Intelligence and Robotics. IGI Global. doi:10.4018/978-1-6684-9999-3.ch014

Reddy, M. A., Gaurav, A., Ushasukhanya, S., Rao, V. C. S., Bhattacharya, S., & Boopathi, S. (2023). Bio-Medical Wastes Handling Strategies During the COVID-19 Pandemic. In Multidisciplinary Approaches to Organizational Governance During Health Crises (pp. 90–111). IGI Global. doi:10.4018/978-1-7998-9213-7.ch006

Reddy. (2023) Development of Programmed Autonomous Electric Heavy Vehicle: An Application of IoT. In: *Proceedings of the 2023 2nd International Conference on Electronics and Renewable Systems, ICEARS 2023*. IEEE. 10.1109/ICEARS56392.2023.10085492

Reddy, M. A., Reddy, B. M., Mukund, C. S., Venneti, K., Preethi, D. M. D., & Boopathi, S. (2023). Social Health Protection During the COVID-Pandemic Using IoT. In F. P. C. Endong (Ed.), (pp. 204–235). Advances in Electronic Government, Digital Divide, and Regional Development. IGI Global. doi:10.4018/978-1-7998-8394-4.ch009

Rene, E. A., & Fokui, W. S. T. (2024). Artificial intelligence-based optimal EVCS integration with stochastically sized and distributed PVs in an RDNS segmented in zones. *Journal of Electrical Systems and Information Technology, 11*(1), 1. doi:10.1186/s43067-023-00126-w

Ren, X., Li, C., Ma, X., Chen, F., Wang, H., Sharma, A., Gaba, G. S., & Masud, M. (2021). Design of multi-information fusion based intelligent electrical fire detection system for green buildings. *Sustainability (Basel), 13*(6), 3405. doi:10.3390/su13063405

Rezoug, M. R., Benaouadj, M., Taibi, D., & Chenni, R. (2021). A New Optimization Approach for a Solar Tracker Based on an Inertial Measurement Unit. Engineering, Technology & *Applied Scientific Research, 11*(5), 7542–7550. doi:10.48084/etasr.4330

Rhee, K.-N., Olesen, B. W., & Kim, K. W. (2017). Ten questions about radiant heating and cooling systems. *Building and Environment, 112*, 367–381. doi:10.1016/j.buildenv.2016.11.030

Riadh, A. D. (2022). Dubai, the sustainable, smart city. *Renewable Energy and Environmental Sustainability, 7*, 3. doi:10.1051/rees/2021049

Riazi, S., Bengtsson, K., Bischoff, R., Aurnhammer, A., Wigstrom, O., & Lennartson, B. (2016). Energy and peak-power optimization of existing time-optimal robot trajectories. *IEEE International Conference on Automation Science and Engineering, 2016-November*. IEEE. 10.1109/COASE.2016.7743423

Ristic, B., Angley, D., Moran, B., & Palmer, J. L. (2017). Autonomous multi-robot search for a hazardous source in a turbulent environment. *Sensors (Basel), 17*(4), 918. doi:10.3390/s17040918 PMID:28430120

Rodrigues, S., Torabikalaki, R., Faria, F., Cafôfo, N., Chen, X., Ivaki, A. R., Mata-Lima, H., & Morgado-Dias, F. (2016). Economic feasibility analysis of small scale PV systems in different countries. *Solar Energy, 131*, 81–95. doi:10.1016/j.solener.2016.02.019

Rodriguez-Martinez, O. F., Andrade, F., Vega-Penagos, C. A., & Luna, A. C. (2023). A Review of Distributed Secondary Control Architectures in Islanded-Inverter-Based Microgrids. *Energies, 16*(2), 878. doi:10.3390/en16020878

Roslan, M. (2020). Particle swarm optimization algorithm-based PI inverter controller for a grid-connected PV system. *PLoS ONE, 15*(12), e0243581. https://doi.org/. pone.0243581 doi:10.1371/journal

Ruppel, P., & Gschwendtner, F. (2011). *Spontaneous, and Privacy-Friendly Mobile Indoor Routing and Navigation*. Second Workshop on Services, Platforms, Innovations and Research for New Infrastructures in Telecommunications, Lübeck, Germany.

Saadh, Almoyad, M. A. A., Arellano, M. T. C., Maaliw, R. R. III, Castillo-Acobo, R. Y., Jalal, S. S., Gandla, K., Obaid, M., Abdulwahed, A. J., Ibrahem, A. A., Sârbu, I., Juyal, A., Lakshmaiya, N., & Akhavan-Sigari, R. (2023a). Long non-coding RNAs: Controversial roles in drug resistance of solid tumors mediated by autophagy. *Cancer Chemotherapy and Pharmacology, 92*(6), 439–453. doi:10.1007/s00280-023-04582-z PMID:37768333

Saadh, Baher, H., Li, Y., chaitanya, M., Arias-Gonzáles, J. L., Allela, O. Q. B., Mahdi, M. H., Carlos Cotrina-Aliaga, J., Lakshmaiya, N., Ahjel, S., Amin, A. H., Gilmer Rosales Rojas, G., Ameen, F., Ahsan, M., & Akhavan-Sigari, R. (2023b). The bioengineered and multifunctional nanoparticles in pancreatic cancer therapy: Bioresponisive nanostructures, phototherapy and targeted drug delivery. *Environmental Research*, *233*, 116490. doi:10.1016/j.envres.2023.116490 PMID:37354932

Saadh, Castillo-Acobo, R. Y., Baher, H., Narayanan, J., Palacios Garay, J. P., Vera Yamaguchi, M. N., Arias-Gonzáles, J. L., Cotrina-Aliaga, J. C., Akram, S. V., Lakshmaiya, N., Amin, A. H., Mohany, M., Al-Rejaie, S. S., Ahsan, M., Bahrami, A., & Akhavan-Sigari, R. (2023b). The protective role of sulforaphane and Homer1a in retinal ischemia-reperfusion injury: Unraveling the neuroprotective interplay. *Life Sciences*, *329*, 121968. doi:10.1016/j.lfs.2023.121968 PMID:37487941

Saadh, Rasulova, I., Almoyad, M. A. A., Kiasari, B. A., Ali, R. T., Rasheed, T., Faisal, A., Hussain, F., Jawad, M. J., Hani, T., Sârbu, I., Lakshmaiya, N., & Ciongradi, C. I. (2024). Recent progress and the emerging role of lncRNAs in cancer drug resistance; focusing on signaling pathways. *Pathology, Research and Practice*, *253*, 154999. doi:10.1016/j.prp.2023.154999 PMID:38118218

Sabarinathan, P., Annamalai, V. E., Vishal, K., Nitin, M. S., Natrayan, L., Veeeman, D., & Mammo, W. D. (2022). Experimental study on removal of phenol formaldehyde resin coating from the abrasive disc and preparation of abrasive disc for polishing application. *Advances in Materials Science and Engineering*, *2022*, 1–8. doi:10.1155/2022/6123160

Safari, M., & Sarvi, M. (2014). Optimal load sharing strategy for a wind/diesel/battery hybrid power system based on imperialist competitive neural network algorithm. *IET Renewable Power Generation*, *8*(8), 937–946.

Sai, Venkatesh, S. N., Dhanasekaran, S., Balaji, P. A., Sugumaran, V., Lakshmaiya, N., & Paramasivam, P. (2023). Transfer Learning Based Fault Detection for Suspension System Using Vibrational Analysis and Radar Plots. *Machines*, *11*(8), 778. doi:10.3390/machines11080778

Sakthi Shunmuga Sundaram, P., Hari Basker, N., & Natrayan, L. (2019). Smart clothes with bio-sensors for ECG monitoring. *International Journal of Innovative Technology and Exploring Engineering*, *8*(4).

Salam Shah, A., Nasir, H., Fayaz, M., Lajis, A., & Shah, A. (2019). A review on Power consumption optimization techniques in IoT-based smart building environments. *Information (Basel)*, *10*(3), 108. doi:10.3390/info10030108

Saleh, M., & Saad, S. (2016). Artificial Immune System based PID Tuning for DC Servo Speed Control. *International Journal of Computer Applications*, *155*(2), 23–26. doi:10.5120/ijca2016912265

Samikannu, R., Koshariya, A. K., Poornima, E., Ramesh, S., Kumar, A., & Boopathi, S. (2022). Sustainable Development in Modern Aquaponics Cultivation Systems Using IoT Technologies. In *Human Agro-Energy Optimization for Business and Industry* (pp. 105–127). IGI Global.

Sampath, B., Sasikumar, C., & Myilsamy, S. (2023). Application of TOPSIS Optimization Technique in the Micro-Machining Process. In IGI:Trends, Paradigms, and Advances in Mechatronics Engineering (pp. 162–187). IGI Global.

Sampath, B. C. S., & Myilsamy, S. (2022). Application of TOPSIS Optimization Technique in the Micro-Machining Process. In M. A. Mellal (Ed.), (pp. 162–187). Advances in Mechatronics and Mechanical Engineering. IGI Global. doi:10.4018/978-1-6684-5887-7.ch009

Sana, F., Azad, N. L., & Raahemifar, K. (2023). Autonomous Vehicle Decision-Making and Control in Complex and Unconventional Scenarios—A Review. In Machines, 11(7). doi:10.3390/machines11070676

Santhosh, M. S., Sasikumar, R., Khadar, S. D. A., & Natrayan, L. (2021). Ammonium Polyphosphate Reinforced E-Glass/Phenolic Hybrid Composites for Primary E-Vehicle Battery Casings -A Study on Fire Performance. *Journal of New Materials for Electrochemical Systems*, *24*(4), 247–253. doi:10.14447/jnmes.v24i4.a03

Santhosh, M. S., Sasikumar, R., Natrayan, L., Senthil Kumar, M., Elango, V., & Vanmathi, M. (2018). Investigation of mechanical and electrical properties of kevlar/E-glass and basalt/E-glass reinforced hybrid composites. *International Journal of Mechanical and Production Engineering Research and Development*, 8(3). Advance online publication. doi:10.24247/ijmperdjun201863

Saravanan, A., Venkatasubramanian, R., Khare, R., Surakasi, R., Boopathi, S., Ray, S., & Sudhakar, M. (2022). POLICY TRENDS OF RENEWABLE ENERGY AND NON. *Renewable Energy*.

Saravanan, K. G., Kaliappan, S., Natrayan, L., & Patil, P. P. (2023). Effect of cassava tuber nanocellulose and satin weaved bamboo fiber addition on mechanical, wear, hydrophobic, and thermal behavior of unsaturated polyester resin composites. *Biomass Conversion and Biorefinery*. Advance online publication. doi:10.1007/s13399-023-04495-0

Saravanan, M., Vasanth, M., Boopathi, S., Sureshkumar, M., & Haribalaji, V. (2022). Optimization of Quench Polish Quench (QPQ) Coating Process Using Taguchi Method. *Key Engineering Materials*, 935, 83–91. doi:10.4028/p-z569vy

Sarker, I. H. (2021). Deep Learning: A Comprehensive Overview on Techniques, Taxonomy, Applications and Research Directions. In SN Computer Science, 2(6). doi:10.1007/s42979-021-00815-1

Sarker, E., Halder, P., Seyedmahmoudian, M., Jamei, E., Horan, B., Mekhilef, S., & Stojcevski, A. (2021). Progress on the demand side management in smart grid and optimization approaches. *International Journal of Energy Research*, 45(1), 36–64. doi:10.1002/er.5631

Satav, S. D., & Lamani, D. K. G., H., Kumar, N. M. G., Manikandan, S., & Sampath, B. (2024). Energy and Battery Management in the Era of Cloud Computing: Sustainable Wireless Systems and Networks. In B. K. Mishra (Ed.), Practice, Progress, and Proficiency in Sustainability (pp. 141–166). IGI Global. doi:10.4018/979-8-3693-1186-8.ch009

Satav, S. D., Hasan, D. S., Pitchai, R., Mohanaprakash, T. A., Sultanuddin, S. J., & Boopathi, S. (2024). Next Generation of Internet of Things (NGIoT) in Healthcare Systems. In B. K. Mishra (Ed.), (pp. 307–330). Practice, Progress, and Proficiency in Sustainability. IGI Global. doi:10.4018/979-8-3693-1186-8.ch017

Sathish, T., Palani, K., Natrayan, L., Merneedi, A., De Poures, M. V., & Singaravelu, D. K. (2021). Synthesis and characterization of polypropylene/ramie fiber with hemp fiber and coir fiber natural biopolymer composite for biomedical application. *International Journal of Polymer Science*, 2021, 1–8.. doi:10.1155/2021/2462873

Satyanarayana, P. V. V., Radhika, A., Reddy, C. R., Pangedaiah, B., Martirano, L., Massaccesi, A., Flah, A., & Jasiński, M. (2023). Combined DC-Link Fed Parallel-VSI-Based DSTATCOM for Power Quality Improvement of a Solar DG Integrated System. *Electronics (Basel)*, 12(3), 505. doi:10.3390/electronics12030505

Scarabelli, G., Oloo, E. O., Maier, J. K. X., & Rodriguez-Granillo, A. (2022). Accurate Prediction of Protein Thermodynamic Stability Changes upon Residue Mutation using Free Energy Perturbation. *Journal of Molecular Biology*, 434(2), 167375. doi:10.1016/j.jmb.2021.167375 PMID:34826524

Sebastian, R. (2011). *Battery energy storage for increasing stability and reliability of an isolated Wind Diesel power system*. Research Gate.

Seeniappan. (2023). *Modelling and development of energy systems through cyber physical systems with optimising interconnected with control and sensing parameters*.

Selvakumar, S., Shankar, R., Ranjit, P., Bhattacharya, S., Gupta, A. S. G., & Boopathi, S. (2023). E-Waste Recovery and Utilization Processes for Mobile Phone Waste. In *Handbook of Research on Safe Disposal Methods of Municipal Solid Wastes for a Sustainable Environment* (pp. 222–240). IGI Global. doi:10.4018/978-1-6684-8117-2.ch016

Selvi. (2023) Optimization of Solar Panel Orientation for Maximum Energy Efficiency. In: *Proceedings of the 4th International Conference on Smart Electronics and Communication, ICOSEC 2023*. IEEE. 10.1109/ICOSEC58147.2023.10276287

Sendrayaperumal, Mahapatra, S., Parida, S. S., Surana, K., Balamurugan, P., Natrayan, L., & Paramasivam, P. (2021). Energy Auditing for Efficient Planning and Implementation in Commercial and Residential Buildings. *Advances in Civil Engineering, 2021*, 1–10. doi:10.1155/2021/1908568

Sengeni, D., Padmapriya, G., Imambi, S. S., Suganthi, D., Suri, A., & Boopathi, S. (2023). Biomedical Waste Handling Method Using Artificial Intelligence Techniques. In *Handbook of Research on Safe Disposal Methods of Municipal Solid Wastes for a Sustainable Environment* (pp. 306–323). IGI Global. doi:10.4018/978-1-6684-8117-2.ch022

Senthil Kumar, M., Mangalaraja, R. V., Senthil Kumar, R., & Natrayan, L. (2019). Processing and characterization of AA2024/Al$_2$O$_3$/SiC reinforces hybrid composites using squeeze casting technique. *Iranian Journal of Materials Science and Engineering, 16*(2). doi:10.22068/ijmse.16.2.55

Senthil Kumar, M., Natrayan, L., Hemanth, R. D., Annamalai, K., & Karthick, E. (2018). Experimental investigations on mechanical and microstructural properties of Al$_2$O$_3$/SiC reinforced hybrid metal matrix composite. In *IOP Conference Series*. Materials Science and Engineering.

Senthil, T. S., Ohmsakthi Vel, R., Puviyarasan, M., Babu, S. R., Surakasi, R., & Sampath, B. (2023). Industrial Robot-Integrated Fused Deposition Modelling for the 3D Printing Process. In R. Keshavamurthy, V. Tambrallimath, & J. P. Davim (Eds.), (pp. 188–210). Advances in Chemical and Materials Engineering. IGI Global. doi:10.4018/978-1-6684-6009-2.ch011

Sepasgozar, S., Karimi, R., Farahzadi, L., Moezzi, F., Shirowzhan, S., Ebrahimzadeh, S. M., Hui, F., & Aye, L. (2020). A systematic content review of artificial intelligence and the internet of things applications in smart home. *Applied Sciences (Switzerland), 10*(9). doi:10.3390/app10093074

Seralathan, S., Chenna Reddy, G., Sathish, S., Muthuram, A., Dhanraj, J. A., Lakshmaiya, N., Velmurugan, K., Sirisamphanwong, C., Ngoenmeesri, R., & Sirisamphanwong, C. (2023). Performance and exergy analysis of an inclined solar still with baffle arrangements. *Heliyon, 9*(4), e14807. doi:10.1016/j.heliyon.2023.e14807 PMID:37077675

Shaahid, S. M., Al-Hadhrami, L. M., & Rahman, M. K. (2014). *Review of economic assessment of hybrid photovoltaic-diesel-battery power systems for residential loads for different provinces of Saudi Arabia* (Vol. 31).

Shah, N. L., & Kurchania, A. K. (2023). Comparative analysis of predictive models for SOC estimation in EV under different running conditions. *E-Prime - Advances in Electrical Engineering. Electronics and Energy, 5*, 100207. doi:10.1016/j.prime.2023.100207

Shahzad, M., & Ma, J. T. (2019). Techno-economic assessment of a hybrid solar- wind-battery system with genetic algorithm. *Energy Procedia, 158*, 6384–6392.

Shair, J., Li, H., Hu, J., & Xie, X. (2021). Power system stability issues, classifications and research prospects in the context of high-penetration of renewables and power electronics. *Renewable & Sustainable Energy Reviews, 145*, 111111. doi:10.1016/j.rser.2021.111111

Shan, Y. (2019). Model Predictive Control of Bidirectional DC-DC Converters and AC/DC Interlinking Converters- A New Control Method for PV-Wind-Battery Micro grids. IEEE Transactions on Sustainable Energy.

Shan, Y., Zheng, B., Chen, L., Chen, L., & Chen, D. (2020). A Reinforcement Learning-Based Adaptive Path Tracking Approach for Autonomous Driving. *IEEE Transactions on Vehicular Technology, 69*(10), 10581–10595. doi:10.1109/TVT.2020.3014628

Sharda, S., Singh, M., & Sharma, K. (2021). Demand side management through load shifting in IoT based HEMS: Overview, challenges and opportunities. *Sustainable Cities and Society, 65*, 102517. doi:10.1016/j.scs.2020.102517

Sharma, P., Said, Z., Kumar, A., Nizetic, S., Pandey, A., Hoang, A. T., Huang, Z., Afzal, A., Li, C., Le, A. T., Nguyen, X. P., & Tran, V. D. (2022). Recent advances in machine learning research for nanofluid-based heat transfer in renewable energy system. *Energy & Fuels, 36*(13), 6626–6658. doi:10.1021/acs.energyfuels.2c01006

Sharma, Raffik, R., Chaturvedi, A., Geeitha, S., Akram, P. S., L, N., Mohanavel, V., Sudhakar, M., & Sathyamurthy, R. (2022). Designing and implementing a smart transplanting framework using programmable logic controller and photo-electric sensor. *Energy Reports, 8*, 430–444. doi:10.1016/j.egyr.2022.07.019

Sharma, S. K. (2021). Mechanical Behavior of Silica Fume Concrete Filled with Steel Tubular Composite Column. *Advances in Materials Science and Engineering, 2021*, 1–9. doi:10.1155/2021/3632991

Shezan, S. A., Kamwa, I., Ishraque, M. F., Muyeen, S. M., Hasan, K. N., Saidur, R., Rizvi, S. M., Shafiullah, M., & Al-Sulaiman, F. A. (2023). Evaluation of different optimization techniques and control strategies of hybrid microgrid: A review. *Energies, 16*(4), 1792. doi:10.3390/en16041792

Shi, Y., Li, W., Tan, B., Yang, F., & Zhang, L. (2023, April). Optimal cost scheduling of virtual power plant for new power system. In *2022 2nd Conference on High Performance Computing and Communication Engineering (HPCCE 2022) (Vol. 12605*, pp. 62-70). SPIE. 10.1117/12.2673283

Shibl, M. M., Ismail, L. S., & Massoud, A. M. (2023). A machine learning-based battery management system for state-of-charge prediction and state-of-health estimation for unmanned aerial vehicles. *Journal of Energy Storage, 66*, 107380. doi:10.1016/j.est.2023.107380

Shimray, B. (2017). A New MLP-GA-Fuzzy Decision Support System for Hydro Power Plant Site Selection. *Arabian Journal for Science and Engineering*.

Shinde, D., Mistry, K. N., Jadhav, G., & Singh, H. (2018, June). Optimization of Automobile Suspension System Using Hybrid GSA Algorithm. []. IOP Publishing.]. *IOP Conference Series. Materials Science and Engineering, 377*(1), 012149. doi:10.1088/1757-899X/377/1/012149

Shi, Y., Xu, B., Wang, D., & Zhang, B. (2018). Using Battery Storage for Peak Shaving and Frequency Regulation: Joint Optimization for Superlinear Gains. *IEEE Transactions on Power Systems, 33*(3), 2882–2894. doi:10.1109/TPWRS.2017.2749512

Shouran, M., & Elgamli, E. (2020). Teaching-Learning based Optimization Algorithm Tuned Fuzzy-PID Controller for Continuous Stirred Tank Reactor (Vol. 7). Research Gate.

Shriyanti Kulkarni. (n.d.). *Vedashree Chaphekar, Md Moin Uddin Chowdhury, Fatih Erden, & Ismail Guvenc*. UAV Aided Search and Rescue Operation Using Reinforcement Learning.

Shu, X., Li, G., Shen, J., Lei, Z., Chen, Z., & Liu, Y. (2020). An adaptive multi-state estimation algorithm for lithium-ion batteries incorporating temperature compensation. *Energy, 207*, 118262. doi:10.1016/j.energy.2020.118262

Singh, T., Solanki, A., & Sharma, S. K. (2021). Role of smart buildings in smart city—components, technology, indicators, challenges, future research opportunities. *Digital cities roadmap: IoT-based architecture and sustainable buildings*, 449-476.

Singh, A. P., Kumar, M. S., Deshpande, A., Jain, G., Khamesra, J., Mhetre, S., Awasthi, A., & Natrayan, L. (2020). Processing and characterization mechanical properties of AA2024/Al$_2$O$_3$/ZrO$_2$/Gr reinforced hybrid composite using stir casting technique. In *Materials Today*. Proceedings.

Singhal, P. (2019). A Case Study on Energy Efficient Green Building with New Intelligent Techniques Used to Achieve Sustainable Development Goal. *2019 20th International Conference on Intelligent System Application to Power Systems (ISAP)*. IEEE. 10.1109/ISAP48318.2019.9065938

Singh, M. (2017). An experimental investigation on mechanical behaviour of siCp reinforced Al 6061 MMC using squeeze casting process. *International Journal of Mechanical and Production Engineering Research and Development*, 7(6). doi:10.24247/ijmperddec201774

Singh, R., Kurukuru, V. S. B., & Khan, M. A. (2023). Advanced Power Converters and Learning in Diverse Robotic Innovation: A Review. *Energies*, 16(20), 7156. doi:10.3390/en16207156

Singh, R., Ren, J., & Lin, X. (2023). A Review of Deep Reinforcement Learning Algorithms for Mobile Robot Path Planning. *Vehicles*, 5(4), 1423–1451. doi:10.3390/vehicles5040078

Singh, S., Kumar, V., Datta, S., Dhanjal, D. S., Sharma, K., Samuel, J., & Singh, J. (2020). Current advancement and future prospect of biosorbents for bioremediation. *The Science of the Total Environment*, 709, 135895. doi:10.1016/j.scitotenv.2019.135895 PMID:31884296

Sivakumar, V., Kaliappan, S., Natrayan, L., & Patil, P. P. (2023). Effects of Silane-Treated High-Content Cellulose Okra Fibre and Tamarind Kernel Powder on Mechanical, Thermal Stability and Water Absorption Behaviour of Epoxy Composites. *Silicon*, 15(10), 4439–4447. Advance online publication. doi:10.1007/s12633-023-02370-1

Sivamayil, K., Rajasekar, E., Aljafari, B., Nikolovski, S., Vairavasundaram, S., & Vairavasundaram, I. (2023). A Systematic Study on Reinforcement Learning Based Applications. In Energies, 16(3). doi:10.3390/en16031512

Song, Q., Wang, S., Xu, W., Shao, Y., & Fernandez, C. (2021). A Novel Joint Support Vector Machine - Cubature Kalman Filtering Method for Adaptive State of Charge Prediction of Lithium-Ion Batteries. *International Journal of Electrochemical Science*, 16(8), 1–15. doi:10.20964/2021.08.26

Soni, S. K., & Bartaria, V. N. (2013). An overview of green building control strategies." *2013 International Conference on Renewable Energy Research and Applications (ICRERA)*. IEEE,. 10.1109/ICRERA.2013.6749837

Soori, M., Arezoo, B., & Dastres, R. (2023). Artificial intelligence, machine learning and deep learning in advanced robotics, a review. In Cognitive Robotics, 3. doi:10.1016/j.cogr.2023.04.001

Srinivas, B., Maguluri, L. P., Naidu, K. V., Reddy, L. C. S., Deivakani, M., & Boopathi, S. (2023). Architecture and Framework for Interfacing Cloud-Enabled Robots: In T. Murugan & N. E. (Eds.), Advances in Information Security, Privacy, and Ethics (pp. 542–560). IGI Global. doi:10.4018/978-1-6684-8145-5.ch027

Stapleton, A. (2017). How solar cells turn sunlight into electricity. *Cosmos magazine*. https://cosmosmagazine.com/technology/how-solar-cells-turn-sunlightinto-electricity

Stavrakas, V., & Flamos, A. (2020). A modular high-resolution demand-side management model to quantify benefits of demand-flexibility in the residential sector. *Energy Conversion and Management*, 205, 112339. doi:10.1016/j.enconman.2019.112339

Stiubiener, U., Carneiro da Silva, T., Federico, B. M. T., & Teixeira, J. C. (2020). PV power generation on hydro dam's reservoirs in Brazil: A way to improve operational flexibility. *Renewable Energy*, 150, 765–776.

Streimikiene, D., Skulskis, V., Balezentis, T., & Agnusdei, G. P. (2020). Uncertain multi-criteria sustainability assessment of green building insulation materials. *Energy and Building*, 219, 110021. doi:10.1016/j.enbuild.2020.110021

Stroe, D. I., Qi, J., Chen, L., Wang, S., Wang, Y., Fan, Y., & Liu, Y. (2023). State of charge estimation strategy based on fractional-order model. In *State Estimation Strategies in Lithium-ion Battery Management Systems* (pp. 191–206). Elsevier. doi:10.1016/B978-0-443-16160-5.00005-6

Stroe, D., Swierczynski, M., Stroe, A.-I., Laerke, R., Kjaer, P. C., & Teodorescu, R. (2016). Degradation Behavior of Lithium-Ion Batteries Based on Lifetime Models and Field Measured Frequency Regulation Mission Profile. *IEEE Transactions on Industry Applications*, 52(6), 5009–5018. doi:10.1109/TIA.2016.2597120

Su, B., Han, W., He, H. Z., Jin, H., Chen, Z., & Yang, S. (2020). A biogas-fired cogeneration system based on chemically recuperated gas turbine cycle. *Energy Conversion and Management, 205*.

Subha, S., Inbamalar, T., Komala, C., Suresh, L. R., Boopathi, S., & Alaskar, K. (2023). A Remote Health Care Monitoring system using internet of medical things (IoMT). *IEEE Explore*, 1–6.

Subramanian, Lakshmaiya, N., Ramasamy, D., & Devarajan, Y. (2022). Detailed analysis on engine operating in dual fuel mode with different energy fractions of sustainable HHO gas. *Environmental Progress & Sustainable Energy, 41*(5), e13850. doi:10.1002/ep.13850

Subramanian, Solaiyan, E., Sendrayaperumal, A., & Lakshmaiya, N. (2022b). Flexural behaviour of geopolymer concrete beams reinforced with BFRP and GFRP polymer composites. *Advances in Structural Engineering, 25*(5), 954–965. doi:10.1177/13694332211054229

Suman. (2023) IoT based Social Device Network with Cloud Computing Architecture. In: *Proceedings of the 2023 2nd International Conference on Electronics and Renewable Systems, ICEARS 2023*. IEEE. 10.1109/ICEARS56392.2023.10085574

Sundaramk, M., Prakash, P., Angalaeswari, S., Deepa, T., Natrayan, L., & Paramasivam, P. (2021). Influence of Process Parameter on Carbon Nanotube Field Effect Transistor Using Response Surface Methodology. *Journal of Nanomaterials, 2021*, 1–9. doi:10.1155/2021/7739359

Suresh, N. S., Thirumalai, N. C., & Dasappa, S. (2019). Modeling and analysis of solar thermal and biomass hybrid power plants. *Applied Thermal Engineering, 160*.

Suresh, V., M, M., & Kiranmayi, R. (2020). Modelling and optimization of an off- grid hybrid renewable energy system for electrification in a rural areas. *Energy Reports, 6*, 594–604. doi:10.1016/j.egyr.2020.01.013

Sutton, R. S., & Barto, A. G. (2017). *Reinforcement Learning: An Introduction* (2nd ed., Vol. 1). MIT Press.

Syafa'ah, L., Fauziyah, L., & Has, Z. (2018). Robust and Accurate Positioning Control of Solar Panel System Tracking based Sun Position Image. *2018 5th International Conference on Electrical Engineering, Computer Science and Informatics (EECSI)*, (pp. 324-329). IEEE. 10.1109/EECSI.2018.8752746

Syamala, M., Komala, C., Pramila, P., Dash, S., Meenakshi, S., & Boopathi, S. (2023). Machine Learning-Integrated IoT-Based Smart Home Energy Management System. In *Handbook of Research on Deep Learning Techniques for Cloud-Based Industrial IoT* (pp. 219–235). IGI Global. doi:10.4018/978-1-6684-8098-4.ch013

Tabak, A., & Duman, S. (2022). Levy flight and fitness distance balance-based coyote optimization algorithm for effective automatic generation control of PV-based multi-area power systems. *Arabian Journal for Science and Engineering, 47*(11), 14757–14788. doi:10.1007/s13369-022-07004-z

Tahir, A., Böling, J., Haghbayan, M. H., Toivonen, H. T., & Plosila, J. (2019). Swarms of Unmanned Aerial Vehicles — A Survey. In Journal of Industrial Information Integration, 16. doi:10.1016/j.jii.2019.100106

Tahir, F., Arshad, M. Y., Saeed, M. A., & Ali, U. (2023). Integrated process for simulation of gasification and chemical looping hydrogen production using Artificial Neural Network and machine learning validation. *Energy Conversion and Management*, *296*, 117702. doi:10.1016/j.enconman.2023.117702

Tamboli, J. A., & Joshi, S. G. (1999). Optimum design of a passive suspension system of a vehicle subjected to actual random road excitations. *Journal of Sound and Vibration*, *219*(2), 193–205. doi:10.1006/jsvi.1998.1882

Tanaka, M. S., Miyanishi, Y., Toyota, M., Murakami, T., Hirazakura, R., & Itou, T. (2017). A study of bus location system using LoRa: Bus location system for community bus Notty. IEEE 6th Global Conference on Consumer Electronics (GCCE), (pp. 1-4). IEEE. 10.1109/GCCE.2017.8229279

Tan, S., Wu, Y., Xie, P., Guerrero, J. M., Vasquez, J. C., & Abusorrah, A. (2020). New challenges in the design of microgrid systems: Communication networks, cyberattacks, and resilience. *IEEE Electrification Magazine*, *8*(4), 98–106. doi:10.1109/MELE.2020.3026496

Tazvinga, H., Xia, X., & Zhang, J. (2013). *Minimum cost solution of photovoltaic–diesel–battery hybrid power systems for remote consumers* (Vol. 96).

Thakre, Pandhare, A., Malwe, P. D., Gupta, N., Kothare, C., Magade, P. B., Patel, A., Meena, R. S., Veza, I., Natrayan L, & Panchal, H. (2023). Heat transfer and pressure drop analysis of a microchannel heat sink using nanofluids for energy applications. *Kerntechnik*, *88*(5), 543–555. doi:10.1515/kern-2023-0034

Thapar, V., Agnihotri, G., & Sethi, V. K. (2011). Critical analysis of methods for mathematical modelling of wind turbines. *Renewable Energy*, *36*(11), 3166–3177. doi:10.1016/j.renene.2011.03.016

Tharakan, R. A., Joshi, R., Ravindran, G., & Jayapandian, N. (2021). Machine Learning Approach for Automatic Solar Panel Direction by using Naïve Bayes Algorithm. *2021 5th International Conference on Intelligent Computing and Control Systems (ICICCS)*, (pp. 1317-1322). IEEE. 10.1109/ICICCS51141.2021.9432114

Tukymbekov, D., Saymbetov, A., Nurgaliyev, M., Kuttybay, N., Nalibayev, Y., & Dosymbetova, G. (2019). Intelligent energy efficient street lighting system with predictive energy consumption, *2019 International Conference on Smart Energy Systems and Technologies (SEST)* (pp. 1-5). IEEE. 10.1109/SEST.2019.8849023

Turek, W., Marcjan, R., & Cetnarowicz, K. (2006). Agent-Based Mobile Robots Navigation Framework. *Lecture Notes in Computer Science*, *3993*, 775–782. doi:10.1007/11758532_101

Türkay, S., & Akçay, H. (2005). A study of random vibration characteristics of the quarter-car model. *Journal of Sound and Vibration*, *282*(1-2), 111–124. doi:10.1016/j.jsv.2004.02.049

Tzafestas, S. G. (2013). Introduction to Mobile Robot Control. In Introduction to Mobile Robot Control. doi:10.1016/C2013-0-01365-5

Ugandar, R. E., Rahamathunnisa, U., Sajithra, S., Christiana, M. B. V., Palai, B. K., & Boopathi, S. (2023). Hospital Waste Management Using Internet of Things and Deep Learning: Enhanced Efficiency and Sustainability. In M. Arshad (Ed.), (pp. 317–343). Advances in Bioinformatics and Biomedical Engineering. IGI Global. doi:10.4018/978-1-6684-6577-6.ch015

Ugle, Arulprakasajothi, M., Padmanabhan, S., Devarajan, Y., Lakshmaiya, N., & Subbaiyan, N. (2023). Investigation of heat transport characteristics of titanium dioxide nanofluids with corrugated tube. *Environmental Quality Management*, *33*(2), 127–138. doi:10.1002/tqem.21999

Usman, M., Khan, M. T., Rana, A. S., & Ali, S. (2017). Techno economic analysis of hybrid solar-diesel-grid connected power generation system. *Journal of Electrical Systems and Information Technology*, 1–10.

Vaishali. (2021). *Guided container selection for data streaming through neural learning in cloud.* International Journal of Systems Assurance Engineering and Management., doi:10.1007/s13198-021-01124-9

Valencia, A., Zhang, W., Gu, L., Chang, N.-B., & Wanielista, M. P. (2022). Synergies of green building retrofit strategies for improving sustainability and resilience via a building-scale food-energy-water nexus. *Resources, Conservation and Recycling, 176,* 105939. doi:10.1016/j.resconrec.2021.105939

Van Le, D., Liu, Y., Wang, R., Tan, R., Wong, Y.-W., & Wen, Y. (2019). Control of air free-cooled data centers in tropics via deep reinforcement learning. *Proceedings of the 6th ACM International Conference on Systems for Energy-Efficient Buildings, Cities, and Transportation,* (pp. 306–315). ACM. 10.1145/3360322.3360845

Vanitha, S. K. R., & Boopathi, S. (2023). Artificial Intelligence Techniques in Water Purification and Utilization. In P. Vasant, R. Rodríguez-Aguilar, I. Litvinchev, & J. A. Marmolejo-Saucedo (Eds.), (pp. 202–218). Advances in Environmental Engineering and Green Technologies. IGI Global. doi:10.4018/978-1-6684-4118-3.ch010

Varga, L., Quintana, V. H., & Miranda, R. (1999). Voltage collapse in the Chilean interconnected system. *IEEE Transactions on Power Systems, 14*(4), 1415–1421. doi:10.1109/59.801905

Vasireddy, S., Ravipati, V., Ravi, T., & Jegan, G. (2016). Wireless sensor based gps mobile application for blind people navigation. *Journal of Engineering and Applied Sciences (Asian Research Publishing Network), 11*(13), 8374–8379.

Veeman, Sai, M. S., Sureshkumar, P., Jagadeesha, T., Natrayan, L., Ravichandran, M., & Mammo, W. D. (2021). Additive Manufacturing of Biopolymers for Tissue Engineering and Regenerative Medicine: An Overview, Potential Applications, Advancements, and Trends. *International Journal of Polymer Science, 2021,* 1–20. doi:10.1155/2021/4907027

Veeranjaneyulu, R., Boopathi, S., Kumari, R. K., Vidyarthi, A., Isaac, J. S., & Jaiganesh, V. (2023). Air Quality Improvement and Optimisation Using Machine Learning Technique. *IEEE- Explore,* (pp. 1–6). IEEE.

Veeranjaneyulu, R., Boopathi, S., Kumari, R. K., Vidyarthi, A., Isaac, J. S., & Jaiganesh, V. (2023). Air Quality Improvement and Optimisation Using Machine Learning Technique. *IEEE- Explore,* 1–6.

Veeranjaneyulu, R., Boopathi, S., Narasimharao, J., Gupta, K. K., Reddy, R. V. K., & Ambika, R. (2023). Identification of Heart Diseases using Novel Machine Learning Method. *IEEE- Explore,* (pp. 1–6). IEEE.

Velmurugan, G., Natrayan, L., Chohan, J. S., Vasanthi, P., Angalaeswari, S., Pravin, P., Kaliappan, S., & Arunkumar, D. (2023). Investigation of mechanical and dynamic mechanical analysis of bamboo/olive tree leaves powder-based hybrid composites under cryogenic conditions. *Biomass Conversion and Biorefinery.* doi:10.1007/s13399-023-04591-1

Vendoti, S., Muralidhar, M., & Kiranmayi, R. (2021). Techno-economic analysis of off-grid solar/wind/biogas/biomass/ fuel cell/battery system for electrification in a cluster of villages by HOMER software. *Environment, Development and Sustainability, 23*(1), 351–372. doi:10.1007/s10668-019-00583-2

Venkateswaran, N., Kumar, S. S., Diwakar, G., Gnanasangeetha, D., & Boopathi, S. (2023a). Synthetic Biology for Waste Water to Energy Conversion: IoT and AI Approaches. In M. Arshad (Ed.), (pp. 360–384). Advances in Bioinformatics and Biomedical Engineering. IGI Global. doi:10.4018/978-1-6684-6577-6.ch017

Venkateswaran, N., Vidhya, K., Ayyannan, M., Chavan, S. M., Sekar, K., & Boopathi, S. (2023). A Study on Smart Energy Management Framework Using Cloud Computing. In P. Ordóñez De Pablos & X. Zhang (Eds.), (pp. 189–212). Practice, Progress, and Proficiency in Sustainability. IGI Global. doi:10.4018/978-1-6684-8634-4.ch009

Venkateswaran, N., Vidhya, R., Naik, D. A., Michael Raj, T. F., Munjal, N., & Boopathi, S. (2023). Study on Sentence and Question Formation Using Deep Learning Techniques. In O. Dastane, A. Aman, & N. S. Bin Mohd Satar (Eds.), (pp. 252–273). Advances in Business Strategy and Competitive Advantage. IGI Global. doi:10.4018/978-1-6684-6782-4.ch015

Vennila, T., Karuna, M. S., Srivastava, B. K., Venugopal, J., Surakasi, R., & B., S. (2023). New Strategies in Treatment and Enzymatic Processes: Ethanol Production From Sugarcane Bagasse. In P. Vasant, R. Rodríguez-Aguilar, I. Litvinchev, & J. A. Marmolejo-Saucedo (Eds.), *Advances in Environmental Engineering and Green Technologies* (pp. 219–240). IGI Global. doi:10.4018/978-1-6684-4118-3.ch011

Vennila, T., Karuna, M., Srivastava, B. K., Venugopal, J., Surakasi, R., & Sampath, B. (2022). New Strategies in Treatment and Enzymatic Processes: Ethanol Production From Sugarcane Bagasse. In Human Agro-Energy Optimization for Business and Industry (pp. 219–240). IGI Global.

Verbrugge, M. W., Liu, P., & Soukiazian, S. (2005). Activated-carbon electric-double-layer capacitors: Electrochemical characterization and adaptive algorithm implementation. *Journal of Power Sources*, *141*(2), 369–385. doi:10.1016/j.jpowsour.2004.09.034

Verma, B. (2020). A Review Paper on STS for PV Power Plant. *International Journal of Engineering Research.* . doi:10.17577/IJERTV9IS020103

Vijayakumar, G. N. S., Domakonda, V. K., Farooq, S., Kumar, B. S., Pradeep, N., & Boopathi, S. (2023). Sustainable Developments in Nano-Fluid Synthesis for Various Industrial Applications. In A. S. Etim (Ed.), (pp. 48–81). Advances in Human and Social Aspects of Technology. IGI Global., doi:10.4018/978-1-6684-5347-6.ch003

Vijayapriya, R. (2022). IoT Based Energy Management System for Net-Zero Energy Building Operation. *Conference Paper, IEEE Delhi Section Conference (DELCON)*. IEEE.

Vijayapriya, R., Umamageswari, A, Bhat, R, Dass, R, & Manikandan, N (2023). *Web-based data manipulation to improve the accessibility of factory data using big data analytics: An industry 4.0 approach*. Data Fabric Architectures: Web-Driven Applications.

Vijayaragavan, Subramanian, B., Sudhakar, S., & Natrayan, L. (2022). Effect of induction on exhaust gas recirculation and hydrogen gas in compression ignition engine with simarouba oil in dual fuel mode. *International Journal of Hydrogen Energy*, *47*(88), 37635–37647. doi:10.1016/j.ijhydene.2021.11.201

Viswanathan, C. (2023). Deep Learning for Enhanced Fault Diagnosis of Monoblock Centrifugal Pumps: Spectrogram-Based Analysis. *Machines*, *11*(9), 874. doi:10.3390/machines11090874

Wallace, N., Kong, H., Hill, A., & Sukkarieh, S. (2019). Energy Aware Mission Planning for WMRs on Uneven Terrains. *IFAC-PapersOnLine*, *52*(30). doi:10.1016/j.ifacol.2019.12.513

Wan, Z., Anwar, A., Hsiao, Y. S., Jia, T., Reddi, V. J., & Raychowdhury, A. (2021). Analyzing and Improving Fault Tolerance of Learning-Based Navigation Systems. *Proceedings - Design Automation Conference, 2021-December*. 10.1109/DAC18074.2021.9586116

Wang, D., Dasari, S., Chambers, M. C., Holman, J. D., Chen, K., Liebler, D. C., Orton, D. J., Purvine, S. O., Monroe, M. E., Chung, C. Y., Rose, K. L., & Tabb, D. L. (2013). Basophile: Accurate Fragment Charge State Prediction Improves Peptide Identification Rates. *Genomics, Proteomics & Bioinformatics*, *11*(2), 86–95. doi:10.1016/j.gpb.2012.11.004 PMID:23499924

Wang, R., Xiong, J., He, M., Gao, L., & Wang, L. (2019). Multi-objective optimal design of hybrid renewable energy system under multiple scenarios. *Renewable Energy*.

Wang, S. L., Tang, W., Fernandez, C., Yu, C. M., Zou, C. Y., & Zhang, X. Q. (2019). A novel endurance prediction method of series connected lithium-ion batteries based on the voltage change rate and iterative calculation. *Journal of Cleaner Production*, *210*, 43–54. doi:10.1016/j.jclepro.2018.10.349

Wang, S., Takyi-Aninakwa, P., Fan, Y., Yu, C., Jin, S., Fernandez, C., & Stroe, D. I. (2022). A novel feedback correction-adaptive Kalman filtering method for the whole-life-cycle state of charge and closed-circuit voltage prediction of lithium-ion batteries based on the second-order electrical equivalent circuit model. *International Journal of Electrical Power & Energy Systems*, *139*, 108020. doi:10.1016/j.ijepes.2022.108020

Wang, S., Takyi-Aninakwa, P., Jin, S., Yu, C., Fernandez, C., & Stroe, D. I. (2022). An improved feedforward-long short-term memory modeling method for the whole-life-cycle state of charge prediction of lithium-ion batteries considering current-voltage-temperature variation. *Energy*, *254*, 124224. doi:10.1016/j.energy.2022.124224

Wang, W., Wang, Y., Wang, D., Zou, Y., & Chen, X. (2023, May). A review of key issues in planning AC/DC distribution systems for renewable energy. []. IOP Publishing.]. *Journal of Physics: Conference Series*, *2503*(1), 012055. doi:10.1088/1742-6596/2503/1/012055

Wang, Y., & Chen, Z. (2020). A framework for state-of-charge and remaining discharge time prediction using unscented particle filter. *Applied Energy*, *260*, 114324. doi:10.1016/j.apenergy.2019.114324

Wang, Y., Liu, C., Pan, R., & Chen, Z. (2017a). Experimental data of lithium-ion battery and ultracapacitor under DST and UDDS profiles at room temperature. *Data in Brief*, *12*, 161–163. doi:10.1016/j.dib.2017.01.019 PMID:28459088

Wang, Y., Liu, C., Pan, R., & Chen, Z. (2017b). Modeling and state-of-charge prediction of lithium-ion battery and ultracapacitor hybrids with a co-estimator. *Energy*, *121*, 739–750. doi:10.1016/j.energy.2017.01.044

Wang, Y., Vatandoost, H., & Sedaghati, R. (2023). Development of a Novel Magneto-Rheological Elastomer-Based Semi-Active Seat Suspension System. *Vibration*, *6*(4), 777–795. doi:10.3390/vibration6040048

Wei, H., Sasaki, H., Kubokawa, J., & Yokoyama, R. (1998). An interior point non linear programming for optimal power flow problems with a novel data structure. *IEEE Transactions on Power Systems*, *13*(3), 870–877. doi:10.1109/59.708745

Wei, L., Nakamura, T., & Imai, K. (2020). Development and optimization of low-speed and high efficiency permanent magnet generator for micro hydro-electrical generation system. *Renewable Energy*, *147*, 1653–1662.

Wilson, D. G., Robinett, R. D., & Eisler, G. R. (2004). Discrete dynamic programming for optimized path planning of flexible robots. *2004 IEEE/RSJ International Conference on Intelligent Robots and Systems (IROS) (IEEE Cat. No.04CH37566)*, *3*, (pp. 2918–2923). IEEE. 10.1109/IROS.2004.1389852

Wu, J., Honglun, W., Huang, Y., Su, Z., & Zhang, M. (2018). Energy Management Strategy for Solar-Powered UAV Long-Endurance Target Tracking. *IEEE Transactions on Aerospace and Electronic Systems*, *PP*, *1*. doi:10.1109/TAES.2018.2876738

Wulandari, D. A., Akmal, M., Gunawan, Y., & others. (2020). *Cooling improvement of the IT rack by layout rearrangement of the A2 class data center room: A simulation study.*

WulfmeierM.ByravanA.HertweckT.HigginsI.GuptaA.KulkarniT.ReynoldsM.TeplyashinD.HafnerR.LampeT.RiedmillerM. A. (2020). Representation Matters: Improving Perception and Exploration for Robotics. *CoRR*, *abs/2011.01758*. https://arxiv.org/abs/2011.01758

Wu, W., Wang, S., Wu, W., Chen, K., Hong, S., & Lai, Y. (2019). A critical review of battery thermal performance and liquid based battery thermal management. *Energy Conversion and Management*, *182*, 262–281. doi:10.1016/j.enconman.2018.12.051

Wu, Y., Zhang, Y., Li, G., Shen, J., Chen, Z., & Liu, Y. (2020). A predictive energy management strategy for multi-mode plug-in hybrid electric vehicles based on multi neural networks. *Energy*, *208*, 118366. Advance online publication. doi:10.1016/j.energy.2020.118366

Xia, G., Cao, L., & Bi, G. (2017). A review on battery thermal management in electric vehicle application. *Journal of Power Sources*, *367*, 90–105. doi:10.1016/j.jpowsour.2017.09.046

Xiao, L., Wang, J., Dong, Y., & Wu, J. (2015). Combined forecasting models for wind energy forecasting: A case study in China. *Renewable & Sustainable Energy Reviews*, *44*, 271–288. doi:10.1016/j.rser.2014.12.012

Xue, Y., Van Cutsem, T., & Ribbons-Pavella, M. (1989). Extended equal area criterion justification, generalizations, applications. *IEEE Transactions on Power Systems*, *4*(1), 44–52. doi:10.1109/59.32456

Ya'u Muhammad, J., Tajudeen Jimoh, M., Baba Kyari, I., Abdullahi Gele, M., & Musa, I. (2019). A Review on Solar Tracking System: A Technique of Solar Power Output Enhancement. *Engineering and Science*, *41*(1), 1–11. doi:10.11648/j.es.20190401.11

Yang, B., Lv, Z., & Wang, F. (2022). Digital Twins for Intelligent Green Buildings. *Buildings*, *12*(6), 856. doi:10.3390/buildings12060856

Yang, J., Zhang, X., Zhang, X., Wang, L., Feng, W., & Li, Q. (2021). Beyond the visible: Bioinspired infrared adaptive materials. *Advanced Materials*, *33*(14), 2004754. doi:10.1002/adma.202004754 PMID:33624900

Yang, T., Zhao, L., Li, W., & Zomaya, A. Y. (2020). Reinforcement learning in sustainable energy and electric systems: A survey. *Annual Reviews in Control*, *49*, 145–163. doi:10.1016/j.arcontrol.2020.03.001

Yang, Z., Chen, M., Liu, X., Liu, Y., Chen, Y., Cui, S., & Poor, H. V. (2021). AI-driven UAV-NOMA-MEC in next generation wireless networks. *IEEE Wireless Communications*, *28*(5), 66–73. doi:10.1109/MWC.121.2100058

Yao, J., Li, X., Zhang, Y., Ji, J., Wang, Y., & Liu, Y. (2022). Path Planning of Unmanned Helicopter in Complex Environment Based on Heuristic Deep Q-Network. *International Journal of Aerospace Engineering*, *1360956*, 1–15. doi:10.1155/2022/1360956

Yap, K. Y., Sarimuthu, C. R., & Lim, J. M.-Y. (2020). Artificial intelligence based MPPT techniques for solar power system: A review. *Journal of Modern Power Systems and Clean Energy*, *8*(6), 1043–1059. doi:10.35833/MPCE.2020.000159

Yatika, G., Kumar, S. R., Patel, P. B., Singh, D. P., Rajkamal, M., & Boopathi, S. (2023). Experimental investigation of RHA biochar and tamarind fibre epoxy composite. *Materials Today: Proceedings*. doi:10.1016/j.matpr.2023.02.439

Yi, X., Lu, T., Li, Y., Ai, Q., & Hao, R. (2023). Collaborative planning and optimization for electric-thermal-hydrogen-coupled energy systems with portfolio selection of the complete hydrogen energy chain. *arXiv preprint arXiv:2311.07891*.

Yi, Y., Zhou, Y., Su, H., Fang, C., Wang, H., Feng, D., & Li, H. (2023). Overview of EV battery testing and evaluation of EES systems located in EV charging station with PV. *Energy Reports*, *9*, 134–144. doi:10.1016/j.egyr.2023.04.075

Yogeshwaran, S., Natrayan, L., Rajaraman, S., Parthasarathi, S., & Nestro, S. (2020a). Experimental investigation on mechanical properties of Epoxy/graphene/fish scale and fermented spinach hybrid bio composite by hand lay-up technique. In *Materials Today*. Proceedings.

Yogeshwaran, S., Natrayan, L., Udhayakumar, G., Godwin, G., & Yuvaraj, L. (2020b). Effect of waste tyre particles reinforcement on mechanical properties of jute and abaca fiber - Epoxy hybrid composites with pre-treatment. In *Materials Today*. Proceedings.

Yogeshwaran, S., Prabhu, R., & Murugan, R. (2015). Mechanical properties of leaf ashes reinforced aluminum alloy metal matrix composites. *International Journal of Applied Engineering Research: IJAER*, *10*(13).

Yoon, C. O., Lee, P. Y., Jang, M., Yoo, K., & Kim, J. (2019). Comparison of internal parameters varied by environmental tests between high-power series/parallel battery packs with different shapes. *Journal of Industrial and Engineering Chemistry*, *71*, 260–269. doi:10.1016/j.jiec.2018.11.034

Yoshimoto, M., Endo, T., Maeda, R., & Matsuno, F. (2018). Decentralized navigation method for a robotic swarm with nonhomogeneous abilities. *Autonomous Robots*, *42*(8), 1583–1599. doi:10.1007/s10514-018-9774-x

Younis, H. A., Ruhaiyem, N. I. R., Ghaban, W., Gazem, N. A., & Nasser, M. (2023). A Systematic Literature Review on the Applications of Robots and Natural Language Processing in Education. In Electronics (Switzerland), 12(13). doi:10.3390/electronics12132864

Yuan, W., Yuan, X., Xu, L., Zhang, C., & Ma, X. (2023). Harmonic Loss Analysis of Low-Voltage Distribution Network Integrated with Distributed Photovoltaic. *Sustainability (Basel)*, *15*(5), 4334. doi:10.3390/su15054334

Yuan, X., Zhou, X., Pan, Y., Kosonen, R., Cai, H., Gao, Y., & Wang, Y. (2021). Phase change cooling in data centers: A review. *Energy and Building*, *236*, 110764. doi:10.1016/j.enbuild.2021.110764

Yu, D., Zhang, T., He, G., Nojavan, S., Jermsittiparsert, K., & Ghadimi, N. (2020). Energy management of wind-PV-storage grid based large electricity consumer using robust optimization technique. *Journal of Energy Storage, 27,* 101054.

Yuvapriya, T., & Lakshmi, P. (2017). Design of fuzzy logic controller for reduction of vibration in full car model using active suspension system. *Asian Journal of Research in Social Sciences and Humanities*, *7*(3), 302–313. doi:10.5958/2249-7315.2017.00172.1

Yuvapriya, T., Lakshmi, P., & Elumalai, V. K. (2022). Experimental validation of LQR weight optimization using bat algorithm applied to vibration control of vehicle suspension system. *Journal of the Institution of Electronics and Telecommunication Engineers*, 1–11.

Yuvapriya, T., Lakshmi, P., & Fahira Haseen, S. (2023). Vibration Control and Ride Comfort Analysis of a Full Car with a Driver Model Using Big-Bang Big-Crunch Optimized FLC. *Journal of the Institution of Electronics and Telecommunication Engineers*, 1–14. doi:10.1080/03772063.2023.2220671

Yuvapriya, T., Lakshmi, P., & Rajendiran, S. (2018). Vibration suppression in full car active suspension system using fractional order sliding mode controller. *Journal of the Brazilian Society of Mechanical Sciences and Engineering*, *40*(4), 1–11. doi:10.1007/s40430-018-1138-0

Yu, W., Dillon, T., Mostafa, F., Rahayu, W., & Liu, Y. (2019). A global manufacturing big data ecosystem for fault detection in predictive maintenance. *IEEE Transactions on Industrial Informatics*, *16*(1), 183–192. doi:10.1109/TII.2019.2915846

Zakariah, A. (2015). Dual-axis Solar Tracking System based on fuzzy logic control and Light Dependent Resistors as feedback path elements. *2015 IEEE Student Conference on Research and Development*, (pp. 139-144). IEEE. 10.1109/SCORED.2015.7449311

Zangeneh, M., Omid, M., & Akram, A. (2012). A Comparative Study Between Parametric and Artificial Neural Networks Approaches for Economical Assessment of Potato Production in Iran. *Spanish Journal of Agricultural Research*, *9*(3), 661–671. doi:10.5424/sjar/20110903-371-10

Zazoum, B. (2023). Lithium-ion battery state of charge prediction based on machine learning approach. *Energy Reports*, *9*, 1152–1158. doi:10.1016/j.egyr.2023.03.091

Zecchino, A. (2021). *Optimal provision of concurrent primary frequency and local voltage control from a BESS considering variable capability curves: Modelling and experimental assessment* (Vol. 190). Electric Power Systems Research.

Zekrifa, D. M. S., Kulkarni, M., Bhagyalakshmi, A., Devireddy, N., Gupta, S., & Boopathi, S. (2023). Integrating Machine Learning and AI for Improved Hydrological Modeling and Water Resource Management. In *Artificial Intelligence Applications in Water Treatment and Water Resource Management* (pp. 46–70). IGI Global. doi:10.4018/978-1-6684-6791-6.ch003

Zeng, J., Ju, R., Qin, L., Hu, Y., Yin, Q., & Hu, C. (2019). Navigation in unknown dynamic environments based on deep reinforcement learning. *Sensors (Basel)*, *19*(18), 3837. doi:10.3390/s19183837 PMID:31491927

Zeng, Y., Zhang, R., & Lim, T. J. (2016). Wireless communications with unmanned aerial vehicles: Opportunities and challenges. *IEEE Communications Magazine*, *54*(5), 36–42. doi:10.1109/MCOM.2016.7470933

Zeraati, M., Hamedani Golshan, M. E., & Guerrero, J. M. (2018). Distributed Control of Battery Energy Storage Systems for Voltage Regulation in Distribution Networks with High PV Penetration. *IEEE Transactions on Smart Grid*, *9*(4), 3582–3593. doi:10.1109/TSG.2016.2636217

Zhang, Q., Meng, Z., Hong, X., Zhan, Y., Liu, J., Dong, J., Bai, T., Niu, J., & Deen, M. J. (2021). A survey on data center cooling systems: Technology, power consumption modeling and control strategy optimization. *Journal of Systems Architecture*, *119*, 102253. doi:10.1016/j.sysarc.2021.102253

Zhang, X.-L., & Zhang, Q. (2021). Optimization of PID Parameters Based on Ant Colony Algorithm. *2021 International Conference on Intelligent Transportation, Big Data & Smart City (ICITBS)*, (pp. 850-853). IEEE. 10.1109/ICITBS53129.2021.00211

Zhang, Y. J. A., Zhao, C., Tang, W., & Low, S. H. (2018). Profit-Maximizing Planning and Control of Battery Energy Storage Systems for Primary Frequency Control. *IEEE Transactions on Smart Grid*, *9*(2), 712–723. doi:10.1109/TSG.2016.2562672

Zhang, Y., Shi, X., Zhang, H., Cao, Y., & Terzija, V. (2022). Review on deep learning applications in frequency analysis and control of modern power system. *International Journal of Electrical Power & Energy Systems*, *136*, 107744. doi:10.1016/j.ijepes.2021.107744

Zhao, Y., Li, T., Zhang, X., & Zhang, C. (2019). Artificial intelligence-based fault detection and diagnosis methods for building energy systems: Advantages, challenges and the future. *Renewable & Sustainable Energy Reviews*, *109*, 85–101. doi:10.1016/j.rser.2019.04.021

Zhou, H., Liu, Q., Yan, K., & Du, Y. (2021). Deep learning enhanced solar energy forecasting with AI-driven IoT. *Wireless Communications and Mobile Computing*, *2021*, 1–11. doi:10.1155/2021/9249387

Zhou, Y. (2022). Advances of machine learning in multi-energy district communities– mechanisms, applications and perspectives. In *Energy and AI* (Vol. 10). Elsevier B.V., doi:10.1016/j.egyai.2022.100187

Zhou, Z., Kang, Y., Shang, Y., Cui, N., Zhang, C., & Duan, B. (2019). Peak power prediction for series-connected LiNCM battery pack based on representative cells. *Journal of Cleaner Production*, *230*, 1061–1073. doi:10.1016/j.jclepro.2019.05.144

Zhu, S., & He, Y. (2023). A Coordinated Control Scheme for Active Safety Systems of Multi-Trailer Articulated Heavy Vehicles. In *28th IAVSD International Symposium on Dynamics of Vehicles on Roads and Tracks*. Ottawa, Canada.

Zhu, Y., Liang, J., Yang, Q., Zhou, H., & Peng, K. (2019). Water use of a biomass direct- combustion power generation system in China: A combination of life cycle assessment and water footprint analysis. *Renewable & Sustainable Energy Reviews, 115*.

Zhu, Y., Li, W., Li, J., Li, H., Wang, Y., & Li, S. (2019). Thermodynamic analysis and economic assessment of biomass-fired organic Rankine cycle combined heat and power system integrated with CO_2 capture. *Energy Conversion and Management, 204*(2).

Zoerr, C., Sturm, J. J., Solchenbach, S., Erhard, S. V., & Latz, A. (2023). Electrochemical polarization-based fast charging of lithium-ion batteries in embedded systems. *Journal of Energy Storage, 72*, 108234. doi:10.1016/j.est.2023.108234

Zou, C., Hu, X., Wei, Z., & Tang, X. (2017). Electrothermal dynamics-conscious lithium-ion battery cell-level charging management via state-monitored predictive control. *Energy, 141*, 250–259. doi:10.1016/j.energy.2017.09.048

About the Contributors

L. Ashok Kumar was a Postdoctoral Research Fellow from San Diego State University. California. He was selected among seven scientists in India for the BHAVAN Fellowship from the Indo-US Science and Technology Forum and also, he received SYST Fellowship from DST, Govt of India. He has 3 years of industrial experience and 22 years of academic and research experience. He has published 173 technical papers in International and National journals and presented 167 papers in National and International Conferences, He has completed 26 Government of India funded projects worth about 15 Crores and currently 9 projects are in progress worth about 12 Crores. He has developed 27 products and out of that 23 products have been technology transferred to industries and for Government funding agencies, He has created Eight Centres of Excellence at PSG Tech in collaboration with Government agencies and Industries namely, Centre for Audio Visual Speech Recognition, Centre for Alternate Cooling Technologies, Centre for Industrial Cyber Physical Systems Research Centre for Excellence in LV Switchgear, Centre for Renewable Energy Systems, Centre for Excellence in Solar PV Systems. Centre for Excellence in Solar Thermal Systems. His PhD work on wearable electronics earned him a National Award from 1STE and he has received 26 awards in the National and in international level. He has guided 92 graduate and postgraduate projects. He has produced 6 PhD Scholars and 12 candidates are doing PhD under his supervision. He has visited many countries for institute industry collaboration and as a keynote speaker. He has been an invited speaker in 345 programs. Also, he has organized 102 events, including conferences, workshops, and seminars. He completed his graduate program in Electrical and Electronics Engineering from University of Madras and his post-graduate from PSG College of Technology, India, and Masters in Business Administration from IGNOU, New Delhi. After completion of his graduate degree he joined as project engineer for Serval Paper Boards Ltd. Coimbatore (now ITC Unit, Kova Presently he is working as a Professor in the Department of EEE, PSG College of Technology. He is also a Certified Chartered Engineer and BSI Certified ISO 50001 2008 Lead Auditor. He has authored 19 books in his areas of interest published by Springer, CRC Press, Elsevier, Nova Publishers. Cambridge University Press. Wiley. Lambert Publishing and IGI Global. He has 11 patents, one Design patent and two Copyrights to his credit and also contributed 18 chapters in various books. He is also the Chairman of Indian Association of Energy Management Professionals and Executive Member in institution of Engineers, Coimbatore Executive Council Member in Institute of Smart Structure and Systems. Bangalore, Associate Member in CODISSIA. He is also holding prestigious positions in various national and International forums and he is a Fellow Member in IET (UK), Fellow Member In IETE, Fellow Member IE and Senior Member in IEEE.

S. Angalaeswari is a faculty member of the School of Electrical Engineering at VIT, Chennai. She possesses over 18 years of teaching experience. Dr. Angalaeswari earned a gold medal in her M.E degree from Anna University and received the Best Project Award for her undergraduate-level project from "The National Council of Engineering – New Delhi." She has authored more than 100 international journal and conference papers, a book titled "Electric Circuit Analysis," as well as patents and book chapters. Furthermore, she has organized over 100 events, including workshops, seminars, faculty development programs, and conferences. Dr. Angalaeswari has played a pivotal role in establishing the MSME certified Center for Industrial Automation Laboratory, valued at Rs.35 lakhs, in collaboration with CDCE Automation Pvt. Ltd, Chennai. She has successfully trained over 70 students in this course. Dr. Angalaeswari has completedto five consultancy projects, working with Daewon India Auto Parts for the Online Vision Inspection Poke Yoke system with unlimited storage, Sekisui DLJM Molding Pvt. Ltd for designing a drilling machine for the 06M Rear Fender, TVS Motors Hosur, Kovai Ortho, and Tonglit Pvt. Ltd. Her research areas include renewable energy sources, integration of distributed sources into the grid, and optimization techniques for controller design in radial distributed systems, among others.

K. Mohana Sundaram received B.E. degree in Electrical and Electronics Engineering from University of Madras in 2000, M.Tech degree in High Voltage Engineering from SASTRA University in 2002 and Ph.D. degree from Anna University, India in 2014.He is senior member in IEEE and presently Vice chair for IEEE PELS Madras section. His research interests include Power electronics Intelligent controllers, and Power systems He has completed funded project of worth Rs.30 .84 lakhs sponsored by DST, Government of India and currently doing a project of worth Rs.22.03 lakhs Sponsored by DST India(2021-23).Presently working as a Professor and Head in EEE department at KPR Institute of Engineering and Technology, India. Under his guidance 09 Ph.D scholars completed PhD from Anna University, Chennai. He received ISTE Faculty National Award 2022 for Innovative work in Engineering and Technology .He has published six books with international publishers and serving as reviewer for IEEE, Springer and Elsevier journals. He is an active member of IEEE, IE, ISTE and IAENG. He has published around 74 papers in International journals such as IEEE, Elsevier, Springer etc.

Ramesh C. Bansal has more than 25 years of diversified experience of scholarship of teaching and learning, accreditation, research, industrial, and academic leadership in several countries. Currently he is a Professor in the Department of Electrical Engineering at University of Sharjah and extraordinary Professor at the University of Pretoria. Previously he was Professor and Group Head (Power) in the ECE Department at University of Pretoria (UP), South Africa. Prior to his appointment at UP, he was employed by the University of Queensland, Australia; University of the South Pacific, Fiji; and BITS Pilani, India. Prof. Bansal has significant experience of collaborating with industry and Government organizations. He has made significant contribution to the development and delivery of BS and ME programmes for Utilities. He has extensive experience in the design and delivery of CPD programmes for professional engineers. He has carried out research and consultancy and attracted significant funding from Industry and Government Organisations. Prof. Bansal has published over 400 journal articles, presented papers at conferences, books and chapters in books. He has Google citations of over 20000 and h-index of 67. He has supervised 25 PhD, 5 Post Docs and currently supervising several PhD students. His diversified research interests are in the areas of Renewable Energy (Wind, PV, DG, Micro Grid) and Smart Grid. Professor Bansal is an editor of several highly regarded journals including IEEE Systems Journal, ECPS,

SGSE. He is a Fellow and Chartered Engineer IET-UK, Fellow Institution of Engineers (India), Fellow SAIEE, and Senior Member of IEEE-USA.

Arunkumar Patil received his M.Tech & PhD in Electrical Engineering from Visvesvaraya Technological University, Belagavi,Karnataka, India, in 2010 and 2017, respectively. Currently, he is working as an Assistant Professor in the Department of Electrical Engineering, School of Engineering, at the Central University of Karnataka. Kalaburagi, Karnataka, India. His areas of interest include Power System Operation and Control, Power System Protection, Wide area monitoring and Control, Alternative Energy sources, Smart Meter Data Analytics, and AI and ML applications in the Power System. He guides research scholars and is involved in sponsored research projects funded by GOI agencies.

***

Sam B completed his undergraduate in Mechanical Engineering and postgraduate in the field of Engineering Design. He completed his Ph.D. from Anna University, Chennai, Tamil Nādu, India.

P Booma Devi is at the Department of Aeronautical Engineering, Sathyabama Institute of Science and Technology, Chennai, Tamil Nadu 600119, India.

Sampath Boopathi is an accomplished individual with a strong academic background and extensive research experience. He completed his undergraduate studies in Mechanical Engineering and pursued his postgraduate studies in the field of Computer-Aided Design. Dr. Boopathi obtained his Ph.D. from Anna University, focusing his research on Manufacturing and optimization. Throughout his career, Dr. Boopathi has made significant contributions to the field of engineering. He has authored and published over 190 research articles in internationally peer-reviewed journals, highlighting his expertise and dedication to advancing knowledge in his area of specialization. His research output demonstrates his commitment to conducting rigorous and impactful research. In addition to his research publications, Dr. Boopathi has also been granted one patent and has three published patents to his name. This indicates his innovative thinking and ability to develop practical solutions to real-world engineering challenges. With 17 years of academic and research experience, Dr. Boopathi has enriched the engineering community through his teaching and mentorship roles. He has served in various engineering colleges in Tamilnadu, India, where he has imparted knowledge, guided students, and contributed to the overall academic development of the institutions. Dr. Sampath Boopathi's diverse background, ranging from mechanical engineering to computer-aided design, along with his specialization in manufacturing and optimization, positions him as a valuable asset in the field of engineering. His research contributions, patents, and extensive teaching experience exemplify his expertise and dedication to advancing engineering knowledge and fostering innovation.

A Prabhu Chakkaravarthy is at the Department of Networking and Communications, School of Computing, College of Engineering and Technology, SRM Institute of Science & Technology, Kattankulathur 603203, India.

Dharani Jaganathan received a Master of Engineering with First Class and Distinction in Computer Science and Engineering and a Bachelor of Technology with First Class in Information Technology Under

Anna University, Chennai in 2013 and 2018, respectively. She is a Full Time Research Scholar at KPR Institute of Engineering and Technology, Coimbatore - 641407. Her Areas of Interest are Reinforcement Learning, Cyber Security, and Intelligent Transportation Systems. She is a lifetime member of the ISTE.

Jegan G received doctorate degree in the field of microwave Engineering in Sathyabama institute of Science and Technology, India in 2019, received bachelor of Engineering, degree in Electronics and Communication Engineering from Government college of Engineering, Bargur, Madras University, India in 2001. He completed his master degree in Sathyabama Institute of Science and Technology, India in 2006. He has totally 16 years of experience in teaching. He was working as assistant Professor in Electronics and communication department in Sathyabama Institute of Science and Technology, Chennai, India. He has published Research articles in national, international Journals and conferences.

Pandiya Rajan G completed B.E Computer Science and Engineering and M.E Computer and Communication. Doing research in the field of Data Analytics and Machine Learning.

Kode Jaya Prakash is at the Department of Mechanical Engineering, VNR Vignana Jyothi Institute of Engineering and Technology, Hyderabad, Telangana 500090, India.

BalaMurugan K is a hard worker in the field of Electric vehicles.

Sathiyasekar K, Professor and Director R&D Department of EEE,K.S.R Institute for Engineering and Technology hails from Erode, Tamilnadu, India. He received Ph.D. in High voltage Engineering from Anna University Chennai, M.Tech ., Degree in High Voltage Engineering from SASTRA University, Tanjore and B.E., Degree in Electrical and Electronics Engineering from University of Madras, Chennai, TN. A total of 14 Research Scholars are pursuing the Ph.D. program under his guidance and supervision in Anna University, Chennai. He is a recipient of the coveted "Certificate of Outstanding Contribution in Reviewing "from International journal of Electrical Power and Energy Systems, Elsevier, Amsterdam, The Netherlands. He has published more than 100 research articles in SCI, Scopus Journals and Conference Proceedings. He is a life member in IE and ISTE and also reviewer for IEEE, Elsevier, Technical Gazette and other reputed journals.

Vishnu Kumar Kaliappan began his career with Konkuk University, Seoul, South Korea as Research Associate, gaining unparalleled experience in the field of development of control algorithms in Unmanned Aerial Vehicle, while also establishing himself as a Head Cyber Physical Systems Group & Office of international relations, KPR Institute of Engineering and Technology, Coimbatore.

Alagar Karthick working as Associate professor in the Electrical and Electronics Engineering department in KPR Institute of Engineering and Technology, Coimbatore, Tamilnadu, India. He has Published more than 100 International Journal and also reviewer for various journals such as Solar Energy, Fuel, Journal of cleaner Production, Heliyon, Building services Engineering research and Technology. Hereceived his Doctor of Philosophy in the field of Building Integrated Photovoltaic (BIPV) from Anna University, Chennai in 2018. He received his Master degree from Energy Engineering and bachelor degree in Electrical and Electronics Engineering. He has received Best Paper Award for his research articles in

Biomass conversion. His research area includes Solar Photovoltaic, Bioenergy, zero energy buildings, Energy with Artificial Intelligence, Machine learning and Deep learning algorithms.

Harpreet Kaur Channi is an Assistant Professor, Department of electrical engineering.,Chandigarh University,Gharuan,Mohali,Punjab India

Smriti Khare is at the Department of Environmental Science, Amity School of Applied Sciences, Amity University, Lucknow, Uttar Pradesh 226010, India.

Venneti Kiran is at the Department of Computer science and engineering (AIML), Aditya College of Engineering, Surampalem, 533437, India

Kiranmayi R. is currently working as a Professor in the Department of Electrical & Electronics Engineering at Jawaharlal Nehru Technological University Anantapur College of Engineering Ananthapuramu, India. She received her B.Tech Degree in JNTU, Hyderabad in 1993. She is the recipient of university gold medal for academic performance. She obtained her M.Tech Degree in 1995 and Ph.D. Degree in 2013 from JNTU, Ananthapuramu, India. She had worked as a Project Engineer (Electrical) at JNTUA CEA and Project Engineer at JNTUH CEH. She has published more than 40 research papers. Her research interests include Electrical Engineering, Control Systems, Power Systems, and Renewable Energy Sources.

D. Ravi Kishore is an professor and hod in EEE department at Godavari Institute of Engineering and and Technology, Rajahmundry. His research areas of energy systems, energy management and control.

N M G Kumar is at the Department of Electrical and Electronics Engineering, Mohan Babu University, Sree Vidyanikethan Engineering College, Tirupathi, Andhra Pradesh 517102, India.

K. Lokeshwaran is at the Department of Computer Science and Engineering (Data Science), Madanapalle Institute of Technology & Science, Madanapalle, Andhra Pradesh 517325, India.

Muthu Kumar M is a Research Scholar/EEE Dept. KPRIET Coimbatore-641407

Muthukannan M is a Professor in the faculty of Engineering at the Kalasalingam Academy of Research and Education (A deemed to be University), India. He has published number of research papers in peer-reviewed International and National journals. He has also several conference presentations to his credit. He has plenty of experience in guiding research scholar for their PhD degrees.

Ramasamy M, received the M.E. (with First Class and Honors) degree in Power Electronics and Drives and obtained Ph.D. degree in Electrical Engineering from Anna University, Chennai, in 2007 and 2013, respectively. He is currently working as an Associate Professor in the Department of Electrical and Electronics Engineering, K.S.R. College of Engineering, Tiruchengode - 637215. He has published 40 research articles in various reputed international journals and presented 26 papers in national and international conferences. His area of interest is power electronics converters, electrical drives, custom power devices, power quality and electrical machines. He is an active member of IEEE and ISTE.

Ebenezar Jebarani M. R. received her doctorate degree in the field of wireless sensor networks in Sathyabama University in 2014, M.E degree in Computer science in 2007 with distinction from Sathyabama University. She has more than 20 years teaching experience. She was working as a Professor in Electronic and communication Department in Sathyabama Institute of Science and Technology. She has published several papers in reputed international/national journal and conferences. She has published around 11 patents and 2 patents got granted. Dr.M.R.Ebenezar Jebarani having the field of interest in Wireless sensor networks, Embedded systems, Python programming, Microprocessors, Microcontrollers, Data communication,Wireless communications and Digital Image Processing. She was the life member of ISTE and IEI. She was acting as a reviewer for various journals

M. Mathiyarasi is at the Department of Aerospace Engineering, Agni College of Technology, OMR, Chennai, Tamil Nadu 600130, India.

Muralidhar M. is currently working as a Director and Professor in the Department of Electrical and Electronics Engineering in Sri Venkateswara College of Engineering & Technology (Autonomous), Chittoor, Andhra Pradesh, India. He received his B.E Degree in the Department of Electrical Engineering, M.E Degree in the Department of Instrumentation & Control Systems, and PhD Degree in the High Voltage Engineering from Sri Venkateswara University, Tirupathi, India. He has published more than 50 research papers. So far, he has guided 06 PhD and 38 M.Tech students. His research interests include Electrical Engineering, Instrumentation & Control Systems, High Voltage Engineering, and Renewable Energy Sources

Akila Muthuramalingam completed her Bachelor's in Computer Science and Engineering from Madurai Kamaraj University, Tamil Nadu and Master's in Computer Science and Engineering from Manonmaniam Sundaranar University, Tamil Nadu. She completed her Ph.D. in Computer Science and Engineering under Anna University, Tamil Nadu. She has 29 years of teaching and industry experience. She is currently working as Professor in the Department of Information Technology, K S Rangasamy College of Technology, Tiruchengode, Tamil Nadu, India. She is a member in various societies like IEEE, IET, ISTE and iGEN. Her research areas include biometrics, pattern matching and soft computing.

Sureshkumar Myilsamy completed his undergraduate in Mechanical Engineering and postgraduate in the field of Engineering Design. He completed his Ph.D. from Anna University, Chennai, Tamil Nādu, India.

Ahalya N is at the Department of Biotechnology, MS Ramaiah Institute of Technology, Bangalore, Karnataka 560054, India.

Loganathan Nachimuthu was born in 1970 in India. He received a B.E degree in Electrical and Electronics Engineering from the Government College of Engineering, Salem in 1999, M.E degree in Power and Energy Systems Engineering from the University Vishveshwaraya College of Engineering, Bangalore, India in 2005, and a Ph.D. degree from Anna University, Chennai in 2013. Previously, he worked as a professor in the Department of Electrical and Electronics Engineering at K.S. Rangasamy College of Technology, Tiruchengode, Tamilnadu, India. Currently working as, a Senior Lecturer, College of Engineering and Technology, Engineering Department at the University of Technology and

Applied Sciences Nizwa, Sultanate of Oman. He has published 8 International Journals including the IEEE Transactions on Dielectrics and Electrical Insulation and 16 National and international conference papers, His research interests include in smart power system protection, High voltage insulation engineering, and Power Systems Engineering.

V Nyemeesha is at the Department of Computer Science and Engineering, VNR VJIET, Hyderabad, Telangana 500090, India.

Kavipriya P obtained her Bachelor's in Engineering (B.E) in the year 2000 from AVC College of Engineering, Bharathidasan University. She received her Master's degree in Applied Electronics, first class with distinction, from Sathyabama University, Chennai in 2005 and also she obtained her Doctorate in the field of wireless networks and communication in the year of 2015. Her current area of interest is Wireless networks, Wireless communication, Wireless Sensor Networks and Circuit Analysis. She joined as a teaching faculty in the year 2001 and presently working as a Professor in ECE, Sathyabama Institute of Science and Technology, Chennai, Tamil Nadu, India. She had published more research articles in reputed Journals and Conference Proceedings. She taught a wide variety of courses for both Under Graduate and Post Graduate levels over a period of twenty years.

Lakshmi P received her B.E. from Government College of Technology, Coimbatore, Masters and Ph.D. degrees from College of Engineering, Guindy, Anna University, Chennai. Presently, she is working as a Professor, EEE Department, College of Engineering, Guindy, Chennai, Tamil Nadu. She published papers in many national and international conferences and national and international journals. Her areas of interest are intelligent controllers, process control and power system stability.

R Pitchai is at the B V Raju Institute of Technology, Telangana 502313, India.

V. Divya Prabha, currently working as Assistant Professor in the Department of Electrical and Electronics Engineering, S.A. Engineering College (Autonomous), pursuing Ph.D. in Electrical and Electronics Engineering from Anna University, Chennai. M.E., Degree in Power Electronics and Drives from Anna University, Chennai., B.E., Degree in Electrical and Electronics Engineering from Anna University, Chennai.Her expertise lies in Renewable Energy, Power Electronics, and Power Quality. She is active member of ISTE.

M. Venkatesan, Principal, K S R Institute for Engineering and Technology, Tiruchengode, Tamil Nadu, India, received his Ph.D from Periyar University, the post graduation in Computer Science from Anna University and under graduation in Electronics from Sri Ramakrishna Mission Vidyalaya College of Arts and Science, Coimbatore. A total of 06 Research Scholars received Ph.D under his guidance and supervision. He has published two books on "Computer Architecture" and "Distributed Systems" and 24 papers in international journals. To the limelight of his career, he has received two times as the Best Principal Award by "Nature Science Foundation", Coimbatore and by "Society for Aerospace & Mechanical Engineering Professionals", Chennai.

R Prasath is at the Department of Computer Science and Engineering, KCG College of Technology, Karapakkam, Chennai, Tamil Nadu 600097, India.

K S Pushpalatha is at the Department of Information Science and Engineering, Acharya Institute of Technology, Acharya Dr. Sarvepalli, Bengaluru 560107, India.

R. Ramya Sri, Assistant Professor, Department of English, Kongu Engineering College, Perundurai, Erode TamilNadu - 638060

K Ramachandra Raju is at the Department of Mechanical Engineering, Bannari Amman Institute of Technology, Erode, Tamil Nadu 638401, India.

Fahira Haseen S received her B.E. Electrical and Electronics Engineering from Alagappa Chettiar Government College of Engineering and Technology, Karaikudi, M.E Control and Instrumentation Engineering in College of Engineering, Guindy, Chennai and She is currently pursuing PhD in EEE department, College of Engineering, Guindy, Chennai, Tamil Nadu. Her areas of interest are control systems, process control and intelligent controllers.

Sankar Ganesh S. completed Ph.D. at Anna University, Chennai in 2013. He has more than 23 years of teaching experience and published many papers in national and international journals. He is presently working in the Department of Artificial Intelligence and Data Science, KPR Institute of Engineering and Technology, Coimbatore. He has profound knowledge and interest in pattern recognition, medical imaging, machine learning, and big data analytics.

Saravanan S is at the Department of Electrical and Electronics Engineering, B V Raju Institute of Technology, Narsapur, Telangana 502 313, India.

Boopa S.B. is at the Department of Textile Technology, MAKAUT, Kolkata, West Bengal 711106, India. sumanta.21394@gmail.com

Devika Sahu is at the Department of Computer Science and Engineering, Government Engineering College, Raipur, Chhattisgarh, 492015, India.

Kaliappan Seeniappan is Professor & Head in the Department of Mechatronics Engineering at KCG College of Technology, Chennai-97, where he has been a faculty member since 2023. He completed his Ph.D. and M.E at Anna University, Chennai and his B.E at R.V.S College of Engineering & Technology, Dindigul. He is having more than 26 years experience . He is recognized supervisor at Anna University in the Department of Mechanical Engineering in the research area Thermal Engineering, Heat Transfer, CFD, Composite Materials. He has served as technical committee head in conferences and workshop and worked as the reviewers and Editorial Board Member for several National and International Journals. He published almost 142 research articles in various national and international journals and conferences, organized STTPs, FDPs, Conferences and other technical events. He holds international Australian Patent Grants and published Indian Patents. He co-authored 18 books . He is Life member in ISTE, IAENG, UAMAE, theIRED and SAEINDIA. He got Best Academician Award for the year 2023 and Young Researcher Award in InSc awards-2021

D Sengeni is at the Department of Electronics and Communication Engineering, CK College of Engineering and Technology, Cuddalore, Tamil Nadu 607003, India.

Kollati Sivaprasad is working as an Assistant Professor at Godavari Institute of Engineering and Technology, Rajahmundry. He is having more than 10 years of experience and published 13 national/international journals. He is an expert of power electronics and drives.

Chalumuru Suresh is at the Department of Computer Science and Engineering, VNR VJIET, Hyderabad, Telangana 500090, India.

Mothilal T, is currently working as Professor and Head in the Department of Automobile in KCG College of Technology, earned Bachelor Degree in Mechanical Engineering from Madras University followed by Master's Degree in Internal Combustion Engineering from Anna University and completed Ph.D. in Thermal science. Having 23 years of teaching, good at thermal related subjects, Research interest lie in thermal Engineering, CFD, composite material, Materials Engineering, Heat Transfer etc, published more than 40 journals,6 patents granted, Throughout the career received numerous accolades like best paper presentation award several times, domain awards in NPTEL, delivered lecture in various engineering college, Life member in Indian Society for Technical Education (LMISTE), International Association of Engineers (IAENG), Institute of Research Engineers and Doctors (theIRED), and also member in Society of Automotive Engineers (SAEINDIA).Recognized research supervisor in Anna University in the Department of Mechanical Engineering, organised several events, workshop, conference etc

Suresh Vendoti is working as an Assistant Professor in the Department of Electrical and Electronics Engineering at Godavari Institute of Engineering and Technology (A), Rajahmundry, Andhra Pradesh, India. I have graduated in Electrical and Electronics Engineering at Narayana Engineering College (JNTUA University), Andhra Pradesh, India. I secured Master of Technology in Power Electronics at P.B.R Visvodaya Institute of Technology & Science (JNTUA University) and Ph.D. in Electrical Engineering at JNTUA University, Andhra Pradesh, India. I have more than 12 years of teaching experience and good publication record includes 3-SCIE indexed journals, 2-Scopus indexed journals, 12-UGC care journals and 12-International/National Conferences. My research area includes Hybrid Electric Vehicles, Renewable Energy Systems, Micro grids and Power Electronics

Dana Victoria is persuing M.Tech in Artificial Intelligence in the Department of Computer Science Engineeringat International School for Technology and Science for Women (A), Rajanagaram, A.P., India. She is Post graduated in Mathematics in V.R. PG institute (Vikrama Simhapuri University), Andhra Pradesh, India. She secured BSC in Aditya Degree college (SV University).

Sanjay B. Warkad obtained his Master's degree in M. E. (EPS) as well as M.B.A. from Amravati and Nagpur University, India. Also obtained his Ph.D. from Visvesvaraya National Institute of Technology (VNIT) Nagpur in Electrical Engineering in 2012. He is working with P. R. Pote (Patil) College of Engineering & Management, Amravati as Professor in Electrical Engineering Department. His major areas of interest are Optimal Power Pricing under Electricity deregulation, Artificial Neural Network applications in Power System, He has guided 14 M.E. research projects, awarded 2 PhDs. and at present pursuing 2 PhDs under him. He has published more than 70 papers at national/international level.

Published Books, Patents and Copyrights. He also awarded Ministry of Energy, Department of Power, Government of India Prize for his research contributions in Power System. He is member of IEEE, Smart grid community, IEEE, IEI, IACSIT, ISTE, IAENG, ISDS Japan. He is also an Editor and reviewer of several International Journals and conferences.

Putchakayala Yanna Reddy is a Research Scholar, Department of Electrical Engineering, National Institute of Technology (NIT), Silchar, Cachar, Assam, 788010

Index

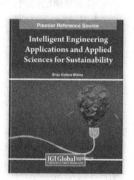
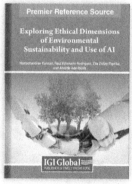

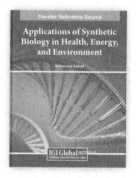

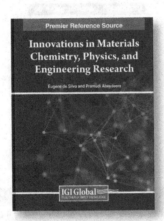

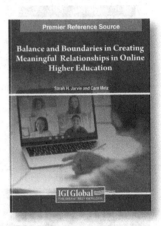

www.igi-global.com

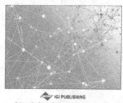

Printed in the United States
by Baker & Taylor Publisher Services